AF509018

Cyber Security of Critical Infrastructures

Cyber Security of Critical Infrastructures

Editors

Leandros Maglaras
Ioanna Kantzavelou
Mohamed Amine Ferrag

MDPI • Basel • Beijing • Wuhan • Barcelona • Belgrade • Manchester • Tokyo • Cluj • Tianjin

Editors
Leandros Maglaras
De Montfort University
UK

Ioanna Kantzavelou
University of West Attica
Greece

Mohamed Amine Ferrag
Guelma University
Algeria

Editorial Office
MDPI
St. Alban-Anlage 66
4052 Basel, Switzerland

This is a reprint of articles from the Special Issue published online in the open access journal *Applied Sciences* (ISSN 2076-3417) (available at: https://www.mdpi.com/journal/applsci/special_issues/Cyber_Security_Critical_Infrastructures).

For citation purposes, cite each article independently as indicated on the article page online and as indicated below:

LastName, A.A.; LastName, B.B.; LastName, C.C. Article Title. *Journal Name* **Year**, *Volume Number*, Page Range.

ISBN 978-3-0365-2726-0 (Hbk)
ISBN 978-3-0365-2727-7 (PDF)

Cover image courtesy of Cyber Security of Critical Infrastructures

Contents

About the Editors

Leandros A. Maglaras is Professor of Cybersecurity in the School of Computer Science and Informatics of De Montfort University. From September 2017 to November 2019, he was the Director of the National Cyber Security Authority of Greece. He obtained the B.Sc. (M.Sc. equivalent) in Electrical and Computer Engineering from the Aristotle University of Thessaloniki, Greece in 1998, M.Sc. in Industrial Production and Management from the University of Thessaly in 2004, and M.Sc. and Ph.D. degrees in Electrical & Computer Engineering from University of Thessaly, in 2008 and 2014, respectively. In 2018, he was awarded a Ph.D. in Intrusion Detection in SCADA systems from the University of Huddersfield. He is featured in Stanford University's list of the world Top 2% scientists. He is a Senior Member of the Institute of Electrical & Electronics Engineers (IEEE) and is an author of more than 160 papers in scientific magazines and conferences.

Ioanna Kantzavelou is an Assistant Professor at the Dept. of Informatics and Computer Engineering in the School of Engineering of the University of West Attica. She received a B.Sc. in Informatics from the Technological Educational Institute of Athens, an MSc by Research in Computer Security from the University College Dublin of the National University of Ireland, and a Ph.D. on Intrusion Detection in Information Technology Security from the University of the Aegean. She has worked in R&D projects funded by the Greek government, the Irish government, and the EU. Her published work includes chapters in books (IOS Press), articles in conferences and journals, recording remarkable citations in her research work. She has joint editorship of three IOS Press collections. She has been a repetitive reviewer in many international conferences, such as ACM SEC, IEEE TrustCom, IFIP SEC, ESORICS, IEEE CIS, and for high ranking journals, of IEEE, Elsevier, Springer, and Emerald. She is a member of the Greek Computer Society (GCS), of the ACM and of the IEEE Computer Society.

Mohamed Amine Ferrag received the Bachelor's, Master's, Ph.D., and Habilitation degrees in computer science from Badji Mokhtar—Annaba University, Annaba, Algeria, in June, 2008, June, 2010, June, 2014, and April, 2019, respectively. Since October 2014, he has been a Senior Lecturer with the Department of Computer Science, Guelma University, Guelma, Algeria. Since July 2019, he has been a Visiting Senior Researcher with the NAU-Lincoln Joint Research Center of Intelligent Engineering, Nanjing Agricultural University, Nanjing, China. His research interests include wireless network security, network coding security, and applied cryptography. He has published over 80 papers in international journals and conferences in the above areas. He has been conducting several research projects with international collaborations on these topics. He was a recipient of the 2021 IEEE TEM BEST PAPER AWARD WINNER. Dr. Ferrag is featured in Stanford University's list of the world's Top 2% scientists for the year 2019. Some of his research findings are published in top-cited journals, such as the IEEE COMMUNICATIONS SURVEYS AND TUTORIALS, IEEE INTERNET OF THINGS JOURNAL, IEEE TRANSACTIONS ON ENGINEERING MANAGEMENT, IEEE ACCESS, IEEE/CAA Journal of Automatica Sinica, MDPI Sensors, Journal of Information Security and Applications (Elsevier), Transactions on Emerging Telecommunications Technologies (Wiley), Telecommunication Systems (Springer), International Journal of Communication Systems (Wiley), Sustainable Cities and Society (Elsevier), and Journal of Network and Computer Applications (Elsevier). He is currently serving on various editorial positions, such as Editorial Board Member in Journals (Indexed SCI and Scopus), such as, ICT Express (JCR IF 4.317), IET Networks (Citescore 4.1),

International Journal of Internet Technology and Secured Transactions (Citescore 1.0), Security and Communication Networks (JCR IF 1.791), and Journal of Sensor and Actuator Networks (Citescore 6.2). He reviewed more than 650 papers for top-cited journals, including, Nature, IEEE transactions, Elsevier, Springer, Wiley journals, etc.

Preface to "Cyber Security of Critical Infrastructures"

Over the years, the digitization of all aspects of life in modern society has come to be considered an acquired advantage. We encounter advanced digital systems and devices for the most trivial things, such as a digital power switch of a household lamp, up to those things that are as absolutely necessary as the fully digitized process of the conversion of wastewater into drinking water. All societies of the developing and developed world have adopted and implemented digital critical infrastructure solutions, or critical infrastructures.

According to the EU directive (2008/114/EC) a critical infrastructure is "a tool, system or section contained in each of the Member States and is final for maintenance of important and sustainable functions, health, safety, financial or social criteria of individuals, and the cessation of its effective operation or its destruction has a significant impact on a Member State, as a direct consequence its failure to maintain the above functions". Within the European context, in fact, when the impact of an attack affects more than one European country, we can talk about European Critical Infrastructure (ECI).

Over the years, the need to shield critical infrastructure becomes more and more urgent. Critical infrastructure must be well-protected from impending attacks and we must continuously improve detection and elimination methods, especially improving the detection and elimination of any attackers as promptly as possible. It is a rather difficult challenge to ensure the security of products or services while simultaneously eliminating cyber-attacks and managing a fully automated production process, all in a timely manner.

To date, it has not been established whether theory can meet practice in the real-word conditions of a critical infrastructure that manages vital products or services. During the production process, different suppliers of devices are used, as well as sensors of low technical specifications and computing power.

In this book, both research and practical aspects of cyber security considerations in critical infrastructures are presented. Aligned with the interdisciplinary nature of cyber security, authors from academia, government, and industry have contributed 13 chapters. The issues that are discussed and analysed include cybersecurity training, maturity assessment frameworks, malware analysis techniques, ransomware attacks, security solutions for industrial control systems, and privacy preservation methods.

Leandros Maglaras, Ioanna Kantzavelou, Mohamed Amine Ferrag
Editors

Editorial

Digital Transformation and Cybersecurity of Critical Infrastructures

Leandros Maglaras [1,*], Ioanna Kantzavelou [2] and Mohamed Amine Ferrag [3]

[1]   School of Computer Science and Informatics, De Montfort University, Leicester LE1 9BH, UK

[2]   Department of Informatics and Computer Engineering, University of West of Attica, 122 43 Athens, Greece; ikantz@uniwa.gr

[3]   Department of Computer Science, Guelma University, Guelma 24000, Algeria; ferrag.mohamedamine@univ-guelma.dz

*   Correspondence: leandros.maglaras@dmu.ac.uk

Citation: Maglaras, L.; Kantzavelou, I.; Ferrag, M.A. Digital Transformation and Cybersecurity of Critical Infrastructures . *Appl. Sci.* **2021**, *11*, 8357. https://doi.org/10.3390/app11188357

Received: 31 August 2021

Accepted: 8 September 2021

Published: 9 September 2021

Publisher's Note: MDPI stays neutral with regard to jurisdictional claims in published maps and institutional affiliations.

Critical infrastructures are vital assets for public safety, economic welfare, and the national security of nations. Vulnerabilities of critical infrastructures have increased with the widespread use of information technologies. As Critical National Infrastructures are becoming more vulnerable to cyberattacks, their protection becomes a significant issue for any organization as well as nation. The risks to continued operations from failing to upgrade ageing infrastructures or not meeting mandated regulatory regimes are considered higher given the demonstrable impact of such circumstances.

Due to the rapid increase in sophisticated cyber threats targeting critical infrastructures with significant destructive effects, cyber security of critical infrastructures has become an agenda item for academics, practitioners, and policy makers. In recent years, cyber attacks, especially those targeting systems that keep or process sensitive information, are becoming more sophisticated. Attacks to such critical systems include penetrations to their network and the installation of malicious tools or programs that can reveal sensitive data or alter the behaviour of specific physical equipment. A holistic view, which covers technical, policy, human, and behavioural aspects, is essential to handle the the cyber security of critical infrastructures effectively.

This editorial presents the manuscripts accepted, after a careful peer-review process, for publication in the Special Issue "Cyber Security of Critical Infrastructures" of the MDPI journal *Applied Sciences*. This Special Issue includes thirteen articles: eleven original research papers describing novel ideas, results, and real-world experiences involving critical infrastructures and two review papers focusing on modern training methods for cybersecurity professionals and privacy preservation methods of cloud-based face recognition methods.

Due to the high volume of cyber attacks that have taken place recently on critical infrastructures, a lot of research regarding cyber-attacks has been conducted. However, there has been a lack of research related to measuring cyber-attacks from the perspective of offensive cybersecurity. Motivated by this, the authors in [1] propose a methodology for quantifying cyber-attacks such that they are measurable rather than abstract. The authors first defined and derived the comprehensive offensive cybersecurity framework and taxonomy; then, they performed a content analysis of public reports of cyber-attacks and identified detailed techniques used in cyber-attacks. They created a systematic scoring model based on the offensive cybersecurity framework and calculated the score results of ten fileless and eight Advanced Persistent Threat (APT) group cyber-attacks. The study presented in this article is the first to be conducted to quantify and score cyber-attacks. The basic finding is that APT cyber-attacks have higher scores than fileless cyber-attacks, due to the APT using various ATT&CK techniques. The main limitation is that the proposed approach cannot analyse real malware, but measuring the score of cyber-attacks is meaningful as an initial research step.

As the human element is generally considered the weakest link in a computer system, professional training is now becoming a necessity, not only for raising the users' awareness, but also for training the technical staff to operate the various protection mechanisms that must be acquired. The authors in [2] try to tackle this issue by combining pedagogical methods that promote skill development and security models that capture the security-related aspects of a process. The proposed methodology tackles the incorporation of educational methods to the overall lifecycle of a complete training programme with the dynamic adaptation of the training process to the trainee's particularities. The trainee starts the learning process by consuming the main teaching material (e.g., lectures, tutorials, videos, etc.) and proceeds to more advance learning procedures, involving hands-on experience on emulated/simulated components. The overall method is integrated in the cyber-ranges platform THREAT-ARREST and the trainee is continuously evaluated.

Focusing on cybersecurity training means and methods, the authors in [3] present an analysis of ten cyber ranges that were recently deployed from universities and organizations. The article presents the current state of the art on testbeds and cyber ranges, analyses the findings of a set of structured interviews with organizations that have a testbed and cyber range, and gives insights of modern cyber ranges. A cyber range system is mainly used for one or more of three main objectives: research, training, and exercises. Based on the analysis of state of the art existing cyber ranges, the authors conclude, among other things, that modern CRs should be enriched with novel features, such as various telecommunication capabilities, emulated Banking systems, hospitals, simulated smart grids, automated vehicles, Virtual Cyber Centres of Operation, and many more, in order to be able to offer realistic and tailored training to cybersecurity professionals.

In order to perform vulnerability testing of web applications, different types of analysis security testing (AST) can be used: static (SAST), dynamic (DAST), or interactive (IAST). Authors in [4] produced an analysis that is the first of its kind—to study the best way to combine the three types of security analysis tools for web applications. They investigate the behaviour of the combination of two static tools, two dynamic tools and two interactive tools using a new methodology. The main finding of this research is that combinations integrated by SAST+DAST+IAST tools as Fortify + Arachni + CCE or Fortify + ZAP + CCE reach very good results for high, medium, and low classifications.

The next article of this Special Issue [5] focuses on Higher Educational Institutes of the UK following a recent JISC report, reaffirming that UHEIs in the UK are not well prepared to defend against, or recover from, cyberattacks. HEIs face a constant challenge of balancing public access in the interest of sharing information, whilst protecting their information assets and could be included in a broader group of critical infrastructures. The work presented in this article proposes a novel Holistic Cybersecurity Maturity Assessment Framework (HCYMAF) for HEIs that can be used in order to conduct a gap analysis against 15 security requirements. Moreover, the proposed framework incorporates several regulations and security best practices into one lightweight online self-assessment guide, producing compliance reports against all regulations that the HEI must be compliant with that can be used in order to design appropriate mitigation plans. The research was based into three pilars: structured interviews with experts in the field, a case study on an HEI, and webinars. The proposed framework could be adjusted in order to be applied to organisations in other sectors, e.g., water or power suppliers.

According to several reports (ENISA, Ventures, etc.), ransomware is one of the Top 10 Cybersecurity Threats in 2021. Ransomware's success is largely owed to the relative simplicity with which an attacker can achieve devastating effects, especially when targeting critical infrastructures. Trying to cope with this issue, the authors in [6] propose a ransomware streaming analytics model by integrating a compact set of 24 static and dynamic traits, a hybrid machine learner, a numeral measurement for ransomware's ancestor family attribution, and a statistic formula for a multi-descent ransomware version via a multi-tiered architecture. In order to showcase the efficiency of the proposed model, the authors conduct extensive experiments on a big dataset consisting of 35,000 ransomware versions of 14 families, 500 versions of 10 malware, and 500 goodware apps aggregated at a different time from different data archives. The proposed solution enriches the accuracy, reduces the mistakes and misclassifications, and shortens the elapsed time versus escalating, big, lifelike, and imbalanced corpora of data.

Dealing with malware that can launch several types of attacks, including ransomware and SolarWinds attacks, is an open issue. The authors in [7] propose an extensible and openly available malware analysis platform entitled Sisyfos. Sisyfos constitutes a significant step in the development of open, modular, and extensible malware platforms that support operational environments, including critical infrastructures. One important aspect that needs to be further investigated is the robustness and fault tolerance of such platforms, especially in boundary cases where samples may lead to system failures.

The next article [8] of the Special Issue proposes an authentication scheme to be employed in rapidly changing variable message format(VMF)-based environments. It is based on the cryptographic hash chain-based authentication technology that includes a time-based one-time password (T-OTP). The proposed lightweight authentication scheme satisfies the demands for a rapidly changing battlefield network and any additional security requirements based on VMF standards. The proposed model could enhance the integrity of tactical message exchanges and reduce unnecessary network transactions and transmission bits for the authentication flow in VMF-based combat network radio (CNR) networks, while ensuring robustness with limited resources.

In most public key infrastructure (PKI) schemes, the public keys are generated by private keys with Rivest–Shamir–Adleman (RSA) and elliptic curve cryptography (ECC). It is now anticipated that quantum computers (QC) will be able to break both RSA and ECC when the technology to manufacture enough quantum nodes becomes available. Paper [9] describes practical ways to generate keys from physical unclonable functions, for both lattice and code-based cryptography and proposes to generate the public–private key pairs by replacing the random number generators with data streams generated from addressable physical unclonable functions (PUFs) to obtain the seeds needed in the post quantum cryptographic (PQC) algorithms. Dissimilar to the key pairs computed by PQC algorithms, the seeds are relatively short, typically 256-bits long.

Devices from the operational technologies (OT) side of this critical infrastructure, which were physically segregated in the past, are now more and more connected to the internet in a series of highly-distributed hierarchical network systems, forming the next generation electric power system or smart grids (SG). SGs are revolutionizing the energy supply sector and this trend is expected to rise in the near future. Unfortunately, SGs have become the target of several serious cyber attacks recently. The authors in [10] focus on such attack scenarios and propose a formal risk assessment framework that is based on threat modelling and probabilistic model checking. The assessment takes into consideration the technological aspects of the SG architecture.

The Industrial Control System (ICS) is an umbrella term that refers to a group of process automation technologies, such as Supervisory Control and Data Acquisition (SCADA) systems and Distributed Control Systems (DCS), which, unfortunately, have been subject to a growing number of attacks in recent years. As they deliver vital services to critical infrastructure—such as communications, manufacturing, and energy among others—hostile intruders mounting attacks represent a serious threat to the day to day running of nation states. Both articles [11,12] focus on detecting vulnerabilities and attacks in an ICS. The former article [11] proposes a pipeline based on existing deep learning models to automatically classify screenshots of ICSs that could be linked to critical infrastructures in order to support the task of detecting vulnerable systems exposed on the internet in real time. The latter [12] introduces interval-valued complex intuitionistic fuzzy relations (IVCIFRs) for recognizing a cyberattack and nullifying its effects.

Advancements in the robotics field have led to the emergence of a diversity of robot-based applications and favoured the integration of robots in the automation of facial recognition tasks. However, this solution faces several security problems. The authors in [13] studied several approaches for robot-secure face recognition in the cloud environment. By using a set of different algorithms to encrypt a set of images, they trained and tested the robot with various deep learning algorithms and evaluated the efficiency in terms of safety, time complexity, and recognition accuracy using the ORL database.

Author Contributions: All the guest editors contributed equally to this editorial. All authors have read and agreed to the published version of the manuscript.

Funding: This research received no external funding.

Institutional Review Board Statement: Not applicable.

Informed Consent Statement: Not applicable.

Data Availability Statement: Not applicable.

Conflicts of Interest: All authors declare no conflict of interest.

References

1. Kim, K.; Alfouzan, F.A.; Kim, H. Cyber-Attack Scoring Model Based on the Offensive Cybersecurity Framework. *Appl. Sci.* **2021**, *11*, 7738. [CrossRef]
2. Hatzivasilis, G.; Ioannidis, S.; Smyrlis, M.; Spanoudakis, G.; Frati, F.; Goeke, L.; Hildebrandt, T.; Tsakirakis, G.; Oikonomou, F.; Leftheriotis, G.; et al. Modern aspects of cyber-security training and continuous adaptation of Programmes to trainees. *Appl. Sci.* **2020**, *10*, 5702. [CrossRef]
3. Chouliaras, N.; Kittes, G.; Kantzavelou, I.; Maglaras, L.; Pantziou, G.; Ferrag, M.A. Cyber ranges and testbeds for education, training, and research. *Appl. Sci.* **2021**, *11*, 1809. [CrossRef]
4. Mateo Tudela, F.; Bermejo Higuera, J.R.; Bermejo Higuera, J.; Sicilia Montalvo, J.A.; Argyros, M.I. On Combining Static, Dynamic and Interactive Analysis Security Testing Tools to Improve OWASP Top Ten Security Vulnerability Detection in Web Applications. *Appl. Sci.* **2020**, *10*, 9119. [CrossRef]
5. Aliyu, A.; Maglaras, L.; He, Y.; Yevseyeva, I.; Boiten, E.; Cook, A.; Janicke, H. A holistic cybersecurity maturity assessment framework for higher education institutions in the United Kingdom. *Appl. Sci.* **2020**, *10*, 3660. [CrossRef]
6. Zuhair, H.; Selamat, A.; Krejcar, O. A Multi-Tier Streaming Analytics Model of 0-Day Ransomware Detection Using Machine Learning. *Appl. Sci.* **2020**, *10*, 3210. [CrossRef]
7. Serpanos, D.; Michalopoulos, P.; Xenos, G.; Ieronymakis, V. Sisyfos: A Modular and Extendable Open Malware Analysis Platform. *Appl. Sci.* **2021**, *11*, 2980. [CrossRef]
8. Kim, D.; Seo, S.; Kim, H.; Lim, W.G.; Lee, Y.K. A Study on the Concept of Using Efficient Lightweight Hash Chain to Improve Authentication in VMF Military Standard. *Appl. Sci.* **2020**, *10*, 8999. [CrossRef]
9. Cambou, B.; Gowanlock, M.; Yildiz, B.; Ghanaimiandoab, D.; Lee, K.; Nelson, S.; Philabaum, C.; Stenberg, A.; Wright, J. Post Quantum Cryptographic Keys Generated with Physical Unclonable Functions. *Appl. Sci.* **2021**, *11*, 2801. [CrossRef]
10. Vallant, H.; Stojanović, B.; Božić, J.; Hofer-Schmitz, K. Threat Modelling and Beyond-Novel Approaches to Cyber Secure the Smart Energy System. *Appl. Sci.* **2021**, *11*, 5149. [CrossRef]
11. Blanco-Medina, P.; Fidalgo, E.; Alegre, E.; Vasco-Carofilis, R.A.; Jañez-Martino, F.; Villar, V.F. Detecting Vulnerabilities in Critical Infrastructures by Classifying Exposed Industrial Control Systems Using Deep Learning. *Appl. Sci.* **2021**, *11*, 367. [CrossRef]
12. Nasir, A.; Jan, N.; Gumaei, A.; Khan, S.U.; Albogamy, F.R. Cybersecurity against the Loopholes in Industrial Control Systems Using Interval-Valued Complex Intuitionistic Fuzzy Relations. *Appl. Sci.* **2021**, *11*, 7668. [CrossRef]
13. Karri, C.; Cheikhrouhou, O.; Harbaoui, A.; Zaguia, A.; Hamam, H. Privacy Preserving Face Recognition in Cloud Robotics: A Comparative Study. *Appl. Sci.* **2021**, *11*, 6522. [CrossRef]

applied sciences

Article

Cyber-Attack Scoring Model Based on the Offensive Cybersecurity Framework

Kyounggon Kim [1,*], Faisal Abdulaziz Alfouzan [2,*] and Huykang Kim [3]

[1] Center of Excellence in Cybercrime and Digital Forensics, College of Criminal Justice, Naif Arab University for Security Sciences, Riyadh 14812, Saudi Arabia
[2] Department of Forensic Sciences, College of Criminal Justice, Naif Arab University for Security Sciences, Riyadh 14812, Saudi Arabia
[3] School of Cybersecurity, Korea University, Seoul 02841, Korea; cenda@korea.ac.kr
* Correspondence: kkim@nauss.edu.sa (K.K.); falfouzan@nauss.edu.sa (F.A.A.)

Citation: Kim, K.; Alfouzan, F.A.; Kim, H. Cyber-Attack Scoring Model Based on the Offensive Cybersecurity Framework. *Appl. Sci.* **2021**, *11*, 7738. https://doi.org/10.3390/app11167738

Academic Editors: Leandros Maglaras, Ioanna Kantzavelou, Mohamed Amine Ferrag and Tiago M. Fernández-Caramés

Received: 11 July 2021
Accepted: 15 August 2021
Published: 23 August 2021

Publisher's Note: MDPI stays neutral with regard to jurisdictional claims in published maps and institutional affiliations.

Abstract: Cyber-attacks have become commonplace in the world of the Internet. The nature of cyber-attacks is gradually changing. Early cyber-attacks were usually conducted by curious personal hackers who used simple techniques to hack homepages and steal personal information. Lately, cyber attackers have started using sophisticated cyber-attack techniques that enable them to retrieve national confidential information beyond the theft of personal information or defacing websites. These sophisticated and advanced cyber-attacks can disrupt the critical infrastructures of a nation. Much research regarding cyber-attacks has been conducted; however, there has been a lack of research related to measuring cyber-attacks from the perspective of offensive cybersecurity. This motivated us to propose a methodology for quantifying cyber-attacks such that they are measurable rather than abstract. For this purpose, we identified each element of offensive cybersecurity used in cyber-attacks. We also investigated the extent to which the detailed techniques identified in the offensive cyber-security framework were used, by analyzing cyber-attacks. Based on these investigations, the complexity and intensity of cyber-attacks can be measured and quantified. We evaluated advanced persistent threats (APT) and fileless cyber-attacks that occurred between 2010 and 2020 based on the methodology we developed. Based on our research methodology, we expect that researchers will be able to measure future cyber-attacks.

Keywords: offensive cybersecurity; cyber-attacks; scoring model; offensive cybersecurity framework

1. Introduction

The development of internet technology has increased the impact of cyber-attacks. The targets of cyber-attacks are steadily changing from traditional systems to cyber–physical systems (CPS). Cyber-attacks on smart mobility and smart homes are steadily increasing at a quickening pace. In 2005, two security researchers revealed the critical vulnerabilities of a self-driving car [1]. They remotely controlled the key features of a self-driving Jeep vehicle and succeeded in stopping the car on a highway. Cyber-attack techniques are becoming more sophisticated and destructive. Not only individual hackers but also state-sponsored hackers are actively entering the field of cyber-attacks. Cyber attackers use offensive cybersecurity technology to perform complex attacks. Offensive cybersecurity refers to a hacking technique that attacks a system, not a defense technology [2]. State-sponsored hackers and the Advanced Persistent Threat (APT) group also use offensive cybersecurity technology.

Attacks on Internet of Things (IoT) and smart homes are also constantly occurring. The Mirai botnet, known as the first IoT malware, attacked the Dyn network server, which is mainly operated in the United States [3]. This prevented Twitter, PayPal, and a significant portion of major online services from providing their services because of the huge amount of network traffic. The attack was not launched from a personal computer such as a zombie

5

botnet. The Dyn server received a massive load of traffic from IoT devices, and a variety of IoT cam companies were compromised by this malware. Korean offensive security researchers found critical vulnerabilities in Z-Wave, a wireless communication protocol that is commonly used in smart homes to communicate between the gateway and small nodes such as door-lock, multi-tab, and gas-lock [4]. A variety of threats to which smart homes are exposed were investigated by using existing offensive cyber security research methods. Much research regarding cyber-attacks has been conducted; however, there is a lack of research regarding systematic measurement of cyber-attacks.

It is necessary to identify the cyber threat actors to measure a cyber-attack. Cyber threat actors include individual hackers, cyber-terrorists, hacktivists, cybercriminals, and state-sponsored hackers. State-sponsored hackers and cybercrime organizations utilize APT that contains various offensive cybersecurity techniques. Table 1 presents some examples of nations and APT groups that have conducted cyber-attacks [5]. Among these APT analysis surveys, many reports use the terms of "sophisticated" attacks. In this paper, we propose an offensive cybersecurity framework as a method to systematically measure a score for the cyber-attacks in each isolated event. To the best of our knowledge, there have been no studies that score cyber-attacks. Hence, we analyze the degree of cyber-attack techniques for APT and fileless cyber-attacks that are using techniques contained in the offensive cybersecurity framework.

Table 1. Names of various Advanced Persistent Threat (APT) groups (sample).

Nations	APT Groups
China	APT1, Common Crew, PLA Unit 61398, Group 3, APT2, PLA Unit 61486, Buckeye, Gothic Panda
Russia	Sofacy, APT28, Sednit, Pawn Storm, Group 74, Fancy Bear, Grizzy Steppe, APT29, Dukes, Group 100, Cozy Duke, Cozy Bear, Cozer
North Korea	Lazarus Group, Labyrinth Chollima, Bureau 121, Whois Hacking Team, Hidden Cobra, DarkHotel, Luder, Karba, APT-C-06, Dubnium, Fallout Team, Tapaoux
Iran	Cutting Kitten, TG-2889, Ghambar, COBALT GYPSY, Magic Hound, Timberworm, Elfin, Refined Kitter, APT33, Holmium, Shamoon2.0

We proposed an offensive cybersecurity framework that systematically organizes techniques used in cyber-attacks and defined offensive cybersecurity taxonomy based on this framework. Then, we described the intention and techniques of cyber-attacks on each offensive cybersecurity module such as encryption, network, web, malicious code and system. We chose fileless cyber-attacks and APT for cyberattack scoring. Fileless cyber-attacks discovered from 2014 to 2018 and APT cyber-attacks assumed to be supported by China, Russia, North Korea, and Iran—known as state-sponsored—were selected. Then, for the selected target, the techniques used in the offensive cybersecurity element were identified. In the case of malicious code, the Cyber Kill Chain (CKC) concept is applied because more detailed attack steps are used. The scoring score was calculated in two steps: (1) the first was to calculate the score for how many Offensive Cybersecurity elements were used in each stage of the CKC; (2) Second, we calculated how many cyber-attack techniques were used in 12 ATT&CK. Finally, the first and second steps were combined to calculate the final score. We utilized published analytical reports for investigating the techniques used.

The main contributions of the proposed scoring model using an offensive cybersecurity framework can be summarized as follows:

1. We defined and derived the comprehensive offensive cybersecurity framework and taxonomy;
2. We performed a content analysis of public reports of cyber-attacks and identified detailed techniques used in cyber-attacks;
3. We provided a systematic scoring model based on the offensive cybersecurity framework;
4. We calculated the score results of ten fileless and eight APT group cyber-attacks.

The remainder of this paper is structured as follows. Section 2 provides the background and presents a literature review of research pertaining to offensive cyber security. Section 3 explains our overall methodology toward offensive cyber security and addresses each element of offensive cyber security in detail. Section 4 discusses the scoring results assigned to cyber-attacks. Finally, Section 5 provides considerations for future work and conclusions.

2. Background and Literature Review

This section presents the literature review used in our proposed cyber-attack scoring model including security for CPS, offensive cybersecurity, and state-sponsored cyber-attacks.

2.1. Security for CPS

Khatoun and Zeadally [6] address concepts, architecture, and research opportunities of smart cities. They consider key components of smart cities to be smart living, smart mobility, smart economy, smart government, and smart people with the core technologies, Internet of Things (IoT), Internet of Data (IoD) and Internet of People (IoP). The major contribution of their paper is an analysis of the detailed elements of smart cities.

Miller and Valasec performed a car attack demonstration that greatly contributed to encouraging research into attacks and critical security vulnerabilities of autonomous cars in smart mobility [1,7,8]. Kim et al. [9] surveyed over 150 papers related to attacks on, and the defense of, autonomous vehicle. Adel et al. [10] summarized the cybersecurity challenges in smart cities in terms of their safety, security and privacy. Their survey showed that the data generated by smart cities emerge from people, homes, transportation, workplaces, schools, commerce and social activities. In addition, their analysis showed that the data in a smart city are not only linear, but have a circular structure that collects and uses the collected data repeatedly. Privacy is an essential aspect of the structure of the smart city. Kim [4] analyzed threats of smart homes based on the STRIDE threat model and constructed a systematic attack tree on the basis of their findings. The authors analyzed smart home communication techniques that usually connect the different nodes with Wi-Fi, ZigBee and Z-Wave. The authors purchased an established Smart Home in which Smart Home techniques had been adopted, and found various vulnerabilities in smart home equipment.

2.2. Offensive Cybersecurity

Dino Dai Zovi addressed the modern history of offensive cyber security [11]. He analysed the history of three generations of offensive cyber security research from 1993 to 2017. The first generation of offensive security was mainly the ventures of underground hackers from 1993 to 1997. During this period, major system vulnerabilities, such as buffer overflow, were investigated. The second generation, from 1997 to 2007, established security companies to provide security consulting and solutions. The third generation of offensive security is that of governments hiring experts who have offensive capability, or academia studying offensive cybersecurity. The Defense Advanced Research Projects Agency (DARPA) hosted the cyber grand challenge (CGC) to find vulnerabilities using automatic machines and artificial intelligence techniques. The USENIX Workshop of Offensive Technology focused on offensive cybersecurity technology from 2007. The movement to promote the necessity of offensive cyber security research has become an inevitable wave.

Richard et al. [12] analyzed the taxonomies of cyber security attacks, which are the first examined known attack type, and they examined actual attack cases based on their classification. The paper classified attack components into seven categories: Attacks, Reconnaissance, Vulnerability, Threats, Exploits, Payloads, and Effects. However, these seven components have substantial mutual conceptual overlap. In addition, the relationship between exploitation and payload is unclear because a payload is the core code part of the exploit code.

Simon et al. [13] classified network and computer attacks. In their paper, the categories in the first dimension of network and computer attacks are viruses, worms, buffer overflows, denial of service attacks, network attacks, physical attacks, password attacks, and information gathering attacks. Then, 15 attacks were classified by adding the second, third, and fourth dimensions. Although they compiled a detailed taxonomy, because their work was published 10 years ago, recent cyber-attack techniques such as web attacks, mobile attacks, and IoT attacks are reflected to a limited extent.

Ben [14] noted that "sophisticated cyber-attacks" have increased dramatically over the last decade. The report identified various institutions, such as financial institutions, telecommunications agencies and state agencies, that have been attacked by "sophisticated cyber-attacks" and the meaning of the terms is not defined; thus, the meaning of "sophistication" is examined more intensively. The authors presented a framework for analyzing sophisticated cyber-attacks that included case studies to improve the formulation of offensive and defensive balance strategies in cyber operations. However, their report is not based on attack elements of the offensive cyber capabilities, but rather on a framework focused on attack methods. In addition, only some attack cases, such as the Stuxnet, were studied. Their study, therefore, does not follow a comprehensive and technical approach.

2.3. State-Sponsored Cyber-Attacks

Edward Snowden, the National Security Agency (NSA) contractor, unveiled an NSA hacking tool to the British newspaper *The Guardian* in 2013. This event drew the world's attention to state-sponsored hackers [15]. In general, Snowden's exposure revealed shocking tactics in the field of cybersecurity, as state-sponsored cyber-attacks were not readily visible to the public. In 2016, the Shadow brokers auctioned another code considered to be an NSA hacking tool [16]. The auction was unsuccessful, but later the NSA hacking tools were released to the world. The code released at the time, known as EternalBlue, exploited the Server Message Block (SMB) vulnerability that was used to create the WannaCry ransomware. Microsoft patched the SMB vulnerability under the name MS17-010 in 2017, necessitated by a widespread attack that was not reported by a whistle-blower in 2017. The United States also reportedly developed Stuxnet in 2010 with Israel with the aim of using cyber-attacks to delay Iran's nuclear development [17].

Another global operation by state-sponsored hackers was the involvement of Russian-sponsored hackers in the US presidential election. Bloomberg news said Russia's cyber-attack during the US presidential election in 2016 was very comprehensive and more powerful than ever before [18]. State-sponsored hacker organizations known to be sponsored by Russia are APT28 and APT29. The latter, also known as the Cozy Bear, is known to be sponsored by the Russian Federal Security Agency (FSB), a key Russian cyber security agency. APT28 is known as the Fancy Bear and is sponsored by the Main Intelligence Directorate (GRU), Russia's secret intelligence agency [19]. Benjamin et al. [20] described how Russian cyber actors disrupt and spy on the digital domain. Cyber tactics form part of the twenty-first century war strategy, and APT28 specifically attacked the Caucasus region and the North Atlantic Treaty Organization (NATO). They also noted that, in the 2016 US presidential election, Russia used psychological espionage to create a psychological impact on American society. Ben and Michael [21] stated that Russia's cyber operations are taking place across a very broad area of the United States. Consequently, the US government then took the Russian cyber-attack seriously and stressed that America should actively prepare for the network.

Various research reports on China's cyber operations are being published. Mandiant [22] researchers published that the PLA unit 61398, a Chinese military organization, has been continuously conducting cyber-attacks on the US government and civilian organizations. These attacks, unlike the one-off attacks by traditional attackers, have been designated APT because they have been continuously occurring for an extensive period of time. Therefore, the People's Liberation Army (PLA) Unit 61398 is named APT1. Antoine et al. [23] surveyed hacker organizations sponsored by China: APT16, APT17 (Aurora Panda), Shell_Crew, APT3

(Gothic Panda), APT15 (Ke3chang), APT12 (IXESHE), APT2 (Putter Panda), and APT30 (Naikon). The main characteristic of China's cyber operation is that it is designed to import industrial secrets from overseas advanced companies, mainly the exploitation of information. Kong et al. [24] investigated the link between North Korean cyber units and cyber operations on the North Korea cyber-attack capabilities. The authors predicted that cyber-attacks could not be expected to occur in only one country, anticipating linkages between political allies. According to a survey paper by Antoine et al. [23], Iran's nuclear weapons program was attacked by a joint operation between the US and Israel, known as the Olympic Games. Iran had been developing a malware named Shamoon two years after the Stuxnet attack and attacked financial and energy companies in the United States and Israel.

State-sponsored hacker groups have been investigated extensively [19]. In particular, after the involvement of Russian hackers in the US presidential election in 2016, additional research has been conducted. The attack techniques used by hacker groups supported by these countries are becoming more sophisticated; however, the level of these sophisticated cyber-attacks has not been measured.

3. Offensive Cybersecurity Framework

We systematized elements of the offensive cybersecurity framework to analyze the purpose and flow of cyber-attacks. Threat actors, which are individual hackers, cybercrime organizations, and nation-state hackers, are conducting cyber-attacks to achieve their goals such as financial gain or system destruction, as shown in Figure 1. Our framework is designed based on the threat actors, internet/network, and targets. The threat actors of cyber-attacks include Individual, Nation-State, and Cybercrime organizations. The Internet and Network are used in the process of accessing the attack target, and Public Network, Proxy, Virtual private network (VPN), and Darknet/Deepweb are used for this. The target of an attack is largely divided into Organizations, and CPS. Within the Organization, there are Web Servers and Web Application Servers (WAS) that are open to the outside, and there are personal computers and mobile phones inside the organization and documents and information in them. Organizations also have internal systems and databases containing important information. Recent attack targets include CPS and Persons. The detailed attack targets of the CPS include Smart Homes, Smart Mobility, Smart Economy, and Smart People. Offensive cybersecurity refers to the attack categories that cyber attackers need to compromise, including encryption, networks, web, malware, and systems.

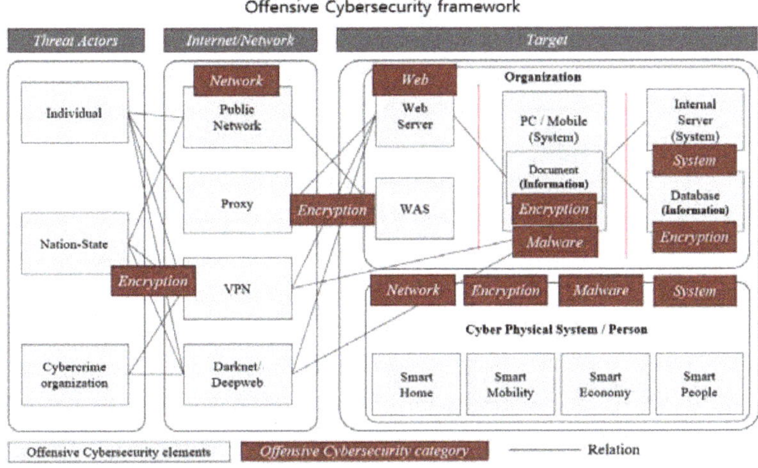

Figure 1. Offensive cybersecurity framework.

Figure 2 shows the taxonomy of Offensive Cybersecurity in detail. Encryption is a method used by cyber attackers to encrypt information, and there are Substitution, Symmetric, Asymmetric, and Hash Algorithms. Examples of network-related cyber-attacks include sniffing, spoofing, and Denial of Service (DoS) and Distributed Denial of Service (DDoS). Web attacks include Injection, Webshell upload, Authentication Access Control, File Download, and Cross-Site Scripting (CSS). Malicious code is more complex, and there is generally key logging, Remote Administration Tools (RAT), DLL injections, script attacks, and so forth, and Obfuscation, Packing, Cryptor and Protector protect the malicious code itself. Lastly, the system has detailed attack targets such as Applications, Services, Operating System (OS)/Kernel and Hypervisor, memory corruption attacks, such as Buffer Overflow (BOF) and Heap Overflow (HOF), and techniques to bypass security functions such as Return-oriented programming (ROP). We analyzed each offensive cybersecurity element to identify detailed techniques used by cyber-attackers.

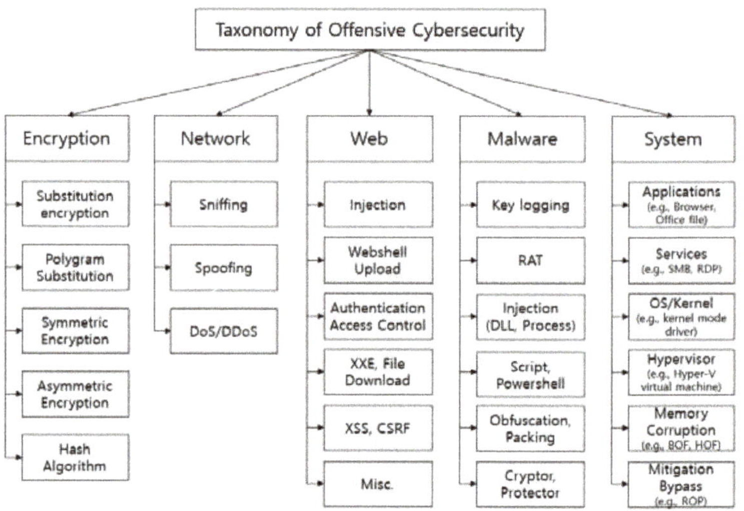

Figure 2. Offensive Cybersecurity Taxonomy.

3.1. Encryption

Encryption is the crucial technique used in cyber-attacks for various purposes. Traditional encryption techniques are used to encrypt confidential information to protect adversary users. The first encryption method was a substitution encryption. Representative encryption methods are the Caesar Cipher (B.C. 500) and the Monoalphabetic Cipher. Then, Polygram substitution, such as the Vigenere cipher (1585~1863) and Enigma cipher (1930), were developed. The creation of modern computing systems was followed by the development of symmetric encryption methods such as the Data Encryption Standard (DES) and Advanced Encryption Standard (AES). These two standards use the same key for encryption and decryption. As a result, an asymmetric encryption method was developed to share the secret key securely. Representative techniques in an asymmetric encryption are the Diffie–Hellman Key Exchange, Rivest–Shamir–Adleman (RSA) and the elliptic-curve cryptography (ECC) method. In the case of developed websites, login credentials, such as passwords, need to be saved in secure databases. In view of this circumstance, hash algorithms (MD5, SHA1 and SHA2) are used to protect users' credentials in the database.

Ransomware uses a symmetric encryption method to encrypt users' valuable files such as images and documents. Ransomware also uses an asymmetric encryption algorithm with the attacker's private key to protect the encryption key that is used in a symmetric encryption [25]. In the era in which state-sponsored hackers are becoming increasingly active, these more elegant attackers are using various encryption methods to hide traces of their activities. After gaining confidential information from a victim, they use various encryption methods to evade being monitored by the victim's system and network. APT28 (sponsored by Russia) uses encrypted POST data to send a command to command and control (C&C) server with an obfuscation base64 (block cipher). The encryption schemes used from an offensive cybersecurity perspective are listed in Table 2, and the cases in which these methods were used in cyber-attacks are compared.

Table 2. Cryptography techniques and cyber-attack cases.

Category	Encryption Method	Cyber-Attack Case
Substantial Encryption	Caesar cipher	Coin Locker
	Rotation 13 (ROT13)	CryptoShield
Symmetric Encryption	DES	-
	3DES	JobCrypter Ransomware
	AES	Ransomware [25], APT29
Asymmetric Encryption	RSA	Ransomware, APT28, APT29
	ECC	OphionLocker Ransomware
Block Encryption	Rivest Cipher (RC)4	ProjectSauron [26], DarkHotel [27]
	RC5	ProjectSauron
	RC6	ProjectSauron
	Base64	APT28, APT29
Hash	Message-Digest (MD)5	-
	Secure Hash Algorithm (SHA)	NetWalker ransomware
Custom	Custom	DarkUniverse

3.2. Network

Most well-known cyber-attacks techniques related to networks are sniffing, spoofing, and Denial of Service. Sniffing is known as a passive network attack because it does not directly attack the target's computer. Spoofing, on the other hand, is an attack that deceives the network protocol and includes Internet Protocol (IP), Domain Name System (DNS), and Address Resolution Protocol (ARP) spoofing. Spoofing is classified as a representative active network attack.

As the Internet evolves, many organizations offer their products and services over the Internet. Network availability has become very important in this process, particularly in the case of CPS. Attackers are using networks to perform denial of service (DoS) attacks and distributed denial of service (DDoS) attacks as a way to attack network availability. DoS and DDoS attacks include Ping of Death, Synchronize (SYN) flooding, and Hypertext Transfer Protocol (HTTP) Get flooding techniques. From the point of view of offensive security, the main objective of a network attack is to conceal the identity and location of the attacker. IP addresses are typically used to identify devices on the network. Attackers hide their source through a proxy or Tor browser to prevent their IP addresses from being revealed.

Hassan et al. [28] and Nazrul et al. [29] classified network cyber-attacks as follows: Information gathering, DoS and DDoS attacks, spoofing, TCP session hijacking, probe, application layer, malformed packet, amplification, and protocol exploit attacks. Dileep et al. [30] classified network attacks based on the network layer. We adopt this concept to construct our offensive framework.

3.3. Web

From the perspective of offensive cybersecurity, the Web is a prime target. Most companies promote their products and services on websites; however, it also makes it easy for attackers to conduct cyber-attacks through a website. Because of this, web hacking frequently occurs and related attack techniques are constantly being studied. We investigated web attack techniques and adopted the Open Web Application Security Project (OWASP) for our methodology. The OWASP organization was formed in 2001 and reports ten dangerous vulnerabilities on its website every three years. The web vulnerability that was consistently ranked number one from 2010 to 2017 is an injection vulnerability. This can be used to fetch database information from a web site or to gain system privileges on the web server. A typical attack among those intended to hack websites, along with injection attacks, is a defacement attack. It is used to steal information, whereas website tampering attacks reveal and hack. Typically, hacktivist organizations hack websites because of their political orientation. Among the attacks used to alter web pages, a typical hacktivist attack uploads a malicious script, known as a web shell, to a web server to alter the web page. We determined that a sizeable proportion of cyber-attacks use compromised websites to spread their malicious code. Table 3 shows the top ten most dangerous web vulnerabilities from 2010 to 2017.

Table 3. OWASP Top 10 (From 2010 to 2017).

Top 10	2010	2013	2017
A1	Injection	Injection	Injection
A2	Cross-Site Scripting (XSS)	Broken Authentication and Session Management	Broken Authentication
A3	Broken Authentication and Session Management	Cross-Site Scripting (XSS)	Sensitive Data Exposure
A4	Insecure Direct Object References	Insecure Direct Object References	XML External Entities (XXE)
A5	Cross-Site Request Forgery (CSRF)	Security Misconfiguration	Broken Access Control
A6	Security Misconfiguration	Sensitive Data Exposure	Security Misconfiguration
A7	Insecure Cryptographic Storage	Missing Function Level Access Control	Cross-Site Scripting (XSS)
A8	Failure to Restrict URL Access	Cross-Site Request Forgery (CSRF)	Insecure Deserialization
A9	Insufficient Transport Layer Protection	Using Known Vulnerable Components	Using Components with Known Vulnerabilities
A10	Unvalidated Redirects and Forwards	Unvalidated Redirects and Forwards	Insufficient Logging and Monitoring

3.4. Malware

A considerable amount of research has been devoted to uncovering malware techniques. Representative research is the form of ATT&CK. Kris and Christian [31] explained malware techniques based on the ATT&CK step. Techniques used during the malware execution step include: "Execution through API (CreateProcessA function)", "using Rundll32", "Command-Line Interface (cmd.exe)", "Service Execution (register or execute as a service)", "PowerShell", and "Windows Management Instrumentation (WMI)". Persistence techniques are "Registry Run Key", "New Service", "Modify Existing Service", "Hooking", "Schedule Task", and "Image File Execution Options Injection". Techniques for Privilege Escalation include Process Injection and Access Token Manipulation. Ekta et al. [32] proposed a malware threat assessment using a fuzzy logic paradigm.

Paul et al. [33] described the behavior of malware as: persistence, configuration, process injection, information stealing and injection, network communications, backconnect, screenshot and video capture, and anti-analysis. Anti-analysis uses obfuscation, packing, cryptor, and protector techniques to confuse analysis malware. In the same way that bullets are important in war, malware is also crucial ammunition in cyber-attacks. Malware uses a wide variety of technologies, and unlike other elements, the stage at which the malicious code is executed is very complicated. According to Paul et al. [33], obfuscation and packing is a technique for hiding malware, and recently, the DLL Side-loading technique has been utilized. We constructed the technology and stages used by malicious code based on the MITRE ATT&CK. At the ATT&CK stage, the technology focused on malicious codes was quantified, and Execution, Persistence, Privileges Escalation, Defense Evasion, Credential Access, Lateral Movement, Command and Control, Exfiltration, and Impact were selected as representative technologies as listed in Table 4.

Table 4. Malicious code techniques.

Stages	Techniques
Execution	Script, Powershell, Cli, Schedule task, Signed binary, User execution, Service execution, Rundll32, Mshta, WMI
Persistence	Registry run keys, Scheduled task, New service
Privilege Escalation	Process injection, Exploit, Access token manipulation, New service
Defense Evasion	Hidden files, Modify registry, Permission modify, Process injection, Packing, Deletion, Obfuscate, Masqurade, deobfuscate, Disable tools, Mshta, Indicator rm
Credential Access	Credential dumping, Brute force, Credential in files, Pass the Hash
Lateral Movement	WA Share, Exploit remote, Remote file copy
Command and Control	Common ports, Multi-hop proxy, Multilayer encryption, Remote file copy, Uncommon ports, Data encoding, Data obfuscation
Exfiltration	Automated exf. Exf. over alt. protocol, Data encrypted, Exf. over C&C
Impact	Disk struct wipe, Encrypt data, Inhibit recovery, Service stop

3.5. System

System attack has been the subject of extensive research. Systems consist of many layers: Application, Services, OS and Kernel, and Hypervisor. The prime vulnerability of systems and applications is a memory corruption. Mitigation techniques have been steadily researched; in addition, mitigation bypass techniques have also been developed continuously. A system is divided into four layers: Applications, Services, OS and Kernel, and Hypervisor for the cloud. Application categories include browsers, Microsoft Office, and Adobe programs. Services represent specific functions that are provided from outside the system and include the SMB and the remote desktop protocol (RDP). The operating system and kernel level are other prevalent attack targets. In the cloud environment, the hypervisor is the basis on which the operating system is run and also a critical target of offensive cyber-attacks. Most types of system vulnerability are in the memory corruption category. These techniques are buffer overflow (BOF), heap overflow (HOF), and integer overflow. A number of mitigation techniques have been developed to defend against system vulnerability. Data execution prevention (DEP) in Windows and the no-execute (NX) bit are designed to defend the execute shell code in the stack area. Address space layer randomization (ASLR), which is also adopted as a defense against memory corruption attacks, changes the stack address after each execution. Offensive techniques to bypass these mitigations are steadily being researched; specifically, return-oriented programming (ROP) is a major attack technique used to bypass a stack address defense.

4. Cyber-Attacks Evaluation

We evaluated each cyber-attack case by modeling offensive cybersecurity. We adopted the proposed methodology by selecting numerous fileless and APT cyber-attack cases. The reason we select fileless and APT is that these kinds of cyber-attacks have advanced the sophistication of cyber-attack techniques.

4.1. Dataset

We used datasets that contain cases of two types of cyber-attacks: fileless cyber-attacks and APT group cyber-attacks. We chose ten recent fileless cyber-attacks listed in Table 5 from the dataset to evaluate our scoring model. The ten selected fileless cyber-attacks were Poweliks, Pozena, Duqu 2.0, Kovter, Petya, Sorebrect, WannaCry, Magniber, Emotet, and GandCrab.

Table 5. Fileless cyber-attack dataset.

No	Cyber-Attack	Year	SHA256
1	Poweliks	2014	-
2	Rozena	2015	c23d6700e93903d05079ca1ea4c1e36151cdba4c5518750dc604829c0d7b80a7
3	Duqu 2.0	2015	52fe506928b0262f10de31e783af8540b6a0b232b15749d647847488acd0e17a
4	Kovter	2016	-
5	Petya	2017	027cc450ef5f8c5f653329641ec1fed91f694e0d229928963b30f6b0d7d3a745
6	Sorebrect	2017	4142ff4667f5b9986888bdcb2a727db6a767f78fe1d5d4ae3346365a1d70eb76
7	WannaCry	2017	ed01ebfbc9eb5bbea545af4d01bf5f1071661840480439c6e5babe8e080e41aa
8	Magniber	2017	c21887eaa1e31b9220d0807d3a7d0f30421ab6f80cfc1c556d534587dd9e6343
9	Emotet	2017	70903a9ef495edd8de01a61f8e9862a037b0dee327d7f92f93ef69e33e461764
10	GandCrab	2018	643f8043c0b0f89cedbfc3177ab7cfe99a8e2c7fe16691f3d54fb18bc14b8f45

We also selected APT Group and Operation from the APT group list [5]. In this study, we first investigated six nations: China, Russia, North Korea, Iran, Israel, and nations in the Middle East. These countries and regions have 87, 20, 9, 9, 2 and 17 cyber-attack groups, respectively, as listed in Table 6. We selected the APT groups of China, Russia, North Korea, and Iran because they are known publicly. Even though China has the most cyber-attack groups, the mere number of cyber-attack groups does not indicate a nation's cyber-attack capabilities. For example, in comparison to China, Israel has only two cyber-attack groups. China has a smattering of small cyber-attack groups, however Israel's Unit 8200 group is known to be the most powerful group.

Table 6. Cyber-attack group dataset.

Nations	Counts	APT Groups Common Name
China	87	Comment Crew, APT2, UPS, IXESHE, APT16, Hidden Lynx, Wekby, Axiom, Winnti Group, Shell Crew, Naikon, Lotus Blossom, APT6, APT26, Mirage, NetTraveler, Ice Fog, Beijing Group, APT22, Suckfly, APT4, Pitty Tiger, Scarlet Mimic, C0d0so, SVCMONDR, Wisp Team, Mana Team, TEMP.Zhenbao, SPIVY, Mofang, DragonOK, Group 27, Tonto Team, TA459, Tick, Lucky Cat, APT40, PassCV, BARIUM, LEAD, Iron Group, Anchor Panda, Big Panda, Electric Panda, Eloquent Panda, Emissary Panda, Foxy Panda, Gibberish Panda, Goblin Panda, Hammer Panda, Hurricane Panda, Impersonating Panda, Judgement Panda, Karma Panda, Keyhole Panda, Kryptonite Panda, Mustang Panda, Night Dragon, Nightshade Panda, Nomad Panda, Pale Panda, Pirate Panda, Poisonous Panda, Predator Panda, Radio Panda, Sabre Panda, Spicy Panda, Stone Panda, Temper Panda, Test Panda, Toxie Panda, Union Panda, Violin Panda, Wet Panda, Calypso, Tropic Trooper, APT41, Poison Carp, AVIVORE, APT-C-01, DarkUniverse, Taskmasters, GALLIUM, RANCOR, ChinaZ, APT-C-37, APT-C-27
Russia	20	Sofacy, APT29, Turla Group, Energetic Bear, Sandworm, FIN7, FIN8, Inception Framework, TeamSpy Crew, BuhTrap, Carberb, FSB 16th & 18th Centers, Cyber Berkut, WhiteBear, GRU GTsST (Main Center for Special Technology), VOODOO BEAR, TEMP.Veles, Zebrocy, SectorJ04, FullofDeep
North Korea	9	Lazarus Group, Group13, DarkHotel, Andariel, Kimsuki, NoName, OnionDog, TEMP.Hermit, Stardust Chollima
Iran	9	Cutting Kitten, Shamoon, Clever Kitten, Madi, Cyber fighters of Izz Ad-Din Al Qassam, Chafer, Prince of Persia, Sima, Oilrig
Israel	2	Unit 8200, SunFlower
Middel East	17	Molerats, AridViper, Volatile Cedar, Syrian Electronic Army (SEA), Cyber Caliphate Army (CCA), Ghost Jackal, Corsair Jackal, Extreme Jackal, Electric Powder, APT-C-23, APT-C-27, Dark Caracal, Tempting Cedar, Sandcat, Group WITRE, ZooPark, APT-C-37
Total	144	

4.2. Investigation

We calculated the cyber-attack score based on the Open Source Intelligence (OSINT) method. Many cybersecurity companies publish an analysis of cyber-attack cases, and certain countries, such as the United States and South Korea, publish reports with analyses of cyber-attack cases. We analyzed these reports to identify the techniques that were being used by the cyber-attack groups.

To identify detailed techniques that were used for each cyber-attack, we selected representative examples of a cyber-attack for each country as well as fileless cyber-attacks. In the first step of our analysis, we used MITRE ATT&CK cyber-attack group artifacts to identify the cyber-attack techniques that were used. Next, we analyzed the cyber-attack techniques in detail based on our proposed offensive cybersecurity framework for each representative cyber-attack as listed in Tables 7 and 8.

Table 7. Techniques used in the fileless cyber-attacks samples.

Fileless Cyber-Attack	Techniques
Poweliks	MS Office Macro vulnerability Inject malicious script into the registry Execution of registry value using Rundll32.exe Execution of registry value encoded using Jscript.Encode Use of Powershell script encoded with Base64 Verification of registry key and path of executed files DLL execution through Powershell schript (injection using dllhost.exe) Deleting files after every operation Dllhost.exe resides Send a user's system information to C&C server through TCP communication
Kovter	Social engineering techniques using email attachments Inject malicious script into the registry Execution of registry value using Mshta.exe Execution of registry value encoded using Jscript.Encode Use of Powershell script encoded with Base64 Injecting codes through Powershell script Deleting files after every operation Regsvr32.exe resides Send a user's system information to C&C server through TCP communication

Table 8. Techniques used in the APT Group cyber-attacks.

Cyber-Attack	Techniques Used
APT29	Network: Multi-hop Proxy, Commonly Used Port, Domain Fronting, Spearphishing Attachment, Spearphishing Link, Standard Non-Application Layer Protocol, User Execution Encryption: Obfuscated Files or Information, Software Packing System: Accessibility Features, Bypass User Account Control, Exploitation for Client Execution, Pass the Ticket Malware: File Deletion, Indicator Removal on Host, PowerShell, Registry Run Keys/Startup Folder, Rundll32, Scheduled Task, Scripting, Shortcut Modification, Windows Management Instrumentation, Windows Management Instrumentation Event Subscription Used tools: CloudDuke, CosmicDuke, CozyCar, GeminiDuke, HAMMERTOSS, meek, Mimikatz, MiniDuke, OnionDuke, PinchDuke, POSHSPY, PowerDuke, PsExec, SDelete, SeaDuke, Tor

Table 8. *Cont.*

Cyber-Attack	Techniques Used
Lazarus Group	Encryption: Custom Cryptographic Protocol, Data Compressed, Data Encoding, Data Encrypted, Obfuscated Files or Information, Standard Cryptographic Protocol Network: Commonly Used Port, Connection Proxy, Exfiltration Over Alternative Protocol, Exfiltration Over Command and Control Channel, Fallback Channels, Multiband Communication, Spearphishing Attachment, Standard Application Layer Protocol, Uncommonly Used Port Web: Drive-by Compromise Malware: Access Token Manipulation, Account Manipulation, Application Window Discovery, Bootkit, Brute Force, Command-Line Interface, Credential Dumping, Data from Local System, Data Staged, Disabling Security Tools, Disk Content Wipe, Disk Structure Wipe, File and Directory Discovery, File Deletion, Hidden Files and Directories, Input Capture, New Service, Process Discovery, Process Injection, Query Registry, Registry Run Keys/Startup Folder, Remote File Copy, Resource Hijacking, Scripting, Service Stop, Shortcut Modification, System Information Discovery, System Network Configuration Discovery, System Owner/User Discovery, System Shutdown/Reboot, System Time Discovery, Timestomp, Windows Admin Shares, Windows Management Instrumentation System: Compiled HTML File, Exploitation for Client Execution, Remote Desktop Protocol, User Execution Used tools: AuditCred, BADCALL, Bankshot, FALLCHILL, HARDRAIN, HOPLIGHT, KEYMARBLE, Mimikatz, netsh, Proxysvc, RATANKBA, RawDisk, TYPEFRAME, Volgmer, WannaCry

4.3. Scoring Procedure

We collected almost 150 reports with analyses of attacks and websites using the links that had already been collected by the APT Group list [5]. Then, we searched each paper for cyber-attack techniques using the keyword-base addressed in Section 3 (Offensive Cybersecurity elements). Finally, we calculated the cyber-attack score based on the offensive cybersecurity elements. Figure 3 and Algorithm 1 shows the overall scoring procedure. Additionally, one cyber-attack group can perform various cyber-attack operations. In this case, we selected representative cyber-attack operations for each cyber-attack group. We used the Lockheed Martin Cyber Kill-Chain process to identify the cyber-attack techniques for each step [34].

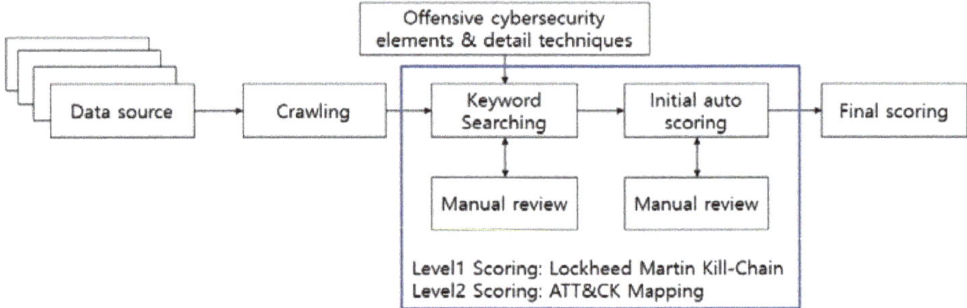

Figure 3. Overall scoring procedure.

MITRE ATT&CK evaluations showed the scoring result for some APT groups such as APT3, APT29, and Carbanak+FIN7. This evaluation was based on the 20 attack stages; however, it only focused on the malware itself rather than mapping the overall chain of the cyber-attack. Thus, our scoring approach differs from that used in the ATT&CK evaluation.

In addition, our approach provides a more comprehensive description of cyber-attack techniques to arrive at a cyber-attack score.

Algorithm 1 Generating score for each cyber-attack.

1: **Input :** Data source-malware analysis report
2: **Output :** Cyber-attack scoring
3: $N \leftarrow$ number of cyber-attack cases
4: *Element[score]* $\leftarrow 0$
5: **for** $k \leftarrow 1$ to N **do**
6:     *keyword* $\leftarrow$ *keyword file*
7:     **if** *content* $==$ *keyword* **then**
8:         *score* $\leftarrow$ *score* $+ 20$
9:         *element*[*score*] $\leftarrow$ *score*
10:     **end if**
11: **end for**
12: Return element[score]

We evaluated cyber-attacks at two levels to calculate the score. Level 1 uses the top offensive cybersecurity elements for each cyber-attack with cyber kill chain phases. We identified the offensive cybersecurity elements that were used in each cyber-attack case. For this purpose, we analyzed the cyber-attacks and mapped them with the Lockheed Martin Cyber Kill-Chain. The Cyber Kill-Chain has seven phases and a total of five cybersecurity high-level elements. We awarded 20 points to each cybersecurity element.

Combining the cyber-attack techniques used in each of the offensive cybersecurity model, the cyber-attack complexity is expressed as follows.

$$Z = Sum \sum_{k=0}^{n} \frac{UOCSM}{TOCSM}. \tag{1}$$

In Equation (1), UOCSM denotes 'Using Offensive Cybersecurity Modules', TOCSM means 'Total Offensive Cybersecurity Modules', and Z represents the utilization of offensive cybersecurity elements. Thus, each phase can earn a maximum of 100 points, which means that the maximum total score (TOCSM) is 700 points.

Level 2 utilized data used by ATT&CK techniques in each element for cyber-attacks. ATT&CK has 12 steps for conducting cyber-attacks. We calculated the sum of technologies used in each ATT&CK phase of the cyber-attack.

4.4. Scoring

4.4.1. Cyber-Attack Scoring Result with Cyber Kill Chain

We analyzed in detail the cyber-attack techniques for each fileless cyber-attack. Figure 4 presents an example of the scoring result for a Poweliks fileless cyber-attack. For Poweriks, in the initial Reconnaissance phase of Cyber Kill Chain, the attackers obtained e-mail addresses of post office workers in the US and Canada. Then, in the Weaponization phase, the attacker created an MS Word document file and inserted malicious code using the Macro vulnerability inside. In the delivery stage, an email containing a malicious program was disguised as a normal email and delivered to the employee's email address. In the Exploitation phase, malicious code was executed using the MS macro function vulnerability. It also inserted a malicious script into the registry and rendered the name of the registrar key unreadable. Additionally, the encoded registry value was executed using the JScript.Encode function. In the installation step, the Base64-encoded PowerShell script was executed. DLL injection was also performed using a PowerShell script, and the malicious DLL is packed with MPRESS. In addition, Poweliks registers malicious code in the automatic program startup registry to perform permanent attacks. In the Command and Control phase, TCP connections are transformed into the two IP addresses that are estimated to be servers. In the Action on objectives step, information about the user's PC is collected and information is transmitted to the attacker's server.

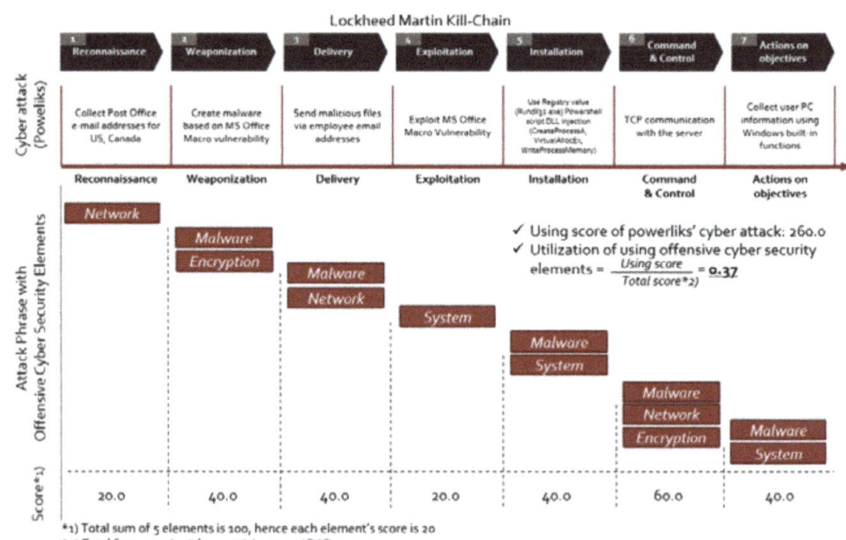

Utilization of using offensive cyber security elements $= \dfrac{Using\ score}{Total\ score\,*2} = 0.37$

Figure 4. Example of scoring using a Poweliks cyber-attack.

For the WannaCry case, it does not appear to have a separate Reconnaissance stage. In the Weaponization phase, the attacker creates a malicious program disguised as the icon of a normal program. In the delivery phase, the attacker uploads a malicious file disguised as a normal file online, and then the victim downloads it. In the Exploitation phase, the MSSecsvc 2.0 service is installed on the victim's PC, and files hidden in the program's resources are dropped and executed. In the installation phase, the dropped program is registered in the registry run key, and it is automatically executed whenever the PC is booted. In the Command and Control phase, communication occurs via the Tor network, and port 9050 is left open should communication with an external server be required. Actions on objectives encrypt all data except for data in a specific file path. The volume shadow file is deleted using Vssadmin on the infected PC. The peculiarity is that the SMB vulnerability also enables the shellcode to be transmitted to a computer on a shared network, and the vulnerability of the PC results in the same process being used to infect the ransomware.

We selected the Lazrus cyber-attack for the APT group from four countries. Figure 5 shows an example of the scoring result for the Lazarus APT cyber-attack. In the Lazarus case, the cyber-attack collected a post-office e-mail address and investigated specific targets with network proving techniques in the reconnaissance step. In the Weaponization step, Lazarus developed malware by exploiting the 0-day vulnerability of Adobe software. Many encryption techniques were adopted during the development of malware. Web, malware, and network techniques were used in the delivery step. In the Exploitation step, Lazarus used various 0-day exploits; thus, we evaluated the system and malware element in the exploitation step. Malware, system, and encryption techniques were used in the Installation step, which used TCP port 443 with some payloads for the implementation of SSL encryption. Actions on the Objectives step in the cyber kill chain were performed by gaining system information, downloading and uploading files, and using the execution command.

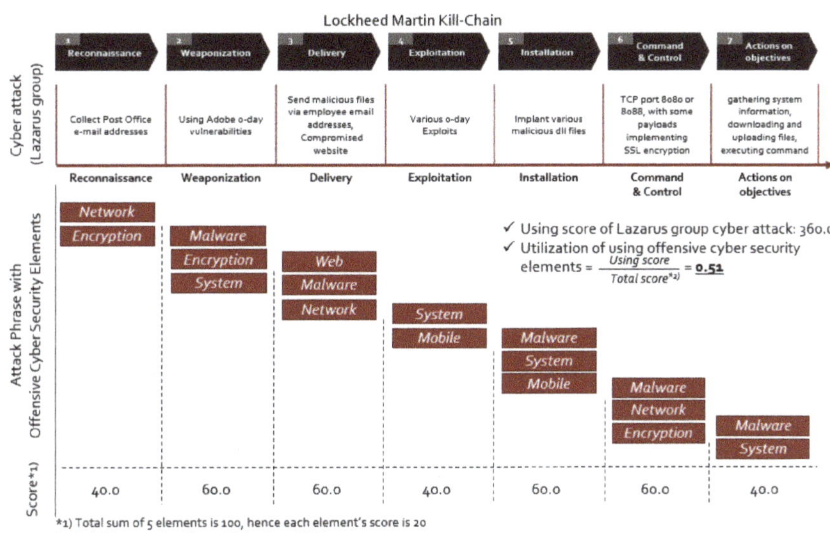

Figure 5. Cyber-attack Score example for Lazarus Group.

Through these methods, we calculated the scoring result of cyber-attacks with Cyber Kill Chain as listed in Table 9. The result using Cyber Kill Chain shows Poweliks (0.3714), Rozena (0.3429), Duqu 2.0 (0.3429), Kovter (0.3429), Petya (0.4286), Sorebrect (0.2857), WannaCry (0.3714), Magniber (0.2857), Emotet (0.3429), and GandCrab (0.3429) for fileless cyber-attacks. For APT cyber-attacks, it shows APT1 (0.4000), Emissary Panda (0.4857), APT29 (0.4286), SectorJ04 (0.4571), Lazarus Group (0.5143), APT38 (0.4286), Chafer (0.4000), MuddyWater (0.4286). The average score for fileless and APT cyber-attacks is 0.3459, 0.4318, respectively. This shows that APT cyber-attacks use more Cyber Kill Chain techniques than fileless cyber-attacks.

Table 9. Scoring result of cyber-attacks using Cyber Kill Chain.

No.	Cyber-Attacks	UOCSM	R	W	D	E	I	C	A	Total
1	Poweliks	260.0	20	40	40	20	40	60	40	0.3714
2	Rozena	240.0	20	40	20	60	40	40	20	0.3429
3	Duqu 2.0	240.0	-	40	40	40	40	40	40	0.3429
4	Kovter	240.0	20	40	40	40	40	20	40	0.3429
5	Petya	300.0	20	60	40	60	60	-	60	0.4286
6	Sorebrect	200.0	-	40	20	40	-	40	60	0.2857
7	WannaCry	260.0	-	40	40	40	40	40	60	0.3714
8	Magniber	200.0	-	40	20	40	40	-	60	0.2857
9	Emotet	240.0	-	40	40	40	40	40	40	0.3429
10	GandCrab	240.0	-	40	20	40	40	40	60	0.3429
11	APT1	280.0	20	60	40	40	60	-	60	0.4000
12	Emissary Panda	340.0	40	40	60	40	60	60	40	0.4857
13	APT29	300.0	40	20	40	40	60	60	40	0.4286
14	SectorJ04	320.0	20	40	40	40	60	60	60	0.4571
15	Lazarus Group	360.0	40	60	60	40	60	60	40	0.5143
16	APT38	300.0	40	40	60	40	40	40	40	0.4286
17	Chafer	280.0	40	40	40	40	40	40	40	0.4000
18	MuddyWater	300.0	20	40	40	40	60	60	40	0.4286

4.4.2. Cyber-Attack Scoring Result with ATT&CK

Owing to the nature of cyber-attacks, increasingly complicated techniques are included in malicious code. Thus, when calculating the cyber-attack score, we conducted a more in-depth analysis of the malicious code. An analyze process was carried out on the 12 stages of the MITRE ATT&CK, and the results are shown in Figure 6.

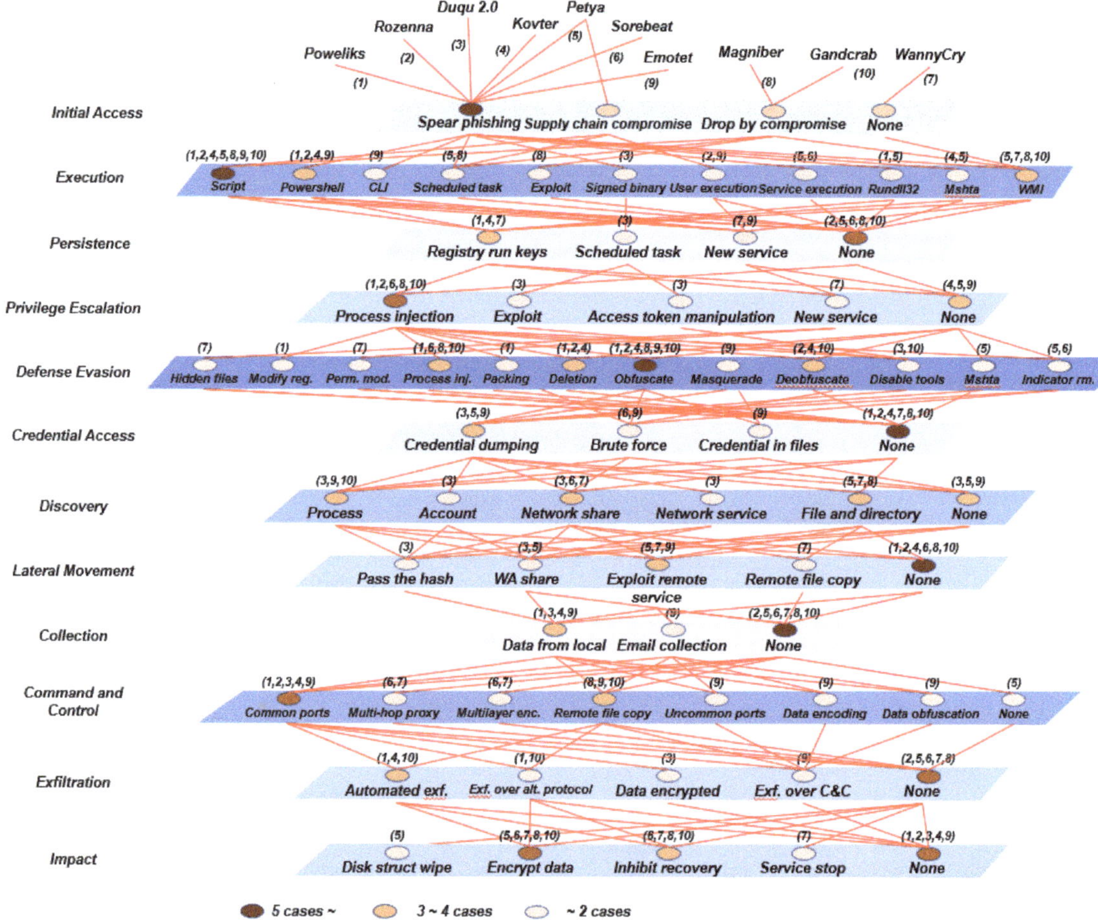

Figure 6. Attack techniques of fileless cyber-attack [35].

In this figure, each layer shows the stages of the cyber-attack in MITRE ATT&CK. The order of layers is the flow of an attack. The red line indicates the connection of a technique used by a malicious code with the previous techniques used. At each stage, the shade of blue plane indicates the cyber-attack techniques. In addition, the color of the circle indicates the number of cyber-attacks that use it. For instance, if the number of cyber-attack types is five or more, the circle's color is dark brown; if it is three or four, it is orange, and if it is two or less, it shows an apricot color.

For example, the result of Duqu 2.0 mapping to ATT&CK is as follows. The initial access step used a spearphishing attachment. Signed binary and proxy execution were used in the Execution step. In the Persistence step, the scheduled task technique was used, and in the Privilege escalation step, exploitation for privilege escalation and access token manipulation techniques were used. In the Defense evasion step, disabling security tools were used, and in the Credential access step, credential dumping was used. In the Discovery step, process discovery, account discovery, network share discovery, and network service scanning were used. Data from local system technology was used in the Collection step, and a commonly used port was used in the Command and Control step. Data encrypted technology was used in the Exfiltration step.

Through these methods, we calculated the scoring result of cyber-attacks with the MITRE ATT&CK as listed in Table 10. The following is a description of the results for Petya, which earned a score of 0.2581. In the initial access step, a spear phishing attachment and supply chain compromise technologies were used. Scripting, mshta, service execution, WMI, rundll32, and schedule tasks were used in the Execution step. In Persistence and Privilege escalation, no special technique was used. In the Defense evasion step, mshta, indicator removal on host technology was used. Credential dumping technology was used in the Credential access step, and file and directory discovery technology was used in the Discovery step. In the Lateral movement step, Petya used Windows admin shares and exploited remote services technology. In the Impact step, disk structure wipe and data encrypted for impact technologies were used.

Table 10. Scoring result of cyber-attacks using ATT&CK phases.

No.	Cyber-Attack	Number of used techniques of MITRE ATT&CK												Total
		I	E	P	P	D	C	D	L	C	C	E	I	
1	Poweliks	1	4	1	1	5	-	-	-	1	1	2	-	0.2581
2	Rozena	1	3	-	1	3	-	-	-	-	1	-	-	0.1452
3	Duqu 2.0	1	2	1	2	1	1	4	2	1	1	1	1	0.2742
4	Kovter	1	4	1	-	5	-	-	-	1	1	1	-	0.2258
5	Petya	2	6	-	-	2	1	1	2	-	-	-	2	0.2581
6	Sorebrect	3	1	-	1	2	1	1	-	-	2	-	2	0.2097
7	WannaCry	-	1	2	1	2	-	2	2	-	2	-	3	0.2419
8	Magniber	1	4	-	1	1	-	1	-	-	1	-	2	0.1774
9	Emotet	1	4	1	-	2	3	1	1	2	5	1	-	0.3387
10	GandCrab	1	3	-	1	4	-	1	-	-	1	2	2	0.2419
11	APT1	-	1	-	-	1	1	6	2	5	-	-	-	0.2581
12	Emissary Panda	1	4	3	5	6	3	6	2	9	2	1	-	0.6774
13	APT29	2	6	2	3	3	-	-	1	-	3	-	-	0.3226
14	Sector04	2	8	-	-	2	2	-	-	-	1	-	1	0.2581
15	Lazarus Group	2	4	3	3	11	2	8	2	6	7	-	7	0.8871
16	APT38	1	1	-	-	4	1	2	-	2	1	-	7	0.3065
17	Chafer	2	4	3	-	1	2	3	2	1	1	-	-	0.3065
18	MuddyWater	2	8	1	1	4	6	6	-	2	6	-	-	0.5806

In contrast, Magniber, with a score of 0.1774, operates as follows. In the initial access step, drive-by compromise technology was used. In the Execution step, scripting, exploitation for client execution, scheduled task, and WMI were used. No technique seems to have been used in the Persistence step. Process injection technology was used in the Privilege escalation step, and obfuscated files or information technology was used in the Defense evasion phase. In the Discovery phase, files and directory discovery technology was used. No technique was used in the Lateral movement and Collection stages. In the Command and Control phase, remote file copy technology was used. In the Impact phase, data encrypted for impact and inhibit system recovery technologies were used.

The technique that was used for Emissary Panda (0.6774), which holds the highest score after Lazarus, is as follows. In the Initial access stage, the drive-by compromise technique was used. In the Execution phase, PowerShell, the windows command shell, Exploitation for client execution, and WMI were used. In the Persistence phase, registry run keys, create or modify system process, and web shell technologies were used. In the

Privilege escalation stage, bypass user access control, exploitation for privilege escalation, hijack execution flow, dll side-loading, and process hollowing techniques were used. In the Defense evasion phase, Windows event logging is disabled, file deletion, network share connection removal and obfuscated files or information were used. In the Credential access phase, OS credential dumping technique for LSA secrets, LSASS memory, and security account manager was used. In the Discovery phase, local account, network service scanning, and query registry methods were used. In the Lateral movement phase, the exploitation of remote services technique is used. In the Collection phase, automated collection is performed. In the Command and Control phase, web protocols, ingress tool transfer are used. In the Exfiltration step, archive via library is used.

Next, we analyzed APT38, which obtained a fairly low score; however, many techniques are used in the Impact step. APT38 used a drive-by compromise technique in the initial access step. In the Execution step, the Windows command shell was used. In the Defense evasion step, indicator removal on host, modify registry, and software-packing techniques were used. In the Credential access step, input capture technique was used. In the Collection step, clipboard data were executed. In the Command and Control step, web protocols and ingress tool transfer were used. In particular, in the Impact step, many techniques were used: data destruction, data encrypted for impact, data manipulation, disk structure wipe, and system shutdown techniques.

4.4.3. Scoring Result Summary

We derived the score for the final cyber-attacks by combining the Cyber Kill Chain score including cybersecurity offensive elements and the score based on MITRE ATT&CK as shown in Figure 7. Each fileless cyber-attack score shows powerliks (0.6295), Rozena (0.4881), Duqu 2.0 (0.6171), Kovter (0.5687), Petya (0.6867), Sorebrect (0.4954), WannaCry (0.6133), Magniber (0.4631), Emotet (0.6816), Gandcrab (0.5848). We can show that Petya has the highest score of the fileless cyber-attacks. Each APT cyber-attack score shows APT1 (0.6581), Emissary Panda (1.1631), APT29 (0.7512), Sectorj04 (0.7152), Lazarus Group (1.4014), APT38 (0.7351), Chafer (0.7065), Muddywater (1.0092). The Lazarus group APT shows the highest score of the cyber-attacks. Overall, APT cyber-attacks usually score higher than fileless cyber-attacks, because he APT use more ATT&CK techniques. The limitation of our approach is that real malware cannot be analyzed. However, we believe that measuring the score of cyber-attacks is meaningful as an initial research step.

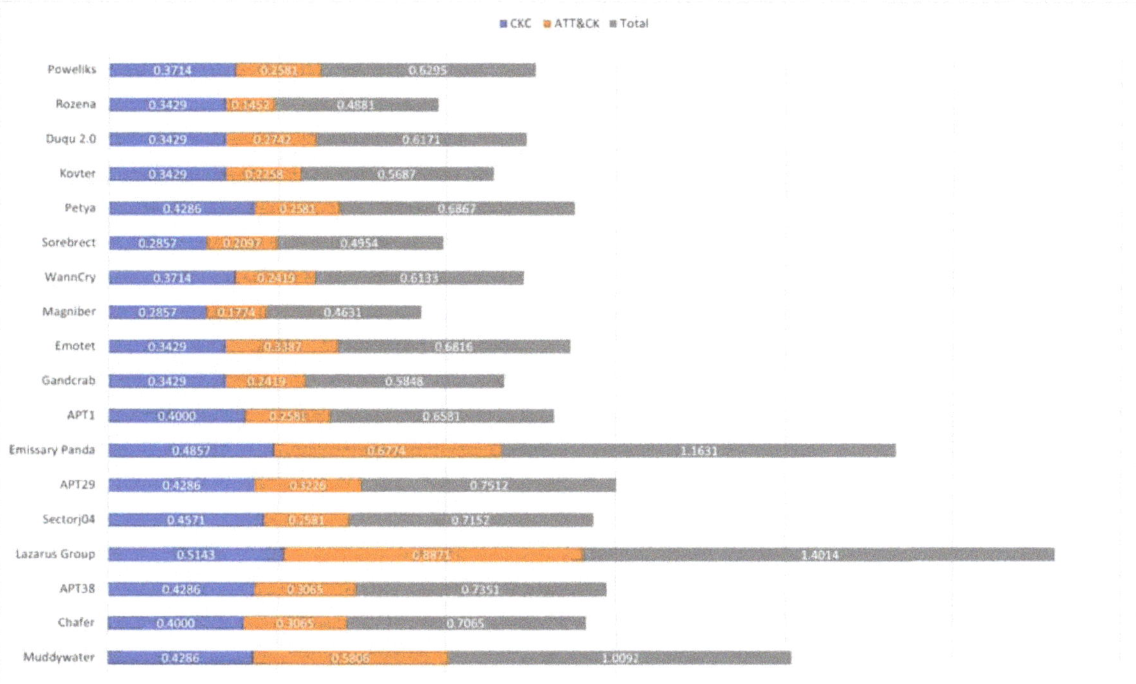

Figure 7. Scoring total result for cyber-attacks.

5. Conclusions

Cyber-attacks are constantly evolving. Starting with simple attacks, such as the defacement of a homepage or the acquisition of personal information, cyber-attacks have been changing to become more complex, such as the theft of national secrets and attacks on national infrastructure. Cyber-attacks include various attack elements. Much security research has addressed the sophistication of cyber-attacks; however, research on scoring attack complexity has been lacking. We conducted research on the complexity of attacks and proposed a model for offensive cybersecurity. In addition, important elements of the offensive cybersecurity model were identified, and detailed descriptions of each element were investigated and described.

Based on this study, we derived scores for fileless and APT group cyber-attacks. The results can be quantified for each of the various elements and techniques of each cyber-attack. The method we investigated was scored on the basis of public cyber-attack reports. This study is the first to be conducted to quantify and score cyber-attacks. We found that APT cyber-attacks have higher scores than fileless cyber-attacks, due to the APT using various ATT&CK techniques. In future research, we will expand to adopt automatic report analysis and gather input from more expert focus groups. In the future, we believe that many researchers are expected to be able to contribute to safeguarding cyberspace from cyber-attacks by researching and developing measurable scoring models for cyber-attacks through our initial research.

Author Contributions: Conceptualization, K.K.; methodology, K.K.; software, K.K.; validation, K.K., F.A.A. and H.K.; formal analysis, K.K.; investigation, K.K. and F.A.A.; resources, K.K., F.A.A. and H.K.; data curation, K.K.; writing—original draft preparation, K.K.; writing—review and editing, F.A.A. and H.K.; visualization, K.K.; supervision, H.K.; project administration, K.K.; funding acquisition, H.K. All authors have read and agreed to the published version of the manuscript.

Funding: For Kim, K, and Alfouzan, F.A, this research received no external funding. For Kim H.K, this research was funded by the Korea government (MSIT).

Acknowledgments: This research was conducted during the work as an Assistant Professor at Naif Arab University for Security Sciences (NAUSS), for Kyounggon Kim and Faisal Abdulaziz Alfouzan. Furthermore, We would like to express our sincere gratitude to Naif Arab University for Security Sciences (NAUSS) and the president of the university for his consistent support and encouragement. For y Huy Kang Kim, this work was supported by Institute of Information & communications Technology Planning & Evaluation (IITP) grant funded by the Korea government (MSIT) (No. 2021-0-00624, Development of Intelligence Cyber Attack and Defense Analysis Framework for Increasing Security Level of C-ITS).

Conflicts of Interest: The authors declare no conflict of interest.

References

1. Miller, C.; Valasek, C. Adventures in automotive networks and control units. In Proceedings of the DEFCON 21 Conference, Las Vegas, NV, USA, 2–4 August 2013; pp. 260–264.
2. Bock, K.; Hughey, G.; Levin, D. King of the hill: A novel cybersecurity competition for teaching penetration testing. In Proceedings of the 2018 USENIX Workshop on Advances in Security Education (ASE 18), Baltimore, MD, USA, 13 August 2018.
3. Antonakakis, M.; April, T.; Bailey, M.; Bernhard, M.; Bursztein, E.; Cochran, J.; Durumeric, Z.; Halderman, J.A.; Invernizzi, L.; Kallitsis, M.; et al. Understanding the mirai botnet. In Proceedings of the 26th USENIX Security Symposium (USENIX Security 17), Vancouver, BC, Canada, 16–18 August 2017; pp. 1093–1110.
4. Gon, K.K.; Hoon, K.S. Using Threat Modeling for Risk Analysis of SmartHome. In Proceedings of Symposium of the Korean Institute of Communications and Information Sciences, November 2015; pp. 378–379. Available online: https://publons.com/journal/418152/proceedings-of-symposium-of-the-korean-institute-o/ (accessed on 18 August 2021).
5. APT Groups and Operations. Available online: https://docs.google.com/spreadsheets/d/1H9_xaxQHpWaa4O_Son4Gx0YOIzlcBWMsdvePFX68EKU (accessed on 18 August 2021)
6. Khatoun, R.; Zeadally, S. Smart cities: Concepts, architectures, research opportunities. *Commun. ACM* **2016**, *59*, 46–57. [CrossRef]
7. Miller, C.; Valasek, C. A survey of remote automotive attack surfaces. *Black Hat USA* **2014**, *2014*, 94.
8. Miller, C.; Valasek, C. Remote exploitation of an unaltered passenger vehicle. *Black Hat USA* **2015**, *2015*, 91.
9. Kim, K.; Kim, J.S.; Jeong, S.; Park, J.H.; Kim, H.K. Cybersecurity for autonomous vehicles: Review of attacks and defense. *Comput. Secur.* **2021**, *103*, 102150. [CrossRef]
10. Elmaghraby, A.S.; Losavio, M.M. Cyber security challenges in Smart Cities: Safety, security and privacy. *J. Adv. Res.* **2014**, *5*, 491–497. [CrossRef] [PubMed]
11. Dai Zovi, D. *A Modern History of Offensive Security Research*; USENIX Association: Baltimore, MD, USA, 2018.
12. Derbyshire, R.; Green, B.; Prince, D.; Mauthe, A.; Hutchison, D. An Analysis of Cyber Security Attack Taxonomies. In Proceedings of the 2018 IEEE European Symposium on Security and Privacy Workshops (EuroS&PW), London, UK, 23–27 April 2018; pp. 153–161.
13. Hansman, S.; Hunt, R. A taxonomy of network and computer attacks. *Comput. Secur.* **2005**, *24*, 31–43. [CrossRef]
14. Buchanan, B. *The Legend of Sophistication in Cyber Operations*; Harvard Kennedy School, Belfer Center for Science and International Affairs: Cambridge, MA, USA, 2017.
15. Poitras, L.; Greenwald, G. NSA Whistleblower Edward Snowden: "I Don't Want to Live in a Society That Does These Sort of Things" [Video]. *The Guardian*, 9 July 2013.
16. Sanger, D. Shadow Brokers Leak Raises Alarming Question: Was the NSA Hacked. *New York Times*, 16 August 2016.
17. Nakashima, E.; Warrick, J. Stuxnet Was Work of US and Israeli Experts, Officials Say. *Washington Post*, 2 June 2012.
18. Riley, M.; Robertson, J. Russian Cyber Hacks on US Electoral System Far Wider Than Previously Known. *Bloomberg*, 13 June 2017.
19. Kim, K.G. *State-Sponsored Hacker and Changes in Hacking Techniques*; NetSec-KR: Seoul, Korea, 2017
20. Jensen, B.; Valeriano, B.; Maness, R. Fancy bears and digital trolls: Cyber strategy with a Russian twist. *J. Strateg. Stud.* **2019**, *42*, 212–234. [CrossRef]
21. Buchanan, B.; Sulmeyer, M. Russia and Cyber Operations: Challenges and Opportunities for the Next US Administration. *Carnegie Endowment for International Peace*, 13 December 2016.
22. Mandiant. Apt1: Exposing One of China's Cyber Espionage Units. 2013. Available online: https://www.fireeye.com/content/dam/fireeye-www/services/pdfs/mandiant-apt1-report.pdf (accessed on 18 August 2021).
23. Lemay, A.; Calvet, J.; Menet, F.; Fernandez, J.M. Survey of publicly available reports on advanced persistent threat actors. *Comput. Secur.* **2018**, *72*, 26–59. [CrossRef]
24. Ji-Young, K.; In, L.J.; Gon, K.K. The All-Purpose Sword: North Korea's Cyber Operations and Strategies. In Proceedings of the 2019 11th International Conference on Cyber Conflict (CyCon), Tallinn, Estonia, 28–31 May 2019; Volume 900, pp. 1–20.
25. Lee, S.; Kim, H.K.; Kim, K. Ransomware protection using the moving target defense perspective. *Comput. Electr. Eng.* **2019**, *78*, 288–299. [CrossRef]

26. GReAT. ProjectSauron: Top Level Cyber-Espionage Platform Covertly Extracts Encrypted Government Comms. 2016. Available online: https://securelist.com/faq-the-projectsauron-apt/75533/ (accessed on 15 August 2021).
27. THE Darkhotel APT a Story of Unusual Hospitality. 2014. Available online: https://media.kasperskycontenthub.com/wp-content/uploads/sites/43/2018/03/08070903/darkhotel_kl_07.11.pdf (accessed on 15 August 2021).
28. Onik, M.M.H.; Al-Zaben, N.; Hoo, H.P.; Kim, C.S. A novel approach for network attack classification based on sequential questions. *arXiv* **2018**, arXiv:1804.00263.
29. Hoque, N.; Bhuyan, M.H.; Baishya, R.C.; Bhattacharyya, D.K.; Kalita, J.K. Network attacks: Taxonomy, tools and systems. *J. Netw. Comput. Appl.* **2014**, *40*, 307–324. [CrossRef]
30. Kumar, D.; Singh, M.K.; Jayanthi, M. *Network Security Attacks and Countermeasures*; IGI Global: Hershey, PA, USA, 2016.
31. Oosthoek, K.; Doerr, C. SoK: ATT&CK Techniques and Trends in Windows Malware. In Proceedings of the International Conference on Security and Privacy in Communication Systems, Orlando, VA, USA, 23–25 October 2019; pp. 406–425.
32. Gandotra, E.; Bansal, D.; Sofat, S. Malware threat assessment using fuzzy logic paradigm. *Cybern. Syst.* **2017**, *48*, 29–48 [CrossRef]
33. Black, P.; Gondal, I.; Layton, R. A survey of similarities in banking malware behaviours. *Comput. Secur.* **2018**, *77*, 756–772 [CrossRef]
34. Martin, L. Cyber Kill Chain®. 2014. Available online: https://www.lockheedmartin.com/content/dam/lockheed-martin/rms/documents/cyber/Gaining_the_Advantage_Cyber_Kill_Chain.pdf (accessed on 18 August 2021).
35. Lee, G.; Shim, S.; Cho, B.; Kim, T.; Kim, K. Fileless cyberattacks: Analysis and classification. *ETRI J.* **2021**, *43*, 332–343. [CrossRef]

Article

Modern Aspects of Cyber-Security Training and Continuous Adaptation of Programmes to Trainees

George Hatzivasilis [1,2,*], Sotiris Ioannidis [1,3], Michail Smyrlis [4,5], George Spanoudakis [5], Fulvio Frati [6], Ludger Goeke [7], Torsten Hildebrandt [8], George Tsakirakis [9], Fotis Oikonomou [10], George Leftheriotis [11] and Hristo Koshutanski [12]

[1]   Foundation for Research and Technology–Hellas, Institute of Computer Science, Vassilika Vouton, 70013 Heraklion, Greece; sotiris@ics.forth.gr

[2]   Department of Electrical and Computer Engineering, Hellenic Mediterranean University (HMU), Estavromenos, 71410 Heraklion, Greece

[3]   Department of Electrical and Computer Engineering, Technical University of Crete, 73100 Chania, Greece

[4]   Innovation Department, Sphynx Technology Solutions AG, 6300 Zug, Switzerland; smyrlis@sphynx.ch

[5]   Research Centre for Adaptive Computing Systems (CeNACS), City, University of London, London EC1V 0HB, UK; g.e.spanoudakis@city.ac.uk

[6]   Department of Computer Science, University of Milan, 20122 Milano, Italy; fulvio.frati@unimi.it

[7]   Innovation Department, Social Engineering Academy, 60322 Frankfurt, Germany; ludger.goeke@social-engineering.academy

[8]   Research Department, SimPlan, 63452 Hanau, Germany; torsten.hildebrandt@simplan.de

[9]   Research and Development Department, ITML, 11525 Athens GR, Greece; gtsa@itml.gr

[10]   Applied Research Department, DANAOS Shipping Company, Limassol CY 3300, Cyprus; drc@danaos.com

[11]   Systems Certification Department, TUV HELLAS (TUV NORD) SA, 15562 Athens GR, Greece; glefthe@tuv-nord.com

[12]   Research Department, ATOS SPAIN SA, 28037 Madrid, Spain; hristo.koshutanski@atos.net

[*]   Correspondence: hatzivas@ics.forth.gr or hatzivas@hmu.gr; Tel.: +30-2810-391600

Received: 6 July 2020; Accepted: 13 August 2020; Published: 17 August 2020

Abstract: Nowadays, more-and-more cyber-security training is emerging as an essential process for the lifelong personnel education in organizations, especially for those which operate critical infrastructures. This is due to security breaches on popular services that become publicly known and raise people's security awareness. Except from large organizations, small-to-medium enterprises and individuals need to keep their knowledge on the related topics up-to-date as a means to protect their business operation or to obtain professional skills. Therefore, the potential target-group may range from simple users, who require basic knowledge on the current threat landscape and how to operate the related defense mechanisms, to security experts, who require hands-on experience in responding to security incidents. This high diversity makes training and certification quite a challenging task. This study combines pedagogical practices and cyber-security modelling in an attempt to support dynamically adaptive training procedures. The training programme is initially tailored to the trainee's needs, promoting the continuous adaptation to his/her performance afterwards. As the trainee accomplishes the basic evaluation tasks, the assessment starts involving more advanced features that demand a higher level of understanding. The overall method is integrated in a modern cyber-ranges platform, and a pilot training programme for smart shipping employees is presented.

Keywords: cyber-ranges; security training; security modelling; serious games; dynamic adaptation; training programmes; computers in education; bloom; STRIDE; smart shipping

1. Introduction

The 4th Industrial Revolution brings the Information Society to the foreground. Every day, highly interconnected systems, utilizing not just the ordinary computer technologies but also the Internet of Things (IoT) and the cloud, exchange high volumes of data and user-related information [1,2]. This complex ecosystem cannot be safeguarded easily, as the attack surface is continuously increasing, while the security of the deployed primitives is not always retained [3–5]. Therefore, successful attacks have been demonstrated by researchers or have been actually performed by hackers, exploiting the underlying vulnerabilities (e.g., [6,7]). The risk still remains high, not only for large organizations, but for small-to-medium enterprises (SMEs) and individuals as well.

As a human is generally considered the weakest link in a computer system, professional training is now becoming a necessity [8,9], not only for raising the users' awareness but also for training the technical staff to operate the various protection mechanisms that must be acquired (e.g., cryptographic protocols, intrusion detection/prevention systems, machine learning and artificially intelligent modules, digital forensics, etc.). Gartner estimates that the global cyber-security awareness and training market will worth around USD 1.5 billion by 2021 [10].

Except from the related academic education that is mainly designed for computer science students, professional programmes are gaining more and more ground, ranging from introductory short courses for non-security persons to highly specialized certifications for security experts. The means to offer such training include (e.g., [11–19]): traditional in-class teaching, on-line training platforms and virtual labs, as well as modern cyber-ranges frameworks that mirror an actual system and provide hands-on experience to the trainee under realistic operational conditions. However, in most cases, these modules target a specific subset of the potential beneficiaries and their educational flexibility is limited. Moreover, the training programmes are designed by technical personnel, who, in most cases, are not aware of the mainstream pedagogical principles. This is a general characteristic of lifelong education that focuses on adult professionals.

In this paper, we try to tackle this issue by combining pedagogical methods that promote skill development and security models that capture the security-related aspects of a process. More specifically, based on the Bloom's taxonomy [20], we categorized the level of difficulty and knowledge maturity that is required in order to learn the underlying training modules for a programme, and based on the Microsoft's STRIDE model [21], we map all these modules in terms of the security aspects that they involve. At first, the trainer organizes the educational content and the learning process for a professional cyber-security certification, by mapping the learning objectives and the training methods with the Constructive Alignment [22] framework. Then, the trainee consumes the teaching material and is continuously evaluated. The assessment starts from the knowledge base and the easiest layers of the Bloom's taxonomy, and if the user is successful, he/she can proceed to the upper layers and the advanced training procedures. The Kolb's learning lifecycle [23] is iteratively performed until the student masters the involved teaching material and accomplishes the learning objectives. The training is finished when the trainee has reached a specific level of understanding for the examined security properties that are included in the STRIDE analysis of this certification. The proposed method is deployed in the THREAT-ARREST cyber-ranges platform [24] as part of the overall trainee and training programme assessment.

The rest paper is organized as: Section 2 refers the related work and the background theory. Section 3 sketches the proposed methodology for the design and evaluation of the cyber-security training programme. Section 4 details the process for establishing a programme for the personnel of a smart shipping company and a preliminary implementation in the THREAT-ARREST cyber-ranges platform. Section 5 summarizes a discussion concerning modern aspects of cyber-security training. Finally, Section 6 concludes and refers future extensions.

2. Materials and Methods

2.1. Modern Cyber-Security Training Platforms

Nowadays, a high variety of research and commercial platforms is available for cyber-security training for both individuals and organizations. A comparison of them with our method is presented in Table 1 and is detailed in [24].

Table 1. Cyber-security training platforms: (A) THREAT-ARREST, (B) BeOne, (C) Kaspersky, (D) ISACA CSX, (E) CyberBit, (F) online training platforms. The following notations are utilized for (Y)es, (N)o, and (P)arial.

Feature	A	B	C	D	E	F
Automatic security vulnerability analysis of a pilot system	Y	N	N	N	N	N
Multi-layer modelling	Y	P	Y	Y	Y	P
Continuous security assurance	Y	N	N	Y	Y	N
Serious gaming	Y	N	Y	Y	N	P
Realistic simulation of cyber systems	Y	P	Y	Y	Y	N
Combination of emulated and real equipment	Y	N	P	Y	N	N
Programme runtime evaluation	Y	N	N	Y	Y	Y
Programme runtime adaptation	Y	N	Y	Y	N	P

Usually, most of the general-purpose e-learning platforms (e.g., Coursera (Mountain View, CA, USA, 2012–2020), Udacity (Mountain View, CA, USA, 2011–2020), edX (MA, USA, 2012–2020), etc.) offer introductory and main educational courses on cyber-security. On the other hand, specialized solutions, such as the SANS (Bethesda, MD, USA, 2000–2020) [11], CyberInternAcademy (MO, USA, 2017–2020) [12], StationX (London, UK, 1996–2020) [13], Cybrary (College Park, MD, USA, 2016–2020) [14], and AwareGO (Reykjavík, Iceland, 2011–2020) [15], support more advance and focused training. In most cases, all these approaches target individuals whose goal is to develop/sharpen new skills. However, they fail to provide hands-on experience on real systems or even cyber-ranges. Modern cyber-ranges platforms, such as BeOne (Hilversum, The Netherlands, 2013–2020) [16], ISACA's CyberSecurity Nexus (CSX) (Rolling Meadows, IL, USA, 1967–2020) [17], Kaspersky (Moscow, Russia, 1997–2020) [18], and CyberBit (Raanana, Israel, 2019–2020) [19], offer more advance features.

THREAT-ARREST combines all modern training aspects of serious gaming [25,26], emulation and simulation in a concrete manner [27], and offers continuous security assurance and programme adaptation based on the trainee's performance and skills (Table 1). The platform [24] offers training on known and/or new advanced cyber-attack scenarios, taking different types of action against them, including: preparedness, detection and analysis, incident response, and post incident response actions. The THREAT-ARREST platform supports the use of security testing, monitoring and assessment tools at different layers in the implementation stack, including:

- Network layer tools (e.g., intrusion detection systems, firewalls, honeypots/honeynet);
- Infrastructure layer tools (e.g., security monitors, passive and active penetration testing tools (e.g., configuration testing, SSL/TLS testing);
- Application layer tools (e.g., security monitors, code analysis, as well as passive and active penetration testing tools such as authentication testing, database testing, session management testing, data validation and injection testing).

The procedure begins by analyzing the organization's system. The Assurance Tool [28] evaluates the current security level and reports the most significant security issues that must drive the following training process. Then, hybrid training programmes are produced, and tailored to the organizational needs and the trainee types. This includes the main training material along with serious games, as well as the simulation and emulation of the cyber range system. THREAT-ARREST also provides continuous evaluation of: (a) the performance of individual trainees in specific training programmes;

and (b) the effectiveness of training programmes across sub-groups of trainees or the entire organization. These evaluations are used to tailor programmes to the needs of individual trainees or alter them at a more macroscopic level.

The whole operation is defined under a methodology called "Cyber Threat and Training Preparation (CTTP) modelling" [24], which determines the learning goals of a training programme, the learning path of the trainee, as well as how to drive the on-demand instantiation of the virtual labs with the advance cyber-ranges features for these programmes and assess the trainee's actions automatically.

This article documents this latest characteristic of the THREAT-ARREST platform and the CTTP modelling concept (see Sections 2.3 and 3). Moreover, the scope of a CTTP programme can be aligned with cyber-security professional specialization programmes, e.g., from ISACA or ISC^2. Therefore, the dynamic adaptation of the training process and the continuous improvement and building of skills constitutes a novel and competitive feature of the THREAT-ARREST solution.

2.2. Teaching Cyber-Security

Surveys concerning cyber-security exercises are reported in [29–31]. ISO-22398 [32] is the international standard that defines several exercise methodologies, such as seminars, simulations, workshops, tabletops and serious games, capture the flag (CTF), red/blue team, etc. These techniques provide hands-on experience to trainees and can assist the development of technical skills. The educational process may involve serious games, simulation with virtual labs, and/or collaboration learning. Although the importance of pedagogical aspects in exercises is recognized in the literature [33], it has not been adequately studied and covered by researchers and practitioners, respectively [33].

To support effectual training, one has to understand how expertise is built and which educational approaches can improve the trainee's performance [29]. Ordinarily, skills' development and behavioral learning start with lecture-oriented teaching. As the trainee's knowledgeable capacity increases, his/her "cognitive learning" is enhanced. Then, deeper knowledge on the subject can be built, by moving to "constructivist learning" approaches that mostly utilize exploratory learning [34,35] (react to learning as a researcher) and problem-based learning [36] (begin by resolving an actual problem and examining the relevant background information). Studies on university students [37] reveal that reaching a high-order of thinking and understanding becomes critical and of great importance in the cyber-security field. Although students successfully complete a relevant course and know (cognitive learning) the main concepts, they usually incorrectly reason about the application of core notions (constructivist learning), such as the differences between confidentiality/integrity or authentication/authorization.

Ericsson defined a well-established Deliberate Practice (DP) theory [38] for the continuous skills' improvement. Thereupon, students require well-specialized goals that improve a specific area of expertise in their field, while on the other hand, they are "not benefitting by tasks which can be completed in an automated fashion". The full achievements of the DP approach can be accomplished when the trainee reaches the highest layers on the Miller's pyramid [39]—an educational method for assessing the trainee's competence based on four levels of: "Knows", "Knows how", "Shows how", and "Does". Cyber Security Exercises (CSEs) [40] is a novel educational methodology for cyber-security that combines the aforementioned pedagogical approaches. An exercise is defined in three phases of: (i) planning the scope and objectives, (ii) implementation, and (iii) evaluation/feedback. This also complies with the relevant phases defined by the MITRE corporation [41] (exercise planning, exercise execution, and post exercise). At the planning stage, the trainer identifies the scope of the exercise, the involved security aspects, and the pedagogical methods, as well as which elements will be simulated during the exercise and the scenario steps. During the implementation stage, the trainer monitors the students and tries to handle events and incidents, driving the students to pass through all learning goals. The process is based on the Boyd's Observe-Orient-Decide-Act (OODA) loop [42]. In the feedback stage, the students and the trainer go through all the main exercise elements. This is the most valuable phase for the individuals as they can ask questions on the underlying concepts, which will hopefully lead to the achievement of the defined learning objectives.

The study in [43] indicates that students can reach competence in cyber-security only via hands-on learning with virtual labs led by an instructor. Therefore, a proper training programme must incorporate a series of good content and tutor interaction, pedagogical framework, and essential virtualized exercises for hands-on interplay. In [44], researchers propose a technology-enhanced pedagogical framework for training with virtual labs. The process starts by applying the Constructive Alignment [22] (map intended learning outcomes with deployed teaching activities) for the design of the curriculum. The learning follows the Kolb's experiential learning cycle [23] (disassembled in four subsequent phases of learning for "Concrete Experience", "Reflective Observation", "Abstract Conceptualization", and "Active Experimentation") and the educational elements are categorized based on the Bloom's Taxonomy [45] (method for the classification of learning objectives into levels of complexity and specificity). Collaborative learning may also be supported for team work. The students are evaluated via on-line quizzes and discussion boards.

Several studies also examine the inclusion of modern gamification techniques in the learning process [46,47]. The implication of serious games is generally considered positive, as the trainee can become familiarized with the involved topics in a more relaxed manner, even in his/her free time.

Another aspect that is usually neglected in cyber-security training programmes is "psychology". This affects both the attacker and the threat model—motivation to devote effort and launch an attack; and the legitimate user-communication/team-working skills, tendency to ignore warnings or defined procedures, etc. These issues are examined in [48]. The "age, sex, or cultural background may make a person more subjectable to some malicious behavior". Thus, despite their familiarization with technology, young people may be at greater risk of being tricked by phishing emails than older ones. Moreover, "different type of trainees has diverse expectations" from a cyber-security course. For instance, computer science students are mostly interested on how an attack can be performed, while psychology students focus more on why someone would exploit a vulnerability and harm a system or a person, and general public may be concerned about the side-effects of a successful hit.

Other challenging issues [49,50] include: (i) the "dynamicity" of the Computer Science, (ii) the "workforce needs" and the requirement for industry standards, and (iii) a "common taxonomy" for threats and the underlying security properties. A modern curriculum design methodology must be able to easily align in the continuous evolving Computer Science and cyber-security fields [49]. Moreover, training programmes should cover the current threat landscape and potentially lead to a professional certification [50]. A common vocabulary across all these aspects must be followed by a well-established programme or body of programmes [50].

The THREAT-ARREST platform supports a model-driven operation based on a methodology called CTTP modelling, which administrates the whole training process. At first, experts examine a piloting system (i.e., for smart shipping, healthcare, and smart energy) and record its main components, user types, etc. The core CTTP sub-model defines how a digital twin of this system can be instantiated on the developed Emulation and Simulation tools. Thereupon, the experts also apply the STRIDE threat model [21] in order to capture the current security status of the piloting system, including the potential threats, vulnerabilities, and the proper deployment of the required defense mechanisms. This information is also part of the core sub-model (a well-structured XML or JSON format [28]) and offers a common and widely-used vocabulary across the whole training experience.

Based on the analysis outcomes, we identify the most critical security aspects for the examined organization and tailor a training programme to its needs. The training perspectives are recorded in the training sub-model. This includes the learning objectives for each trainee type and the organization as a whole, as well as the dynamic adaptation and skill development features that are presented in this article (Sections 3 and 4).

The trainer defines complete training programmes with ordinary training material (e.g., lectures, tutorials, etc.), serious games, and virtual labs (emulated and simulated scenarios). The learning path for a programme is consisted by a series of CTTP models. Each model defines which of these modules will be activated and their correlation with the learning objectives (Constructive Alignment). The model-driven approach enables us to provide a high variety of CTTP models where different scenarios of escalated difficulty are activated based on the trainee's type, expectations, and performance. The variations of a model are mapped in the Bloom's taxonomy. The trainee begins the training by building the basis of the cognitive learning and then proceeding to constructivist learning and high-order thinking. Multi-user CTTP models are also supported (i.e., red/blue team and advance CTF scenarios), offering also collaborative learning opportunities. Thus, the successful learning of a security (or other) topic is performed in several iterations based on the Kolb's learning cycle. Moreover, the programmes curriculum can correlated with professional specification bodies, such as those from ISACA and ISC^2, and learning outcomes of the models and the programme as a whole are mapped based on the Constructive Alignment methodology.

2.3. The Building-Blocks of the THREAT-ARREST Cyber-Security Training Framework

The operation of the THREAT-ARREST cyber-ranges platform is driven by the CTTP models. In this subsection, we briefly introduce the CTTP-modelling features and how we establish a training programme. More details can be found in [24].

2.3.1. CTTP Modelling

The THREAT-ARREST modelling approach is consisted of four main stages (see Figure 1): (i) analysis of a pilot system, (ii) establishment of the training programme, (iii) training and user feedback, and (iv) post-training monitoring and security evaluation.

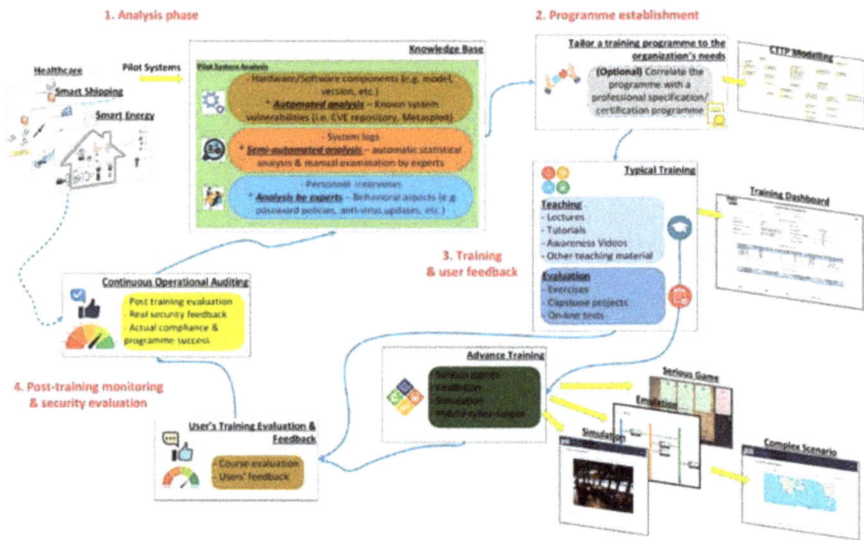

Figure 1. The THREAT-ARREST lifecycle.

Initial Analysis of a Pilot System

At first, we analyze the customer organization system based on the STRIDE method and build the knowledge base for the training programme. The goal is to estimate the current security status and identify the weak points (e.g., system or behavioral vulnerabilities). The platform's Assurance Tool [28] deploys monitoring modules in the piloting system that disclose its technical features (such as the type and version of the running software or the installed hardware components) and check if it operates securely. Then, it searches to widely-known security repositories (i.e., CVE) and automatically discovers the active vulnerabilities of the system (e.g., if a server uses MSQL 5.5.35, then it is vulnerable to buffer overflow attacks based on the CVE-2014-0001). The vulnerabilities set is assessed in a semi-automated fashion by the experts, who identify the most significant of them for the evaluated organization. Based on this information, we define the core assurance sub-model.

Experts also interview the organizations personnel and record the followed operational procedures (e.g., password-update policy, anti-virus updates, etc.). The training programme is designed afterwards based on the overall outcomes of this initial analysis.

Moreover, during this phase, the experts gather real-operational log or other data files from the piloting system. This knowledge is further processed in order to enhance the advance training procedures of the THREAT-ARREST platform. At first, we perform statistical analysis on the original data to disclose the statistical patterns of each file. This is performed either through manual examination by experts or via an automatic statistical analysis module. The goal is to produce synthetic events (i.e., a series with legitimate and/or phishing emails) or other data (i.e., a database's content with dummy but realistic entries) via our Data Fabrication Tool that will be later used in order to provide advance training under realistic conditions.

CTTP Programme Establishment

Then, based on the initial analysis results, we tailor a CTTP programme to the organization's special needs, which could also be combined and cover the training for a professional certification programme (e.g., Certified Information Security Manager (CISM) by ISACA or Certified Information Systems Security Professional (CISSP) by ISC^2), in order to increase the THREAT-ARREST's efficiency. Therefore, we define the main parameters of the "Training Programme", such as the programme's goals, actuators, trainee rules, etc.

Afterwards, we gather the relevant teaching material for the typical training (e.g., lectures, tutorials, awareness videos, etc.) and model the advance training scenarios based on "simulation sub-model", "emulation sub-model", and "gamification sub-model", as well as the "data fabrication sub-model" for the required synthetic data. The resulted training ingredients and exercises are classified based on the Bloom's taxonomy. Henceforth, we can map the desired security learning outcomes of the STRIDE modelling with the developed training elements, based on the Constructive Alignment [22] technique.

Training and User Feedback

Once the trainee has completed the basic training for a learning unit, the accompanied CTTP models are activated in the Dashboard and the trainee can now proceed with the advanced training. The CTTP models describe a virtual system and how to instantiate it via the Emulation, Simulation, and Gamification Tools, respectively.

These virtual labs and digital twins, which could resemble the organization's actual system and followed procedures, offer hands-on experience to the trainees/personnel. Thus, they can test and evaluate new technologies and policies, break-down the system, restore the default state and start over again, without affecting the actual system. The trainees begin the programme, consume the teaching material and are assessed against the desired learning goals. The CTTP models can be adjusted dynamically at runtime in order to be adapted to each individual trainee's needs. The goal is to continuously adapting the difficulty level throughout the various iterations of the games and virtual labs and the phases of the Kolb's lifecycle.

After the completion of the training, the platform displays the results for each trainee and the programme as a whole. This process indicates the scores of the trained personnel and their achievements regarding the educational processes. Discussion sessions can follow in collaboration with the trainers in order to revise the main learning topics and explain potential open issues or unresolved tasks to the trainees. Finally, the trainees can also complete questionnaires and provide feedback to the THREAT-ARREST operator, e.g., for the platform modules, the programme, etc., in order to update and improve our system. All these could form ordinary characteristics of a training platform.

Post-Training Monitoring and Security Evaluation

However, the successful completion of a programme does not always reflect to the improvement of the pilot organization's security in a straightforward manner. The security level is increased only when the trainees apply what they have learnt in the actual system. The evaluation of this phase is one of the THREAT-ARREST's novelties in comparison with other alternative solutions.

Thus, our platform continues to audit the pilot system for a determined period after the training phases. The deployed controls from the initial phase (Section 2.3.1) continuously assure the organization's security-sensitive components. The goal is to capture if the trainees really applied what they were toughed.

For example, in the analysis phases we discover that the trainees do not update their email passwords in a regular basis, i.e., by examining the log-file of the mailing server (assurance sub-model). Thus, we tailor a programme to include the learning topic of password management (Training Programme and simulation, emulation, gamification, and data fabrication sub-models). When the programme is finished, we inspect the server's log and check if the password-update entries have been increased or not.

The confirmation that the personnel adheres with the learned features, and thus the system's security is really improved, constitutes the actual validation that the programme was successful. This process is facilitated by the Assurance Tool and the relevant model. Feedback is collected from this phase in order to improve the THREAT-ARREST's operation for future training iterations and new programmes.

3. Results

3.1. Modelling of the Learning Process

This section presents the main educational and pedagogical aspects of the proposed framework. This includes the incorporation of the STRIDE threat model for the analyses of the cyber-security aspects that are involved in the programme, the Bloom's taxonomy for the classification of the learning elements, the Kolb's learning lifecycle, and the Constructive Alignment, along with the integration of these methods in the cyber-ranges platform THREAT-ARREST [24]. In this article, we extend the CTTP models and embody the aforementioned methods in our training framework. The goal is to enrich the overall model-driven approach in an attempt to accomplish continuous and dynamic adaptation of the training process to the trainee's particularities and enhance the skills' development operations.

3.1.1. Security Modelling

During the initial analysis phase of the THREAT-ARREST lifecycle (Section 2.3.1), we analyze the piloting environment based on the STRIDE methodology [21]. STRIDE is a widely-known security model for defining threats, which was designed by Microsoft. The name is the abbreviation of the six threat categories that it analyzes: (i) Spoofing, (ii) Tampering, (iii) Repudiation, (iv) Information disclosure, (v) Denial of Service (DoS), and (vi) Elevation of privilege. Each of one of them reflects a potential violation of a desired security property in the system, i.e., authentication, integrity, non-repudiation, confidentiality, availability, and authorization, respectively. Figure 2 depicts this mapping between threats and security goals.

The threat model assesses the detailed system design. Data-flow diagrams (DFDs) identify the involved entities, events, and the boundaries of the system. The model has been successfully applied to cyber-only and cyber-physical environments. While Microsoft no longer maintains STRIDE, the model is part of the Microsoft Security Development Lifecycle (SDL) and implemented within the Threat Modeling Tool, which is still available.

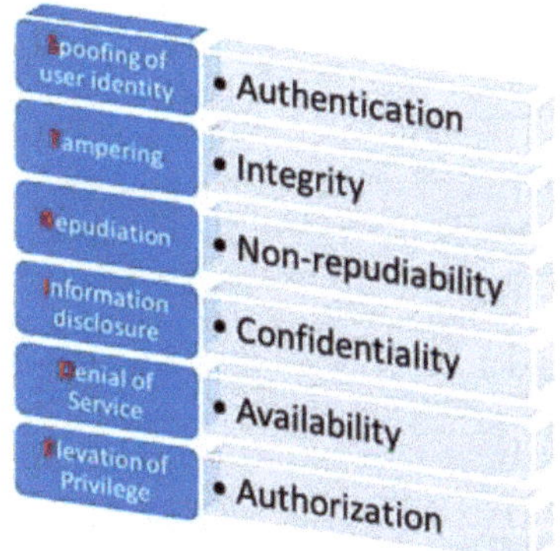

Figure 2. The STRIDE threat model.

Today, there are several threat modelling techniques [51], including Attack Tress, Security Cards, the MITRE ATT&CK framework, etc. STRIDE is a mnemonic method that focuses on assets. We choose this primitive as it can be easily understood and applied by a trainer during the design of the training programme and the underlying training procedures, correlating also threats with respective defensive countermeasures.

3.1.2. Training Programme Preparation

Training programmes are established during the second THREAT-ARREST modelling phase (Section 2.3.1). At first the trainer must design the lifecycle for a training programme. The preparation of the learning procedure is important and resolves the problem of teaching a learning topic in the determined time limits of the programme. The trainer can sketch the learning evolution and becomes more confident in the class (or virtual class). Problematic issues are foreseen and avoided while the timely preparation helps in saving time and reveals the potentials of the educational content.

Learning is a cyclic process and involves the four Kolb's stages [23]. At first, the trainee based on his/her knowledge and experience faces new problems, takes decisions, acts, and applies what he/she has learnt in practice. Then, the trainee proceeds, copes with real conditions and acquires new experiences. The gained experiences are examined via several perspectives, the results are processed, their significance is understood, and conclusions are drawn. Finally, these experiences are grouped, linked to scientific data and/or theoretic approaches, general principles are drawn, and action guidelines are formed. These phases are repeated in a cyclic manner, as they are depicted in Figure 3.

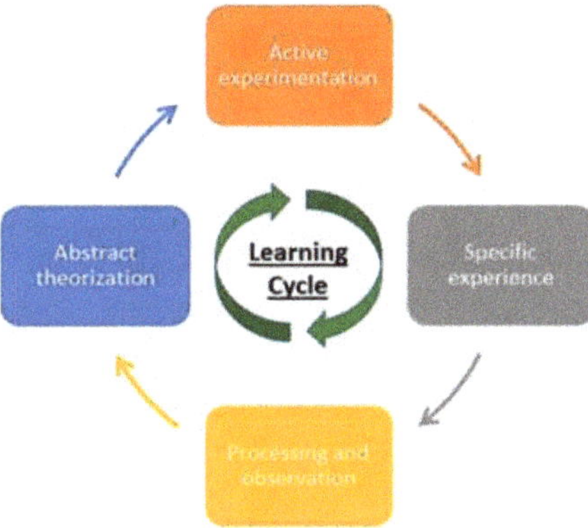

Figure 3. The 4 phases of the learning cycle.

The trainee's evaluation has to be continuous, systematic, methodic, pedagogical, and multi-factor in terms of what has been taught, learned, and is capable of doing. Thus, an effective training procedure must be able in adapting to each individual trainee's needs and capabilities, and continually contribute to their improvement.

Benjamin Bloom was one of the first scientists who systematically categorized the educational objectives and the related educational goals [20]. The so called "Bloom's taxonomy" is one of the main principles of the educational sciences, which has been revised and updated in the last years [45]. In general, the taxonomy forms a hierarchical model for the classification of educational learning objectives into levels of specificity and complexity. The overall method tries to enhance the communication between educators on the design of curricula, exercises, and examinations. It has been adopted by related teaching philosophies that lean more on skills rather than on content.

It consists of 6 layers, with the 3 bottom levels (remembering, understanding, and applying) denoting the basic understanding of the examined topic, while the coverage of the 3 top ones (analyzing, evaluating, and creating) reveals that the trainee has achieved a higher-order of thinking. Thus, the learning procedure is built from bottom-up, as the trainee goes through the cognitive, affective, and sensory learning domains [45]. Starting from the lowest maturity layer, where the trainee needs only to know the basic learning material, the process may reach at the highest point, where the trainee must have fully understood the overall learning concept. Figure 4 illustrates the main features for the latest revisited Bloom's taxonomy [52].

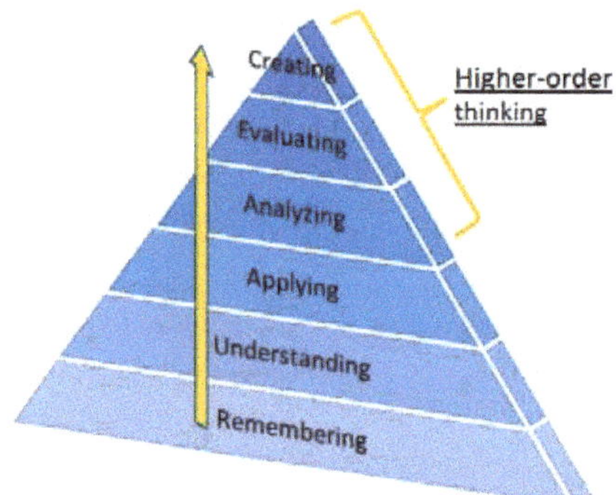

Figure 4. The revisited Bloom's taxonomy.

The first three layers assess the trainee's knowledge about the teaching content while skill development is promoted with "higher-order thinking". This also forms the final aim of the Bloom's taxonomy—building a culture of thinking.

The Blooms taxonomy was chosen for the scope of our study (instead of other candidate ones like the Miller's pyramid [39]) as: (i) it fully covers the educational objectives for cyber-security training, (ii) it is a well-established pedagogical methodology and widely-known among tutors, and (iii) it offers a good balance between simplicity and completeness for the categorization of the learning elements.

3.1.3. Continuous Trainee Assessment and Dynamic Adaptation of the Training Process

The trainee is taught the teaching material and then he/she is evaluated (in a single or several learning cycles). Afterwards, the results are surveyed and feedback is provided (both to the trainee and the trainer). During the evaluation phase, the overall process chooses the involved learning goals that will be evaluated (based on the teaching material which has been consumed by the specific trainee so far) and records the trainee's achievements. The process selects these goals based on the Bloom's revisited taxonomy, starting from the bottom (base of the knowledge pyramid) to the top (advanced knowledge and hands-on capabilities/experiences). As the trainee accomplishes the lower-level goals, he/she proceeds to the upper/layers. Denoting also the increment of the training difficulty.

When the accomplishment ratio for the goals of a specific maturity layer goes beyond a threshold (i.e., 85%), we consider that the trainee has "cover" this layer. Thus, four "professional certification levels" are determined for each educational phase, based on the layers of the Bloom's taxonomy:

- Foundation: the trainee has covered the first layer. He/she knows the main theoretic background of the educational topic.
- Practitioner: the trainee proceeds and accomplishes the layers 2–3. He/she has practical knowledge regarding the application and operation of the underlying concepts.
- Intermediate: the trainee reaches the layers 4–5. He/she has hands-on experience and technical knowledge regarding the deployment and management correlation of the various learning subjects.
- Expert: the trainee reaches the top layer 6. He/she has complete knowledge of the educational topic and is able in designing, developing, and administrating all aspects of the involved subject.

The absolute completion of a topic (100%) presents that the trainee has successfully learned all the underlying learning goals. Moreover, various trainee types with divert expectations and skill development needs could target a different level of certification.

4. The Smart Shipping Use Case

This section details the implementation of our educational method in the THREAT-ARREST platform, as well as the application of the adaptive learning for the real case of a smart shipping organization. The overall programme preparation and evaluation is composed of eight main phases: (i) description of the training programme, (ii) learning outcome of the training module, (iii) teaching and learning strategies, iv) student participation, (v) overview of assessments and training levels, (vi) study plan (learning schedule), (vii) resources required to complete the training, and viii) bench marking of the module. The phases are detailed in the following subsections.

4.1. Description of the Training Programme

The maritime sector is under an on-going process of digitalization in all aspects of operation. For a long period, seafarers have been trained with computer-based training programs on-board according to regulated training models. These days they are consuming training courses offered by sophisticated e-learning platforms. No doubt that maritime personnel are considered skilled enough to navigate properly in a web environment.

A typical topology of the on-board information technology (IT) and operation technology (OT) infrastructure [53] which is exposed to cyber threats and to risks in the format of environmental, crew safety or financing negative uncertainties is portrayed below in Figure 5.

Ships are becoming more and more integrated with shore-side operations because digital communication is being used to conduct business, manage operations, and stay in touch with head office. Furthermore, critical ship systems essential to the safety of navigation, power and cargo management have been increasingly digitalized and connected to the Internet to perform a wide variety of legitimate functions (e.g., updates, versioning upgrades, remote maintenance, voyage or ship performance monitoring from ashore, etc.). The ship–shore interface is conducted with several communication methodologies and protocols whistle cyber threats could be applicable to the full range of networking.

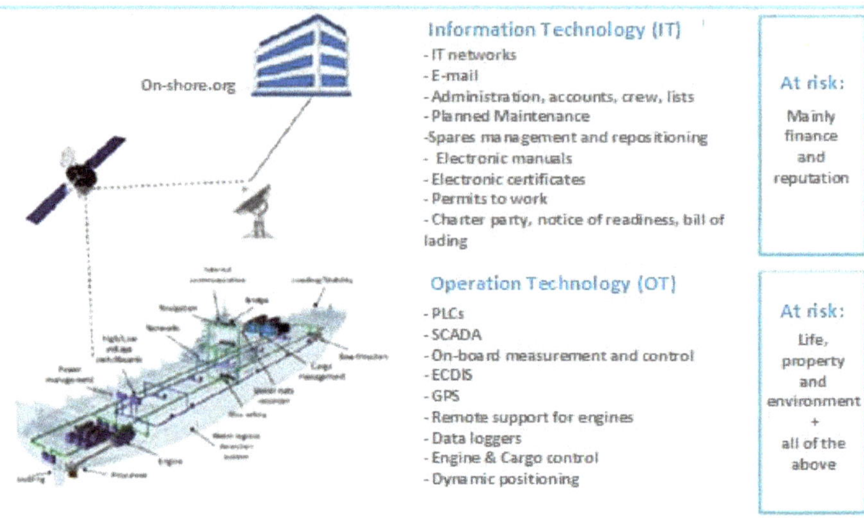

Figure 5. Smart shipping system architecture.

A schematic approach on the aforementioned networking for consumption of services between two distinct partners (shore and ship, supplier and vessel, third-party OS system provider and vessel, etc.) is following. The next figure is displaying and describing the configuration of DANAOS' communication protocols (web services, emails, telco, calls etc.) and security protections. Firewalls applied at each side of junctions between network components and data protection is secured with not storing data in centralized repositories but with controlling from a tailor-made and internally developed service platform (DANAOSone platform [54]). The overall platform modules are depicted in Figure 6.

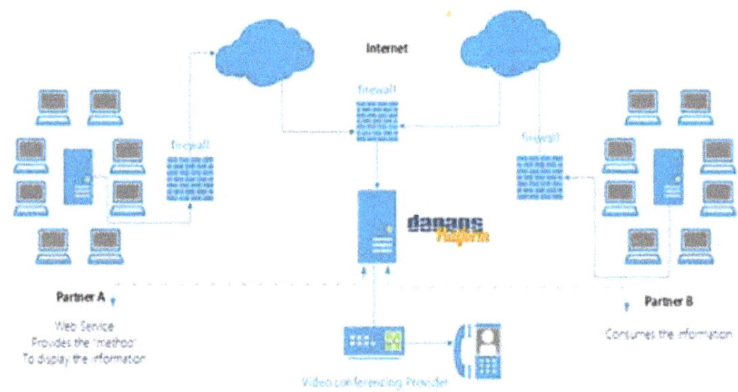

Figure 6. DANAOS configuration of communication protocols.

Cyber threats are raised where vulnerabilities in the system exist. A cyber-attack involves the attacker who in turn is motivated to trigger the attack in order to achieve a certain objective and the victim, who in turn faces the consequences of the attack. Protective barriers either in the form of technical protection or human awareness are set forward to prevent attack from impacting the system network components and cause negative consequences [55]. A schematic flow of cyber threat mechanism is given in Figure 7.

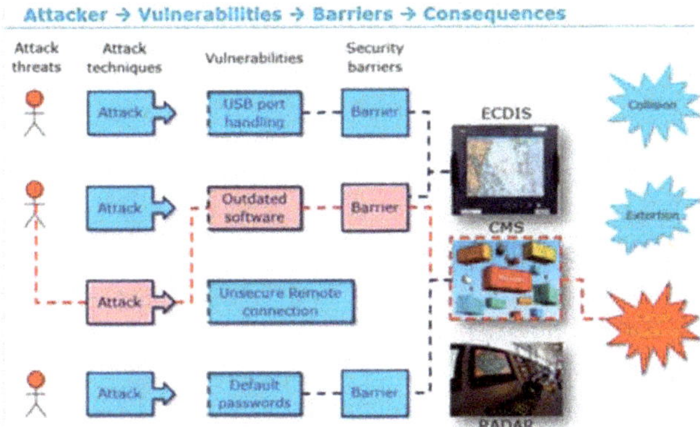

Figure 7. Flow of cyber threat mechanism.

Along that cyber threat mechanism, training and awareness is the key supporting element and an important barrier along with technical and physical protection to an effective approach to cyber safety and security.

4.2. Learning Outcome of the Training Module

Shipping Company's staff have a key role in protecting IT and OT systems. Training and awareness should be tailored to the appropriate levels for:

- On-board personnel including the master, officers and crew.
- Shore-side personnel, who support the management and operation of the ship.

An awareness or training framework should be in place for all personnel, covering at least the following risk factors and awareness aspects:

1. Risks related to emails and how to behave in a safe manner (examples are phishing attacks where the user clicks on a link to a malicious site);
2. Risks related to Internet usage, including social media, chat forums and cloud-based file storage where data movement is less controlled and monitored;
3. Risks related to the use of own devices (these devices may be missing security patches and controls, such as anti-virus, and may transfer the risk to the environment to which they are connected to);
4. Risks related to installing and maintaining software on company hardware using infected hardware (removable media) or software (infected package);
5. Risks related to poor software and data security practices where no anti-virus checks or authenticity verifications are performed;
6. Safeguarding user information, passwords and digital certificates;
7. Cyber risks in relation to the physical presence of non-company personnel, e.g., where third-party technicians are left to work on equipment without supervision;
8. Detecting suspicious activity or devices and how to report if a possible cyber incident is in progress (examples of this are strange connections that are not normally seen or someone plugging in an unknown device on the ship network);
9. Awareness of the consequences or impact of cyber incidents to the safety and operations of the ship.

Applicable personnel should be able to identify the signals when a system has been compromised. For example, training scenarios should trigger and evaluate user awareness aiming at the effective and efficient identification of hidden threats between applicable signs such as:

- An unresponsive or slow to respond system;
- Unexpected password changes or authorized users being locked out of a system;
- Unexpected errors in programs, including failure to run correctly or programs running; unexpected or sudden changes in available disk space or memory;
- Emails being returned unexpectedly;
- Unexpected network connectivity difficulties;
- Frequent system crashes;
- Abnormal hard drive or processor activity;
- Unexpected changes to browser, software or user settings, including permissions.

4.3. Teaching and Learning Strategies

At first, we begin by establishing a training programme that is tailored to the organization's particularities. Then, we model the overall "learning path" (from basic to advance training) and the trainee starts the process. He/she is continuously evaluated, and the learning procedures are adapted to his/hers needs. Figure 8 sketches the overall process, which is further detailed in the following subsections.

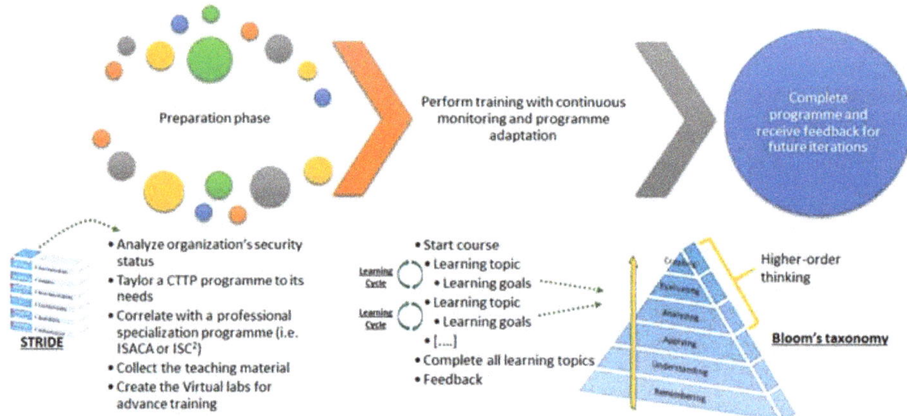

Figure 8. Training Programme lifecycle and the Learning path.

Initially, security experts interview the personnel of the evaluated organization (i.e., DANAOS shipping company). Then, we execute the Assurance Tool of the THREAT-ARREST platform [24,28] with the specifications of the pilot system (e.g., software modules, hardware equipment, network topology, business processes, etc.). With this Tool, we can: (i) export the system's security vulnerabilities and threats, (ii) conduct a risk analysis to identify the most significant of them, and (iii) perform statistical analysis on the various system log-files in order to produce realistic synthetic logs (i.e., with the platform's Data Fabrication Tool). Afterwards, these logs are utilized by the CTTP models and can be processed by the Gamification, Emulation, and/or Simulation Tools [25–27].

After the initial analysis, we define which are the main user/trainee types (e.g., simple users, operators, administrators, security experts, business managers and general personnel, CISOs, etc.), the security-related features (based on the STRIDE model), and the learning goals that we want to achieve (based on the Bloom's taxonomy). Furthermore, we determine the involved learning topics that have to be taught to the organization's personnel for the basic training procedure (e.g., information systems security, network security, cryptography, social-engineering, password management, etc.). For the advance training procedures, several valuable scenarios are also designed (e.g., serious games, emulated and/or simulated settings, potential synthetic logs, etc.).

The outcome is a tailored training programme for the specific needs of the evaluated user types. The programme specifies the learning topics and the advance evaluation scenarios for each trainee type, along with the correlated learning goals.

4.4. Student Participation

The main users involve the backend employees (e.g., office or administrative personnel, security experts, CSO, etc.), as well as, the captain and the crew of a smart vessel, who must be in position to face cyber threats even in the case where the communication with the backend systems/experts is not feasible. In general, the captain is a valuable actuator and he is the person in charge with the responsibility to take decisions for a potential ongoing cyber security incident in the vessel. Although he/she is not a security expert, he/she ought to possess sufficient knowledge in order to take the correct actions. On the other hand, the crew is ordinarily considered as users with low security awareness.

Shipping Company's staff have a key role in protecting Information Technology (IT) and Operational Technology (OT) systems. Training and awareness should be tailored to the appropriate levels for:

- On-board personnel including the master, officers and crew.
- Shore-side personnel, who support the management and operation of the ship.

Applicable personnel should be able to identify the signals when a system has been compromised. The objective is to increase the security awareness in shipping ICT systems' operators, and security attacks and help towards identifying new threats which jeopardize the operations of ICT systems in the Shipping Management industry.

A secure network depends on the IT/OT set up onboard the ship, and the effectiveness of the company policy based on the outcome of the risk assessment.

Special attention should be given when there has been no control over who has access to the on-board systems. This could, for example, happen during dry-docking, layups or when taking over a new or existing ship.

Cyber Security protection measures may be technical and focused on ensuring that on-board systems are designed and configured to be resilient to Cyber Attacks. Protection measures may also be procedural and should be covered by company policies, safety management procedures, security procedures and access controls.

Implementation of Cyber Security controls should be prioritized, focusing first on those measures, or combinations of measures, which offer the greatest benefit.

The guidelines for preventing deliberate attacks on ships and port facilities is defined in the International Ship and Facility Security Code ISPS adopted by the International Maritime Organization (IMO) in 2002 [56]. DANAOS is also following the guidelines of the Center of Internet security (CIS) [57] to apply critical security controls to equipment and data onboard vessels.

4.5. Overview of Assessments and Training Levels

In the aforementioned context of risk awareness framework and signal identification, THREAT-ARREST develops an advanced training programme incorporating emulation, simulation, serious gaming and visualization capabilities to adequately train and evaluate crew users with different types of responsibility and levels of expertise in recognizing signals of possible cyber-attacks, raising awareness on impact and consequences of attacks while following the necessary corrective actions to defend high-risk cyber systems. This also includes the design of several cyber-ranges scenarios, as are illustrated in Figure 9.

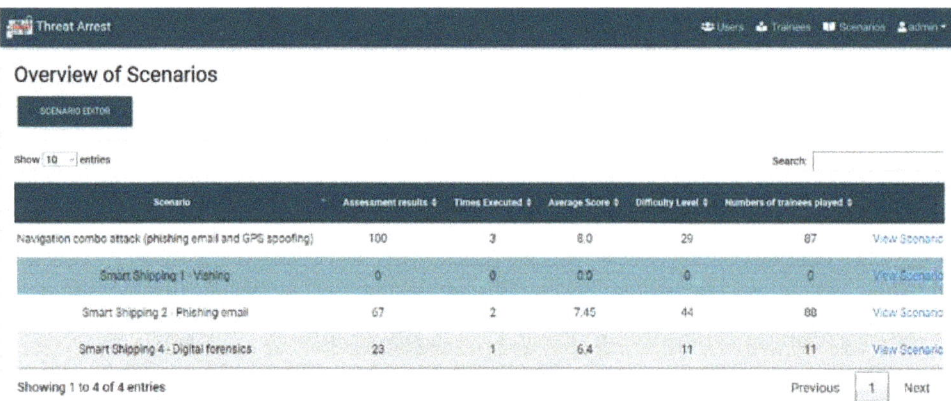

Figure 9. Scenarios overview.

From the previous figure, Table 2 summarizes the four main cyber-ranges exercises that have been implemented so far. Notice that due to the model-driven approach of the THREAT-ARREST platform, we can easily produce a high volume of variations of these four scenarios and the related CTTP models (as depicted in Figure 9), supporting the dynamic adaptation features based on the pedagogical methods that were described in the previous sections. Moreover, the same models can be applied in other application domains (e.g., smart energy, healthcare, etc.) with slight changes.

Table 2. Main Smart Shipping Scenarios.

#	Description	Trainee Type	Security Expertise	Platform Tools
1	Navigation combo attack (phishing email and GPS spoofing)	Captain	Highly-privilege actuator with low/moderate security knowledge	• Emulation • Simulation • Gamification
2	Vishing (social engineering)	Crew/Offshore officers	Non-security actuators with low access privileges	• Training • Gamification
3	Attacks on the Offshore system	IT Administrators of the shipping company	Highly-privilege actuators with moderate/high security knowledge	• Emulation • Assurance Tool
4	Digital Forensics	The organization's security engineers	Security experts	• Emulation • Simulation • Data Fabrication

The smart shipping pilot is based on the system of the DANAOS shipping company. This mainly includes the backend system at the organization's premises, along with the DANAOS communication platform (DANAOSone), as well as the systems on the smart vessels and their communication with the main system. Figure 10 depicts the pilot's architecture and main components.

For the deployment of the main Virtual Labs under THREAT-ARREST, the backend system and the system of smart vessels are emulated. The operational behavior of the vessels' on-board equipment (e.g., navigation modules, smart devices, etc.) is simulated.

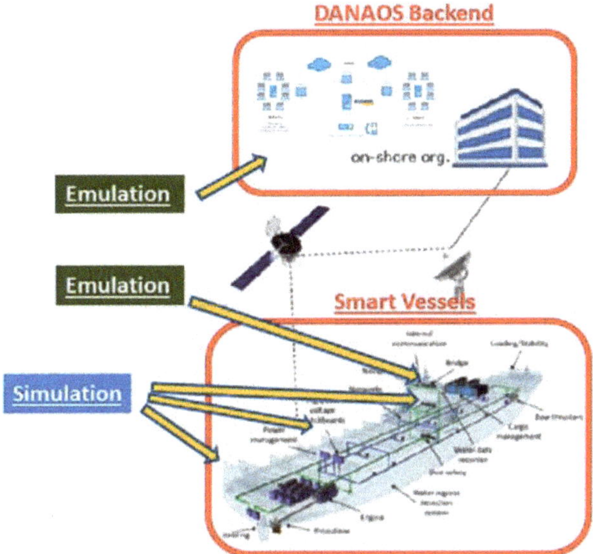

Figure 10. The Smart Shipping pilot architecture and Virtual Lab deployment.

Application Example of the STRIDE Model and the Bloom Taxonomy

In this subsection, we will describe the application of the STRIDE methodology for the modelling of the security aspects of the social engineering scenario. The trainee type is the captain of the vessel (valuable actuator with moderate security knowledge). He/she must start a (simulated) journey from the Heraklion port to Piraeus, which will be designated by the backend office via an email to the captain. All legitimate emails are digitally signed with PGP.

The programme involves the security aspects of "Tampering" and "Spoofing". During the basic training, the trainee gets familiar with the main cryptographic primitives (Remembering), practices cryptography via related tools, i.e., CryptTool-2, (Understanding), and signs/verifies emails with PGP (Applying). Moreover, the trainee is touch the general concepts of social engineering and phishing attacks (Remembering), reviews specific examples of attacks and plays a PROTECT game with a social engineering card-deck (Understanding), and tries to classify email examples as legitimate or malicious (Applying).

For the advance training (Analyzing/Evaluating) as the emulated scenario starts, a faulty (but legitimate) email, commanding the captain to go to the Thessaloniki port, is sent. The email contains the details of another journey and was sent to the trainee by mistake: (i) the trainee identifies that this is a legitimate email, (ii) since the destination port was Piraeus, the trainee understands that this email was sent to him/her by mistake, and (iii) the trainee ignores the email and reports it back to the backend office. Then, the trainee receives a malicious (phishing) email, alerting him/her that a bad weather condition will take place, thus, he/she needs to go to another port to make a stop: (i) the trainee identifies that this is a phishing email and ii) ignores the email and reports it to the backend office. Lastly, the captain receives a legitimate email with the weather forecast, denoting that the weather is good, and the destination is the Piraeus port: (i) the trainee understands that this is a legitimate email and (ii) starts the journey in the Simulation Tool (where CTTP simulation sub-models can be activated with on-ship attacks for more complex scenarios, i.e., GPS spoofing).

If the trainee succeeded in all steps and has learnt the underlying concepts, he/she can act as the trainer and create the emails (legitimate, faulty, or malicious) that will be sent to other trainees during the emulation scenario (Create). Table 3 summarizes the modelling steps for the social engineering scenario of Table 2 and Figure 9. The overall accomplishments of the trainee disclose his/her level of understanding concerning the tampering and spoofing perspectives of social engineering attacks and the usage of the relevant countermeasures that would assure integrity and authentication, respectively.

Table 3. Modelling of a Social Engineering Scenario.

STRIDE Property	Bloom Taxonomy Layer	Description
Tampering/Integrity	Remembering	Introductory lesson to cryptography
	Understanding	Exercises with the educational Crypt Tool 2
	Applying	Practice with PGP (sign/verify emails)
	Analyzing/Evaluating	Emulated scenario where the trainee has to verify emails' integrity with PGP and send signed responses to the back office
	Creating	Act as the back office employee or the attacker and send the emails of the emulated scenario to other trainees
Spoofing/Authentication	Remembering	Lesson for social engineering and phishing attacks
	Understanding	Review of actual phishing email examples and play a tailored PROTECT game
	Applying	Classify email examples as legitimate or malicious
	Analyzing/Evaluating	Emulated scenario where the trainee must audit emails (e.g., the sender's email address, the email's content, PGP verification, etc.) and justify if they are legitimate, faulty, or malicious.
	Creating	Act as the back office employee or the attacker and send the emails of the emulated scenario to other trainees

4.6. Study Plan (Learning Schedule)

4.6.1. Basic Training

The basic training involves the Training and the Gamification Tools [25]. The trainees are registered and we compound their training sessions.

For the preparation of the Training Tool, we gather the content for the basic training (e.g., lectures, awareness videos, tutorials, and other educational material) and map it to the programmes for these specific trainees.

Then, the users start the training process by consuming the related teaching material. After completing a training section, the trainee's knowledge can be evaluated by exercises, capstone projects, and/or online tests (e.g., questionnaires).

Meanwhile, the trainee can practice his/her knowledge by playing serious games that are related with the learning material which has been consumed by the specific trainee. Each game has a pool of gaming ingredients, such as cards, set of questions, scenarios, etc. In each round, the trainee is given one of these ingredients and tries to find the correct action. For the THREAT-ARREST, an ingredient has also a tag-list that contains the learning topics which are related to the ingredient. For example, a card in the PROTECT game [25] for phishing attacks is correlated with training for information systems security and social-engineering (see Figure 11).

When a trainee starts a game, the Training Tool collects the learning topics that have been consumed by the specific trainee and sends them to the Gamification Tool. Then, the game selects randomly a set of the underlying ingredients from the pool that contain the learning topics in their tag-list. The trainee plays the game and the score is maintained within the game. Once it is over, the overall evaluation is sent back to the Training Tool and the trainee's profile is updated.

The basic training is considered successful when the trainee:

- Has consumed the main teaching material;
- Has passed the training evaluation (e.g., exercises, exams, etc.) with an adequate score;
- Has passed a game, which contains all the involved learning topics of the learning unit, with an adequate score.

Once a good level of understanding has been accomplished by the trainee, he/she can proceed with the related advance training scenarios, which are modelled in the form of CTTP models.

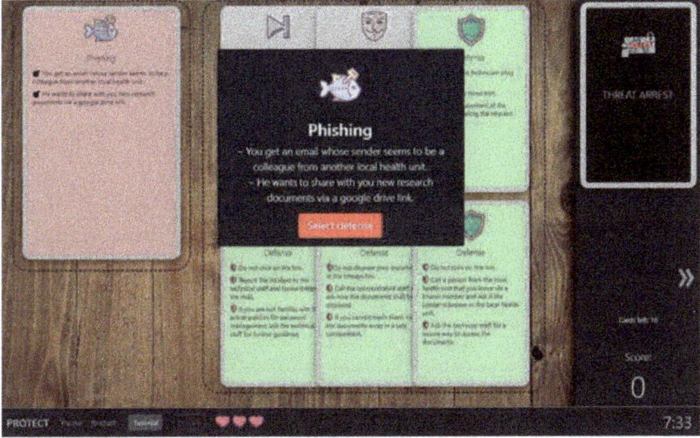

Figure 11. Game view of the serious game PROTECT.

4.6.2. Advance Training

The advance training involves emulated and/or simulated scenarios (see Figure 9 and Table 2). Once the trainee has completed the basic training for a learning unit, the accompanied CTTP models are activated in the Training Tool's Dashboard and the trainee can now proceed with the advanced training. The CTTP models describe a virtual system and how to instantiate it via the Emulation and Simulation Tools. In most cases, this virtual system will resemble the pilot system of the evaluated organization.

The trainee chooses one of the available/active CTTP models from the Dashboard. Then, the Training Tool parses the CTTP model and identifies the underlying emulated/simulated components, exports the instantiation scripts for each of these emulated/simulated components, and deploys the components sequentially, based on a designated instantiation order which is defined in the CTTP model. Specifically, the Training Tool sends the script for each component to the relevant Tool, receives an acknowledgement that the component is up and running, and proceeds to the next component. When all components are set correctly and are operative, the trainee is notified in the dashboard and can begin interacting with them (see Figure 12).

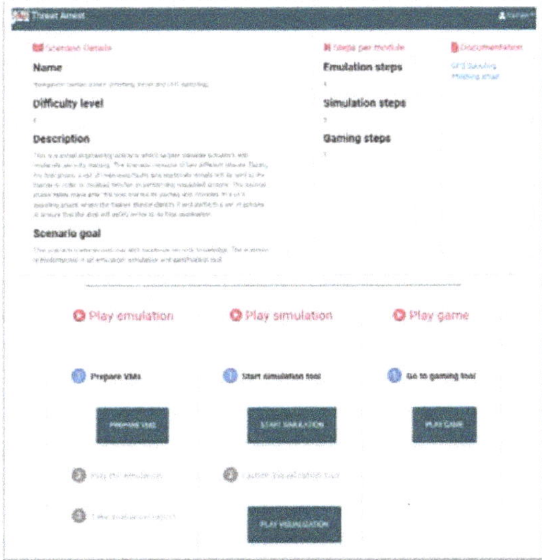

Figure 12. Training scenario details.

4.7. Resources Required to Complete the Training

THREAT-ARREST platform includes mechanisms that have been deployed with respect to the aggregated scoring of trainees in the various training scenarios, in order to provide real-time assessment information through the interface of the Training Tool. The evaluation process is briefly depicted in Figure 13.

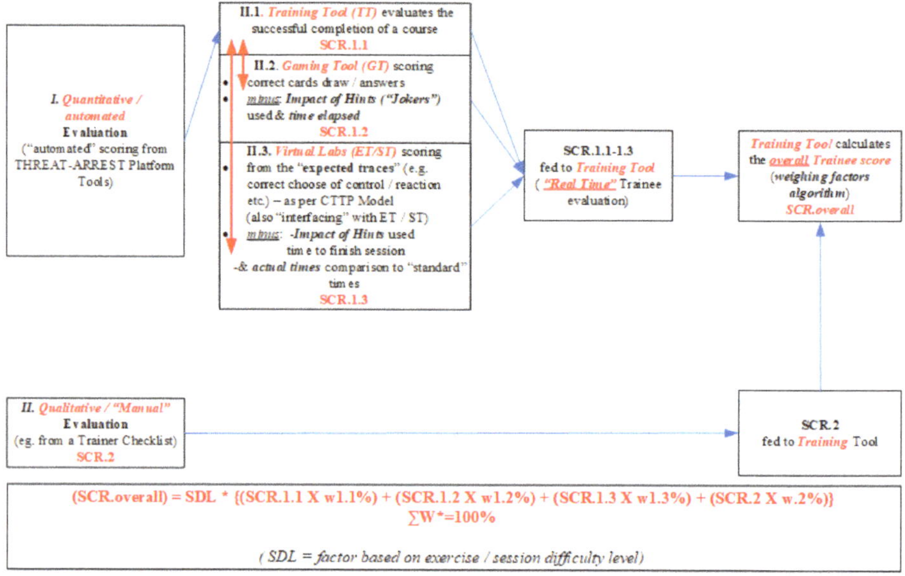

Figure 13. Scoring method for trainees' performance assessment.

Based on that, two complementary basic scoring "sources" are being used:

1   A quantitative (automated) scoring based on the TREAT-ARREST platform's tools and the relevant information derived from the CTTP Models. The first one can be divided to three sub-scores stemmed from:

    a   The Training Tool;
    b   The Gamification Tool;
    c   And the virtual labs with the Emulation and Simulation Tools.

2   And a qualitative (manual) scoring, e.g., when the trainee answers a questionnaire.

The overall score is calculated through the formula presented at the bottom of Figure 13, with the weights of each score to be defined by the administrator or the trainer. The exact algorithm and weights are pre-defined, based on a specific scenario/exercise and the CTTP Programmes standardization/certification associations.

Additionally, to the evaluation of the individual progress of each trainee, we also need a way to evaluate a CTTP Programme for an organization. Thus, aggregated metrics are also utilized to capture the success of an organization's trainees. In the main form, the min and max scores will be used from all the pilot trainees to disclose the deviation of the training among trainees of the same category (e.g., administrators) as well as the mean value and regression analysis to reflect the overall achievement and the generic security posture of the examined organization.

4.8. Benchmarking of the Module

After each iteration, the trainee's scores are updated (see Figure 14).

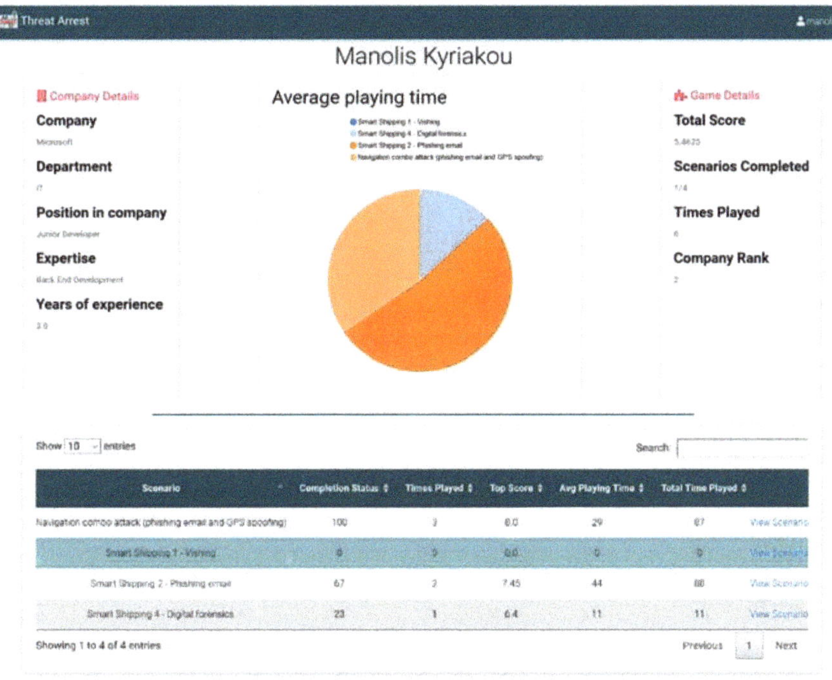

Figure 14. Trainee's scores.

The organization can also review the progress of all its trainees along with the evaluation metrics for the programme as a whole (see Figure 15).

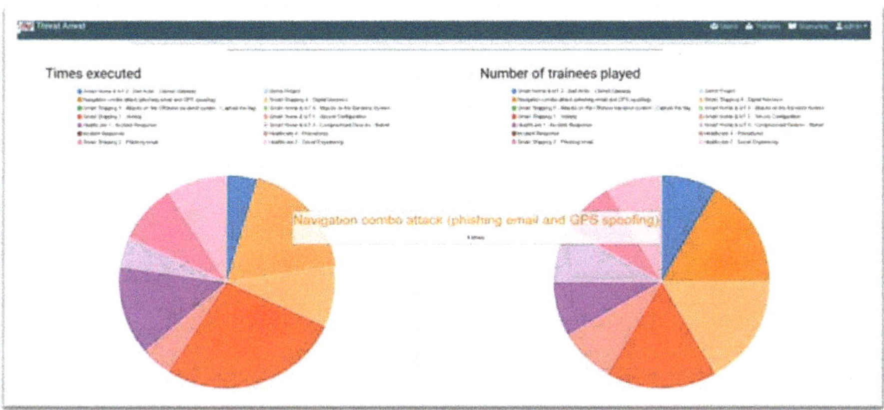

Figure 15. Smart shipping trainees' scores and overall programme evaluation graphs.

DANAOS capitalizes on the THREAT-ARREST platform which delivers security training, based on a model-driven approach where CTTP models, specifying the potential attacks, the security controls of cyber systems against them, and the tools that may be used to assess the effectiveness of these controls while driving the training process, and align it (where possible) with operational cyber system security assurance mechanisms to ensure the relevance of training.

The THREAT-ARREST's maritime pilot objective is to increase the security awareness in shipping Information and Communications Technology (ICT) systems' operators, and security attacks and help towards identifying new threats which jeopardize the operations of ICT systems in the Shipping Management industry.

5. Discussion

Cyber-security training is always important for the general public and can be even imperative for some economic sectors. The evolving digitalization of our daily activities is expected to bring more and more cyber-security in the foreground. Although there is a plethora of training platforms with advance technical features, the focus to the pedagogical aspects is expected to gain more focus for the next generation of these platforms.

The European Cyber Security Organization (ECSO) along with the European Cybersecurity Competence Network Pilot projects published a concrete report for 2019–2020 [58,59], concerning the modern features and aspects that novel cyber-security training platforms have to support. The overall THREAT-ARREST approach supports several of the modern educational features that an innovative training environment has to support, such as virtual labs, serious games, collaborating exercises, discussion sessions, the human in the training loop, etc. Moreover, the maritime sector is identified among the important economic sectors that require advance cyber-security training programmes.

This paper tackles the incorporation of educational methods to the overall lifecycle of a complete training programme with the dynamic adaptation of the training process to the trainee's particularities. In the current version, these procedures are more-or-less predefined to some degree by the trainer or the programme designer. Therefore, we can support different difficulty levels for different trainee types (ranging from main security for the general public to advance training for security experts), as well as, the gradual building of the Bloom's knowledge pyramid for each one of them. The model-driven operation enables us to easily generate a high variety of training models and cope with the dynamicity of the training requirements.

One important aspect which should be offered by a modern platform based on the surveys from ECSO [58,59], is the adaptation of the training based on machine learning and artificial intelligence. Therefore, the dynamicity will be mostly supported by an intelligent system and will be further adapted to a person's behavior. The goal is to make the training even more human-centric. The THREAT-ARREST platform does not support this functionality. Nevertheless, the benchmarking of the training modules (Section 4.8) could act as a training dataset for the potential machine learning proposals. Improvements can be suggested regarding the time that is required for specific trainee groups to complete an exercise, the use of assistive hints throughout the exercise, as well as the assessment of the mapping between the training modules and the learning objectives. Furthermore, a model-driven design, such as the one developed by THREAT-ARREST, could make the implementation of this vision feasible.

As aforementioned, expansion to other economic sectors and industries should also be considered [58]. Now, we are in progress of providing targeted training scenarios for healthcare and smart energy piloting systems. Videos with demos for the main platform tools as well as a set of training scenarios can be found on our YouTube channel at www.youtube.com/channel/ UCBUClnDkE6cjYtw7cEgP0vQ. The platform is currently under evaluation and actual training sessions with real employees from the shipping company are to be conducted this summer.

6. Conclusions

This paper proposes an educational methodology for the dynamic adaptation of cyber-security training programmes. A training session is disassembled into learning topics, which are then categorized based on the revisited Bloom's taxonomy and are mapped to the STRIDE security model. The trainee starts the learning process by consuming the main teaching material (e.g., lectures, tutorials, videos, etc.) and proceeds to more advance learning procedures, involving hands-on experience on emulated/simulated components. The trainee is continuously evaluated. The assessment begins from learning topics that cover the knowledge basis of the examined teaching unit (modelled in the Bloom's taxonomy), and if the trainee is successful, he/she can proceed to the correlated modules for advanced training. The beneficiary aims to develop his/her skills and earn a professional certification on specific cyber-security fields, based on the four specialization levels that are offered (foundation, practitioner, intermediate, and expert). The overall method is integrated in the cyber-ranges platform THREAT-ARREST and a preliminary application is presented, where a training programme for smart shipping personnel is established.

As a future extension, we consider the further evaluation of the method based on feedback that we receive by trainers and/or other users of the platform. Artificial Intelligence empowered by machine learning for the adaptation of the training to the trainee's skills is also an interesting approach that can be implemented when a sufficient volume of trainee profiles has been collected from future iterations. Moreover, we are now planning new training programmes for the cases of healthcare and smart energy organizations.

Author Contributions: Conceptualization, all co-authors; methodology, G.H., F.O., and G.L.; software, M.S., G.T., F.F., L.G. and T.H.; validation, F.O., G.H., H.K. and G.L.; resources, F.O.; data curation, M.S. and G.T.; writing—original draft preparation, G.H.; writing—review and editing, G.H. and M.S.; visualization, T.H., G.T. and L.G.; Supervision and Project administration, S.I. and G.S. All authors have read and agreed to the published version of the manuscript.

Funding: This work has received funding from the European Union Horizon's 2020 research and innovation programme H2020-DS-SC7-2017, under grant agreement No. 786890 (THREAT-ARREST).

Acknowledgments: This work has received funding from the European Union Horizon's 2020 research and innovation programme H2020-DS-SC7-2017, under grant agreement No. 786890 (THREAT-ARREST).

Conflicts of Interest: "The authors declare no conflict of interest".

References

1. Lin, J.; Yu, W.; Zhang, N.; Yang, X.; Zhang, H.; Zhao, W. A survey of Internet of Things: Architecture, enabling technologies, security and privacy, and applications. *IEEE Internet Things J.* **2017**, *4*, 1125–1142. [CrossRef]
2. Hatzivasilis, G.; Fysarakis, K.; Soultatos, O.; Askoxylakis, I.; Papaefstathiou, I.; Demetriou, G. The Industrial Internet of Things as an enabler for a Circular Economy Hy-LP: A novel IIoT Protocol, evaluated on a Wind Park's SDN/NFV-enabled 5G Industrial Network. In *Computer Communications—Special Issue on Energy-Aware Design for Sustainable 5G Networks*; Elsevier: Amsterdam, The Netherlands, 2018; Volume 119, pp. 127–137.
3. Habibi, J.; Midi, D.; Mudgerikar, A.; Bertino, E. Heimdall: Mitigating the Internet of Insecure Things. *IEEE Internet Things J.* **2017**, *4*, 968–978. [CrossRef]
4. Hatzivasilis, G.; Soultatos, O.; Ioannidis, S.; Verikoukis, C.; Demetriou, G.; Tsatsoulis, C. Review of Security and Privacy for the Internet of Medical Things (IoMT). In Proceedings of the 1st International Workshop on Smart Circular Economy (SmaCE), Santorini Island, Greece, 30 May 2019; pp. 1–8.
5. Hatzivasilis, G.; Soultatos, O.; Ioannidis, S.; Spanoudakis, G.; Katos, V.; Demetriou, G. MobileTrust: Secure Knowledge Integration in VANETs. *ACM Trans. Cyber-Phys. Syst.* **2020**, *4*, 1–25. [CrossRef]
6. Khandelwal, S. United airlines hacked by sophisticated hacking group. *The Hacker News*, 30 July 2015.
7. Hirschfeld, J.D. Hacking of government computers exposed 21.5 million people. *The New York Times*, 9 July 2015.
8. Santa, I. *A Users' Guide: How to Raise Information Security Awareness*; ENISA: Heraklion, Greece, 2010; pp. 1–140.
9. Manifavas, C.; Fysarakis, K.; Rantos, K.; Hatzivasilis, G. DSAPE—Dynamic Security Awareness Program Evaluation. In *International Conference on Human Aspects of Information Security, Privacy, and Trust*; Springer: Cham, Switzerland, 2014; pp. 258–269.
10. Kish, D.; Carpenter, P. Forecast Snapshot: Security Awareness Computer-Based Training, Worldwide. 2017. Gartner Research, ID G00324277, March 2017. Available online: https://www.gartner.com/en/documents/3629840/forecast-snapshot-security-awareness-computer-based-trai (accessed on 24 July 2020).

11. SANS: Online Cyber Security Training. 2000–2020. Available online: https://www.sans.org/online-security-training/ (accessed on 24 July 2020).

12. CYBERINTERNACADEMY: Complete Cybersecurity Course Review on Cyberinernacademy. 2017–2020. Available online: https://www.cyberinternacademy.com/complete-cybersecurity-course-guide-review/ (accessed on 24 July 2020).

13. StationX: Online Cyber Security & Hacking Courses. 1996–2020. Available online: https://www.stationx.net/ (accessed on 24 July 2020).

14. Cybrary: Develop Security Skills. 2016–2020. Available online: https://www.cybrary.it/ (accessed on 24 July 2020).

15. AwareGO: Security Awareness Training. 2011–2020. Available online: https://www.awarego.com/ (accessed on 24 July 2020).

16. BeOne Development: Security Awareness Training. 2013–2020. Available online: https://www.beonedevelopment.com/en/security-awareness/ (accessed on 24 July 2020).

17. ISACA: CyberSecurity Nexus (CSX) Training Platform. 1967–2020. Available online: https://cybersecurity.isaca.org/csx-certifications/csx-training-platform (accessed on 24 July 2020).

18. Kaspersky: Kaspersky Security Awareness. 1997–2020. Available online: https://www.kaspersky.com/enterprise-security/security-awareness (accessed on 24 July 2020).

19. CyberBit: Cyber Security Training Platform. 2019–2020. Available online: https://www.cyberbit.com/blog/security-training/cyber-security-training-platform/ (accessed on 24 July 2020).

20. Bloom, B. Taxonomy of educational objectives: The classification of educational goals. In *Handbook I: Cognitive Domain*; David McKay Company: New York, NY, USA, 1956.

21. Johnstone, M.N. Threat modelling with STRIDE and UML. In Proceedings of the 8th Australian Information Security Management Conference (AISM), Perth Western, Australia, 30 November 2010; pp. 18–27.

22. Biggs, J. *Teaching for Quality Learning at University: What the Student Does*, 4th ed.; Open University Press: Maidenhead, UK, 2011; pp. 1–416.

23. Sims, R. R: Kolb's Experiential Learning Theory: A Framework for Assessing Person-Job Interaction. *Acad. Manag. Rev.* **1983**, *8*, 501–508. [CrossRef]

24. Othonas, S.; Fysarakis, K.; Spanoudakis, G.; Koshutanski, H.; Damiani, E.; Beckers, K.; Wortmann, D.; Bravos, G.; Ioannidis, M. The TREAT-ARREST Cyber-Security Training Platform. In Proceedings of the 1st Model-driven Simulation and Training Environments for Cybersecurity (MSTEC), Luxembourg, 27 September 2019.

25. Goeke, L.; Quintanar, A.; Beckers, K.; Pape, S. PROTECT—An Easy Configurable Serious Game to Train Employees Against Social Engineering Attacks. In Proceedings of the 1st Model-Driven Simulation and Training Environments for Cybersecurity (MSTEC), Luxembourg, 27 September 2019.

26. Beckers, K.; Pape, S.; Fries, V. HATCH: Hack and trick capricious humans—A serious game on social engineering. In Proceedings of the 30th International BCS Human Computer Interaction (HCI) Conference Fusion, Bournemouth, UK, 11–15 July 2016; pp. 1–3.

27. Braghin, C.; Cimato, S.; Damiani, E.; Frati, F.; Mauri, L.; Riccobene, E. A model driven approach for cyber security scenarios deployment. In Proceedings of the 1st Model-Driven Simulation and Training Environments for Cybersecurity (MSTEC), Luxembourg, 27 September 2019.

28. Somarakis, I.; Smyrlis, M.; Fysarakis, K.; Spanoudakis, G. Model-driven Cyber Range Training—The Cyber Security Assurance Perspective. In Proceedings of the 1st Model-Driven Simulation and Training Environments for Cybersecurity (MSTEC), Luxembourg, 27 September 2019.

29. Hautamäki, J.; Karjalainen, M.; Hämäläinen, T.; Häkkinen, P. Cyber security exercise: Literature review to pedagogical methodology. 13th annual International Technology. In Proceedings of the Education and Development Conference, Valencia, Spain, 11–13 March 2019; pp. 3893–3898.

30. McDaniel, L.; Talvi, E.; Hay, B. Capture the Flag as Cyber Security Introduction. In Proceedings of the 49th Hawaii International Conference on System Sciences (HICSS), Koloa, HI, USA, 5–8 January 2016; pp. 5479–5486.

31. James, J.E.; Morsey, C.; Phillips, J. Cybersecurity education: A holistic approach to teaching security. In *Issues in Information Systems*; Maria, E.C., Ed.; IACIS: Leesburg, VA, USA, 2016; Volume 17, pp. 150–161.

32. ISO 22398: Societal Security—Guidelines for Exercises. Available online: https://www.iso.org/standard/50294.html (accessed on 24 July 2020).

33. Arabo, A.; Serpell, M. Pedagogical Approach to Effective Cybersecurity Teaching. In *Transactions on Edutainment XV*; Springer: Berlin/Heidelberg, Germany, 2019; Volume 11345, pp. 129–140.

34. Freitas, S.; Oliver, M. How can exploratory learning with games and simulations within the curriculum be most effectively evaluated? *Comput. Educ.* **2006**, *46*, 249–264. [CrossRef]

35. Israel, M.; Lash, T. From classroom lessons to exploratory learning progressions: Mathematics + computational thinking. *Interact. Learn. Environ.* **2019**, *28*, 362–382. [CrossRef]

36. Mann, L.; Chang, R.L.; Chandrasekaran, S.; Coddington, A.; Daniel, S.; Cook, E.; Crossin, E.; Cosson, B.; Turner, J.; Mazzurco, A.; et al. From problem-based learning to practice-based education: A framework for shaping future engineers. *Eur. J. Eng. Educ.* **2020**, 1–21. [CrossRef]

37. Scheponik, T.; Sherman, A.T.; Delatte, D.; Phatak, D.; Oliva, L.; Thompson, J.; Herman, G.L. How Students Reason about Cybersecurity Concepts. In Proceedings of the Frontiers in Education Conference (FIE), Erie, PA, USA, 12–15 October 2016; pp. 1–5.

38. Ericsson, K.A. Deliberate practice and acquisition of expert performance: A general overview. *Acad. Emerg. Med.* **2008**, *15*, 988–994. [CrossRef] [PubMed]

39. Miller, G.E. The assessment of clinical skills/competence/performance. *Acad. Med.* **1990**, *65*, 63–67. [CrossRef] [PubMed]

40. Karjalainen, M.; Kokkonen, T.; Puuska, S. Pedagogical Aspects of Cyber Security Exercises. In Proceedings of the IEEE European Symposium on Security and Privacy Workshops, Stockholm, Sweden, 17–19 June 2019; pp. 103–108.

41. Kick, J. Cyber Exercise Playbook. The MITRE Corporation. Available online: https://www.mitre.org/sites/default/files/publications/pr_14-3929-cyber-exercise-playbook.pdf (accessed on 24 July 2020).

42. Lif, P.; Sommestad, T.; Granasen, D. Development and evaluation of information elements for simplified cyber-incident reports. In Proceedings of the International Conference On Cyber Situational Awareness, Data Analytics and Assessment (Cyber SA), Glasgow, UK, 11–12 June 2018; pp. 1–10.

43. Said, S.E. Pedagogical Best Practices in Higher Education National Centers of Academic Excellence/Cyber Defense Centers of Academic Excellence in Cyber Defense. Ph.D. Thesis, Union University, Tennessee, TN, USA, May 2018.

44. Athauda, R.; AlKhaldi, T.; Pranata, I.; Conway, D.; Frank, C.; Thorne, W.; Dean, R. Design of a Technology-Enhanced Pedagogical Framework for a Systems and Networking Administration course incorporating a Virtual Laboratory. In Proceedings of the Frontiers in Education Conference (FIE), San Jose, CA, USA, 3–6 October 2018; pp. 1–5.

45. Pohl, M. *Learning to Think—Thinking to Learn: Models and Strategies to Develop a Classroom Culture of Thinking*, 1st ed.; Hawker Brownlow Education: Cheltenham, Australasia, 2000; pp. 1–98.

46. Švábenský, V.; Vykopal, J.; Čermák, M.; Laštovička, M. Enhancing cybersecurity skills by creating serious games. In Proceedings of the 23rd Annual ACM Conference on Innovation and Technology in Computer Science Education (ITiCSE), Larnaca, Cyprus, 2–4 July 2018; pp. 194–199.

47. Jin, G.; Tu, M. Evaluation of Game-Based Learning in Cybersecurity Education for High School Students. *J. Educ. Learn.* **2018**, *12*, 150–158. [CrossRef]

48. Taylor-Jackson, J.; McAlaney, J.; Foster, J.; Bello, A.; Maurushat, A.; Dale, J. Incorporating Psychology into Cyber Security Education: A Pedagogical Approach. In Proceedings of the AsiaUSEC'20, Financial Cryptography and Data Security (FC), Sabah, Malaysia, 14 February 2020; pp. 1–15.

49. Shah, V.; Kumar, A.; Smart, K. Moving Forward by Looking Backward: Embracing Pedagogical Principles to Develop an Innovative MSIS Program. *J. Inf. Syst. Educ.* **2018**, *29*, 139–156.

50. Knapp, K.J.; Maurer, C.; Plachkinove, M. Maintaining a Cybersecurity Curriculum: Professional Certifications as Valuable Guidance. *J. Inf. Syst. Educ.* **2017**, *28*, 101–114.

51. Shafiq, H.; Kamal, A.; Ahmad, S.; Rasool, G.; Iqbal, S. Threat modelling methodologies: A survey. *Sci. Int.* **2014**, *26*, 1607–1609.

52. Anderson, L.W.; Krathwohl, D.R.; Airasian, P.W.; Cruikshank, K.A.; Mayer, R.E.; Pintrich, P.R.; Raths, J.; Wittrock, M.C. *A Taxonomy for Learning, Teaching, and Assessing: A Revision of Bloom's Taxonomy of Educational Objectives*; Reference and Research Book News: Dublin, OH, USA, 2001; Volume 16, pp. 1–336.

53. Bird, J.; Kim, F. Survey on application security programs and practices. In *A SANS Analyst Survey*; SANS Institute: Bethesda, MD, USA, 2014; pp. 1–24.

54. DANAOS Shipping Company: DANAOSone Platform. DANAOS Management Consultants S.A. Available online: https://web2.danaos.gr/maritime-software-solutions/danaosone-platform/ (accessed on 24 July 2020).

55. Trustwave. *Security Testing Practices and Priorities: An Osterman Research Survey Report*; Osterman Research: Seattle, WA, USA, 2016; pp. 1–15.

56. IMO. *SOLAS Chapter XI-2—International Ship and Port Facility Security Code (ISPS Code)*; International Maritime Organization (IMO): London, UK, 2004.
57. CIS: Center of Internet Security. Available online: https://www.cisecurity.org/ (accessed on 24 July 2020).
58. ESCO: Results of Simulation-Based Competence Development Survey. European Cyber Security Organisation, 2019–2020 Report. Available online: https://echonetwork.eu/report-results-of-simulation-based-competence-development-survey/ (accessed on 24 July 2020).
59. Aaltola, K.; Taitto, P. Utilising Experiential and Organizational Learning Theories to Improve Human Performance in Cyber Training. *Inf. Secur. Int. J.* **2019**, *43*, 123–133. [CrossRef]

applied sciences

Article

Cyber Ranges and TestBeds for Education, Training, and Research

Nestoras Chouliaras [1], George Kittes [1], Ioanna Kantzavelou [1], Leandros Maglaras [2,*] and Grammati Pantziou [1,*] and Mohamed Amine Ferrag [3]

[1] Department of Informatics and Computer Engineering, University of West of Attika, 12241 Athens, Greece; nchouliaras@uniwa.gr (N.C.); cscyb19010@uniwa.gr (G.K.); ikantz@uniwa.gr (I.K.)
[2] School of Computer Science and Informatics, De Montfort University, Leicester LE1 9BH, UK
[3] Department of Computer Science, Guelma University, Guelma 24000, Algeria; ferrag.mohamedamine@univ-guelma.dz
[*] Correspondence: leandros.maglaras@dmu.ac.uk (L.M.); pantziou@uniwa.gr (G.P.)

Abstract: In recent years, there has been a growing demand for cybersecurity experts, and, according to predictions, this demand will continue to increase. Cyber Ranges can fill this gap by combining hands-on experience with educational courses, and conducting cybersecurity competitions. In this paper, we conduct a systematic survey of ten Cyber Ranges that were developed in the last decade, with a structured interview. The purpose of the interview is to find details about essential components, and especially the tools used to design, create, implement and operate a Cyber Range platform, and to present the findings.

Keywords: testbeds; cyber ranges; cyber exercises; education; training; research

Citation: Chouliaras, N.; Kittes, G.; Kantzavelou, I.; Maglaras, L.; Pantziou, G.; Ferrag, M.A. Cyber Ranges and TestBeds for Education, Training, and Research. *Appl. Sci.* **2021**, *11*, 1809. https://doi.org/10.3390/app11041809

Academic Editor: David Megías

Received: 23 December 2020
Accepted: 9 February 2021
Published: 18 February 2021

Publisher's Note: MDPI stays neutral with regard to jurisdictional claims in published maps and institutional affiliations.

1. Introduction

In recent years, cyber attacks, especially those targeting systems that keep or process sensitive information, are becoming more sophisticated. Critical National Infrastructures are the main targets of cyber attacks since essential information or services depend on their systems, and their protection becomes a significant issue that is concerning both organizations and nations [1–4]. Attacks to such critical systems include penetrations to their network and installation of malicious tools or programs that can reveal sensitive data or alter the behaviour of specific physical equipment.

Following this increase in cyber attacks, the need for professionals will also continue to increase in the upcoming years. According to predictions from Cybersecurity Ventures, an estimated 3.5 million cybersecurity jobs will be available and eventually unfilled by 2021. While global Cybercrime damages are predicted to reach $6 Trillion annually by 2021 [5], 61% of companies find most of the cybersecurity applicants unqualified [6]. The majority of chief information security officers around the world are worried about the cybersecurity skills gap, with 58% of CISOs believing the problem of not having an expert cyber staff will worsen [7].

Gartner Inc. [8] delivered its first-ever forecast report titled: "Forecast Analysis: Container Management (Software and Services) Worldwide", for the software container management software and services market, stating that adoption of the technology will be widespread. Software containers have enjoyed massive growth in recent years. Popular with developers, they provide a way for applications to be built once and run in any kind of computing environment, helping make enterprises much more agile. Gartner reckons that software containers will become the "default choice for 75% of new customer enterprise applications" by 2024. As a result, 15% of all applications will be running in containers by then, up from just 5% today.

Training activities and environments that can support challenging situations, followed by concrete guidance, procedures, and tools are needed. These platforms can help individuals to react in different, unpredictable situations in a collective and collaborative way. This environment should blend simulations and emulations of real components and systems, embedding different attack and defense mechanisms [9] and must be able to adapt to a variety of different incidents, in order to be cost-effective and attractive for organizations and educational institutes. Experiential learning is an educational technique that proposes the active involvement of the participants in order to help them learn through experience—an efficient method for delivering experiencing learning exercises as part of serious games. Cyber ranges are exercising environments that contain both physical and virtual components and can be used to represent realistic scenarios for training [10].

In recent years, cyber ranges have been offering additional features/capabilities from a simple simulation environment. Chandra [11] proposed that efficiency may be achieved by harnessing operating system container technology. Carnegie Mellon University [12] has developed by SEI open-source software tools to create secure and realistic cyber simulations. These tools recreate the real world and make training exercises more realistic.

In this article, we present the current state of the art on testbeds and cyber ranges that are used for training and research purposes. A systematic review of the literature on cyber range systems was carried out and the study revealed that there is a variety of implementations with different approaches that have been developed in different environments, using real, virtual, or hybrid equipment. Moreover, in order to better understand what the important components of a modern cyber range (CR) are, we conducted structured interviews with technical directors that have developed and used recently cyber ranges and present the findings.

The contributions of the article are:

- It presents the current state of the art on testbeds and cyber ranges.
- It presents the findings of a set of structured interviews with organizations that have a testbed and cyber range.
- It discusses the findings and gives insights of modern cyber ranges.

The findings of the research will be a guide for the effort to design, develop and implementation of a Cyber Range platform for the University of West Attica (UNIWA) but can also be a guide for other cyber ranges that are under development. UNIWA was founded in March 2018, from the merging process of two Technological Institutes. It operates with high educational and research standards and strives to respond to the ever-increasing demands of modern society for the creation of executives that have attained a solid scientific and technological background. UNIWA is the third-largest in Greece in terms of student numbers, approximately 52,000 undergraduates, 1150 postgraduates, and 210 doctoral students. The aim of a modern cyber range should be to enhance courses with hands-on experience of participants. In addition, it will enhance the research goals of the university through using a more complex and realistic environment than it currently has. UNIWA has a cybersecurity team (INSSec) with active participation in national and international cybersecurity exercises over the last decade as well as CTF competitions such as UniCTF 2019 and UniCTF 2020. In addition, it organized the CTF competition [13], UniwaCTF 2019, a competition between Greek universities. A Cyber Range system will enhance the realism of CTF contests, allowing UNIWA to organize more complex cyber exercises, such as the blue vs. red team..

The remainder of the paper is organized as follows: Section 2 discusses related surveys and showcases the value of this article. Section 3 introduces the key concepts and the overall architecture of current testbeds and cyber ranges. Section 4 presents the findings of the questionnaire. Section 6 discusses the findings and concludes the paper.

2. Related Surveys

During this literature review conducted from March to June 2020, several cyber ranges and testbeds were identified in different domains, such as Educational, CTF, Industrial Control Systems, Cyber Physical, and SCADA.

Davis and Magrath (2013) [14] conduct a survey of Cyber Ranges and classified their findings into three categories: Modeling and Simulation, Ad-hoc or Overlay, and Emulations. Specifically, their survey had the purpose of assisting organizations to select and build their desired CR capability. Hence, they surveyed the available options for constructing and managing a CR, for monitoring and analysis, training scenarios, communities for collaboration, and commercial offerings. They categorized CR using a two-level model. Firstly, they distinguished the CRs by their type, as Simulation, Ad-hoc or Overlay and Emulation. They also named the fourth category as Analytics without actually using it. Following previously defined methodologies, they categorized a CR as simulation when utilizing software models of real cases, as overlay if they use the real production equipment, and as emulation in the case of running the real applications on separate equipment. The second-level criteria of their categorization have been the sector the CR supports and the categories have been academic, military or commercial. The survey makes interesting points about the above-mentioned categories. Simulation CRs are sterilized, emulation ones have more realistic behaviour, but they are expensive, while overlays are only a small minority. According to the survey, the emulation CRs are the best category, especially when using virtualization. Moreover, the survey states that the main use of CRs is training, leaving far behind cybersecurity testing and research and development. This survey is quite broad as it covers almost 30 CRs, and it fulfills its aim. It refers widely to military developed and operated cases. This is expected as, at the time, military implementations had quite a few operating CRs. However, this survey is already seven years old, meaning that a lot of things have changed since. Moreover, it overlooks the cases where several categories are combined in hybrid cross-category environments.

Holm (2015) [15] surveyed 30 ICS testbeds. This survey has been a part of a study about critical infrastructures and eventually refers specifically to Industrial Control Systems (ICS). The study was motivated by the increasing vulnerability of ICS to cyber-attacks. It was titled "Virtual Industrial Control System Testbed" and was performed for FOI, the Swedish Defense Research Agency. The main purpose of the study was to specify the way to create a high-fidelity Virtual Industrial Control System (VICS) and the first step had been surveying the existing relevant testbeds through five Research Questions. The expected outcome would be the creation of a new testbed (CRATE). The survey collected information from 30 ICS testbeds in 12 countries. The study covers several testbed characteristics like the three methods that can be used to implement ICS in testbeds (virtualization, simulation, and hardware), including relevant subcategories (Operating System virtualization, Programming Language virtualization, Library virtualization) and categorization of these testbeds' objectives into 11 categories (Vulnerability analysis, Education, Tests of defense mechanisms, Power system control tests, Performance analysis, Creation of standards, Honeynet, Impact analysis, Test robustness, Tests in general, Threat analysis). Furthermore, the survey presents per category how the reviewed 30 testbeds implement their control center, communication architecture, field devices, and observed/controlled processes. The available categories are again Virtualization, Simulation, Emulation, and Hardware. However, this survey leaves room for hybrid methods. In addition, the survey states Fidelity, Repeatability, Measurement Accuracy, and Safe execution of tests as the basic requirements that testbeds should comply. It is clarified though that these requirements are not a product of the survey itself, but they pre-existed. The survey concludes that none of the questioned testbeds implements an overlay model (enables executing a real field device inside a virtual/emulated container). The complexity of ICS accounts for this conclusion. Finally, it distinguishes vulnerabilities as Policy and Procedure Vulnerabilities, Platform Vulnerabilities, and Network Vulnerabilities. Finally, the survey describes the architecture and functionality of a designed testbed (CRATE). This survey follows a stable methodology, approaching the testbeds from various different angles. Moreover, the analysis has taken into account a satisfactory amount of 30 testbeds. However, its main focus is the industrial (ICS) testbeds, and, eventually, the results are narrowed to this specific category of testbeds. In addition, since the time of the survey (2015), ICS systems have become more connected and have revealed more surface to the attackers. Unavoidably, the survey and its vulnerability analysis haven't taken into account the evolved and interconnected situation nowadays.

Yamin [16] presents a survey of Cyber Ranges and security testbeds and provide a taxonomy and an architectural model of a generic Cyber Range. Their work begins with the definition of a cyber exercise where they define the stages of such an exercise as well as the teams involved (white, blue, red). They identify a gap in existing surveys as they characterize them as sectorial or outdated. The chosen methodology of this survey has been the systematic literature review which consists of eight stages (Statement of purpose, protocol establishment, a search of the sources, screening of the literature, assessment, data extraction, synthesis of the outcome, review). During this process, they produce an initial taxonomy where a CR consists of five basic pillars (scenario, monitoring, teaming, scoring, management). Indicative of the width of the survey is the variety of cyber exercise teams/roles they have identified (red, blue, white, orange, purple, yellow, green, autonomous). An outcome of the survey is a classification of the capabilities and functionalities of modern CRs as well as a new taxonomy based on the information gathered, with six pillars (scenario, monitoring, learning, management, teaming, environment) has been produced. The survey has researched and recorded a multitude of simulation, emulation, hardware, management, monitoring, traffic generation, and other relevant tools and solutions implemented in contemporary CRs. In addition, the functional architecture of a generic Cyber Range is described. Based on the surveyed CRs, the survey attempts to predict the future shape of the Cyber Range environment. This survey is, by all means, an impressive work that firstly analyses and then combines data from multiple papers mainly for the period 2015–2017. The survey performs a wide approach and analysis of the literature. However, the survey concludes in a rather conservative manner, and the predicted future cyber ranges don't quite differ from the present ones.

Kucek (2020) [17] investigates the underlying infrastructures and CTF environments, specifically open-source CTF environments, and examined eight open-source CTF environments. The survey aims to be used as a valuable reference for whoever is involved in CTF challenges. Starting from 28 platforms, the survey shortlisted 12 environments that are open-source and finally managed to examine eight of them (CTFd, FacebooookCTF, HackTheArch, Mellivora, Pedagogic-CTF, PicoCTF, RootTheBox, WrathCTF), and to extract valuable conclusions and comparison data. The study was motivated by the popularity of CTF events combined with the lack of studies that examine the underlying infrastructure and configuration of real-time cyber exercises like CTFs. Once more, it starts with a questionnaire of four Research Questions (RQs). The survey distinguished the open-source CTF environments and attempted empirical research of them. They followed an organized methodology of five comprehensive steps (general review, shortlist of open-source CTFs, install, configure challenges, conclusions). In order to empirically examine each of the eight shortlisted environments, the survey conducted 16 different challenges categorized in five CTF types (quiz, jeopardy, Attack-defense, Mixtures, King of the Hill). Some interesting results include the architecture of the platforms. Some of them run on a certain O/S, while others run on any O/S. The next (higher) layer above the O/S is either the container layer or the virtualization one. The CTF challenges are configured on top of these layers. The survey concludes that the examined environments differ in some features they support and the respective configurations that are available. All the examined platforms have some generic features (participant registration, challenge provision, user manual, scoring methodology). The platforms differ in the specifics and the available options of the mentioned features. The survey has been both original and ambitious to deepen the performed analysis. However, its main objective is the CTF implementations and, consequently, it is narrowed to this specific category of testbeds. Moreover, the actual research is limited to eight CTF environments. Starting from around 30 candidate Cyber Ranges, they finally realized the empirical study on eight of them because of various reasons (proprietary environments, lack of adequate documentation, etc.).

Ukwandu [18] present a survey of Cyber Ranges and security testbeds. In this very recent survey, only publications from selected databases and only from the last five years (2015–2020) are examined. A taxonomy is developed to provide a broader comprehension of the future of Cyber Ranges and testbeds. The paper makes multiple references to the smart-everything technological transformation which must be taken into account when assessing or training in cybersecurity. Once more, the followed approach has been the chain: plan, select, extract, execute. The survey is presented as an overview of the Cyber Ranges and Test Beds which can be found in the literature and 44 CRs are identified. These instances are categorized in multiple ways, initially based on their application (Military/Defense/intelligence, Academic, Commercial, Law Enforcement, etc.) and their type (Private, Public, Federated). In addition, the teaming options are presented. The survey presents a classification of the found CRs according to their implementation method (Emulation, Simulation, Overlay, Live). The survey describes in fair detail the architecture and interconnection of CR building blocks. The survey provides a definition of a CR scenario and then different scenario options and differentiation factors (design, validation, deployment) are described. The stages that a training testbed should include are presented in an impressively simple but straightforward plan. The different approaches to training are described (gamification, Mock Attack Training, Role-Based Training, exercises). The survey argues in favor of the differentiation between Cyber Ranges and Test Beds. It presents Cyber Ranges as far more complicated than Testbeds. This argument concludes with the need for different taxonomies, respectively. Finally, according to the survey, the future shape of Cyber Ranges and Test Beds is going to combine real-time, intelligent implementations featuring mobility, automatic configuration, and integration of different technologies, applications, and appliances. Throughout this extensive analysis, the survey doesn't avoid some minor contradictions. Moreover, our survey integrates a structured interview that has been performed on a selected group of representative cyber ranges.

As shown in Table 1, we classify the surveys according to the following criteria:

- Focus area: We categorize surveys in relation to their scope.
- Method: this category indicates the method of collection and analysis of the data that are related to the CRs.

Most of the surveys, including ours, have a broad scope, while only two of them were focused on a specific area of research, ICS and CTFs. The main difference of our survey as compared to the previous ones is the use of mixed data collection methods that included both literature review and structured interviews with Universities and agencies that have deployed and run such CRs. This method helped us cover the lack of published information in terms of architecture, topology and tools.

Table 1. Related surveys on Cyber Ranges and TestBeds.

Survey	Reference	Systems Studied	Focus Area	Year	Method
Davis-Magrath et al.	[14]	30	Broad	2013	Literature Review
Holm et al.	[15]	30	ICS	2015	Literature Review
Yamin et al.	[16]	100	Broad	2019	Literature Review
Kucek et al.	[17]	28	CTFs	2020	Empirical Review
Ukwandu et al.	[18]	44	Broad	2020	Literature Review
Chouliaras et al.	Our survey	25	Broad	2021	Literature Review Structured Interviews

3. Background

Among many cyber incidents that have occurred in the last decade, two of them can be considered as major triggers for the development of Cyber Ranges—firstly the attack against the nuclear program of Iran. This attack that was revealed in 2010 used the computer worm Stuxnet and specifically targeted the programmable logic controllers (PLCs) used to automate machine processing systems. Since then, the malware has been mutated and discovered in other industrial and energy installations. Secondly, on 23 December 2015 via a series of cyber-attacks, cyber attackers remotely controlled the Ukrainian power grid, specifically the SCADA distribution management system, and eventually caused a significant power outage to the Ukrainian constituency. The above mentioned incidents have been more than persuasive of the vulnerability of industrial systems. This resulted in widely opening the way for the development of cyber ranges.

Initially, an up-to-date survey of the present situation of Cyber Range systems was conducted. This survey has revealed multiple useful outcomes. Some of them are the characteristics of modern cyber ranges and testing beds, the various development platforms used, the tools and methods which are implemented, how fast do the implementations occur, how are the exercises conducted and executed, how are the relevant scenarios created and implemented, etc.

Apart from the need to test and evaluate the cybersecurity aspect of applications, tools, and systems, cyber ranges are extremely useful for the capacity building of cyber experts. They must develop and possess several abilities like being deeply technically skilled, capable of recognizing and responding to complicated and urgent situations, able to assess risks and vulnerabilities, to handle uncertainty, to solve problems to provide explanations to think adversarial. In a nutshell, today's security experts must possess a "security mindset" as described in [19].

Various definitions of cyber ranges have been given in the relevant literature and publications. The definition given in NIST one-pager [20] has been chosen as the first among equals. Thus, according to NIST, cyber ranges are interactive, simulated representations of an organization's local network, system, tools, applications that are connected to a simulated Internet level environment. They provide a safe, legal environment to gain hands-on cyber skills and a secure environment for product development and security posture testing.

The research performed reveals that the environment of cyber ranges in terms of their development can be categorized into three main types: simulation, emulation, and hybrid. A Simulation involves using a model, a virtual instance in order to recreate a complex network environment based on the real network components' behaviour. Emulation is when the cyber range runs on the dedicated physical network infrastructure of the CR. Hybrid emerges from a customized combination of any of the above types. An additional category refers to overlay cyber ranges which are the instances that run in parallel with the actual production systems on the real equipment and infrastructure.

Recently, attention to Cyber Ranges has been growing. Cyber range systems are predominantly used for three main objectives: Research, Training and Exercise.

- Research (testing implementations including methods, tools, building blocks and systems)
- Training/Education (academia, specialized security courses and cyber-security certifications)
- Exercises/Competitions on Cyber Security (security training by means of cyber security exercises like Capture the Flag or Cyber Defense Exercises).

Research demands for environments that are fully controlled and isolated but at the same time complex to develop and test a new tool, or to design new attack techniques or methods. The training serves cybersecurity practice and education. Trainees have the opportunity to practice various cyber range scenarios, according to their specific training needs. The third and maybe most popular category of cyber range use nowadays is for cyber exercises. Here, the users compete in cyber contests, capture the flag competitions, hack the box challenges, and attack/defense games.

We can also categorize cyber ranges based on their operator. The main players for the development of Cyber Ranges and similar testbeds have been universities, government agencies, military research centers, international organizations, and their affiliates. While the details of some Cyber Ranges are publicly available, there also exist cyber ranges that are funded by the military and governments throughout the world and their details are eventually classified. Throughout the recent development and widening of the cyber range constituency, the concept of a federation of cyber ranges has emerged. The concept of federation relies on the consideration that a single cyber range would have enormous costs and would be extremely complicated if it was to have all the necessary features and functionalities, the whole package. Therefore, it would be better organized, and also modular and in effect realistic, if multiple cyber ranges, each within a specific area of expertise, could collaborate in order to offer to their users a wide variety of use cases and different scenarios. For example, some cyber ranges simulate social media networks or publicly available internet resources while other cyber ranges may be specialized in simulating industrial control systems or critical infrastructures. The combination of the capabilities of different cyber ranges would result in the development of a much broader simulation environment available for their end-users, while at the same time the overall cost would remain unchanged. Following this concept, several cyber range federations are being developed. Such an example is the Cyber Ranges Federation project which aims at building an EU-wide cyber range. Participants of this federation include eleven EU member states, the European Space Agency (ESA) as well as the European Defence Agency (EDA). Another relevant initiative is the CyberSec4Europe project which refers to designing, testing and demonstrating potential governance structures for a future European Cybersecurity Competence Network. One more example is the ECHO project (European network of Cybersecurity centers and competence Hub for innovation and Operations) launched by the European Commission with the vision to establish and operate a Cybersecurity Competence Network.

The Deployment models of cloud computing are categorized into four commonly used categories. Private Cloud, Public Cloud, Community Cloud and Hybrid Cloud. Additionally, there are three Services models of Cloud Computing: Infrastructure, Software, and Platform as a Service (IaaS, SaaS, PaaS). In the SaaS model, a software provider sells a software application that can be used on-demand. In the IaaS, the provider offers as service computing resources like storage, server or peripherals. The users can have a virtual server in a very short time, and they pay only for the resources they use. The PaaS model represents an abstraction layer between the IaaS and SaaS and its target group includes deployers and developers. Infrastructure platforms and tools include OpenStack [21], Opennebula [22], Proxmox [23], VMware [24], Public cloud (AWS), Minimega [25] and KVM [26].

Infrastructure as code (IaC) is another step ahead towards infrastructure agility and flexibility. With IaC, the management of infrastructure (networks, virtual machines, load balancers, and connection topology) is realized in a descriptive model. Some Infrastructure as code (IaC) tools that we came across in our survey include Chef [27], Puppet [28], Ansible [29], SaltStack [28], Terraform [30], and Vagrant [27].

In the present paragraph, some terms that are necessary for the forthcoming analysis are defined. When we talk about deployment, we refer to the process of putting a new application, or a new version of an application, to run on a prepared application server. Orchestration is the arrangement or coordination of multiple systems that are designed to cooperate. Provisioning (used by DEVOps) refers to getting computers or virtual hosts to use and installing needed libraries or services to them. Configuration management (CM) is a system engineering process for the establishment and maintenance of a product's performance, functional, and physical attributes with its requirements, design and operational information. Configuration management aims at bringing consistency in the infrastructure. The above-mentioned tools (Chef, Puppet, Ansible, SaltStack) are all "configuration management" tools, which means they are designed to install and manage software on existing servers, whereas Terraform is an "orchestration tool", meaning that it is designed to provision the servers themselves, leaving the configuration of these servers to other tools. These two categories are not mutually exclusive, as most configuration management tools can do some degree of provisioning and most orchestration tools can do some degree of configuration management.

Using the cyber range background and environment as described in the previous paragraphs, we now move forward to explain the features of the cyber ranges we found in our survey. We analyze 25 CRs, and we write about features that they use (see Table 2) like objective, environment, supporting sector, etc. Research (R), Training (T), Exercise (E), Education (ED), Operations (O), Testing (TE), Academic (A), Military (M), Government (G), Private Enterprise (PE), Industry (I), Demonstrations (DM), Development (DV),Testing (TS), Emulation (EM), Simulation (S), Hybrid/Cyber Physical (HCP), VMWARE (VW), Openstack (O), Minimega (MN), TerraForm (TR), Public cloud AWS (AW), QEMU / KVM (Q), Virtualbox (VB), Custom (C), Yes (Y), No (N), Not Available (N/A), Docker (D), Instructors (IN), Provided on demand (OD), In house (IH), On-Premise (OP), Online (ON) and On-Site (OS). Then, based on these findings, we select the ten most representative cyber-ranges, and we moved forward with the structured interview (Tables 3 and 4).

Table 2. Summary of cyber ranges and testBeds.

Operator	Objective	Sector	Environment	Infrastructure Platform(s)	Dataset
NATO Cyber Range [31]	T, E	M	EM	VW	N/A
Masaryk University (KYPO) [32,33]	R, T, E, ED	A	EM	O	Y
Florida Cyber Range [34]	ED, R, T, O	M	N/A	N/A	N/A
Sandia National Laboratories (Cyber Scorpion) [25]	T	G	N/A	MN	N/A
Virginia Tech [35,36]	R, T, E	A	S	AW	N/A
De Montfort University [10]	R, T, E	A	HCP	Q	OD
Royal Military Academy [37,38]	R, T	A, M	S	VB, C	N
AIT Austrian Institute of Technology [39–41]	R, T, E	A,G, M PE	HCP	O, TR	N
Naval Postgraduate School [42,43]	T, E, ED	A, G, M	S	D	IN
Norwegian University of Science and Technology (NCR) [44]	R, T, E, Ts	A, G, M, PE	EM, S, HCP	O, VB, VM, D	OD
Università degli Studi di Milano [21]	T	R	EM, S	O	No
JAMK University of Applied Sciences (JYVSECTEC) [45,46]	R, T, E	A, G, M, PE	EM, S, HCP	N/A	IH
Swedish Defence Research Agency (CRATE) [47,48]	R, T, E	G, M	HCP	VB	ON
Michigan Cyber Range [49]	T	A	N/A	N/A	Yes
Silensec [50]	T	I	N/A	N/A	ON
CYBERIUM (fujitsu) [51]	T	I	N/A	N/A	ON
DECIDE (NUARI) [52]	R, T, E	A	N/A	N/A	ON
Georgia Cyber Range [53]	ED, T, R, DM, DV	A	N/A	N/A	ON
IBM X-Force Command C-TOC [54]	T, E	I	N/A	N/A	ON
Cybexer [55]	T, E	I	N/A	N/A	ON
Airbus Cyber Range [56]	R, T, E	I	S	N/A	ON
Raytheon Cyber Range [57]	T ,E	M, A, I	N/A	N/A	ON
hns-platform [58]	T, E	I	HCP, S	N/A	ON
Cyberbit Cyber Range [59]	T, E	I	S	N/A	ON, OP
Cyber Warfare Range [60]	T, E	I	S	N/A	OS, OP

Table 3. Cyber Ranges' features.

Operator	Security Challenges	Courses	Access	Roles	Teams	Events
De Montfort University	W, E, AP	DF, SS, IC	OP	SC, NC, CR	B, RD, G, YL, WT, P	EV, WS, EX
Royal Military Academy	W, F, DD, AP	DF, NS, WS	RA	N/A	B	EX
Masaryk University	W, F, E, SI, MA	DF, NS	OP, RA	CR, CS	B, RD, G, YL, WT, P	EV, WS, EX
AIT Austrian Institute of Technology	W, F, E, DD, AP, R, MA	NS, WS, OS	OP, RA	SC, CR, CS, CO, IT, LG	B, RD, G, YL, WT, P	EV, WS, EX
Naval Postgraduate School	W, C, E, DD, SI, MA, RE	C, SS, NS, WS	L	Various	N/A	EI
Norwegian University of Science and Technology	W, C, F, E, S, DD, AP, R, SI, MA, RE, RM, ISE, CM, CP	C, DF, HS, SS, NS, CS, WS, CM	OP, RA	SC, NC, CR, CS, M, CV	B, RD, WT P	EV, WS, EX
Virginia Tech	W, C, F, E, S, SI, RE	C, DF, SS, NS, WS	RA	N/A	N/A	EV, WS, EX, EI
Università degli Studi di Milano	W, F, SI, MA, RB	DF, WS	OP, RA	N/A	B, RD, G	N/A
JAMK University of Applied Sciences	W, C, F, E, S, DD, AP, R, SI, MA, RE	DF, HS, SS, NS, CS, WS	OP, RA	SC, NC, CR, CS	B, RD, G, YL, WT, P	EV, WS, EX
Swedish Defence Research Agency	W, F, DD, R, MA	SS, NS	OP, RA	SC, NC, CR	B, RD, G, WT	EV, WS, EX

Table 4. Cyber Ranges' tools.

Operator	VM	Network	Scoring	Scenarios	Manage	Monitor	Traffic	User Behavior
De Montfort University	MS	PR	Y	MF	PR	OS, SN, SU, N, W, M	D, B	B
Royal Military Academy	A	VB	N/A	J	C	SN, N	N/A	G
Masaryk University	A	O	C	JS, YM, C	C	NG	N/A	C
AIT Austrian Institute of Technology	A	O	N/A	JS	N/A	OS, W	N/A	C
Naval Postgraduate School	DC	LDT	AG	LDT	N/A	A	N/A	N/A
Norwegian University of Science and Technology	A, V, MS	O	J	YM	XT	OS	OF, D	GS
Virginia Tech	P	CL	CC	N/A	AP	CW	N/A	N/A
Università degli Studi di Milano	OH	O	IT	X	IT	Y	IT	N/A
JAMK University of Applied Sciences	N/A	N/A	N/A	N/A	N/A	N/A	N/A	N/A
Swedish Defence Research Agency	S	VB, VX	N/A	N/A	N/A	N/A	N/A	AI, BT

4. Analysis of Results

Due to the lack of several features that are not mentioned in the publications but also to have a better picture of the systems used, a structured questionnaire [61] (see Appendix A) was created and sent to selected universities and research centers that develop and maintain such systems (see Tables 3 and 4).

Table 3 has the following analysis: Web (W), Cryptography (C), Forensics (F), Exploitation (E), Steganography (S), DDoS (DD), APT (AP), Ransomware (R), SQL Injections (SI), Malware Analysis (MA), Reverse Engineering (RE), Risk Management (RM), Information Security Economics (ISE), Cyber Crisis Management (CM), Cyber Policy Analysis (CP), Digital Forensics (DF), Software Security (SS, ICS Security (IC), Custom (CU), Request Base (RB), Digital Forensics (DF), Network security (NS), Web Security (WS), Software Security (SS), ICS Security (IC), OT Security (OS), Hardware Security (HS), Cloud Security (CS), Data-driven cybersecurity management (CM), On Premise (OP), Remote Access (RA), Local (L), SOC (SC), NOC (NC), CERT (CR), CSIRT (CS), CISO (CO), IT-Team (IT), Legal (LG),

Managers (M), C-levels (CV), BLUE (B), RED (RD), GREEN (G), YELLOW (YL), WHITE (WT), PURPLE (P),Event (EV), Workshop (WS), Exercise (EX), and Educational Institutions (EI).

Table 4 has the following analysis: Manual Scripting (MS), Ansible (A), Docker containers (DC), Vagrant (V), Packer (P), Openstack Heat (OH), PROXMOX (PR), Virtualbox (VB), Openstack (O), Cloudformation (CL), VXLAN (VX), Labtainers designer tool (LDT), Custom (C), Artifacts Gathered (AG), Jeopardy Board (J), CloudCTF (CC), Internal Tools (IT), JSON (JS), YAML (YM), Multiple Formats (ML), XML (X), Automatic (A), Xentop (XT), API (AP), OSSIM (OS), Snort (SN), Suricata (SU), Netflow (N), Wireshark (W) , MALCOM (M), Nagios (NG), Cloudwatch (CW), DNP3 (D), Bespoke (B), OpenFlow (OF), GHOSTS (G), AutoIT (AI), Bot(BT), Yes (Y), and Not Available (N/A).

The motivation for the questionnaire was, despite a large number of published works and surveys [14–18], the lack of data on the tools used for the development and management of Cyber Ranges, when used to organize cybersecurity exercises and provide a data-set for further research. At first, it was checked to see if there are cyber range systems in universities and research centers in Greece. The limited number of existing systems that are located in Greece led us to broaden the search in Europe, Asia and the rest of the world.

The questionnaire was addressed to technical directors or managers who were directly involved with the Cyber Range. The survey was conducted from 01/06/2020 until 04/08/2020. The results of the research were produced by 10 different systems located in nine different countries and two continents. The countries are the USA, the United Kingdom, Italy, Norway, Sweden, Finland, the Czech Republic, Belgium and Austria.

The first question was about the objective of the Cyber Range and, as expected, participants answered that their main objective is training.

The largest percentage of the participants use CR systems for research, training and security exercises [16,18]. No participant has developed their system exclusively to cover a single objective, and, more specifically, 80% of participants cover at least two, as shown in Figure 1.

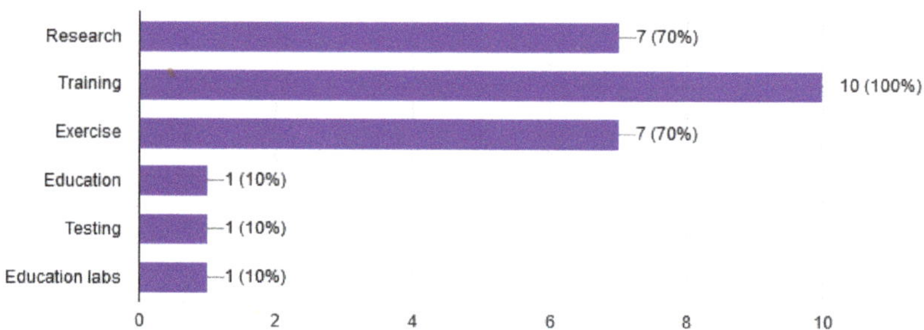

Figure 1. Objectives of the Cyber Range.

Question 2. The questionnaire was sent to the CR system providers covering all four key areas [14] Academic, Government, Military, and Private Enterprise. We have covered this requirement due to the feedback from all areas, Figure 2.

Of course, the majority of the answers as shown in the figure supporting sector are mainly from the Academic sector. This is because military and Private Enterprise providers do not disclose details about their systems due to confidentiality, and the existing literature is limited. However, we have managed to cover all areas, even for the military and Private Enterprise sectors, and draw useful conclusions about technologies, implementations, and development tools as shown in the next questions.

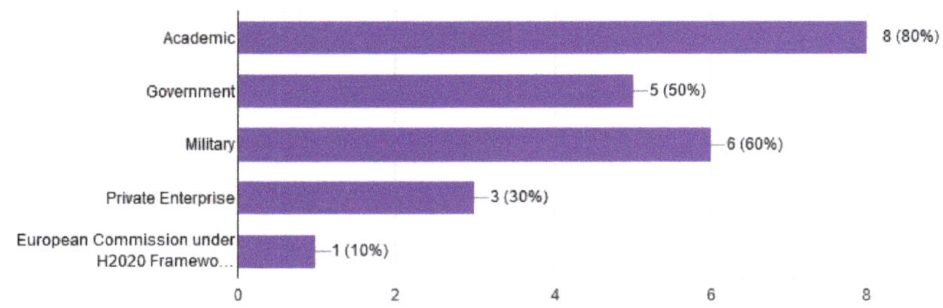

Figure 2. Sectors of the Cyber Range.

In question 3, we have another categorization of a cyber range, which is the domain that the systems operate. Another area that is flourishing is the conduct of cybersecurity exercises [62–64]. As expected, the results of the domain cybersecurity competition are very high, Figure 3, about 80%, as well as in SCADA, reach 60%. An interesting conclusion from the analysis of the results is that 30% of the systems are focused only on conducting security exercises, and 20% only on SCADA.

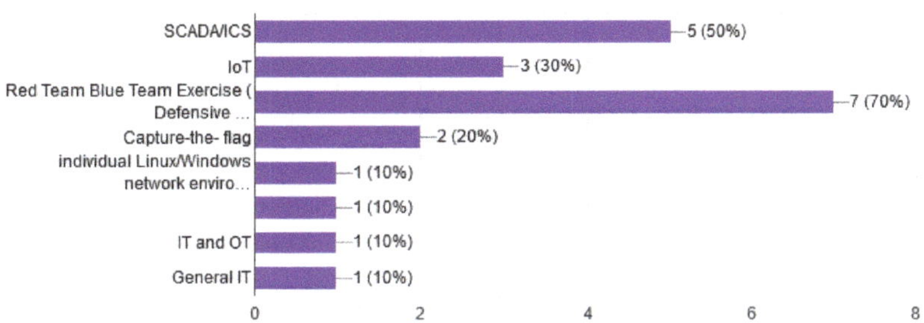

Figure 3. Domains of the Cyber Range.

Mainly after the incident of Iran's nuclear program, and the attack of the Ukraine power grid, a great development in cyber range systems aimed at improving the security of SCADA and ICS and OT generally was observed. By correlating questions 3, 4, and 5, we observe that cyber ranges do not focus only on only one domain as before but have evolved by adding new components and managed to cover many domains like business, banking, telecom, health, and transport.

Question 4 describes the security challenges that occur in Cyber Range platforms. The most popular challenge is web security that is provided by all responders. In addition, as shown in Figure 4, the Forensics come first with 80% and Exploitation and Malware analysis follows with 70%. Additionally, one of the responders stated that they can create any challenge based on specific demands.

The content of security challenges [17] varies and depends on the type of cybersecurity competition or curriculum of the university/research center. Cyber security exercises allow students to gain hands-on experiences while immersed in environments that mimic real-world operational systems. Highly realistic training allows students to gain valuable experience that employers are looking for [65]. A very interesting approach is the inclusion of challenges like Risk Management, Information Security Economics, Cyber Crisis Management, and Cyber Policy Analysis. These are hot areas and we suggest other universities to add these kinds of challenges to their cyber range platforms.

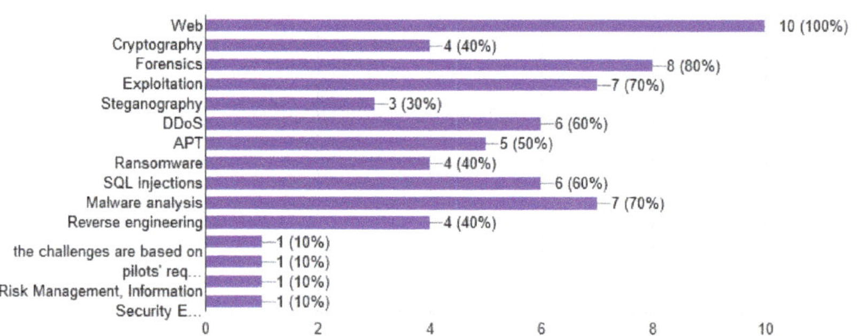

Figure 4. Security challenges of the Cyber Range.

A key motivation of our research is the development and implementation of a CR platform for the University of West Attica that covers three areas of research, education and conducting security exercises. Wanting to more deeply cover the educational side, we sought to find out if the CR platform is also used for educational purposes. All responders answered positively. According to Beveridge [65], injecting realism into cybersecurity training and education is beneficial to rapidly train qualified, skilled and experienced cybersecurity professionals. Additionally, we asked which courses they use for the CR platform. The most popular courses as shown in Figure 5 are network security by 80%, followed by web security and digital forensics by 70%, and software security by 60%.

Universities are linked to the educational curriculum courses related to emerging technologies such as cloud security, OT security and Data-driven cybersecurity management. Cyber ranges can combine security courses and hands-on experience and give cybersecurity experts the mentality, the problem solving capability and the appropriate technical tools for capacity building.

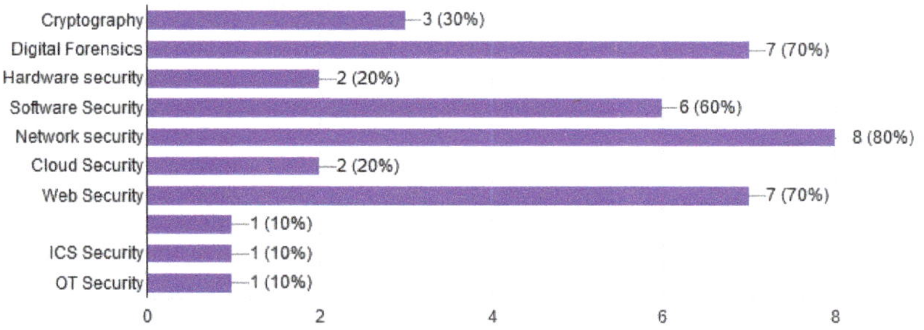

Figure 5. Educational courses of the Cyber Range.

Another categorization of Cyber Ranges is the type of environment. Davis [14] in 2013 categorize CR and security testbeds in three main categories emulation, simulation, and Ad-hoc or Overlay. In our questionnaire, we asked the participants to identify the environment also in three categories—the first is emulation: testbed built with real hardware or software, the second is a simulation: testbed built with software virtualization, and the last is Hybrid/Cyber-Physical: virtual testbeds connected with real hardware. Apart from one participant who had developed an emulated environment and two participants who have developed a simulation environment, all responders have chosen a mixed type of environment, as shown in Figure 6.

The rapid virtualization growth helps create complex environments, thus managing to achieve the highest possible accuracy, fidelity, scalability and flexibility while reducing implementation costs. Additionally, by using a simulation/hybrid environment, a university can develop a CR [35,66–68], while, before 2010, CR was developed for military purposes only (Emulab [69], NCR, StealthNet, and LARIAT [70]) mainly due to high development and maintenance costs.

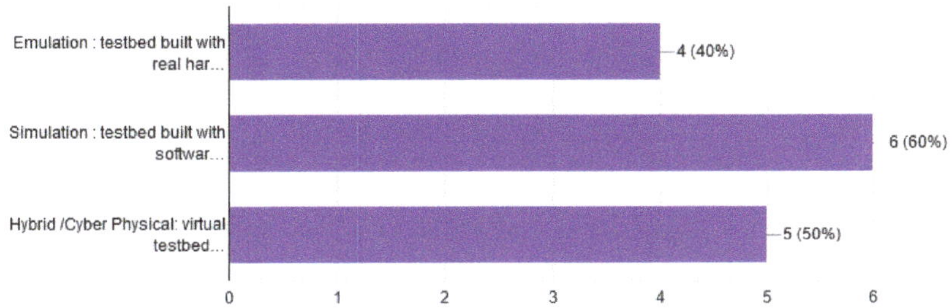

Figure 6. Types of environment.

In question 7, we discuss which type of virtualization technology is chosen for the development of CR, and, according to ECSO [71], there are two types, conventional and cloud virtualization. Conventional virtualization uses hypervisor-based technology and containers, mostly Docker. A list of both types of hypervisors containVirtualbox, Vmware, XenServer, Hyper-V, QEMU, etc. Cloud virtualization is divided into three types, public, private, and hybrid. The best advance of the cloud is the sharing of resources, great capabilities for automation and minimization of cost reduction [29]. OpenNebula, CloudStack, and OpenStack [27] are mostly used to deploy cloud virtualization [21–23]. The finding of questionnaires, as shown in Figure 7, says that up 50% uses the cloud, both Openstack and AWS, and 40% use traditional technology. In addition, we conclude that OpenStack is the main tool (44%) used to deploy cloud infrastructure.

The development of cloud computing has opened new horizons for the evolution of cyber ranges. Cloud environments constitute internet-based platforms to be used for computer technology. The technology used to develop the CR platforms is mainly open source and the use of commercial tools is partial. We found that the use of container technology has little impact on the systems we analyzed. We believe that there should be greater development through container technology since they improve realism and user behaviour [12].

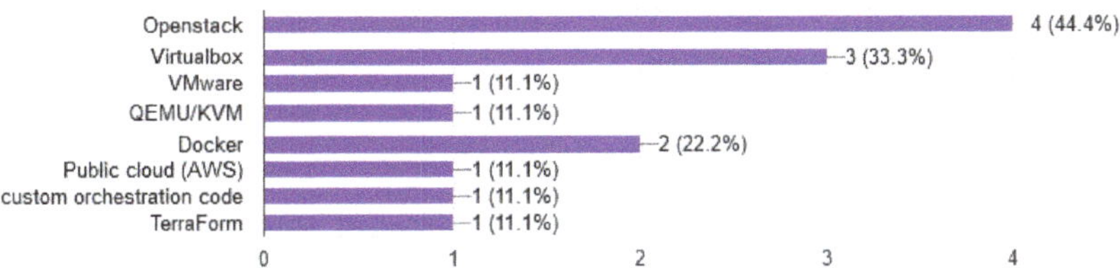

Figure 7. Infrastructure platform.

Question 8 is about the type of access that CRs can provide to platform participants. As presented in Figure 8, these are on-premises 70%, remote access 80% and 10% local. Moreover, 60% of CRs can provide both types of access, on-premises and remote access. In addition, finally, one platform can provide only on-premises access. The advantage [65] of providing remote access to participants is important for conducting distance learning courses, or long-distance security competitions.

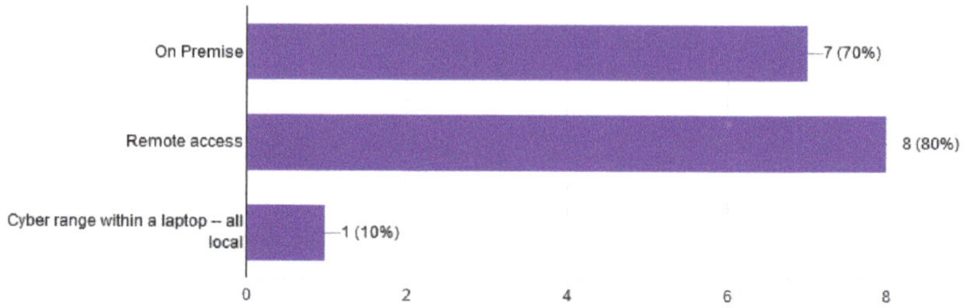

Figure 8. Type of access.

Question 9 is one of the most important questions we asked in the questionnaire. When searching in the literature to find out how to implement a Cyber Range system, the result was disappointing and the findings were negligible, especially regarding military and commercial systems. With the main motivation of discovering the design technology and the implementation tools, we proceeded to compile this question. As shown in Figure 9, the technology of CRs is dominated by the use of Infrastructure as code (IaC) tools [27–30] and especially Ansible with 40%, Vagrant, and Packer. In addition, in a small percentage, where obviously there is no cloud infrastructure, the configuration of virtual machines is done with the use of manual scripting with an imprint in the speed of implementation and in the flexibility of configuration.

Today, IaC is the process of managing and provisioning computer data centers through machine-readable definition files, rather than physical hardware configuration or interactive configuration tools. IaC tools are used to configure systems, deploy software and updates, and orchestrate. The biggest advantage is the speed and ease of their use as opposed to manual scripting.

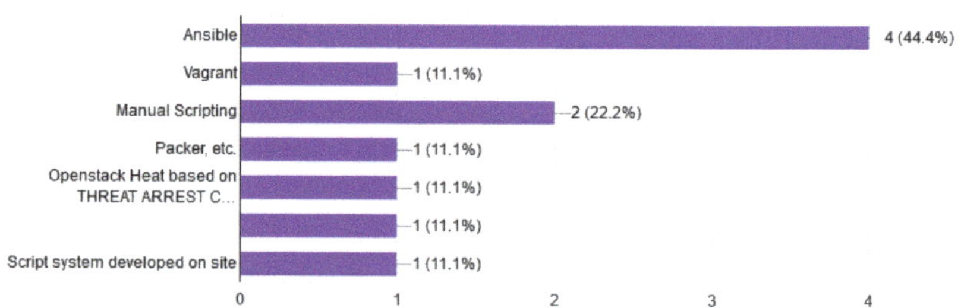

Figure 9. Set up VMs.

The tools used for the network topology are shown in Figure 10. Network tools provided by the infrastructure platform are mainly used. This can guide researchers/developers to invest in network tools that can be adopted by other CR systems.

In order to keep scoring during cybersecurity competitions like cyber security exercises or CTFs, several tools and mechanisms are provided. These tools are responsible for counting the flags in CTF [17] and awarding points, or artifacts from a CDX. As shown in Figure 11, the majority of scoring tools are custom made and depending on challenge, architecture of exercises, and infrastructure platforms.

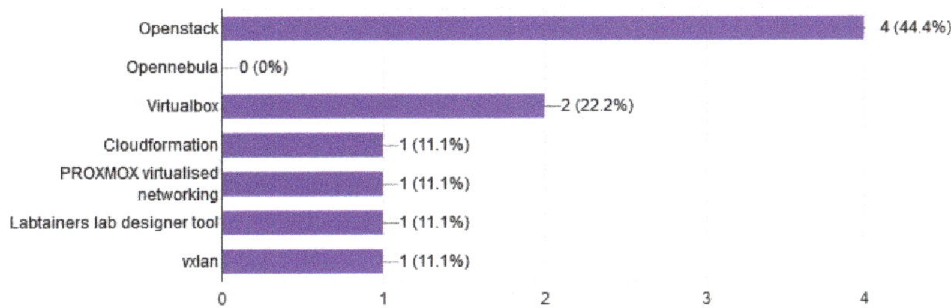

Figure 10. Network topology.

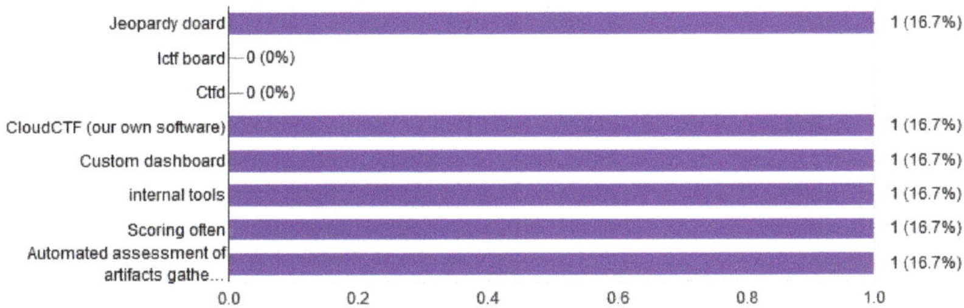

Figure 11. Scoring tools for Cyber Ranges.

JSON and YAML are the main scripting language that is used as shown in Figure 12, for designing a CTF or CDX. In addition, with the use of scripting language, it became possible to create dynamic scenarios. Planning an exercise requires a script. The scenario was initially static and required the configuration of all parameters during the development of each exercise. This resulted in complex development and management of exercises, required high management costs, and demanded long development times recently, with the development of dynamic scripts [4,72] based on scripting languages such as JSON, YAML and XML or IaC [30] Tools.

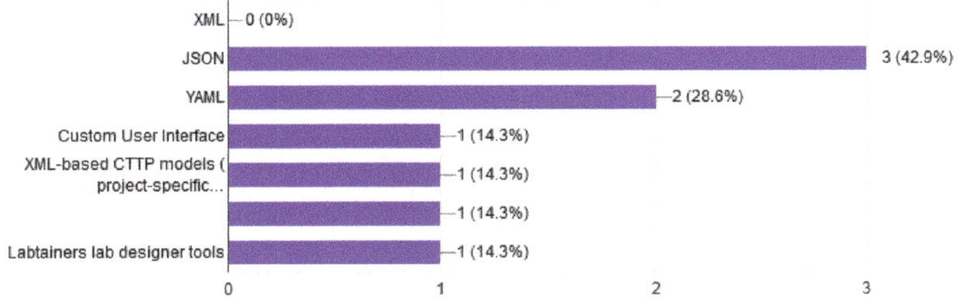

Figure 12. Tools to create cyber security scenarios.

A CR platform should have the right tools for managing users and groups as shown in Figure 13. Moreover, the CR must have a graphical user interface (GUI), capable of managing resources [28] like memory, usage, performance, reports, error logs, alert, etc. The responders identified that most use tools that are provided by the platform (OpenStack, Proxmox, AWS) or developed their own tools.

Dynamic scenarios require minimal administrative effort and in less time (from seconds to a few minutes) that could include new environments with different network topologies.This may be an opportunity for researchers/developers to produce tools that can be used by other systems.

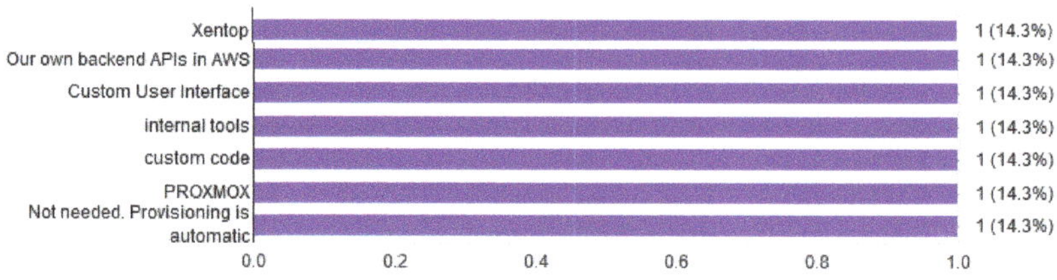

Figure 13. Tools to manage.

The CR platform must be able to monitor data. It must have all the necessary components for supervision, whether they are exercise training, research, or testing a system. The tools deploy depending on the type of exercise or field of the research. The responders answered that they are mostly used for monitoring purposes and open-source tools (see Figure 14), mainly SIEM tools such as OSSIM or Nagios. IDS tools such as Snort or Suricata are also used.

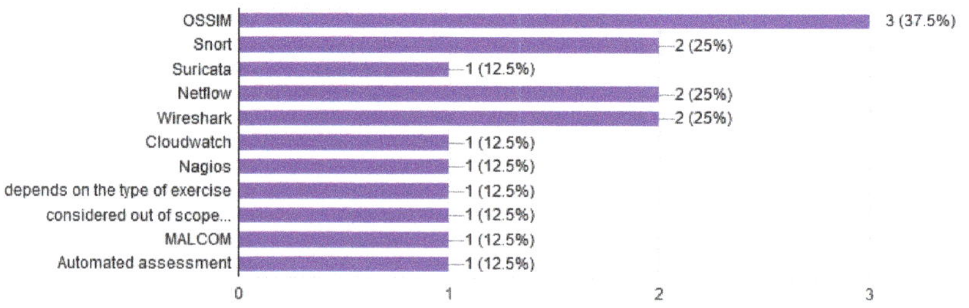

Figure 14. Tools to monitor.

CR platforms use tools [73–75] for monitor data. OpenFlow and DNP3 have been used by the responders in several occasions, but mainly in-house tools or scripts are used, as shown in Figure 15. Testing of security tools [76] should take place under conditions that are as realistic as possible. Network traffic of the testing infrastructure should approach a real network of a company or a university [77]. Based on the answers, we don't find a tool that has a high level of acceptance yet.

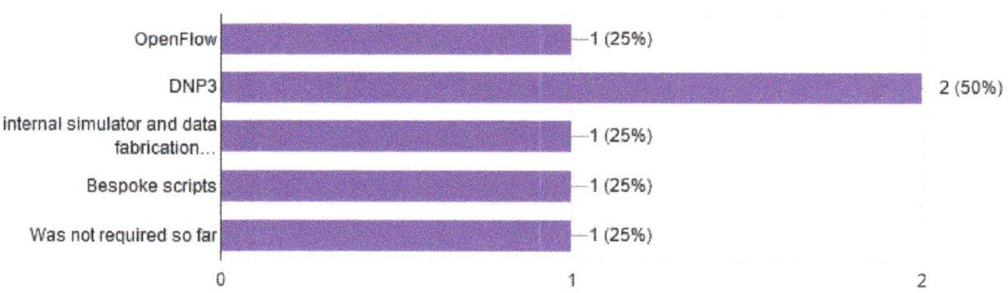

Figure 15. Network traffic.

Another example of an automation user/team is the automation of the red team in conducting cybersecurity attacks. The use of such an automated team covers the need to find qualified cybersecurity experts with knowledge of attacking systems, which is very difficult. There are published papers describing how to create such red teams mostly in the military domain such as K0ala from Lincoln Laboratory [70] and SVED from FOI [48] that used for automating the behaviour of a red team. GHOSTS as shown in Figure 16, a tool developed by the SEI, creates non-player characters (NPCs) that behave realistically without human intervention in order to help build complex cyber simulations. GHOSTS create NPCs that behave like real people to generate context-driven traffic. As a result, creators of simulations can challenge participants in blue or red teams with engaging content that helps them develop elite skill sets [12,78] and red team automation. From the answers, we notice that systems have used the GHOSTS tool [12] that develops SEI and provided through GITHUB, while the other platforms have developed their own tools.

In general, scripting languages are capable of creating complex environments, including realistic user behaviour, thus improving realism. In such a use case scenario, an automated user can send or receive emails, browse the internet site, open office documents or print them, etc., resembling a typical office user that works in a company working environment. Realistic user behaviour is an important part of creating complex cybersecurity exercises.

Figure 16. User behaviour.

In question 10, we identify how many groups can participate in an exercise. The answers were quite different and related not only to the implementation of the CR but also to the capacity of the infrastructure of the environment that supports it. The answers varied from systems that support only groups with one user to systems with a capacity of thousands of groups. However, on average, systems support up to 10 groups. Moreover, we examined the total number of participants that varies from one to thousands of simultaneous users. The average of users falls in the range between 50 and 100. Another point of measurement of the analysis and complexity of the exercises [79] is the number of different teams [16] that participate. As expected, the teams [80] that mainly participate are the blue 80% and the red 70%. In addition, apart from two participants who did not inform us about the teams, at least half of the participants stated that blue, red, yellow, purple, green, and white teams take part in the exercises as shown in Figure 17.

One main purpose of question 10 was also to identify the complexity of the exercises and the capacity of the cyber ranges. The roles of the participants are also very important, since they support, as shown in Figure 18, the development of security teams such as SOC, NOC, CERT, and CSIRT. It is also interesting that, in some cases, some other roles were used from CRs such as Managers, C-level executives, and legal representatives.

Figure 17. Cyber Security Teams.

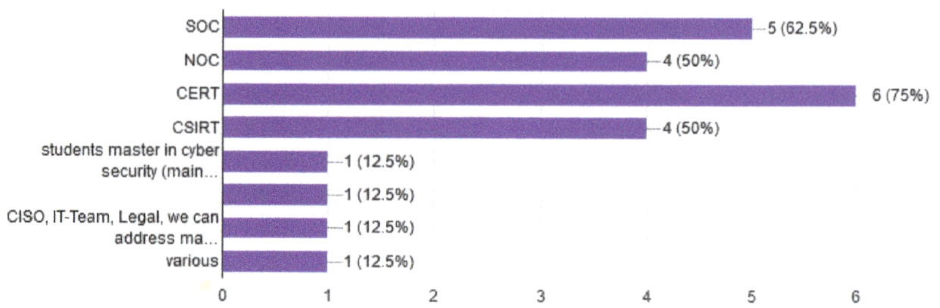

Figure 18. Roles of participants.

In question 11, we asked the participants if the CR platform has already been used. As shown in Figure 19, 90% of the respondents answered positively. In many cases, a system is created for research purposes, such as a research program that has an expiration date. The CR systems analyzed in this questionnaire are already used for educational, research, or CDX and presented in a public event.

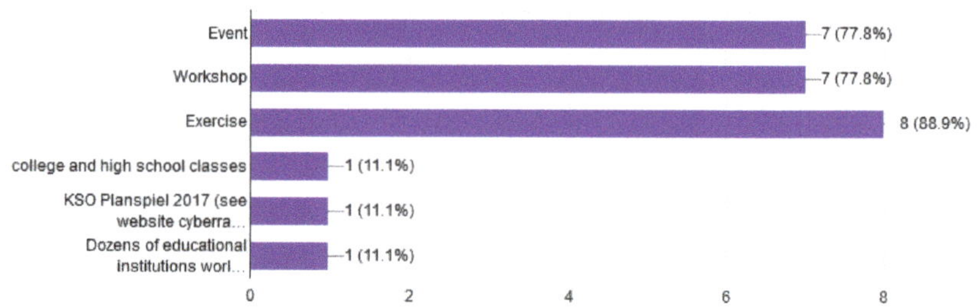

Figure 19. What type of event.

The last question is about datasets. An important element of datasets is whether they contain measurable data. Researchers using datasets can evaluate the performance of IDSs, measuring their accuracy, false positives, and overall efficiency. In Figure 20, the results showed that a large percentage, around 60%, of the systems produce datasets or this action is included in the upcoming plans.

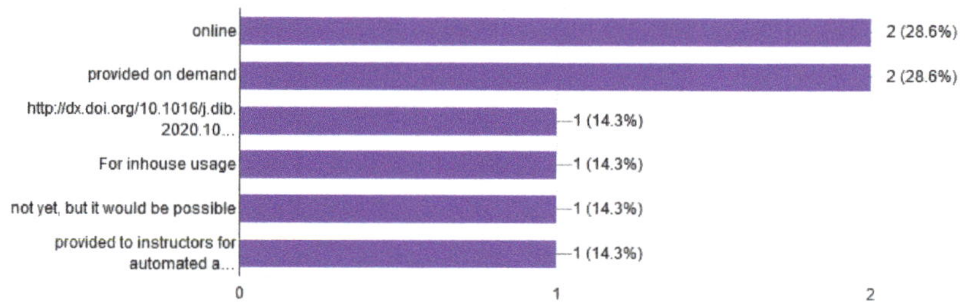

Figure 20. Dataset.

The creation of a dataset that contains capture network traces, from cybersecurity exercises, can enhance or produce new sophisticated methods on detection techniques for cybersecurity attacks (see Figure 21).

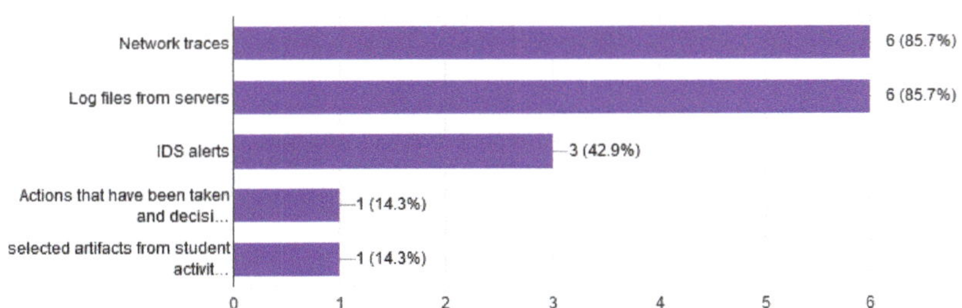

Figure 21. Dataset.

5. Challenges and Future Directions

CR research teams should be focusing on improving various aspects of their testbeds. In addition, modern CRs should be enriched with novel features, such as various telecommunication capabilities, emulated Banking systems, hospitals [81], simulated smart grids, automated vehicles [82], Virtual Cyber Centres of Operation, wireless sensor networks, real time Intrusion Detection Systems [83], honeypots [84], novel authentication mechanisms [85], mobile security scenarios, and several privacy mechanisms. By adding these features, new attack scenarios can be easily deployed on a testbed, revealing vulnerabilities of the various systems and thus giving the researchers the opportunity of developing innovative defence mechanisms. Moreover, any novel CR should be built in a way that could be easily used for research purposes inside EU projects. This could be accomplished if the CRs are capable of being connected to various real-world devices to the network, making it that way ideal for launching attacks and testing the defence mechanisms of various systems. One other important aspect that should be taken into account is the capability of modern CRs to create measurable data in a semi automated way with limited human intervention.

Modern CRs should include a portable version for demonstration purposes and for easy deployment as a modern teaching instrument in various cyber security events that take place around Europe. Moreover, research teams should also be working towards the capability of their CRs to provide remote access to researchers. Via such a federated model, researchers all around the world will be given the opportunity to implement various protocols and study their behaviour in custom tailor-made environments. Finally, the need for moving from traditional cyber ranges to digital twins is a trend that is going to become dominant in the near future, especially for replicating critical infrastructures.

6. Conclusions

In this paper, we present a systematic survey of ten Cyber Ranges with a structured interview. The purpose of the questionnaire is to examine key components that consist of a Cyber Range platform, and particularly the tools used to design, create, implement, and operate a cyber range platform. As analysed in Section 4, most of the current cyber ranges are moving towards more realistic and competitive scenarios that can help the users receive focused experiential learning. The combination of emulated and simulated into hybrid environments can help a cyber range to be more adaptive, expandable, and thus efficient. One important aspect of a modern cyber range is the datasets that are produced and how these can be shared with other scholars in order to help them test new security mechanisms.

The findings of the research will be a guide for the effort to design, develop and implementation a Cyber Range platform for the University of West Attica (UNIWA) but can also be a guide for other cyber ranges that are under development.

Author Contributions: Conceptualization, N.C., G.K. and L.M.; Methodology, N.C., I.K. and L.M.; Software, M.A.F., G.P. and G.K.; Validation, G.P., I.K. and L.M.; formal analysis, I.K., M.A.F. and G.K.; investigation, N.C., G.K., I.K. and L.M.; resources, M.A.F., G.K. and N.C.; data curation, N.C., G.P., I.K. and L.M.; writing—original draft preparation, N.C., G.K. and M.A.F.; writing—review and editing, I.K., G.P. and L.M.; visualization, M.A.F., N.C., G.K. and L.M.; supervision, I.K., G.P. and L.M. All authors have read and agreed to the published version of the manuscript.

Funding: This research received no external funding.

Institutional Review Board Statement: Not applicable.

Informed Consent Statement: Not applicable.

Conflicts of Interest: All authors declare no conflict of interest.

Appendix A. Questionnaire

Cyber Range Questionnaire [61]

1. What is the objective of the Cyber Range? (select all that apply)
2. What is the supporting sector of the Cyber Range? (select all that apply)
3. What is the domain that is emulated or replicated in the operational environment? (select all that apply)
4. What type of security challenges are provided? (select all that apply)
5. Is the Cyber Range used for educational purposes?
6. What is the type of Cyber Range environment?
7. Which infrastructure platform(s) is (are) used to develop the Cyber Range?
8. What type of access does it provide to participants? (select all that apply)
9. What tools are used to

    i. Set up Vms?
    ii. Set up network topology?
    iii. Keep scoring? (flag dashboards, log analyzers, etc.)
    iv. Create cyber security scenarios?
    v. Manage the Cyber Range? (resources)
    vi. Monitoring the exercises ? (SIEM, IDS, etc.)
    vii. Generate network traffic?
    viii. Generate user behaviour?
    ix. Other functions?

10. Teams, Roles and Participants

    i. How many teams can participate at the same time?
    ii. Total number of active participants?
    iii. PARTICIPANTS: What are the roles/functions? (select all that apply) iv. Roles

11. Has the Cyber Range been used already?
12. Has the Cyber Range provided any dataset?

    i. if yes the dataset is?
    ii. What type of information does the dataset contain?

References

1. Maglaras, L.A.; Kim, K.H.; Janicke, H.; Ferrag, M.A.; Rallis, S.; Fragkou, P.; Maglaras, A.; Cruz, T.J. Cyber security of critical infrastructures. *ICT Express* **2018**, *4*, 42–45. [CrossRef]
2. Ferrag, M.A. EPEC: An efficient privacy-preserving energy consumption scheme for smart grid communications. *Telecommun. Syst.* **2017**, *66*, 671–688. [CrossRef]
3. Ferrag, M.A.; Nafa, M.; Ghanemi, S. EPSA: An efficient and privacy-preserving scheme against wormhole attack on reactive routing for mobile ad hoc social networks. *Int. J. Secur. Netw.* **2016**, *11*, 107–125. [CrossRef]
4. Braghin, C.; Cimato, S.; Damiani, E.; Frati, F.; Mauri, L.; Riccobene, E. A Model Driven Approach for Cyber Security Scenarios Deployment. In *Computer Security*; Fournaris, A.P., Athanatos, M., Lampropoulos, K., Ioannidis, S., Hatzivasilis, G., Damiani, E., Abie, H., Ranise, S., Verderame, L., Siena, A., et al., Eds.; Springer International Publishing: Cham, Switzerland, 2020; pp. 107–122.
5. Chung, M. Signs your cyber security is doomed to fail. *Comput. Fraud. Secur.* **2020**, *2020*, 10–13. [CrossRef]

6. Crumpler, W.; Lewis, J.A. *Cybersecurity Workforce Gap*; Center for Strategic and International Studies (CSIS): Washington, DC, USA, 2019.
7. Angafor, G.N.; Yevseyeva, I.; He, Y. Bridging the Cyber Security Skills Gap: Using Tabletop Exercises to Solve the CSSG Crisis. In Proceedings of the IFIP Joint International Conference on Serious Games, Stoke-on-Trent, UK, 19–20 November 2020; Springer: Berlin/Heidelberg, Germany, 2020; pp. 117–131.
8. Gartner, I. *Forecast Analysis: Container Management (Software and Services)*; Gartner, Inc.: Stamford, CT, USA, 2020.
9. Stewart, B.; Rosa, L.; Maglaras, L.A.; Cruz, T.J.; Ferrag, M.A.; Simoes, P.; Janicke, H. A novel intrusion detection mechanism for scada systems which automatically adapts to network topology changes. *EAI Endorsed Trans. Ind. Networks Intell. Syst.* **2017**, *4*. [CrossRef]
10. Hallaq, B.; Nicholson, A.; Smith, R.; Maglaras, L.; Janicke, H.; Jones, K. CYRAN: A hybrid cyber range for testing security on ICS/SCADA systems. In *Cyber Security and Threats: Concepts, Methodologies, Tools, and Applications*; IGI Global: Hershey, PA, USA, 2018; pp. 622–637.
11. Chandra, Y.; Mishra, P.K. Design of Cyber Warfare Testbed. In *Software Engineering*; Hoda, M.N., Chauhan, N., Quadri, S.M.K., Srivastava, P.R., Eds.; Springer: Singapore, 2019; pp. 249–256.
12. Updyke, D.; Dobson, G.; Podnar, T.; Osterritter, L.; Earl, B.; Cerini, A. *GHOSTS in the Machine: A Framework for Cyber-Warfare Exercise NPC Simulation*; Technical Report CMU/SEI-2018-TR-005; Software Engineering Institute, Carnegie Mellon University: Pittsburgh, PA, USA, 2018.
13. UNIWA. UNIWA CTF. Available online: http://www.pdsn.uniwa.gr/profile/inssec/ (accessed on 17 January 2021).
14. Davis, J.; Magrath, S. *A Survey of Cyber Ranges and Testbeds Executive*; Cyber Electronic Warfare Division DSTO (Defence Science and Technology Organisation): Edinburgh, Australia, 2013.
15. Holm, H.; Karresand, M.; Vidström, A.; Westring, E. A Survey of Industrial Control System Testbeds. In *Secure IT Systems*; Buchegger, S., Dam, M., Eds.; Springer International Publishing: Cham, Switzerland, 2015; pp. 11–26.
16. Yamin, M.M.; Katt, B.; Gkioulos, V. Cyber ranges and security testbeds: Scenarios, functions, tools and architecture. *Comput. Secur.* **2020**, *88*, 101636. [CrossRef]
17. Kucek, S.; Leitner, M. An Empirical Survey of Functions and Configurations of Open-Source Capture the Flag (CTF) Environments. *J. Netw. Comput. Appl.* **2020**, *151*, 102470. [CrossRef]
18. Ukwandu, E.; Farah, M.A.B.; Hindy, H.; Brosset, D.; Kavallieros, D.; Atkinson, R.; Tachtatzis, C.; Bures, M.; Andonovic, I.; Bellekens, X. A review of cyber-ranges and test-beds: Current and future trends. *Sensors* **2020**, *20*, 7148. [CrossRef] [PubMed]
19. Dark, M. Thinking about Cybersecurity. *IEEE Secur. Priv.* **2015**, *13*, 61–65. [CrossRef]
20. NIST. *Cyber Ranges*; NIST: Gaithersburg, MD, USA.
21. Braghin, C.; Cimato, S.; Damiani, E.; Frati, F.; Riccobene, E.; Astaneh, S. Towards the Monitoring and Evaluation of Trainees' Activities in Cyber Ranges. In *International Workshop on Model-Driven Simulation and Training Environments for Cybersecurity*; Springer: Berlin/Heidelberg, Germany, 2020; pp. 79–91.
22. Eichler, Z. Cloud-Based Security Research Testbed: A DDoS Use Case. In Proceedings of the 2014 IEEE Network Operations and Management Symposium (NOMS), Krakow, Poland, 5–9 May 2014.
23. Goldman, R. *Learning Proxmox VE*; Packt Publishing Ltd.: Birmingham, UK, 2016.
24. Østby, G.; Berg, L.; Kianpour, M.; Katt, B.; Kowalski, S.J. A Socio-Technical Framework to Improve cyber security training: A Work in Progress. In Proceedings of the fifth Workshop on Socio-Technical Perspective in IS development, Stockholm, Sweden, 10 June 2019; pp. 81–96.
25. Raybourn, E.M.; Kunz, M.; Fritz, D.; Urias, V. A Zero-Entry Cyber Range Environment for Future Learning Ecosystems. In *Cyber-Physical Systems Security*; Springer: Berlin/Heidelberg, Germany, 2018; pp. 93–109.
26. Pham, C.; Tang, D.; Chinen, K.I.; Beuran, R. Cyris: A cyber range instantiation system for facilitating security training. In Proceedings of the Seventh Symposium on Information and Communication Technology, Ho Chi Minh, Vietnam, 8–9 December 2016; pp. 251–258.
27. Luchian, E.; Filip, C.; Rus, A.B.; Ivanciu, I.; Dobrota, V. Automation of the infrastructure and services for an openstack deployment using chef tool. In Proceedings of the 2016 15th RoEduNet Conference: Networking in Education and Research, Bucharest, Romania, 7–9 September 2016; pp. 1–5.
28. Kostromin, R. Survey of Software Configuration Management Tools of Nodes in Heterogeneous Distributed Computing Environment. Available online: http://ceur-ws.org/Vol-2638/paper15.pdf (accessed on 17 January 2021).
29. Tkachuk, R.V.; Ilie, D.; Tutschku, K. Orchestrating Future Service Chains in the Next, Generation of Clouds. In Proceedings of the 15th SNCNW 2019, Lulea, Sweden, 4–5 June 2019; pp. 18–22.
30. Brikman, Y. Why We Use Terraform and not Chef, Puppet, Ansible, Saltstack, or Cloudformation. 2016. Available online: https://lsi.vc.ehu.eus/pablogn/docencia/AS/Act7%20Admin.%20centralizada%20infrastructure-as-code,%20Configuration%20Management/Terraform%20Chef%20Puppet%20Ansible%20Salt.pdf (accessed on 17 January 2021).
31. Pernik, P. *Improving Cyber Security: NATO and the EU*; International Center for Defence Studies: Tallinn, Estonia, 2014.
32. Vykopal, J.; Ošlejšek, R.; Čeleda, P.; Vizvary, M.; Tovarňák, D. Kypo Cyber Range: Design and Use Cases. 2017. Available online: https://is.muni.cz/publication/1386573/en/KYPO-Cyber-Range-Design-and-Use-Cases/Vykopal-Oslejsek-Celeda-Vizvary (accessed on 17 January 2021).

33. Vykopal, J.; Vizvary, M.; Oslejsek, R.; Celeda, P.; Tovarnak, D. Lessons learned from complex hands-on defence exercises in a cyber range. In Proceedings of the 2017 IEEE Frontiers in Education Conference (FIE), Indianapolis, IN, USA, 18–21 October 2017; pp. 1–8.
34. Range, F.C. Florida Cyber Range. Available online: https://floridacyberrange.org/ (accessed on 24 November 2020).
35. Range, V.C. About the Virginia Cyber Range. Available online: https://www.virginiacyberrange.org/ (accessed on 25 November 2020).
36. Darwish, O.; Stone, C.M.; Karajeh, O.; Alsinglawi, B. Survey of Educational Cyber Ranges. In Proceedings of the Workshops of the International Conference on Advanced Information Networking and Applications, Caserta, Italy, 15–17 April 2020; Springer: Berlin/Heidelberg, Germany, 2020; pp. 1037–1045.
37. Debatty, T.; Mees, W. Building a Cyber Range for training CyberDefense Situation Awareness. In Proceedings of the 2019 International Conference on Military Communications and Information Systems (ICMCIS), Budva, Montenegro, 14–15 May 2019; pp. 1–6.
38. Llopis, S.; Hingant, J.; Pérez, I.; Esteve, M.; Carvajal, F.; Mees, W.; Debatty, T. A comparative analysis of visualisation techniques to achieve cyber situational awareness in the military. In Proceedings of the 2018 International Conference on Military Communications and Information Systems (ICMCIS), Warsaw, Poland, 22–23 May 2018; pp. 1–7.
39. Leitner, M.; Frank, M.; Hotwagner, W.; Langner, G.; Maurhart, O.; Pahi, T.; Reuter, L.; Skopik, F.; Smith, P.; Warum, M. AIT Cyber Range: Flexible Cyber Security Environment for Exercises, Training and Research. In Proceedings of the European Interdisciplinary Cybersecurity Conference (EICC), Rennes, France, 18 November 2020; pp. 18–19.
40. Frank, M.; Leitner, M.; Pahi, T. Design Considerations for Cyber Security Testbeds: A Case Study on a Cyber Security Testbed for Education. In Proceedings of the 2017 IEEE 15th Intl Conf on Dependable, Autonomic and Secure Computing, 15th Intl Conf on Pervasive Intelligence and Computing, 3rd Intl Conf on Big Data Intelligence and Computing and Cyber Science and Technology Congress(DASC/PiCom/DataCom/CyberSciTech), Orlando, FL, USA, 6–10 November 2017; pp. 38–46. [CrossRef]
41. Kucek, S.; Leitner, M. Training the Human-in-the-Loop in Industrial Cyber Ranges. In *Digital Transformation in Semiconductor Manufacturing*; Keil, S., Lasch, R., Lindner, F., Lohmer, J., Eds.; Springer International Publishing: Cham, Switzerland, 2020; pp. 107–118.
42. Irvine, C.E.; Thompson, M.F.; McCarrin, M.; Khosalim, J. Live Lesson: Labtainers: A Docker-based Framework for Cybersecurity Labs. In Proceedings of the 2017 USENIX Workshop on Advances in Security Education (ASE 17), Vancouver, BC, Canada, 5 August 2017; USENIX Association: Vancouver, BC, Canada, 2017.
43. Thompson, M.F.; Irvine, C.E. Individualizing Cybersecurity Lab Exercises with Labtainers. *IEEE Secur. Priv.* **2018**, *16*, 91–95. [CrossRef]
44. Kianpour, M.; Kowalski, S.; Zoto, E.; Frantz, C.; Øverby, H. Designing Serious Games for Cyber Ranges: A Socio-technical Approach. In Proceedings of the 2019 IEEE European Symposium on Security and Privacy Workshops (EuroS PW), Stockholm, Sweden, 17–19 June 2019; pp. 85–93.
45. Kokkonen, T.; Hämäläinen, T.; Silokunnas, M.; Siltanen, J.; Zolotukhin, M.; Neijonen, M. Analysis of Approaches to Internet Traffic Generation for Cyber Security Research and Exercise. In *Internet of Things, Smart Spaces, and Next, Generation Networks and Systems*; Balandin, S., Andreev, S., Koucheryavy, Y., Eds.; Springer International Publishing: Cham, Switzerland, 2015; pp. 254–267.
46. Karjalainen, M.; Kokkonen, T.; Puuska, S. Pedagogical Aspects of Cyber Security Exercises. In Proceedings of the 2019 IEEE European Symposium on Security and Privacy Workshops (EuroS PW), Stockholm, Sweden, 17–19 June 2019; pp. 103–108.
47. Gustafsson, T.; Almroth, J. Cyber Range Automation Overview with a Case Study of CRATE. Available online: https://www.researchgate.net/profile/Tommy_Gustafsson2/publication/346559585_Cyber_range_automation_overview_with_a_case_study_of_CRATE/links/5fc73339299bf188d4e8f40b/Cyber-range-automation-overview-with-a-case-study-of-CRATE.pdf (accessed on 17 January 2021).
48. Holm, H.; Sommestad, T. SVED: Scanning, Vulnerabilities, Exploits and Detection. In Proceedings of the MILCOM 2016—2016 IEEE Military Communications Conference, Baltimore, MD, USA, 1–3 November 2016; pp. 976–981. [CrossRef]
49. Rege, A.; Adams, J.; Parker, E.; Singer, B.; Masceri, N.; Pandit, R. Using cybersecurity exercises to study adversarial intrusion chains, decision-making, and group dynamics. In Proceedings of the European Conference on Cyber Warfare and Security, Dublin, Ireland, 29–30 June 2017; Academic Conferences International Limited: Cambridge, MA, USA, 2017; pp. 351–360.
50. Silensec. Silensec. Available online: https://www.silensec.com/about-us/cyberranges (accessed on 24 November 2020).
51. Hara, K. Cyber Range CYBERIUM for Training Security Meisters to Deal with Cyber Attacks. *Fujitsu Sci. Tech. J.* **2019**, *55*, 59–63.
52. Nuari. Nuari. Available online: https://nuari.net/ (accessed on 24 November 2020).
53. Center, G.C. Georgia Cyber Center. Available online: https://www.gacybercenter.org/ (accessed on 24 November 2020).
54. IBM. IBM Xforce. Available online: https://exchange.xforce.ibmcloud.com/ (accessed on 25 November 2020).
55. Cybexer. Cybexer. Available online: https://cybexer.com/ (accessed on 25 November 2020).
56. Airbus. Airbus Cyber Range. Available online: https://airbus-cyber-security.com/products-and-services/prevent/cyberrange/ (accessed on 25 November 2020).
57. Raytheon. Raytheon Cyber Range. Available online: https://www.raytheon.com/cyber/capabilities/range (accessed on 25 November 2020).
58. DIATEAM. Hns-Platform Cyber Range. Available online: https://www.hns-platform.com/ (accessed on 25 November 2020).

59. Cyberbit. Cyberbit Cyber Range. Available online: https://www.cyberbit.com/platform/cyber-range/ (accessed on 25 November 2020).
60. Range, C.W. Cyber Warfare Range. Available online: https://www.azcwr.org/ (accessed on 25 November 2020).
61. Chouliaras, N. Cyber Range Questionnaire. Available online: https://docs.google.com/forms/d/e/1FAIpQLSek34D2Ks4laS4 AmajwHZAGqWGOrQxCOGIM3Lcmyaof2xyd2w/viewform?usp=sf_link (accessed on 15 September 2020).
62. Seker, E.; Ozbenli, H.H. The Concept of Cyber Defence Exercises (CDX): Planning, Execution, Evaluation. In Proceedings of the 2018 International Conference on Cyber Security and Protection of Digital Services (Cyber Security), Oxford, UK, 3–4 June 2018; pp. 1–9.
63. Forero, C.A.M. Tabletop Exercise For Cybersecurity Educational Training; Theoretical Grounding in addition, Development. Master's Thesis, University of Tartu Institute of Computer Science, Tartu, Estonia, 2016.
64. Kick, J. *Cyber Exercise Playbook*; Technical Report; MITRE CORP: Bedford, MA, USA, 2014.
65. Beveridge, R. Effectiveness of Increasing Realism Into Cybersecurity Training. *Int. J. Cyber Res. Educ. (IJCRE)* **2020**, *2*, 40–54. [CrossRef]
66. Technology, J.J.S. RGCE Organizational Environments. Available online: https://jyvsectec.fi/2018/01/rgce-organizational-environments/ (accessed on 25 November 2020).
67. Teknisk-Naturvitenskapelige Universitet, N. Om Norwegian Cyber Range. Available online: https://www.ntnu.no/ncr (accessed on 25 November 2020).
68. University, M. KYPO Cyber Range Platform. Available online: https://crp.kypo.muni.cz/ (accessed on 25 November 2020).
69. Nussbaum, L. Testbeds Support for Reproducible Research. In Proceedings of the Reproducibility Workshop, 2017; pp. 24–26.
70. Braje, T.M. *Advanced Tools for Cyber Ranges*; Technical Report; MIT Lincoln Laboratory: Lexington, KY, USA, 2016.
71. (ECSO), E.C.S.O. Understanding Cyber Ranges: From Hype to Reality. Available online: https://ecs-org.eu/documents/publications/5fdb291cdf5e7.pdf (accessed on 25 November 2020).
72. Russo, E.; Costa, G.; Armando, A. Building Next, Generation Cyber Ranges with CRACK. *Comput. Secur.* **2020**, *95*, 101837. [CrossRef]
73. Behal, S.; Kumar, K. Characterization and Comparison of DDoS Attack Tools and Traffic Generators: A Review. *IJ Netw. Secur.* **2017**, *19*, 383–393.
74. Patil, B.R.; Moharir, M.; Mohanty, P.K.; Shobha, G.; Sajeev, S. Ostinato—A Powerful Traffic Generator. In Proceedings of the 2017 2nd International Conference on Computational Systems and Information Technology for Sustainable Solution (CSITSS), Bangalore, India, 21–23 December 2017; pp. 1–5.
75. Botta, A.; Dainotti, A.; Pescapè, A. A tool for the generation of realistic network workload for emerging networking scenarios. *Comput. Netw.* **2012**, *56*, 3531–3547. [CrossRef]
76. Erlacher, F.; Dressler, F. How to Test an IDS? GENESIDS: An Automated System for Generating Attack Traffic. In Proceedings of the 2018 Workshop on Traffic Measurements for Cybersecurity, Budapest, Hungary, 20–24 August 2018; pp. 46–51.
77. Berk, V.H.; de Souza, I.G.; Murphy, J.P. Generating realistic environments for cyber operations development, testing, and training. In *Sensors, and Command, Control, Communications, and Intelligence (C3I) Technologies for Homeland Security and Homeland Defense XI*; Carapezza, E.M., Ed.; International Society for Optics and Photonics, SPIE: Bellingham, WA, USA, 2012; Volume 8359, pp. 51–59. [CrossRef]
78. Applebaum, A.; Miller, D.; Strom, B.; Korban, C.; Wolf, R. Intelligent, automated red team emulation. In Proceedings of the 32nd Annual Conference on Computer Security Applications, Los Angeles, CA, USA, 5–9 December 2016; pp. 363–373.
79. Kokkonen, T.; Puuska, S. Blue team communication and reporting for enhancing situational awareness from white team perspective in cyber security exercises. In *Internet of Things, Smart Spaces, and Next, Generation Networks and Systems*; Springer: Berlin/Heidelberg, Germany, 2018; pp. 277–288.
80. NCCDC. National CCDC. Collegiate Cyber Defense Competition. Available online: https://www.nationalccdc.org/ (accessed on 25 September 2020).
81. Evans, M.; He, Y.; Maglaras, L.; Janicke, H. HEART-IS: A novel technique for evaluating human error-related information security incidents. *Comput. Secur.* **2019**, *80*, 74–89. [CrossRef]
82. Kosmanos, D.; Prodromou, N.; Argyriou, A.; Maglaras, L.A.; Janicke, H. MIMO techniques for jamming threat suppression in vehicular networks. *Mob. Inf. Syst.* **2016**, *2016*, 8141204. [CrossRef]
83. Ferrag, M.A.; Maglaras, L.; Ahmim, A.; Derdour, M.; Janicke, H. Rdtids: Rules and decision tree-based intrusion detection system for internet-of-things networks. *Future Internet* **2020**, *12*, 44. [CrossRef]
84. Doubleday, H.; Maglaras, L.; Janicke, H. SSH Honeypot: Building, Deploying and Analysis. 2016. Available online: https://dora.dmu.ac.uk/handle/2086/12079 (accessed on 17 January 2021).
85. Papaspirou, V.; Maglaras, L.; Ferrag, M.A.; Kantzavelou, I.; Janicke, H.; Douligeris, C. A novel Two-Factor HoneyToken Authentication Mechanism. *arXiv* **2020**, arXiv:2012.08782.

Article

On Combining Static, Dynamic and Interactive Analysis Security Testing Tools to Improve OWASP Top Ten Security Vulnerability Detection in Web Applications

Francesc Mateo Tudela [1], **Juan-Ramón Bermejo Higuera** [1,*], **Javier Bermejo Higuera** [1], **Juan-Antonio Sicilia Montalvo** [1,*] and **Michael I. Argyros** [2]

[1] Escuela Superior de Ingeniería y Tecnología, Universidad Internacional de La Rioja, Avda. de La Paz 137, 26006 Logroño, La Rioja, Spain; fmateo@gmail.com (F.M.T.); javier.bermejo@unir.net (J.B.H.)

[2] Department of Computing and Technology, Cameron University, Lawton, OK 73505, USA; Michael.Argyros@cameron.edu

* Correspondence: juanramon.bermejo@unir.net (J.-R.B.H.); juanantonio.sicilia@unir.net (J.-A.S.M.)

Received: 1 December 2020; Accepted: 18 December 2020; Published: 20 December 2020

Featured Application: This document provides a complete comparative study of how different types of security analysis tools, (static, interactive and dynamic) can combine to obtain the best performance results in terms of true and false positive ratios taking into account different degrees of criticality.

Abstract: The design of the techniques and algorithms used by the static, dynamic and interactive security testing tools differ. Therefore, each tool detects to a greater or lesser extent each type of vulnerability for which they are designed for. In addition, their different designs mean that they have different percentages of false positives. In order to take advantage of the possible synergies that different analysis tools types may have, this paper combines several static, dynamic and interactive analysis security testing tools—static white box security analysis (SAST), dynamic black box security analysis (DAST) and interactive white box security analysis (IAST), respectively. The aim is to investigate how to improve the effectiveness of security vulnerability detection while reducing the number of false positives. Specifically, two static, two dynamic and two interactive security analysis tools will be combined to study their behavior using a specific benchmark for OWASP Top Ten security vulnerabilities and taking into account various scenarios of different criticality in terms of the applications analyzed. Finally, this study analyzes and discuss the values of the selected metrics applied to the results for each n-tools combination.

Keywords: web application; security vulnerability; analysis security testing; static analysis security testing; dynamic analysis security testing; interactive analysis security testing; assessment methodology; false positive; false negative; tools combination

---

1. Introduction

In recent years, the use of web applications has increased in many types of organizations, such as public and private, government, critical infrastructures, etc. These applications have to be continuously developed in the shortest time possible to face the competitors. Therefore, developers make programing security vulnerabilities or use third-party modules or components that are vulnerable. Sometimes they have limited budgets. These cases often make them forget an essential component within the development life cycle—security. In addition, companies find that developers do not have enough security knowledge, which increases the risk of producing insecure developments.

Taking into account the different types of analysis security testing (AST) [1] and the great amount of security vulnerabilities that web applications have in their code and configurations, it seems necessary that the security analyst making a security analysis should select the best static (SAST), dynamic (DAST) and interactive (IAST) analysis security testing tools in combination. The results of several studies [2–4] confirm that the web applications analyzed did not pass the OWASP Top Ten project [5]. SQL injection (SQLI) and cross site scripting (XSS) vulnerabilities continue to be the most frequent and dangerous vulnerabilities.

Several tools of the same type can be combined to achieve a better performance in terms of true and false positives [6–9]. Several works have also shown that better ratios of true and false positives can be obtained by combining different types of techniques to leverage the synergies that different type of tools can have [10–12]. These works show how a tools combination can reduce false positives (detected vulnerabilities that do not really exist) and false negatives (real vulnerabilities not found). Analyzed works concluded that each security vulnerability included in an AST tool report, including manual reviews, is required to be verified. False positives are not really a danger and can be fixed by the security analyst. However, a false negative is more difficult to find if the tool has not previously detected it, causing a real danger. These techniques include the use of static white box security analysis (SAST), dynamic black box security analysis (DAST) or interactive white box security analysis (IAST) tools. Manual analysis requires highly specialized staff and time. To carry out an analysis of web application security, using any method, it is necessary to cover the entire attack surface accessing all parts and application layers and using tools to automate security analysis as much as possible. There are some questions to investigate:

- How is each type of AST tool's average effectiveness, considering each type of vulnerability without combination?
- How is each type of AST tool's combinations average effectiveness, considering each type of vulnerability and the number of tools in a combination?
- How is each type of AST tool's average effectiveness at detecting OWASP Top Ten security category vulnerabilities without combination?
- How is the n-tools combination's average effectiveness at detecting OWASP Top Ten security vulnerabilities computing different metrics?
- Which are the best combinations for analyzing the security of web applications with different levels of criticality?

It is necessary to establish in the organizations a Software Development Life Cycle (SSDLC), as defined in the work of Vicente et al. [13], in order to standardize the use of SAST, DAST and IAST tools with the objective of deploying in production web applications as secure as possible.

This work aims to be the first of its kind—to study the best way to combine the three types of security analysis tools for web applications. Therefore, the first goal of this study is to investigate the behavior of the combination of two static tools (Fortify SCA by Microfocus, Newbury, United Kingdom, and FindSecurityBugs, OWASP tool created by Philippe Arteau, licensed under LGPL), two dynamic tools (OWASP ZAP open source tool with Apache 2 licenseand Arachni open source tool with public source License v1.0 created by Tasos Laskos) and two interactive tools (Contrast Community Edition by Contrasst Security, Los Altos, EE.UU. and CxIAST by Chekcmarx, Raman Gan, Israael) using a new specific methodology and a test bench for Java language with test cases for the main security vulnerabilities. We investigate specifically the security performance of two and three AST tools combinations, selecting adequate metrics. The second main goal of this study is to select the best combinations of tools taking into account different security criticality scenarios in the analyzed applications. To investigate the best combinations of AST tools according to each of the three levels that are criticality established, different metrics are selected considering the true and false positive ratios and the time available to perform a subsequent audit to eliminate the possible false positives obtained.

The tools are executed against the OWASP Benchmark project [14] based on the OWASP Top Ten project [5] to obtain the results of n-tools combination effectiveness. Well-known metrics are selected for the execution results to obtain a strict rank of combination tools. Finally, the paper gives some practical recommendations about how to improve their effectiveness using the tools in combination.

Therefore, the contributions of this paper can be summarized as follows:

- A concrete methodology using the OWASP Benchmark project to demonstrate its feasibility evaluating n-tools combinations by the detection of web vulnerabilities using appropriate criteria for benchmark instantiation.
- An analysis of the results obtained by n-tools in combination using a defined comparative method to classify them according to different degree levels of web applications importance using carefully selected metrics.
- An analysis of results of leading commercial tools in combination for allowing the practitioners to choose the most appropriate tools to perform a security analysis of a web application.

The outline of this paper is as follows: Section 2 reviews background in web technologies security focusing on vulnerabilities, SAST, DAST and IAST tools, security web application benchmarks and related work. In Section 3 the OWASP Benchmark project to evaluate security analysis tools (AST) tools is presented. Section 4 describes the steps of the comparative methodology proposal designed by enumerating the steps followed to rank the SAST tools in combination using the selected benchmark. Finally, Section 4 contains the conclusions and Section 5 sketches the future work.

2. Background and Related Work

This section presents the background on web technologies security, benchmarking initiatives, security analysis tools, as well as a review and analysis of different security analysis tools combination results in previous comparatives.

2.1. Web Applications Security

The OWASP Top Ten project brings together the most important vulnerability categories. There are several works that confirm web applications tested did not pass the OWASP Top Ten project [2–4]. Web applications in organizations and companies connected through the Internet and Intranets imply that they are used to develop any type of business, but at the same time they have become a valuable target of a great variety of attacks by exploiting the design, implementation or operation vulnerabilities, included in the OWASP Top Ten project, to obtain some type of economic advantage, privileged information, denial, extortion, etc.

Today there are a great variety of web programing languages such as .NET framework languages (C# or Visual Basic), swift for iOS platforms, or PHP. Java is the most used language, according to several analyses [15,16]. NodeJs, Python and C ++ are among the most frequently chosen today. Modern web applications use asynchronous JavaScript language and XML (AJAX), HTML5, flash technologies [17,18] and Javascript libraries such as Angular, Vue, React, Jquery, Bootstrap, etc. Also. Vaadin is a platform for building, collaborative web applications for Java backends and HTML5 web user interfaces.

Developers should have training in secure code development to prevent security vulnerabilities in web application source code [1]. Another form of prevention is to use secure languages that do type and memory length checks at compile time. C#, Rust and Java are some of these languages [1]. Designers and developers have to use security benchmarks for all configurations of navigators, application and database servers. Complementary and specific measures of online protection are necessary as the deployment of a Web Application Firewall [19–21].

2.2. Analysis Security Testing

There are different types of testing techniques that a security auditor or analyst can select to perform a security analysis of a web application, static white box security analysis (SAST), dynamic black box security analysis (DAST) or interactive white box security analysis (IAST) techniques [1]. The OWASP Security Testing Guide Methodology v4.1 [22] suggests that to perform a complex web application security analysis it is necessary to automate as possible using static, dynamic and interactive analysis testing tools, including manuals checking to find more true positives, and to reduce the number of false positives.

2.2.1. Static Analysis Security Testing

A good technique to avoid security vulnerabilities in source code is prevention [1,23]. Vulnerabilities can be prevented if developers are trained in secure web application development to avoid making "mistakes" that could lead to security vulnerabilities [24]. Obviously, prevention will avoid some of the vulnerabilities, but programing mistakes will always be made in despite of applying preventive secure best practice. Therefore, other subsequent security analysis techniques are necessary once the source code is developed.

SAST tools performs a white box security analysis. It analyzes both source code and executable, as appropriate. SAST tools start with a problem because of the act of determining if a program reaches its final state, or not [25]. Despite this problem, security code analysis can reduce the review code effort [26]. SAST tools are considered the most important security activity within a SSDLC [13].

Security analysts need to improve on recognizing all types of vulnerabilities in the source code for a particular programing language [27]. In the study of Díaz and Bermejo [28], tools such as Fortify SCA, Coverity, Checkmarx or Klocwork are good examples of tools that provide a trace information for eliminating false positives. SAST tools interfaces can be more or less "friendly" in terms of the error trace facilities to audit a security vulnerability.

SAST tools analyze the entire application covering all attack surface. They can also revise the configuration files of the web application. For this reason, static analysis requires a final manual audit of the results to discard the false positives and find the false negatives (much more complicated). However, several works confirm that different SAST tools have distinct algorithm designs as Abstract Interpretation [29–31], Taint Analysis [32], Theorem Provers [33], SAT Solvers [34] or Model Checking [35,36]. Therefore, combining SAST tools can find different types of vulnerabilities and therefore obtain a better combination result [6,7].

2.2.2. Dynamic Analysis Security Testing

DAST tools are black box tools that allow the analyses of a running application attacking all external source inputs of a web application [37]. In a first phase they try to discover (crawling) the attack surface of the web application, that is, all the possible source inputs of the application. The crawling phase must be manual using the tool as an intercept proxy, while also performing an automatic crawling that includes some information of the web application such as programing languages, application server, database server, authentication or session methods. After a crawling phase, the tools perform a recursive attack with malicious payload to all source inputs of web application discovered. Next each http response is syntactically analyzed to check if a security vulnerability exits. Finally, the vulnerability report must be manually revised to discard false positives and discover possible false negatives. Unlike the white box tools, here the source code of the application is not known. Tests are launched against the interface of the running web application. DAST tools usually discover less true positives and also has less false positives than SAST tools [10,38].

DAST tools allow the detection of vulnerabilities in the deployment phase of software development. The behavior of an attacker is simulated to obtaining results for analysis. In addition, these tools that can be executed independently of the language used by the application. DAST tools evolve over time, incorporating new authentication methods (JWT), attacks vectors (XML, JSON, etc.) and techniques to automate the detection of vulnerabilities [1], among which they stand out fuzzing techniques, based on tests on the application trying to make it fail, such as modifying the entries on forms.

2.2.3. Interactive Analysis Security Testing

Interactive analysis, the ability to monitor and analyze code as it executes, has become a fundamental tool in computer security research. Interactive analysis is attractive because it allows us to analyze the actual executions, and thus can perform precise security analysis based upon runtime information. IAST tools are white box tools and they are an evolution of the SAST tools [39]. They allow code analysis, but unlike the SAST tools, they do it in real time and in an interactive way similar to the DAST tools. The main difference is that the tool runs directly on the server and has to be integrated into the application. Therefore, it is recommended to have a testing environment to perform all the tests and detect the greatest number of vulnerabilities. They are based on the execution of an agent on the server side. This agent is responsible for monitoring the behavior of the application being integrated into all layers of it, providing a broader control of the application flow and data flow, creating possible attack scenarios. The main characteristics of IAST tools are [40]:

- It runs on the server as an agent, obtaining the results of the behavior generated by the end user on the published application. As they are agents that run on the server side, they must be compatible in the language that the application is designed. They can produce a relative runtime overhead in the application server.
- They can allow the sanitization of the entries, making the information that comes from the client side clean, eliminating possible injections or remote executions.
- The data correlation between SAST and IAST tools is usually more accurate as they are white box tools.
- They generate fewer false positives and they are not be able to detect client-side vulnerabilities.

One of the techniques used by IAST tools is taint analysis. The purpose of dynamic taint analysis is to track information flow between sources and sinks. Any program value whose computation depends on data derived from a taint source is considered tainted. Any other value is considered untainted. A taint policy determines exactly how taint flows as a program executes, what sorts of operations introduce new taint, and what checks are performed on tainted values [41–47]. Another technique used by IAST tools is symbolic execution. It allows the analyses of the behavior of a program on many different inputs at one time, by building a logical formula that represents a program execution. Thus, reasoning about the behavior of the program can be reduced to the domain of logic. One of the advantages of forward symbolic execution is that it can be used to reason about more than one input at once [48–51].

2.3. Related Work

In this section, we review the main and recent studies about the combination of different types of web applications security analysis tools with the main objectives of discovering more vulnerabilities and reducing the number of false positives. Several works combine static analysis tools with machine learning techniques for automatic detection of security vulnerabilities in web applications reducing the number of false positives [52,53]. Other approximations are based in attacks and anomalies detection using machine learning techniques [54].

2.3.1. SAST Tools Comparisons Studies

In the work of Diaz and Bermejo [28], nine SAST tools for C language comparison is accomplished to rank them according a set of selected metrics using SAMATE tests suites for C language. This comparison includes several leaders' commercial tools. It is very useful to the practitioners to select the best tool to analyze the source code security.

The work of Nunes et al. [9] present an approach to design benchmarks for evaluating SATS tools having into account distinct levels of criticality. The results of selected SAST tools show that the metrics could be improved to balance the rates of the true positives (TP) and false positives (FP). However, this approach could improve their representative with respect to security vulnerabilities coverage for other types, besides SQLI and XSS.

The work of Algaith et al. [6] is based on a previously published data set that resulted from the use of five different SAST tools to find SQL injections (SQLI) and cross-site scripting (XSS) vulnerabilities in 132 WordPress Content Management System (CMS) plugins. This work could be improved using a benchmark with more vulnerability categories and including leader commercial tools.

Finally, a paper [8] can be examined that presents a benchmarking approach [55] to study the performance of seven static tools (five commercial tools) with a new methodology proposal. The benchmark is representative and it is designed for the vulnerability categories included in the known standard OWASP Top Ten project for SAST tools evaluations.

2.3.2. DAST Tools Comparisons Studies

A study of current DAST tools [56] provides the background needed to evaluate and identify the potential value of future research in this area. It evaluates eight well-known DAST against a common set of sample applications. The study investigates what vulnerabilities are tested by the scanners and their effectiveness of. The study shows that DAST tools are adept at detecting straightforward historical vulnerabilities.

In the work [57], a comprehensive evaluation is performed on a set of open source DAST tools. The results of this comparative evaluation highlighted variations in the effectiveness of security vulnerability detection and indicated that there are correlations between different performance properties of these tools (e.g., scanning speed, crawler coverage, and number of detected vulnerabilities). There was a considerable variance on both types and numbers of detected web vulnerabilities among the evaluated tools.

Another study [58] evaluates five DAST tools against seven vulnerable web applications. The evaluation is based on different measures such as the vulnerabilities severity level and types, numbers of false positive and the accuracy of each DAST tool. The evaluation is conducted based on an extracted list of vulnerabilities from OWASP Top Ten project. The accuracy of each DAST tool was measured based on the identification of true and false positives. The results show that Acunetix and NetSparker have the best accuracy with the lowest rate of false positives.

The study of Amankwah el al. [59] proposes an automated framework to evaluate DAST tools vulnerability severity using an open-source tool called Zed Attach Proxy (ZAP) to detect vulnerabilities listed by National Vulnerability Database (NVD) in a Damn Vulnerable Web Application (DVWA). The findings show that the most prominent vulnerabilities, such as SQL injection and cross-site scripting found in modern Web applications are of medium severity.

2.3.3. SAST-DAST Comparisons Studies

The study of Antunes and Vieira [10], compares SAST vs DAST tools against web services benchmarks and show that the SAST tools usually obtain better true positive ratios and worse false positive ratios than DAST tools. Additionally, these works confirm that SAST and DAST can find different types of vulnerabilities.

2.3.4. SAST Tools Combinations Studies

The work of Xypolytos et al. [60] proposes a framework for combining and ranking tool findings based on tool performance statistics. The initial result shows potential capabilities of the framework to improve tool effectiveness and usefulness. The framework weights the performance of SAST Tools per defect type and cross-validates the findings between different SAST tools reports. An initial validation shows the potential benefits of the proposed framework.

The work of Tao Ye et al. [61], performed a comparison study of commercial (Fortify SCA and Checkmarx) and open source (Splint) SAST tools focusing on the Buffer Overflow vulnerability. In their work they describe the comparison of these tools separately, analyzing 63 open source projects including 100 bugs of this type. In their study, they include the combination of the tools in pairs (Fortify + Checkmarx, Checkmarx + Splint and Fortify + Splint).

In another paper [7], the problem of combining diverse SAST tools is investigated to improve the overall detection of vulnerabilities in web applications, considering four development scenarios with different criticality. It tested five SAST tools against two benchmarks, one with real WordPress plugins and another with synthetic test cases. This work could be improved using test benchmarks with more vulnerability categories than SQLI and XSS and including leaders' commercial tools.

2.3.5. SAST-DAST and SAST-IAST Hybrid Prototypes

The goal of the proposed approach in [62] is to improve penetration testing of web applications by focusing on two areas where the current techniques are limited: identifying the input vectors of a web application and detecting the outcome of an attempted attack. This approach leverages two recently developed analysis techniques. The first is a static analysis technique for identifying potential input vectors, and the second is a dynamic analysis to automate response analysis. The empirical evaluation included nine web applications, and the results show that the solution test the targeted web applications more thoroughly and discover more vulnerabilities than other two tools (Wapiti and Sqlmap), with an acceptable analysis time.

There are several recent works that combine SAST tools with IAST tools to runtime monitor attacks with the information of static analysis. The work of Mongiovi et al. [63] presented a novel data flow analysis approach for detecting data leaks in Java apps, which combines static and interactive analysis in order to perform reliable monitoring while introducing a low overhead in target apps. It provides an implementation as the JADAL tool, which is publicly available. The analysis shows that JADAL can correctly manage use cases that cannot be handled by static analysis, e.g., in presence of polymorphism. Additionally, it has a low overhead since it requires monitoring only a minimal set of points.

Another study presents a new approach to protect Java Enterprise Edition web applications against injection attacks as XSS [64]. The paper first describes a novel approach to taint analysis for Java EE. It explains how to combine this method with static analysis, based on the Joana IFC framework. The resulting hybrid analysis will boost scalability and precision, while guaranteeing protection against XSS.

Another paper [65] proposes a method where firstly vulnerabilities are detected through static code analysis. Then, code to verify these detected vulnerabilities is inserted into the target program's source code. Next, by actually executing the program with the verification code, it is analyzed whether the detected vulnerabilities can be used in a real attack. So, the proposed method comprises of vulnerability detection stage through static code analysis, instrumentation stage to link static code analysis and dynamic analysis, dynamic analysis stage of injecting an error and monitoring the impact on a program.

3. Method Proposal to Analyze AST n-Tools Combinations

In this section we present a new methodology repeatable to combine, compare and rank the SAST, DAST and IAST tools combinations (see Figure 1).

- Select the OWASP Benchmark project.
- Select the SAST, DAST and IAST tools. In this concrete instantiation of the methodology, we choice six commercial and open source tools according to the analysis of the related works in Section 2.3 and official lists of SAST, DAST and IAST tools.

- Run the selected tools against the OWASP Benchmark project with the default configuration for each tool.
- Select appropriate metrics to analyze results according three different levels of web applications criticality.
- Metrics calculation.
- Analysis, discussion and ranking of the results.

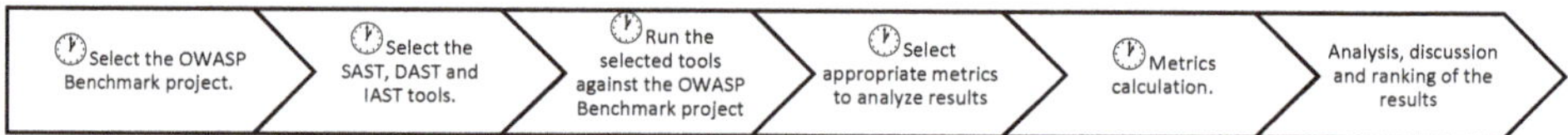

Figure 1. Method proposal to analyze analysis security testing (AST) n-tools combinations.

The following figure presents the proposed method in graphic form:

3.1. Benchmark Selection

An adequate test bench must be credible, portable, representative, require minimum changes and be easy to implement, and the tools execution must be under the same conditions [10]. We have investigated several security benchmarks for web applications as Wavsep used in the comparisons of [66,67]; Securebench Micro Project used in the works of [68,69]; Software Assurance Metrics And Tool Evaluation (SAMATE) project of National Institute of Standards and Technology (NIST) used in several works [9,28,70–73]; OWASP benchmark project [14]; Delta-bench by Pashchenko et al. [74] and OWASP Top Ten Benchmark [55] adapted for OWASP Top Ten 2013 and 2017 vulnerability categories projects and designed for static analysis only used in the work of Bermejo et al. [8].

Taking into account statistics of security vulnerabilities reported by several studies [2–4], the most adequate test bench for using SAST, DAST and IAST tools is OWASP benchmark project [14]. This benchmark is an open source web application in Java language deployed in Apache Tomcat. It meets the properties required for a benchmark [10] and it covers dangerous security vulnerabilities of web applications according to OWASP Top Ten 2013 and OWASP Top Ten 2017 projects. It contains exploitable test cases for detecting true and false positives, each mapped to specific CWEs, which can be analyzed by any type of application security testing (AST) tool, including SAST, DAST (like OWASP ZAP), and IAST tools. We have randomly selected 320 test cases from OWASP benchmark project and distributed them equally according to the number and type of vulnerabilities represented in the test bed. Of these cases half are false positives and half are true positives. It is easily portable as a Java project and it does not require changes with any tool. Table 1 shows the test cases distribution for each type of security vulnerability.

Table 1. OWASP Benchmark Project Test Cases.

CWE	Vulnerability Types	Number of Test Cases
78	Command Injection	40
643	Xpath Injection	20
79	Cross Site Scripting (XSS)	40
90	LDAP Injection	20
22	Path Traversal	40
89	SQL Injection	40
614	Secure Cookie flag	20
501	Trust Boundary Violation	20
330	Weak Randomness	40
327	Weak Cryptographic	20
328	Weak Hashing	20
	Total	320

3.2. SAST, DAST and IAST Tools Selection

- SAST and IAST tools are selected according with Java 2 Platform, Enterprise Edition (J2EE), the most used technology in web applications development, the programing language used by J2EE is Java, one of the labeled as more secure [75]. The next step is the selection of two (2) SAST tools for source code analysis that can detect vulnerabilities in web applications developed using the J2EE specification; two (2) IAST tools for detecting vulnerabilities in runtime and finally two (2) DAST tools:
- With the premises of the above comparatives and analyzing the availability of commercial and open source tools are selected three commercial and three open source meaning tools. Selected tools:

  - Fortify SCA (SAST Commercial tool by Microfocus, Newbury, United Kingdom) supports 18 distinct languages, the most known OS platforms and offers SaaS (Software as a service) and it detects more than 479 vulnerabilities.
  - FindSecurityBugs. (SAST OWASP open source tool, created by Philippe Arteau and licensed under LGPL). It has plugins are available for SonarQube, Eclipse, IntelliJ, Android Studio and NetBeans. Command line integration is available with Ant and Maven.
  - OWASP ZAP (DAST OWASP open source tool with Apache 2 license). Popular tool developed by the active OWASP Foundation community. It is a widely used tool in the field of penetration testing. It supports a great number of plugins, features and input vectors.
  - Arachni (DAST tool with public source License v1.0 created by Tasos Laskos). Open source vulnerability scanner highly valued by the community, and which allows the assessment of vulnerabilities such as SQLi, XSS, Command Injection and other included in OWASP Top Ten project.
  - Contrast Community Edition (IAST free version of the commercial tool for 30 days by Contrasst Security, Los Altos, EE.UU.). It allows the analyses of the application in Java language in an interactive way, making an agent in charge of reporting the vulnerabilities to the server. It provides the complete functionality of our paid platform solutions, Contrast Assess, and Contrast Protect. Contrast Community Edition's main limitations from the paid platform are language support (Java, .NET Core framework) and only one (1) application can be onboarded.
  - CxIAST (IAST commercial tool, by Chekcmarx, Raman Gan, Israael). It is an IAST engine for Java language. It inspects custom code, libraries, frameworks, configuration files, and runtime data flow. It supports Java, .Net framework and NodeJs languages. It can integrate with CxSAST (SAST tool) to correlate and improve results regarding true and false positives.

3.3. Metrics Selection

The selected metrics are widely accepted in others works [8–10,28,73,76] and in the work of Antunes and Vieira [77] that analyzes distinct metrics for different levels of the criticality of web applications. The metrics used in this methodology are:

- Precision (1). Proportion of the total TP detections:

$$TP \ / \ (TP + FP) \tag{1}$$

where TP (true positives) is the number of true vulnerabilities detected in the code and FP (false positives) is the number of vulnerabilities detected that, really, do not exist.

- True positive rate/Recall (TPR) (2). Ratio of detected vulnerabilities to the number that really exists in the code:

$$TP \ / \ (TP + FN) \tag{2}$$

where TP (true positives) is the number of true vulnerabilities detected in the code and FN (false negatives) is the total number of existing vulnerabilities not detected in the code.

- False positive rate (FPR) (3). Ratio of false alarms for vulnerabilities that do not really exists in the code:

$$TN \,/\, (TN + FP) \tag{3}$$

where TN (true negatives) is the number of not detected vulnerabilities that do not exist in the code and FP (false positives) is the total number of detected vulnerabilities that do not exist in the code.

- F-measure (4) is harmonic mean of precision and recall:

$$\frac{(2 \; x \; precision \; x \; recall)}{(precision + recall)} \tag{4}$$

- F_β-Score (5) is a particular F-measure metric for giving more weight to recall or precision. For example, a value for β of 0.5 gives more importance to precision metric, however a value or 1.5 gives more relevance to recall precision:

$$\left(1 + \beta^2\right) x \; \frac{precision \; x \; recall}{((\beta^2 \; x \; precision) + recall)} \tag{5}$$

- Markedness (6) has as a goal to penalize the ratio of true positives considering the ratio of false positives and the ratio of false negatives.

$$\frac{TP * TN - FP * FN}{\sqrt{(TP + FP) * (TP + FN) * (TN + FP) * (TN + FN)}} \tag{6}$$

- Informedness (7) computes all positive and negative cases found are reported, and for each true positive the result is increased in a ratio of 1/P while if not detected it is decreased in a ratio of 1/N, where P is the number of positives (test cases with vulnerability) and N is the number on Negatives (test cases with no vulnerability).

$$\frac{TP}{TP + FP} + \frac{TN}{FN + TN} - 1 = \frac{TP}{TP + FP} - \frac{FN}{FN + TN} \tag{7}$$

3.4. Metrics Calculation

In this section, the selected tools run against the OWASP Benchmark project test cases. We obtain the true positive and false positive results for each type of vulnerability. Next, the metrics selected in Section 3.4 are applied to obtain the most appropriate good interpretation of the results and draw the best conclusions.

To calculate all metrics, we have developed a C program to automatize the computation of metrics of each tool. Once executed each tool against OWASP Benchmark project the results are manually revised and included in a file that the C program reads to obtain the selected metrics.

3.4.1. 1-Tool Metrics Calculation

For each tool, Recall (Rec), FPR, Precision (Prec), F-measure (F-Mes), $F_{0.5}$Score ($F_{0.5}$S), $F_{1.5}$Score ($F_{1.5}$S), Markedness (Mark) and Informedness (Inf) metrics are computed. In Table 2 false positives (FP), false negatives (FN), true negatives (TN) and true positives (TP) are computed. Additionally, in Table 3, all selected metrics are computed for each tool. The abbreviations used for some of the tools are: FindSecurityBugs (FSB), Contrast Community Edition (CCE).

Table 2. Metrics applied to individual tools.

	FP	FN	TN	TP
CCE	0	0	160	160
FSB	107	3	53	157
CxIAST	0	15	160	145
Fortify	92	21	68	139
ZAP	8	118	152	42
Arachni	0	124	160	36

Table 3. Metrics applied to individual tools.

	Rec	FPR	Prec	F-Mes	$F_{0.5}S$	$F_{1.5}S$	Mark	Inf
CCE	1	0	1	1	1	1	1	1
FSB	0.99	0.67	0.59	0.74	0.65	0.82	0.56	0.32
CxIAST	0.91	0	1	0.95	0.98	0.93	0.91	0.91
Fortify	0.87	0.58	0.6	0.71	0.64	0.76	0.37	0.29
ZAP	0.29	0.01	0.98	0.45	0.67	0.37	0.56	0.29
Arachni	0.23	0	1	0.37	0.59	0.3	0.56	0.23

A first classification order according to the TPR score from left to right are showed in Table 3 for 1-tool in isolation.

Table 4 shows the execution times of each tool against OWASP Benchmark project. DAST tools (ZAP and Arachni) usually spend more time because they attack the target application externally by performing a penetration test.

Table 4. Execution times of each tool against OWASP Benchmark project.

Tool	Elapsed Time
FSB	22 min
Fortify	19 min and 86 s
ZAP	36 h
Arachni	13 h and 10 min
CCE	2 min and 43 s
CxIAST	4 min and 10 s

3.4.2. 2-Tools and 3-Tools Metrics Calculation

The rules showed in Table 5 have been considered for computing TP, TN, FP and FN metrics in n-tools combinations.

Table 5. Logic applied to the combination of two tools.

	Tool A		Tool N	N-Tools
Positives cases (P)	TP	or	TP	TP
	TP	or	FN	TP
	FN	or	TP	TP
	FN	and	FN	FN
Negative cases (N)	FP	or	FP	FP
	FP	or	TN	FP
	TN	or	FP	FP
	TN	and	TN	TN

We use the 1-out-of-N (1ooN) strategy for combining the results of the AST tools. The process proposed to calculate the combined results for two or more tools is based on a set of automated steps. 1ooN SAST for true positives detection: Any "TP alarm" from any of the n-tools in a test case for TP detection will lead to a TP alarm in a 1ooN system. If no tool in a combination detects a TP in a test case, it is a FN result for the 1ooN system.

Additionally, we use the 1-out-of-N (1ooN) strategy for combining the results of the AST tools. The process proposed to calculate the combined results for two or more tools is based on a set of automated steps. 1ooN SAST for true positives detection: Any "FP alarm" from any of the n-tools in a test case for FP detection will lead to an FP alarm in a 1ooN system. If no tool in a combination detects a FP in a test case, it is a TN result for the 1ooN system.

Table 6 shows the metrics calculations for 2-tools combinations. In any 2-tools combination can be included tools of the same type.

Table 6. Application of the metrics to 2-tools combinations.

	Rec	FPR	Prec	F-Mes	FPR	$F_{0.5}S$	$F_{1.5}S$	Mark	Inf
Arachni + CCE	1	0	1	1	0	1	1	1	1
CCE + CxIAST	1	0	1	1	0	1	1	1	1
ZAP + CCE	1	0.05	0.95	0.98	0.05	0.96	0.98	0.95	0.95
Fortify + CCE	1	0.58	0.63	0.78	0.58	0.68	0.85	0.63	0.43
FSB + CCE	1	0.67	0.6	0.75	0.67	0.65	0.83	0.6	0.33
FSB + CxIAST	1	0.67	0.6	0.75	0.67	0.65	0.83	0.6	0.33
FSB + Fortify	1	0.78	0.56	0.72	0.78	0.62	0.81	0.56	0.23
Forti + CxIAST	0.99	0.58	0.63	0.77	0.58	0.68	0.84	0.6	0.41
FSB + ZAP	0.98	0.67	0.59	0.74	0.67	0.65	0.82	0.54	0.31
FSB + Arachni	0.98	0.67	0.59	0.74	0.67	0.65	0.82	0.54	0.31
Fortify + Arachni	0.94	0.58	0.62	0.75	0.58	0.66	0.81	0.49	0.36
Fortify + ZAP	0.93	0.58	0.62	0.74	0.58	0.66	0.8	0.47	0.35
Ar + CxIAST	0.92	0	1	0.96	0	0.98	0.94	0.92	0.92
ZAP + CxIAST	0.91	0.05	0.95	0.93	0.05	0.94	0.92	0.86	0.86
ZAP + Arachni	0.3	0.05	0.86	0.44	0.05	0.63	0.38	0.43	0.25

Table 7 shows the metrics calculations for 3-tools combinations. In any 3-tools combination tools of the same type can be included.

Table 7. Application of the metrics to 3-tools combinations.

	Rec	FPR	Prec	F-Mes	$F_{0.5}S$	$F_{1.5}S$	Mark	Inf
Ar + CCE + CxIAST	1	0	1	1	1	1	1	1
ZAP + Ararchi + CCE	1	0.05	0.95	0.98	0.99	0.98	0.95	0.95
ZAP + CCE + CxIAST	1	0.05	0.95	0.98	0.99	0.98	0.95	0.95
Fortify + ZAP + CCE	1	0.58	0.63	0.77	0.9	0.85	0.63	0.42
Fortify + Ararchi + CCE	1	0.58	0.63	0.78	0.9	0.85	0.63	0.43
Fortify + CCE + CxIAST	1	0.58	0.63	0.78	0.9	0.85	0.63	0.43
FSB + ZAP + CCE	1	0.67	0.6	0.75	0.88	0.83	0.6	0.33
FSB + ZAP + CxIAST	1	0.67	0.6	0.75	0.88	0.83	0.6	0.33
FSB + Ararchi + CCE	1	0.67	0.6	0.75	0.88	0.83	0.6	0.33
FSB + Ararchi + CxIAST	1	0.67	0.6	0.75	0.88	0.83	0.6	0.33
FSB + CCE + CxIAST	1	0.67	0.6	0.75	0.88	0.83	0.6	0.33
FSB + Fortify + ZAP	1	0.78	0.56	0.72	0.87	0.81	0.56	0.23
FSB + Fortify + Ararchi	1	0.78	0.56	0.72	0.87	0.81	0.56	0.23
FSB + Fortify + CCE	1	0.78	0.56	0.72	0.87	0.81	0.56	0.23
FSB + Fortify + CxIAST	1	0.78	0.56	0.72	0.87	0.81	0.56	0.23
Fortify + ZAP + CxIAST	0.99	0.58	0.63	0.77	0.89	0.84	0.6	0.41
Fortify + Arachni+ CxIAST	0.99	0.58	0.63	0.77	0.89	0.85	0.62	0.42
FSB + ZAP + Ararchi	0.98	0.67	0.59	0.74	0.87	0.82	0.54	0.31
Fortify + ZAP + Ararchi	0.94	0.58	0.62	0.74	0.85	0.81	0.49	0.36
ZAP +Arachni + CxIAST	0.92	0.05	0.95	0.93	0.92	0.93	0.87	0.87

3.5. Analysis and Discussion

Next, in this section, the main research questions formulated in the Introduction Section 1 are going to be analyzed.

3.5.1. What s Each Type of AST Tool's Average Effectiveness Considering Each Type of Vulnerability without Combination?

Figure 2 shows the average effectiveness considering each type of AST tool and each type of vulnerability. SAST tools obtain very good ratios of Recall/TPR between 0.80 and 1 scores and worse FPR results which are too high in several vulnerability types as CMDi, LDAPi, SQLi, Trust Boundary Violation or Xpathi. DAST tools have a very good ratio of TPR for all vulnerability types, however they have a wide improving margin with respect to Recall/TPR where the best score is 0.70 for XSS and 0.60 for SQLi vulnerabilities but the scores are very low for the rest of vulnerability types. IAST tools obtain excellent TPR and FPR scores for all vulnerability types. Therefore, combining two IAST tools or combining IAST tools with DAST or SAST tools can be a very good choice. Combining IAST tools with SAST tools the TPR ratio can be improved. However, these combinations can obtain a worse FPR combined ratio. Combining IAST tools with DAST tools can improve the TPR ratio.

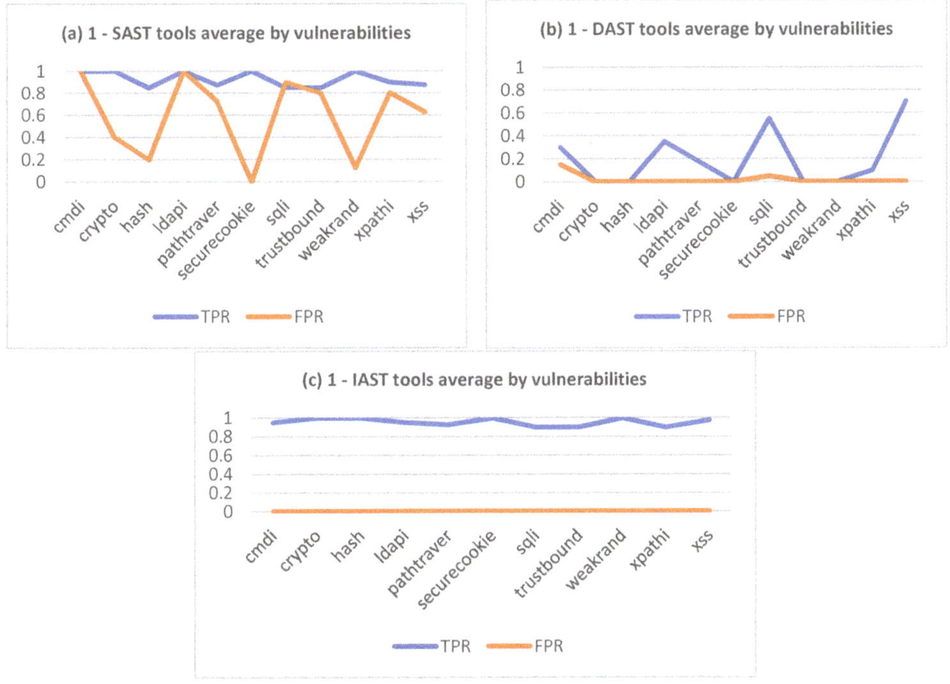

Figure 2. Effectiveness of each type of AST tool considering each type of vulnerability without combination.

3.5.2. What Is That Each Type of AST Tool's Combinations Average Effectiveness Considering Each Type of Vulnerability and the Number of Tools in a Combination?

The graphics of Figure 3 show the average effectiveness of the tools in combination. The TPR ratio increases as the number of tools in the combination increases, reaching practically the value 1 in all vulnerabilities. The FPR ratio increases mainly in several vulnerability types as CMDi, LDAPi, SQLi, Trust Boundary Violation or Xpathi.

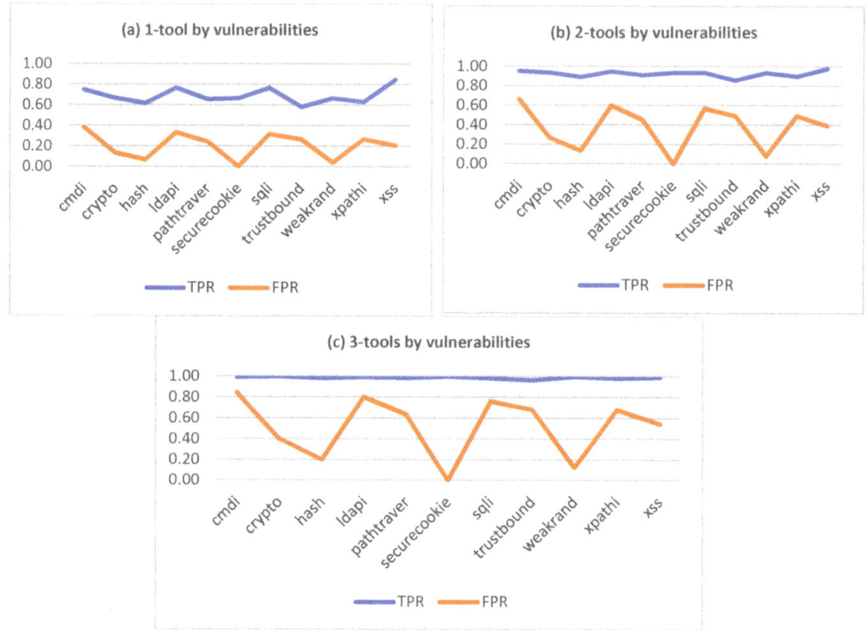

Figure 3. Average effectiveness considering n-tools combinations for each type of vulnerability.

3.5.3. What Is Each Type of AST Tool's Average Effectiveness Detecting OWASP Top Ten Security Category Vulnerabilities without Combination?

In the Figure 4 the results for each type of AST tool effectiveness detecting the OWASP Top Ten security category vulnerabilities without combination is shown. IAST tools obtain the best results in all metrics near the best score (1) and also for FPR where they obtain the best score (0). SAST tools obtain good results in general but they have a great improving margin with respect to FPR. DAST tools obtain good results with respect to FPR but they have a great improving margin with respect to Recall metric. The combination of Recall and TPR supposes a very good precision metric (>0.90). Depends on the specific vulnerability findings of each type of tool in a DAST-SAST tools combination, a concrete combination will be able to obtain better or worse results metrics. Combinations of IAST tools with SAST tools will obtain more false positives due to high FPR ratio of SAST tools.

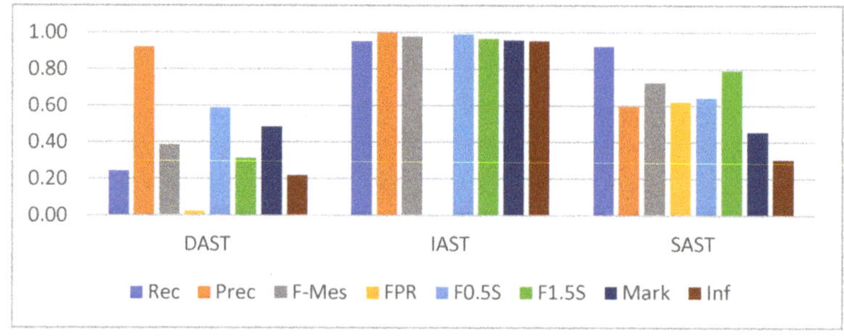

Figure 4. Tools metrics comparison without combination.

Combinations of IAST tools with DAST tools can obtain better metrics results due to DAST tools have a low ratio of FPR and they can find some distinct vulnerabilities than IAST tools.

3.5.4. What Is the n-Tools Combinations Average Effectiveness Detecting OWASP Top Ten Security Vulnerabilities Computing Different Metrics?

In Figures 5 and 6 the results for each type of AST n-tools combinations effectiveness detecting OWASP Top Ten security category vulnerabilities without combination is shown. Figure 5 shows the 2-tools combinations results for each combination and Figure 3 shows the 3-tools combinations results for each combination. In each combination can be included tools of the same type.

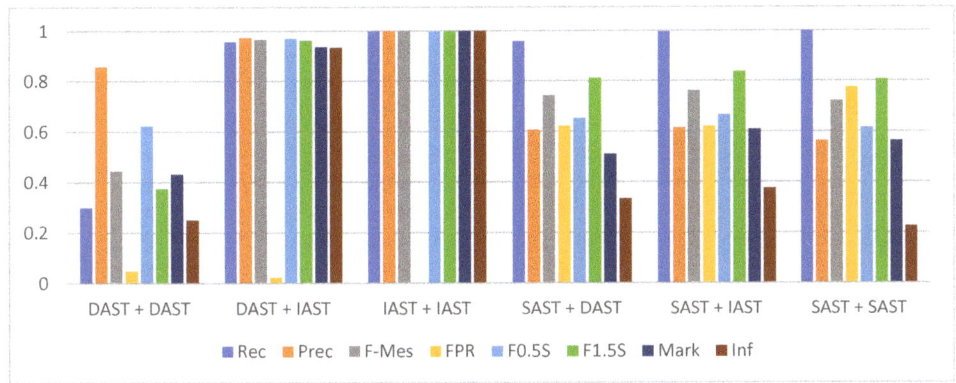

Figure 5. Metric results comparison of 2-tools combinations.

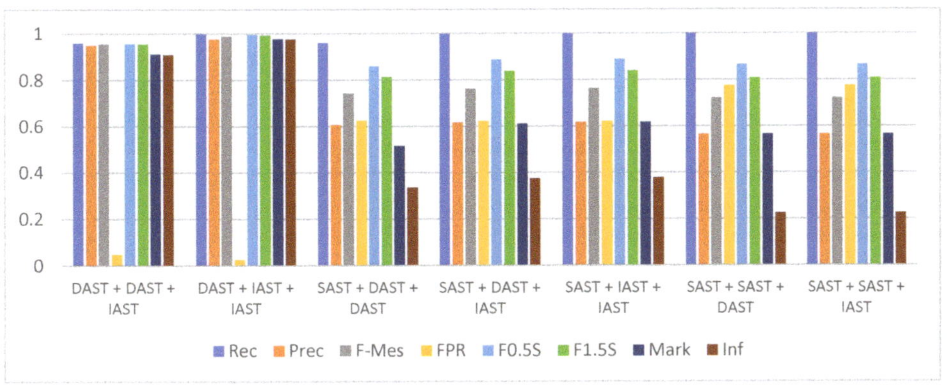

Figure 6. Metric results comparison of 3-tools combinations.

The results of 2-tools combinations show that the combination of two IAST tools obtain the maximum results for all metrics (1) and (0) for FPR metric. DAST+IAST tools combinations is the second-best combination with very good results for all metrics (>0.90) and also very good result for FPR metric. The IAST+SAT tools combinations is the third-best combination with excellent results for Recall metric but with high score for FPR, which make the other metric be worse. SAST+DAST combinations obtain better precision than SAST+SAST combinations, however SAST+SAST obtain better results with respect to Recall than SAST+DAST combinations. Depending on the objective of an analysis, the auditor will choose a combination type. This choice is going to be analyzed in the next study's questions—considering distinct types of web applications criticality. In all cases, the combination of two tools improves the metrics results obtained by each tool in isolation.

Figure 6 shows that the conclusions about metrics results obtained by 3-tools combinations are similar to the ones 2-tools combinations but it is important to note that the combination of three tools improves the metrics results obtained by 2-tools combinations and the ones obtained by each tool in isolation. The IAST+IAST+DAST combinations obtain excellent results for all metrics followed by IAST+DAST+DAST combinations. Recall/TPR is excellent for all combinations, between 0.95 and 1 scores. For the rest of the combinations, the fact of increasing the FPR ratio increasing the FPR ratio implies a decrease in the values of the other metrics. The higher the FPR ratio, the worse the values of metrics precision and $F_{0.5}$-Score (SAST+SAST+DAST and SAST+SAST+IAST combinations). Similar metrics values are obtained by SAST+DAST+DAST, SAST+DAST+IAST, SAST+IAST+IAST.

3.5.5. Which Are the Best Combinations for Analyzing the Security of Web Applications with Different Levels of Criticality?

Each organization should adapt the security criticality levels of their applications according to the importance they consider appropriate. The criticality levels have been established according to the data stored by the applications, based on the importance of the exposure of sensitive data today, data governed by data protection regulations, but extendable to any type of application. Three distinct levels of web applications criticality, high, medium and low are proposed:

- High criticality—a high criticality environment would be those environments with very sensitive data that, if violated, can cause very serious damage to the company.
- Medium criticality—a medium critical environment would be those environments with especially sensitive data.
- Low criticality—a low criticality environment would be those environments that do not carry any risk for the disclosure of information published in it.

For high criticality web applications, the metrics that provide a greater number of true positive results have been taken into account, that is, FP and TP are detected. It has been considered that being a critical application, any alert detected by the tools should be reviewed. It is also important to ensure that there are no undetected vulnerabilities, i.e., FNs. The FN ratio should be low. TPR must be considered for this type of application, as the closer the value is to 1, the lower the FN. On the other hand, it does not matter that there is a high number of FP, since as it is a critical application, it is considered necessary to audit the vulnerability results to discard false positives.

We have considered for applications with medium criticality that those metrics provide a higher number of positive detections including a medium false positive ratio, as well as being as accurate as possible. That is why if you have to take into account both metrics, the best is to use F-Measure, which provides the harmonic between both. Later, because of its criticality, it is advisable to give priority to TPR and finally to Precision. This relationship will help you to choose the best tool for this purpose.

Finally, in the low criticality applications, the aim is to choose those metrics that provide the highest number of TP, having a very low number of FP, even if this means that there is some vulnerability that is not detected. Therefore, it is advisable to use the Markedness metric, which takes into account the number of TP and FP detections that have been correctly detected. It is considered in these cases that there is no much time to audit the report of vulnerabilities to discard the possible false positives.

In Table 8 are shown the selected metrics considering the distinct types of web application criticality.

Table 8. Selection of metrics by web application criticality.

Metric—Web Application Criticallity	High	Medium	Low
Recall	1	2	
Precision (Prec)		3	1
F-Measure (F-Mes)		1	
$F_{0.5}$Score ($F_{0.5}$S)			2
$F_{1.5}$Score ($F_{1.5}$S)	2		
Markedness (Mark)			3
Informedness (Inf)	3		

According to classification metrics included in Table 2, a ranking of n-tools combinations is going to be established for each degree of web applications criticality. This ranking is established considering three metrics for each level of criticality and the order established for each metric and for each level of criticality in Table 7.

Table 9 shows the classification for 2-tools combinations.

Table 9. Ranking of 2-tools combinations according to distinct level of web application criticality.

HIGH	Rec	$F_{1.5}$S	Info	MEDIUM	F-Mes	Rec	Prec	LOW	Prec	$F_{0.5}$S	Mark
Arachni + CCE	1.00	1.00	1.00	Arachni + CCE	1.00	1.00	1.00	Arachni + CCE	1.00	1.00	1.00
CCE + CxIAST	1.00	1.00	1.00	CCE + CxIAST	1.00	1.00	1.00	CCE + CxIAST	1.00	1.00	1.00
ZAP + CCE	1.00	0.98	0.95	ZAP + CCE	0.98	1.00	0.95	Arachni + CxIAST	1.00	0.98	0.92
Fortify + CCE	1.00	0.85	0.43	Arachni + CxIAST	0.96	0.92	1.00	ZAP + CCE	0.95	0.96	0.95
FSB + CCE	1.00	0.83	0.33	ZAP + CxIAST	0.93	0.91	0.95	ZAP + CxIAST	0.95	0.94	0.86
FSB + CxIAST	1.00	0.83	0.33	Fortify + CCE	0.78	1.00	0.63	ZAP + Arachni	0.86	0.63	0.43
FSB + Fortify	1.00	0.81	0.23	Fortify + CxIAST	0.77	0.99	0.63	Fortify + CCE	0.63	0.68	0.63
Fortify + CxIAST	0.99	0.84	0.41	FSB + CCE	0.75	1.00	0.60	Fortify + CxIAST	0.63	0.68	0.60
FSB + ZAP	0.98	0.82	0.31	FSB + CxIAST	0.75	1.00	0.60	Fortify + Arachni	0.62	0.66	0.49
FSB + Arachni	0.98	0.82	0.31	Fortify + Arachni	0.75	0.94	0.62	Fortify + ZAP	0.62	0.66	0.47
Fortify + Arachni	0.94	0.81	0.36	Fortify + ZAP	0.74	0.93	0.62	FSB + CCE	0.60	0.65	0.60
Fortify + ZAP	0.93	0.80	0.35	FSB + ZAP	0.74	0.98	0.59	FSB + CxIAST	0.60	0.65	0.60
Arachni + CxIAST	0.92	0.94	0.92	FSB + Arachni	0.74	0.98	0.59	FSB + ZAP	0.59	0.65	0.54
ZAP + CxIAST	0.91	0.92	0.86	FSB + Fortify	0.72	1.00	0.56	FSB + Arachni	0.59	0.65	0.54
ZAP + Arachni	0.30	0.38	0.25	ZAP + Arachni	0.44	0.30	0.86	FSB + Fortify	0.56	0.62	0.56

The ranking of Table 9 confirms that the benchmark contains vulnerability types that AST tools can detect and permits to establish an adequate order with respect to the tool effectiveness in terms of the selected metrics. IAST tools combinations obtain the best results for high, medium and low critical web applications. Another conclusion is that IAST and DAST tools combinations have very good results for medium and low critical applications. IAST and SAST tools combinations have very good results for high, applications and also good results for medium and low critical web applications. ZAP+Arachni obtain the worse results for high and medium web applications critical levels.

Figure 7 includes three graphics to show the rankings of 2-tools combinations for each criticality level taking into account the first metric used for each classification. High (a): Recall; Medium (b): F-measure and Low (c): $F_{0.5}$ Score. ZAP+ Arachni combination obtains a better result for low level due to DAST tools have a low false positives ratio and the $F_{0.5}$ Score metrics favors on a good FPR metric. However, FSB + Fortify combination obtains a worse result for low level due to SAST tools have a high ratio of false positives. Additionally, FSB + Fortify combination obtains a worse result for medium level due to SAST tools have a high ratio of false positives.

The rankings of Tables 9 and 10 confirm that the benchmark contains vulnerability types that AST tools can detect but it permits to establish an adequate order with respect to the tool effectiveness in terms of the selected metrics. Table 10 shows that IAST tools combinations obtain the best results. Another conclusion is that there is IAST and DAST tools combinations have very good results. Combinations with SAST tools generally obtain good results if spite of having a higher ratio of false positives than IAST and DAST tools they have a very good ratio of true positives. SAST tools have to be considered to include them in a security analysis of a web application because find distinct vulnerabilities and the false positives can be eliminated in the necessary subsequent audit of the vulnerability reports. The Combinations integrated by SAST+DAST+IAST tools as Fortify + Arachni + CCE or Fortify + ZAP + CCE obtain a very good result in the HIGH, MEDIUM and LOW classifications. In the three 3 classifications the two first 2-tools combinations are the same due to the presence of one or two IAST tools in the combination, but from the third position the three criticality classifications are quite different. In the three 3 classifications the three first 3-tools combinations are the same due to the presence of one or two IAST tools in the combination, but from the fourth position the three criticality classifications are quite different. For high level of criticality, the results reached for the 3-tools combinations are between 0.92 and 1 for recall metric.

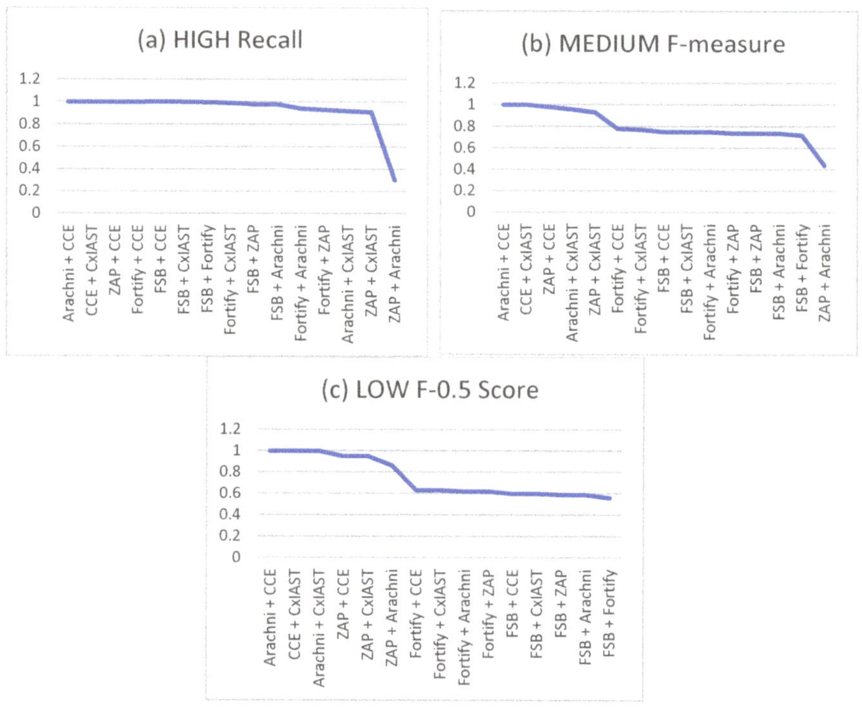

Figure 7. Rankings for (**a**) high, (**b**) medium and (**c**) low criticality levels for 2-tools combinations.

Table 10. Ranking of 3-tools combinations according to distinct level of web application criticality.

HIGH	Rec	$F_{1.5}S$	Info	MEDIUM	F-Mes	Rec	Prec	LOW	Prec	$F_{0.5}S$	Mark
Arachni + CCE + CxIAST	1.00	1.00	1.00	Arachni + CCE + CxIAST	1.00	1.00	1.00	Arachni + CCE + CxIAST	1.00	1.00	1.00
ZAP + Arachni + CCE	1.00	0.98	0.95	ZAP + Arachni + CCE	0.98	1.00	0.95	ZAP + Arachni + CCE	0.95	0.99	0.95
ZAP + CCE + CxIAST	1.00	0.98	0.95	ZAP + CCE + CxIAST	0.98	1.00	0.95	ZAP + CCE + CxIAST	0.95	0.99	0.95
Fortify + Arachni + CCE	1.00	0.85	0.43	ZAP + Arachni + CxIAST	0.93	0.92	0.95	ZAP + Arachni + CxIAST	0.95	0.92	0.87
Fortify + CCE + CxIAST	1.00	0.85	0.43	Fortify + Arachni + CCE	0.78	1.00	0.63	Fortify + Arachni + CCE	0.63	0.90	0.63
Fortify + ZAP + CCE	1.00	0.85	0.42	Fortify + CCE + CxIAST	0.78	1.00	0.63	Fortify + CCE + CxIAST	0.63	0.90	0.63
FSB + ZAP + CCE	1.00	0.83	0.33	Fortify + ZAP + CCE	0.77	1.00	0.63	Fortify + Arachni + CxIAST	0.63	0.89	0.62
FSB + ZAP + CxIAST	1.00	0.83	0.33	Fortify + Arachni + CxIAST	0.77	0.99	0.63	Fortify + ZAP + CCE	0.63	0.90	0.63
FSB + Arachni + CCE	1.00	0.83	0.33	Fortify + ZAP + CxIAST	0.77	0.99	0.63	Fortify + ZAP + CxIAST	0.63	0.89	0.60
FSB + Arachni + CxIAST	1.00	0.83	0.33	FSB + ZAP + CCE	0.75	1.00	0.60	Fortify + ZAP + Arachni	0.62	0.85	0.49
FSB + CCE + CxIAST	1.00	0.83	0.33	FSB + ZAP + CxIAST	0.75	1.00	0.60	FSB + ZAP + CCE	0.60	0.88	0.60
FSB + Fortify + ZAP	1.00	0.81	0.23	FSB + Arachni + CCE	0.75	1.00	0.60	FSB + ZAP + CxIAST	0.60	0.88	0.60
FSB + Fortify + Arachni	1.00	0.81	0.23	FSB + Arachni + CxIAST	0.75	1.00	0.60	FSB + Arachni + CCE	0.60	0.88	0.60
FSB + Fortify + CCE	1.00	0.81	0.23	FSB + CCE + CxIAST	0.75	1.00	0.60	FSB + Arachni + CxIAST	0.60	0.88	0.60
FSB + Fortify + CxIAST	1.00	0.81	0.23	Fortify + ZAP + Arachni	0.74	0.94	0.62	FSB + CCE + CxIAST	0.60	0.88	0.60
Fortify + Arachni + CxIAST	0.99	0.85	0.42	FSB + ZAP + Arachni	0.74	0.98	0.59	FSB + ZAP + Arachni	0.59	0.87	0.54
Fortify + ZAP + CxIAST	0.99	0.84	0.41	FSB + Fortify + ZAP	0.72	1.00	0.56	FSB + Fortify + ZAP	0.56	0.87	0.56
FSB + ZAP + Arachni	0.98	0.82	0.31	FSB + Fortify + Arachni	0.72	1.00	0.56	FSB + Fortify + Arachni	0.56	0.87	0.56
Fortify + ZAP + Arachni	0.94	0.81	0.36	FSB + Fortify + CCE	0.72	1.00	0.56	FSB + Fortify + CCE	0.56	0.87	0.56
ZAP + Arachni + CxIAST	0.92	0.93	0.87	FSB + Fortify + CxIAST	0.72	1.00	0.56	FSB + Fortify + CxIAST	0.56	0.87	0.56

Figure 8 includes three graphics to show the rankings of 3-tools combinations for each criticality level having into account the first metric used for each classification. High (a): Recall; Medium (b): F-measure and Low (c): $F_{0.5}$ Score. For high level the combinations that include 2 DAST tools have a wort results for recall metric. For high and medium levels, combinations between DAST and IAST tools obtain the best results.

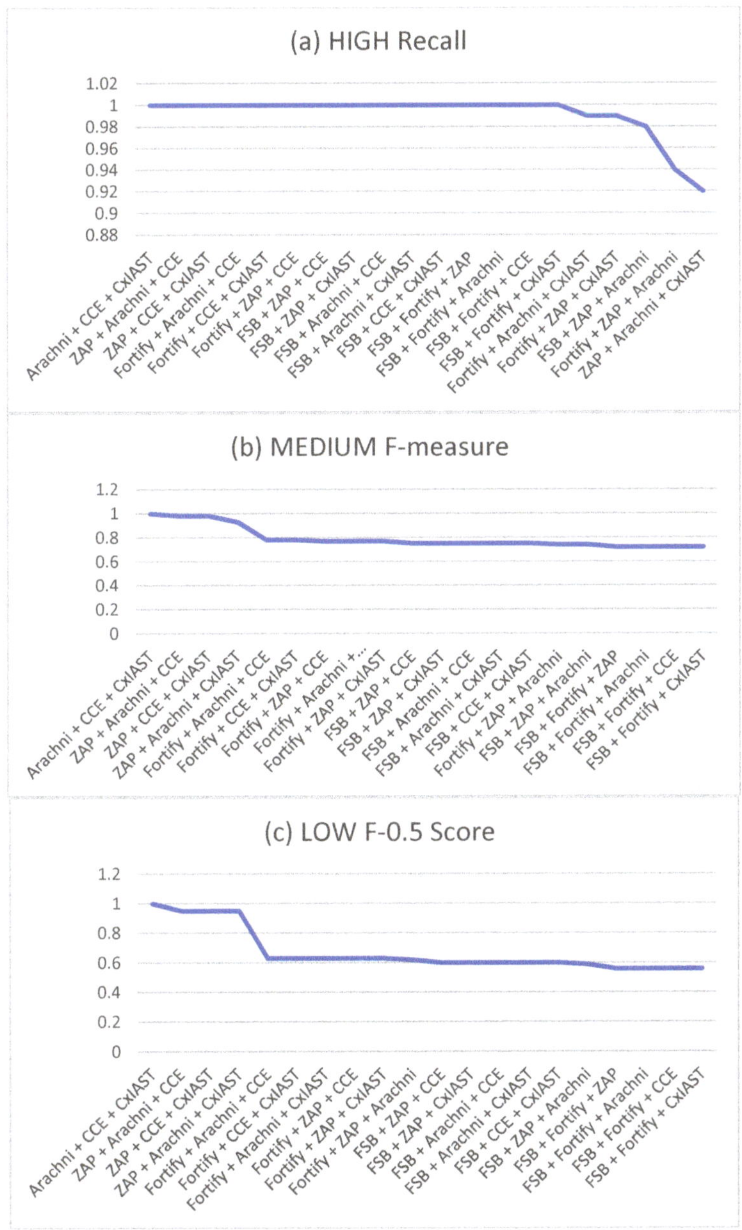

Figure 8. Rankings for (**a**) high, (**b**) medium, and (**c**) low criticality levels for 3-tools combinations.

4. Conclusions

Vulnerability analysis is still for many organizations, the great unknown. In others, it is a natural way of working, providing the necessary security required to keep the product free of vulnerabilities. Using different types of tools to detect distinct security vulnerabilities helps both developers and organizations to release applications securely, reducing the time and resources that must later be provided to fix initial errors. It is during the secure software development life cycle of an application where vulnerability detection tools help to naturally integrate security into the web application. It is necessary to use SAST, DAST and IAST tools in the development and test phases to develop a secure web application within a secure software development life cycle auditing the vulnerability reports to eliminate false positives and correlating the results obtained for each tool in a concrete combination of different types of tools.

In recent years, different combinations of tools have been seen to improve security, but until the writing of this study, it has not been possible to combine the three types of tools that exist on the market. It has been proven how, in a remarkable way, such a combination helps to obtain a very good ratios of true and false positives in many of the cases. IAST tools have obtain very good results metrics, the use of an IAST tool in a combination can improve the web applications security of an organization. Three tools included in this study are commercial, it is very interesting that the auditors and practitioners have the possibility of knowing how the security performance and the behavior of a commercial tool is as part of a n-tools combination.

To reach the final results we have used a methodology applied against a selected representative benchmark for OWASP Top Ten vulnerabilities to obtain results from different types of tools. Next, the results have been applied to a set of carefully selected metrics to perform the comparison of tools combinations to find out which of them are part of the best option when choosing these tools considering various levels of criticality in web applications to be analysed in an organization.

IAST tools combinations obtain the best results. Another conclusion is that IAST and DAST tools combinations have very good results. Combinations with SAST tools obtain generally good results if spite of having a higher ratio of false positives than IAST and DAST tools. However, SAST tools have a very good ratio of true positives. Therefore, SAST tools have to be considered to include them in a security analysis of a web application because find distinct vulnerabilities than DAST tools and the false positives can be eliminated in the necessary subsequent audit of the vulnerability reports. The combinations integrated by SAST+DAST+IAST tools as Fortify + Arachni + CCE or Fortify + ZAP + CCE obtain a very good result in the high, medium and low classifications.

The correlation of results between tools of different type is still an aspect that is not very widespread. It is necessary to develop a methodology or a custom-made software that allows in an automatic or semiautomatic way for the evaluation and correlation of the results obtained by several tools of different types.

It is very important to develop representative test benchmarks for the set of vulnerabilities included in the OWASP Top Ten categories that allow proper assessment and comparison of combinations of tools including SAST, DAST and IAST tools. It is also necessary to evolve them to include more test cases designed for more types of vulnerabilities included in the OWASP Top Ten categories. The rankings established in Tables 8 and 9 confirm that the benchmark contains vulnerability types that AST tools can detect but it permits to establish an adequate order with respect to the tool effectiveness in terms of the selected metrics.

5. Future Work

As future work we are going to work on the elaboration of a representative benchmark for web applications including a wide set of security vulnerabilities that permits an AST tools comparison. The main objective is that the evaluation of each combination of different tools can be the most effective possible using the methodology proposed in this work, allowing one to distinguish the best combinations of tools considering several levels of criticality in the web applications of an organization.

Author Contributions: Conceptualization, F.M.T., J.B.H. and J.-R.B.H.; methodology, J.B.H., F.M.T.; validation, J.-R.B.H., M.I.A. and J.-A.S.M.; investigation, J.B.H., F.M.T., J.-R.B.H. and J.-A.S.M.; resources, J.B.H., F.M.T., J.-R.B.H. and J.-A.S.M.; writing—original draft preparation, J.B.H., and F.M.T.; writing—review and editing, J.B.H., F.M.T., J.-R.B.H. and J.-A.S.M.; visualization, J.B.H., F.M.T., J.-R.B.H., M.I.A. and J.-A.S.M.; supervision, J.B.H., F.M.T., J.-R.B.H. and J.-A.S.M.; project administration, J.-R.B.H., M.I.A. and J.-A.S.M.; software M.I.A. All authors have read and agreed to the published version of the manuscript.

Funding: This research received no external funding.

Conflicts of Interest: The authors declare no conflict of interest.

References

1. Felderer, M.; Büchler, M.; Johns, M.; Brucker, A.D.; Breu, R.; Pretschner, A. Security Testing: A Survey. In *Advances in Computers*; Elsevier: Cambridge, MA, USA, 2016. [CrossRef]
2. Homaei, H.; Shahriari, H.R. Seven Years of Software Vulnerabilities: The Ebb and Flow. *IEEE Secur. Priv. Mag.* **2017**, *15*, 58–65. [CrossRef]
3. Barabanov, A.; Markov, A.; Tsirlov, V. Statistics of software vulnerability detection in certification testing. In *International Conference Information Technologies in Business and Industry 2018*; IOP Publishing: Tomsk, Russia, 2017. [CrossRef]
4. Sołtysik-Piorunkiewicz, A.; Krysiak, M. The Cyber Threats Analysis for Web Applications Security in Industry 4.0. In *Towards Industry 4.0—Current Challenges in Information Systems*; Studies in Computational Intelligence; Springer: Cham, Switzerland, 2020. [CrossRef]
5. OWASP Foundation. OWASP Top Ten 2017. Available online: https://www.owasp.org/index.php/Top_10_2017-Top_10 (accessed on 1 December 2020).
6. Algaith, A.; Nunes, P.; Fonseca, J.; Gashi, I.; Viera, M. Finding SQL injection and cross site scripting vulnerabilities with diverse static analysis tools. In Proceedings of the 14th European Dependable Computing Conference, IEEE Computer Society, Iasi, Romania, 10–14 September 2018. [CrossRef]
7. Nunes, P.; Medeiros, I.; Fonseca, J.C.; Neves, N.; Correia, M.; Vieira, M. An empirical study on combining diverse static analysis tools for web security vulnerabilities based on development scenarios. *Computing* **2018**, *101*, 161–185. [CrossRef]
8. Bermejo, J.R.; Bermejo, J.; Sicilia, J.A.; Cubo, J.; Nombela, J.J. Benchmarking Approach to Compare Web Applications Static Analysis Tools Detecting OWASP Top Ten Security Vulnerabilities. *Comput. Mater. Contin.* **2020**, *64*, 1555–1577. [CrossRef]
9. Nunes, P.; Medeiros, I.; Fonseca, J.C.; Neves, N.; Correia, M.; Vieira, M. Benchmarking Static Analysis Tools for Web Security. *IEEE Trans. Reliab.* **2018**, *67*, 1159–1175. [CrossRef]
10. Antunes, N.; Vieira, M. Assessing and Comparing Vulnerability Detection Tools for Web Services: Benchmarking Approach and Examples. *IEEE Trans. Serv. Comput.* **2015**, *8*, 269–283. [CrossRef]
11. Monga, M.; Paleari, R.; Passerini, E. A hybrid analysis framework for detecting web application vulnerabilities. In Proceedings of the 2009 ICSE Workshop on Software Engineering for Secure Systems, Vancouver, BC, Canada, 19 May 2009; pp. 25–32. [CrossRef]
12. Higuera, J.B.; Aramburu, C.A.; Higuera, J.-R.B.; Sicilia, M.-A.; Montalvo, J.A.S. Systematic Approach to Malware Analysis (SAMA). *Appl. Sci.* **2020**, *10*, 1360. [CrossRef]
13. Mohino, J.D.V.; Higuera, J.B.; Higuera, J.-R.B.; Montalvo, J.A.S.; Higuera, B.; Mohino, D.V.; Montalvo, J.A.S. The Application of a New Secure Software Development Life Cycle (S-SDLC) with Agile Methodologies. *Electronics* **2019**, *8*, 1218. [CrossRef]
14. OWASP Foundation. OWASP Benchmark Project. Available online: https://www.owasp.org/index.php/Benchmark (accessed on 1 December 2020).
15. Nanz, S.; Furia, C.A. A comparative study of programming languages in rosetta code. In Proceedings of the 37th International Conference on Software Engineering 2015, Florence, Italy, 16–24 May 2015; Volume 1, p. 778. [CrossRef]
16. Aruoba, S.B.; Fernández-Villaverde, J. A comparison of programming languages in macroeconomics. *J. Econ. Dyn. Control.* **2015**, *58*, 265–273. [CrossRef]
17. Beasley, R.E. Ajax Programming. In *Essential ASP.NET Web Forms Development*; Apress: Berkeley, CA, USA, 2020. [CrossRef]

18. Moeller, J.P. *Security for Web Developers: Using Javascript, HTML and CSS*; O'Reilly Media: Sebastopol, Russia, 2016.
19. Razzaq, A.; Hur, A.; Shahbaz, S.; Masood, M.; Ahmad, H.F.; Abdul, R. Critical analysis on web application firewall solutions. In Proceedings of the 2013 IEEE Eleventh International Symposium on Autonomous Decentralized Systems (ISADS), Mexico City, Mexico, 6–8 March 2013; pp. 1–6. [CrossRef]
20. Holm, H.; Ekstedt, M. Estimates on the effectiveness of web application firewalls against targeted attacks. *Inf. Manag. Comput. Secur.* **2013**, *21*, 250–265. [CrossRef]
21. Tekerek, A.; Bay, O. Design and implementation of an artificial intelligence-based web application firewall model. *Neural Netw. World* **2019**, *29*, 189–206. [CrossRef]
22. OWASP Foundation. OWASP Testing Guide, 2020. Available online: https://owasp.org/www-project-web-security-testing-guide/ (accessed on 1 December 2020).
23. Huth, M.; Nielsen, F. *Static Analysis for Proactive Security. Computing and Software Science. Lecture Notes in Computer Science*; Springer: Cham, Switzerland, 2019. [CrossRef]
24. Al-Amin, S.; Ajmeri, N.; Du, H.; Berglund, E.Z.; Singh, M.P. Toward effective adoption of secure software development practices. *Simul. Model. Pr. Theory* **2018**, *85*, 33–46. [CrossRef]
25. Sipser, M. *Introduction to the Theory of Computation*, 2nd ed.; Thomson Course Technology: Boston, MA, USA, 2006.
26. Singh, D.; Sekar, V.R.; Stolee, K.T.; Johnson, B. Evaluating How Static Analysis Tools Can Reduce Code Review Effort. In Proceedings of the IEEE Symposium on Visual Languages and Human-Centric Computing, Raleigh, NC, USA, 11–14 October 2017; pp. 101–105. [CrossRef]
27. Yang, J.; Tan, L.; Peyton, J.; Duer, K.A. Towards better utilizing static application security testing. In Proceedings of the 41st International Conference on Software Engineering: Software Engineering in Practice, Montreal, QC, Canada, 25–31 May 2019; IEEE Computer Society: Montreal, QC, Canada, 2019. [CrossRef]
28. Díaz, G.; Bermejo, J.R. Static analysis of source code security: Assessment of tools against SAMATE tests. *Inf. Softw. Technol.* **2013**, *55*, 1462–1476. [CrossRef]
29. Fromherz, A.; Ouadjaout, A.; Miné, A. Static Value Analysis of Python Programs by Abstract Interpretation. In *NASA Formal Methods. NFM 2018. Lecture Notes in Computer Science*; Dutle, A., Muñoz, C., Narkawicz, A., Eds.; Springer: Cham, Switzerland, 2018; Volume 10811. [CrossRef]
30. Urban, C.; Ueltschi, S.; Müller, P. Abstract interpretation of CTL properties. In *SAS '18. LNCS*; Springer: Berlin/Heidelberg, Germany, 2018; Volume 11002, pp. 402–422. [CrossRef]
31. Oortwijn, W.; Gurov, D.; Huisman, M. An Abstraction Technique for Verifying Shared-Memory Concurrency. *Appl. Sci.* **2020**, *10*, 3928. [CrossRef]
32. Ferrara, P.; Olivieri, L.; Spoto, F. BackFlow: Backward Context-Sensitive Flow Reconstruction of Taint Analysis Results. In *Verification, Model Checking, and Abstract Interpretation. VMCAI 2020. Lecture Notes in Computer Science*; Beyer, D., Zufferey, D., Eds.; Springer: Berlin/Heidelberg, Germany, 2020; Volume 11990. [CrossRef]
33. Khan, W.; Kamran, M.; Ahmad, A.; Khan, F.A.; Derhab, A. Formal Analysis of Language-Based Android Security Using Theorem Proving Approach. *IEEE Access* **2019**, *7*, 16550–16560. [CrossRef]
34. Biere, A.; Kröning, D. SAT-Based Model Checking. In *Handbook of Model Checking*; Clarke, E., Henzinger, T., Veith, H., Bloem, R., Eds.; Springer: Cham, Switzerland, 2018. [CrossRef]
35. Beyer, D.; Gulwani, S.; Schmidt, D.A. Combining Model Checking and Data-Flow Analysis. In *Handbook of Model Checking*; Clarke, E., Henzinger, T., Veith, H., Bloem, R., Eds.; Springer: Cham, Switzerland, 2018. [CrossRef]
36. Nielson, F.; Nielson, H.R.; Zhang, F. Multi-valued Logic for Static Analysis and Model Checking. In *Models, Mindsets, Meta: The What, the How, and the Why Not? Lecture Notes in Computer Science*; Margaria, T., Graf, S., Larsen, K., Eds.; Springer: Cham, Switzerland, 2019; Volume 11200. [CrossRef]
37. Mirjalili, M.; Nowroozi, A.; Alidoosti, M. A survey on web penetration test. *Adv. Comput. Sci.* **2014**, *3*, 107–121.
38. Vega, E.A.A.; Orozco, L.S.; Villalba, L.J.G. Benchmarking of Pentesting Tools. *Int. J. Comput. Electr. Autom. Control Inf. Eng.* **2017**, *11*, 602–605.
39. Pan, Y. Interactive Application Security Testing. In Proceedings of the 2019 International Conference on Smart Grid and Electrical Automation (ICSGEA), Xiangtan, China, 10 August 2019; pp. 558–561. [CrossRef]
40. Bermejo, J.R. Assessment Methodology of Web Applications Automatic Security Analysis Tools for Adaptation in the Development Life Cycle. Ph.D. Thesis, UNED, Madrid, Spain, 2014. Available online: http://e-spacio.uned.es/fez/view/tesisuned:IngInd-Jrbermejo (accessed on 1 December 2020).

41. Ren, Y.; Dong, W.; Lin, J.; Miao, X. A Dynamic Taint Analysis Framework Based on Entity Equipment. *IEEE Access* **2019**, *7*, 186308–186318. [CrossRef]

42. Zhao, J.; Qi, J.; Zhou, L.; Cui, B. Dynamic Taint Tracking of Web Application Based on Static Code Analysis. In Proceedings of the 2016 10th International Conference on Innovative Mobile and Internet Services in Ubiquitous Computing (IMIS), Fukuoka, Japan, 6–8 July 2016; pp. 96–101. [CrossRef]

43. Karim, R.; Tip, F.; Sochurkova, A.; Sen, K. Platform-Independent Dynamic Taint Analysis for JavaScript. *IEEE Trans. Softw. Eng.* **2018**. [CrossRef]

44. Kreindl, J.; Bonetta, D.; Mossenbock, H. Towards Efficient, Multi-Language Dynamic Taint Analysis. In Proceedings of the 16th ACM SIGPLAN International Conference on Managed Programming Languages and Runtimes, MPLR 2019, Athens, Greece, 21–22 October 2019; Association for Computing Machinery: New York, NY, USA, 2019; pp. 85–94. [CrossRef]

45. Cho, S.; Kim, G.; Cho, S.-J.; Choi, J.; Park, M.; Han, S. Runtime Input Validation for Java Web Applications Using Static Bytecode Instrumentation. In Proceedings of the International Conference on Research in Adaptive and Convergent Systems, RACS'16, Odense, Denmark, 11–14 October 2016; Association for Computing Machinery: New York, NY, USA, 2016.

46. Wang, R.; Xu, G.; Zeng, X.; Li, X.; Feng, Z. TT-XSS: A novel taint tracking based dynamic detection framework for DOM cross-site scripting. *J. Parallel Distrib. Comput.* **2018**, *118*, 100–106. [CrossRef]

47. Bichhawat, A.; Rajani, V.; Garg, D.; Hammer, C. Information Flow Control in WebKit's JavaScript Bytecode. In *Principles of Security and Trust, POST 2014*; Lecture Notes in Computer Science; Abadi, M., Kremer, S., Eds.; Springer: Berlin/Heidelberg, Germany, 2015; Volume 8414. [CrossRef]

48. Joseph, P.N.; Jackson, D. Derailer: Interactive security analysis for web applications. In Proceedings of the 29th ACM/IEEE International Conference on Automated Software Engineering (ASE '14), Vasteras, Sweden, 15–19 September 2014; ACM: New York, NY, USA, 2014; pp. 587–598.

49. Duraibi, S.; Alashjaee, A.M.; Song, J. A Survey of Symbolic Execution Tools. *Int. J. Comput. Sci. and Secur. (IJCSS)* **2019**, *13*, 244.

50. Baldoni, R.; Coppa, E.; D'Elia, D.C.; Demetrescu, C.; Finocchi, I. A survey of symbolic execution techniques. *ACM Comput. Surv.* **2018**, *51*, 50. [CrossRef]

51. Balasubramanian, D.; Zhang, Z.; McDermet, D.; Karsai, G. Dynamic symbolic execution for the analysis of web server applications in Java. In Proceedings of the 34th ACM/SIGAPP Symposium on Applied Computing, SAC '19, Limassol, Cyprus, 8–12 April 2019; pp. 2178–2185. [CrossRef]

52. Pistoia, M.; Tripp, O.; Lubensky, D. Combining Static Code Analysis and Machine Learning for Automatic Detection of Security Vulnerabilities in Mobile Apps. In *Application Development and Design: Concepts, Methodologies, Tools, and Applications*; IGI Global: Hershey, PA, USA, 2018. [CrossRef]

53. Pereira, J.D.; Campos, J.R.; Vieira, M. An Exploratory Study on Machine Learning to Combine Security Vulnerability Alerts from Static Analysis Tools. In Proceedings of the 2019 9th Latin-American Symposium on Dependable Computing (LADC), IEEE, Natal, Brazil, 19–21 November 2019. [CrossRef]

54. Riera, T.S.; Higuera, J.-R.B.; Higuera, J.B.; Martínez-Herráiz, J.-J.; Montalvo, J.A.S. Prevention and Fighting against Web Attacks through Anomaly Detection Technology. A Systematic Review. *Sustainability* **2020**, *12*, 4945. [CrossRef]

55. Bermejo, J.R. OWASP Top Ten-Benchmark. Available online: https://github.com/jrbermh/OWASP-Top-Ten-Benchmark (accessed on 1 December 2020).

56. Bau, J.; Bursztein, E.; Gupta, D.; Mitchell, J.C. State of the Art: Automated Black-Box Web Application Vulnerability Testing. In Proceedings of the IEEE Symposium on Security and Privacy, Berkeley/Oakland, CA, USA, 16–19 May 2010.

57. Alsaleh, M.; Alomar, N.; Alshreef, M.; Alarifi, A.; Al-Salman, A. Performance-based comparative assessment of open source web vulnerability scanners. *Secur. Commun. Netw.* **2017**, *2017*, 6158107. [CrossRef]

58. Qasaimeh, M.; Shamlawi, A.; Khairallah, T. Black box evaluation of web applications scanners: Standards mapping aproach. *J. Theor. Appl. Inf. Technol.* **2018**, *96*, 4584–4596.

59. Amankwah, R.; Chen, J.; Kudjo, P.K.; Agyemang, B.K.; Amponsah, A.A. An automated framework for evaluating open-source web scanner vulnerability severity. *Serv. Oriented Comput. Appl.* **2020**, *14*, 297–307. [CrossRef]

60. Xypolytos, A.; Xu, H.; Vieira, B.; Ali-Eldin, A.M.T. A Framework for Combining and Ranking Static Analysis Tool Findings Based on Tool Performance Statistics. In Proceedings of the 2017 IEEE International Conference on Software Quality, Reliability and Security Companion (QRS-C), Prague, Czech Republic, 25–29 July 2017; pp. 595–596. [CrossRef]

61. Ye, T.; Zhang, L.; Wang, L.; Li, X. An Empirical Study on Detecting and Fixing Buffer Overflow Bugs. In Proceedings of the 2016 IEEE International Conference on Software Testing, Verification and Validation (ICST), Chicago, IL, USA, 11–15 April 2016; pp. 91–101. [CrossRef]

62. Halfond, W.G.J.; Choudhary, S.R.; Orso, A. Improving penetration testing through static and dynamic analysis. *Softw. Test. Verif. Reliab.* **2011**, *21*, 195–214. [CrossRef]

63. Mongiovi, M.; Giannone, G.; Fornaia, A.; Pappalardo, G.; Tramontana, E. Combining static and dynamic data flow analysis: A hybrid approach for detecting data leaks in Java applications. In Proceedings of the 30th Annual ACM Symposium on Applied Computing, Salamanca, Spain, 13–17 April 2015; ACM: New York, NY, USA, 2015; pp. 1573–1579. [CrossRef]

64. Loch, F.D.; Johns, M.; Hecker, M.; Mohr, M.; Snelting, G. Hybrid taint analysis for java EE. In Proceedings of the 35th Annual ACM Symposium on Applied Computing, Brno, Czech Republic, 30 March–3 April 2020; ACM: New York, NY, USA, 2020. [CrossRef]

65. Kim, S.; Kim, R.Y.C.; Park, Y.B. Software Vulnerability Detection Methodology Combined with Static and Dynamic Analysis. *Wirel. Pers. Commun.* **2015**, *89*, 777–793. [CrossRef]

66. Alavi, S.; Bessler, N.; Massoth, M. A comparative evaluation of automated vulnerability scans versus manual penetration tests on false-negative errors. In Proceedings of the Third International Conference on Cyber-Technologies and Cyber-Systems, IARIA, Athens, Greece, 18–22 November 2018.

67. Idrissi, S.E.; Berbiche, N.; Sbihi, M. Performance evaluation of web application security scanners for prevention and protection against vulnerabilities. *Int. J. Appl. Eng. Res.* **2017**, *12*, 11068–11076.

68. Livshits, B.V.; Lam, M.S. Finding security vulnerabilities in java applications with static analysis. In Proceedings of the 14th Conference on USENIX Security Symposium USENIX Association, Berkeley, CA, USA, 31 July–5 August 2005.

69. Martin, B.; Livshits, B.; Lam, M.S. Finding application errors and security flaws using PQL: A program query language. In Proceedings of the 20th Annual ACM Conference on Object-Oriented Programming, Systems, Languages, and Applications, San Diego, CA, USA, October 2005. [CrossRef]

70. Krishnan, R.; Nadworny, M.; Bharill, N. Static analysis tools for security checking in code at Motorola. *ACM SIGAda Ada Lett.* **2008**, *28*, 76–82. [CrossRef]

71. Cifuentes, C.; Scholz, B. Parfait–designing a scalable bug checker. In Proceedings of the 2008 Workshop on Static Analysis, SAW '08, Tucson, AZ, USA, 12 June 2008; ACM: New York, NY, USA, 2008. [CrossRef]

72. Shrestha, J. Static Program Analysis. Ph.D. Thesis, Uppsala University, Uppsala, Sweden, 2013.

73. Goseva-Popstojanova, K.; Perhinschi, A. On the capability of static code analysis to detect security vulnerabilities. *Inf. Softw. Technol.* **2015**, *68*, 18–33. [CrossRef]

74. Pashchenko, I.; Dashevskyi, S.; Massacci, F. Delta-bench: Differential benchmark for static analysis security testing tools. In Proceedings of the 11th ACM/IEEE International Symposium on Empirical Software Engineering and Measurement, IEEE Computer Society, Toronto, ON, Canada, 9–10 November 2017. [CrossRef]

75. Long, F.; Mohindra, D.; Seacord, R.C.; Sutherland, D.F.; Svoboda, D. *Java™ Coding Guidelines: 75 Recommendations for Reliable and Secure Programs*; Pearson Education: Boston, MA, USA, 2014.

76. Heckman, S.; Williams, L. A systematic literature review of actionable alert identification techniques for automated static code analysis. *Inf. Softw. Technol.* **2011**, *53*, 363–387. [CrossRef]

77. Antunes, N.; Vieira, M. On the metrics for benchmarking vulnerability detection tools. In Proceedings of the 45th Annual IEEE/IFIP International Conference on Dependable Systems and Networks, Rio de Janeiro, Brazil, 22–25 June 2015. [CrossRef]

Publisher's Note: MDPI stays neutral with regard to jurisdictional claims in published maps and institutional affiliations.

Article

A Holistic Cybersecurity Maturity Assessment Framework for Higher Education Institutions in the United Kingdom

Aliyu Aliyu, Leandros Maglaras *, Ying He *, Iryna Yevseyeva, Eerke Boiten, Allan Cook and Helge Janicke

School of Computer Science and Informatics, De Montfort University, Leicester LE1 9BH, UK;
P17243308@my365.dmu.ac.uk (A.A.); iryna@dmu.ac.uk (I.Y.); eerke.boiten@dmu.ac.uk (E.B.);
allan.cook@dmu.ac.uk (A.C.); helge.janicke@cybersecuritycrc.org.au (H.J.)
* Correspondence: leandros.maglaras@dmu.ac.uk (L.M.); ying.he@dmu.ac.uk (Y.H.)

Received: 15 April 2020; Accepted: 18 May 2020; Published: 25 May 2020

Abstract: As organisations are vulnerable to cyberattacks, their protection becomes a significant issue. Capability Maturity Models can enable organisations to benchmark current maturity levels against best practices. Although many maturity models have been already proposed in the literature, a need for models that integrate several regulations exists. This article presents a light, web-based model that can be used as a cybersecurity assessment tool for Higher Education Institutes (HEIs) of the United Kingdom. The novel Holistic Cybersecurity Maturity Assessment Framework incorporates all security regulations, privacy regulations, and best practices that HEIs must be compliant to, and can be used as a self assessment or a cybersecurity audit tool.

Keywords: assessment framework; cybersecurity; GDPR; PCI-DSS; DSPT; NISD

---

1. Introduction

In an age of information growth, technology plays a key role in shaping all aspects of human life. In the education sector, teachers and students can make use of the ever-expanding resources available, creating a diverse learning experience that caters for many teaching and learning styles. However, with this adoption of technology, Higher Education Institutions (HEIs) are finding themselves the targets of malicious cyberactivities, with a recent JISC report [1] reaffirming that UHEIs in the UK are not well prepared to defend against, or recover from cyberattacks.

Due to their nature, HEIs hold a significant amount of information and accumulated knowledge. As a result, they are attractive to threat actors who target research findings, financial data, and computing resources. Katz [2] identified that HEIs are under continual risk of cyberattacks. Consequently, HEIs face a constant challenge of balancing public access in the interest of sharing information, whilst protecting their information assets.

A study of businesses students in New England was conducted by Kim [3] on the attitude of students regarding Information Security Awareness (ISA). It was evident in the findings that students who participated found the ISA training important and necessary in improving their knowledge in cybersecurity. Studies in 2013 by the Kaspersky Lab [4] showed over the period of a year that 91% of organisations surveyed reported their IT infrastructure had been the victim of at least one cyberattack. Additionally stated in the report, there was an increase in cybercrime such as email phishing, unauthorised network access, malware, and theft of mobiles in 2013 compared to 2012. The study focused on corporate IT infrastructures and highlighted that for years, IT infrastructures such as those in HEIs had been deficient in terms of security and had always been a target for threat actors.

In the market, there are currently many frameworks available for organisations to adopt to improve the effectiveness of their cybersecurity. These frameworks support action at both an individual and organisational level. Aloul [5] highlights that for the success and security of any security improvement program adopted by an institution, it is important that students and staff are given training and education in information security awareness. This should be made part of the risk/security assessment plan adopted by all levels of administration, from students, to teachers, and all administrative employees, as teaching the front-end users will serve as the first line of defence against attackers [6].

To build a secure environment, providing a relevant security awareness program is the initial step. There should be constant training and education provided to equip students, staff, and employees to deal with the latest cyberthreats and modern prevention methods [7]. There should also be effectiveness metrics that the institution can measure and monitor. Changes to management and audits can be adopted by the institution to strengthen the level of cybersecurity [8]. One important set of tools that HEIs can use in order to measure their cybersecurity readiness and compliance levels is maturity models [9].

Matthew J. Butkovic [10] defined the maturity model as "a set of characteristics, attributes, indicators, or patterns that represent progression and achievement in a particular domain or discipline". The artefacts that make up the model are typically agreed on by the domain or discipline, which validates them through application and iterative recalibration. In order to make maturity models more effective, the measurable transitions between levels should be based on empirical data that have been validated in practice. This means each level in the model should be more mature than the previous level. In essence, what constitutes mature behaviours must be characterised and validated. This can be challenging to achieve unambiguously in many maturity models.

Our proposed Holistic Cybersecurity Maturity Assessment Framework (HCYMAF) is based on a process methodology called a Capability Maturity Model (CMM) [11]. CMMs were originally developed by the Carnegie Mellon University Software Engineering Institute (CMU/SEI) to improve the management of software development, and have been subsequently used in many other domains, such as cybersecurity. A maturity model defines a set of metrics for measuring organisational competency or maturity in terms of a set of recognised best practices, skills, or standards. Metrics are organised into categories and quantified on a performance scale. Using specific rating criteria, organisations can measure their performance against these maturity levels.

This paper makes the following contributions,

- It proposes a novel Holistic Cybersecurity Maturity Assessment Framework (HCYMAF) for HEIs that can be used in order to conduct a gap analysis against 15 security requirements.
- The proposed framework incorporates several regulations and security best practices into one lightweight online self-assessment guide.
- It produces compliance reports against all regulations that the HEI must be compliant with in order to facilitate mitigation plans.
- It can be adapted and expanded in order to be used on other critical sectors of the UK and abroad.

The rest of the paper is organised as follows: In Section 2, we present related work, while in Section 3 we describe our system framework. In Section 3.4, we present the validation procedure. Finally, in Section 5, we conclude this paper and present future work.

2. Related Work

2.1. Essential Components of a Maturity Model

A maturity model should follow a structure to ensure its consistency. It typically includes the components, levels, attributes, appraisal and scoring methods, and model domains. Levels represent the measurement aspect of a maturity model, however, if the scaling is inaccurate or incomplete, we may not be able to validate the model and the results produced may not be accurate or consistent.

Attributes represent the main content of the model and are classified by domains and levels. Attributes are defined at the intersection of a domain and a maturity level, which are typically based on observed practices, standards, or other expert knowledge. These can be expressed as characteristics, indicators, practices, or processes. In capability models, attributes also express qualities of organisational maturity (e.g., planning and measuring) for supporting process improvement regardless of the process being modelled.

Appraisal and scoring methods are used to facilitate the assessment. They can be formal or informal, expert-led or self-applied. Scoring methods are algorithms devised by the community to ensure consistency of appraisals and they are common standards for measurement. Scoring methods can include weighting (so that important attributes are valued over less important ones) or can value different types of data collection in different ways (e.g., providing higher marks for documented evidence than for interview-based data).

Model domains essentially define the scope of a maturity model. Domains are a means for grouping attributes into an area of importance for the subject matter and intent of the model. In capability models, the domains are often (but not necessarily) referred to as process areas as they are a collection of processes that make up a larger process or discipline (e.g., software engineering). Depending on the model, users may be able to focus on improving a single domain or a group of domains.

2.2. Maturity Model Types

Caralli [12] classified maturity models into three different types—progression models, capability models, and hybrid models. Progression models represent a simple progression or scaling of a characteristic, indicator, attribute, or pattern in which the movement through the maturity levels indicates some progression of attribute maturity. Progression models typically place their focus on the evolution of the model's core subject matter (such as practices or technologies) rather than attributes that define maturity (such as the ability and willingness to perform a practice, the degree to which a practice is validated, etc.). In other words, the purpose of a progression model is to provide a simple road map of progression or improvement as expressed by increasingly better versions (for example, more complete, more advanced) of an attribute as the scale progresses [10]. For capability models such as CMM, the dimension that is being measured is a representation of organisational capability around a set of characteristics, indicators, attributes, or patterns—often expressed as processes. A CMM measures more than the ability to perform a task; it also focuses on broader organisational capabilities that reflect the maturity of the culture and the degree to which the capabilities are embedded (or institutionalised) in the culture [10]. Hybrid models merge two abilities; the ability to measure maturity attributes and the ability to measure evolution or progression in progressive models. This type of model reflects transitions between levels that are similar to capability model levels (i.e., that describe capability maturity) but also account for the evolution of attributes in a progression model [10].

2.3. Existing Work on Maturity Models

Evaluation of maturity capability was developed in 1986 by the US Department of Defense for assessing maturity capabilities of Software Engineering processes of the software companies they worked with [13]. This model was later adopted by different domains including cybersecurity.

Various cybersecurity maturity models were developed according to the needs of organisations. Currently, the most popular and widely used maturity models are incorporated into (inter)national standards. For instance, ISO/IEC 27001 [14,15] and NIST [16]—European and American standards for cybersecurity, respectively. ISO/IEC 27001 was developed based on the British Standard BS7799 and ISO/IEC 17799 to provide requirements as well as to maintain and improve Information Security

Management System (ISMS) [13]. ISO/IEC 27001 defines ISMS as a part of the overall management system, which "establish, implement, operate, monitor, review, maintain and improve information security" [14,15].

Sabillon et al. [17] proposed a Cybersecurity Audit Model (CSAM) in order to improve cybersecurity assurance. The CSAM was designed to be used for conducting cybersecurity audits in organisations and Nation States. CSAM evaluates and validates audit, preventive, forensic, and detective controls for all organisational functional areas. The CSAM was then tested, implemented, and validated along with the Cybersecurity Awareness TRAining Model (CATRAM) in a Canadian higher education institution. Adler et al. [18] created a Dynamic Capability Maturity Model for improving cybersecurity. It extends an existing cybersecurity CMM into a dynamic performance management framework. It is a software-based framework that enables organisations to create, test, validate, or refine plans to improve their cybersecurity maturity levels. Almuhammadi et al. identified the gaps of the NIST Cyber Security Framework for Critical Infrastructure (NIST CSF) by comparing it to the COBIT, ISO/IEC 27001, and ISF frameworks, and then proposed an Information Security Maturity Model (ISMM) to fill in the gaps and measure NIST CSF implementation progress [19]. Miron et al. reviewed Cybersecurity Capability Maturity Models for providers of critical infrastructure, and provided recommendations on employing capability maturity models to measure and communicate readiness [20].

Akinsanya et al. investigated the effective assessment of healthcare cybersecurity maturity models for healthcare organisations using cloud computing [21]. The findings showed that the assessment practices are sometimes considered ineffective since the measurements of individual IS components were not capable of depicting the overall security posture within a healthcare organisation. The effects of cloud computing technology in healthcare were also not taken into account.

The existing maturity models offer a manageable approach for assessing the security level of a system or organisation, however, it is difficult to establish sound security models and mechanisms for protecting the cyberspace, as the definitions and scopes of both cyberspace and cybersecurity are still not well defined [22]. Most of the existing maturity models provide a minimum compliance model rather than an aspired cybersecurity model that can address emerging threat landscape. The model should allow multiusers including management, security experts, and practitioners to assess the overall security status of the organisation/system and take measures to address the weaknesses identified from the assessment. Most of the existing models are measured by qualitative metrics/processes, however quantitative metrics should be essential for security assessment [22,23].

2.4. Selected Existing Models Adopted for HEIs Maturity Assessment

Existing models were reviewed for their applicability for HEIs maturity assessment. The basis of our maturity model was formed according to the CMMI [24]. The CMMI was used as it provides an evolutionary path to performance improvement.

The starting point of a cybersecurity assessment is the definition of requirements for an Information Security Management System (ISMS) of an organisation. ISO/IEC 27001 Information Security Management [14,15] is the best-known standard for providing a set of necessary requirements and this was used in our framework.

In addition to the evaluation of maturity, our model provides a set of cybersecurity actions and controls to be implemented to close the existing gaps in HEI cybersecurity. For this, we reviewed a number of well-established models and selected the most critical ones to be used for HEIs' protection from known cyberattack vectors. The CIS Controls [25] are specifically technical controls that can be used to mitigate from specific attacks. ENISA's guidelines on assessing DSP security and OES compliance with the NISD security requirements [26] provided insight into the self-assessment/management framework for the DSP security against the security requirements. The cybersecurity evaluation tool provided a systematic approach for evaluating an organisation's security posture by assessing operational resilience, cybersecurity practices, organisational management of external dependencies, and other key elements of a robust cybersecurity framework.

Except the above models, Citigroup's Information Security Evaluation Model (CITI-ISEM) [20], Computer Emergency Response Team / CSO Online at Carnegie Mellon University (CERT/CSO), and the U.S. Cybersecurity Capability Maturity Model (C2M2) and its National Initiative for Cybersecurity Education's Capability Maturity Model (NICE-CMM) were also reviewed [20]. These models were reviewed in order to check that we did not miss any important security controls from incorporating them into our framework.

The work of Mbanaso et al. titled Conceptual Design of a Cybersecurity Resilience Maturity Measurement (CRMM) Framework [27] was also reviewed and it provided insight into measuring the effectiveness and efficiency of organisation's controls with respect to cybersecurity resilience, and also the steps that can be taken to improve resilience maturity. Lastly, the work of Butkovic and Caralli titled Advancing Cybersecurity Capability Measurement using the CERT-RMM Maturity Indicator Level Scale [28] provided insight into how the CMMI maturity levels can be utilised to show incremental improvement in maturity.

Recently, ENISA [29] published a report that presents a mapping of the main security objectives between the NISD and General Data Protection Regulation (GDPR) in order to support organisations in their process of identifying appropriate security measures. At the same time, ISO issued the ISO 27701 Standard [30] in order to help organisation establish, implement, maintain, and continually improve a Privacy Information Management System by combining the ISMS with the privacy framework and principles defined in ISO/IEC 29100. NIST has also published the Privacy Framework [31] that follows the structure of the Framework for Improving Critical Infrastructure Cybersecurity (the Cybersecurity Framework) in order to facilitate the use of both frameworks together. It is obvious that all major security organisations and authorities have identified the need for mapping cybersecurity requirements from different frameworks, but until now, only initial works that map GDPR with NIST and NISD have been published.

Apart from the lack of a security maturity model tailored for HEIs, the other identified gaps in the review of these maturity models occur in the aspect of adoption: the maturity models are either too complicated to implement, or they require the organisation's processes to be refined to suit their implementation. HEIs have more fluid and less controllable environments, which render many of ISO controls nonapplicable or introduce too significant barriers for HE to manage effectively.

A holistic framework that incorporates all regulations and can be used either offline or online with easily followed and understood maturity assessment metrics was needed for the HEIs. The proposed framework incorporates several regulations and security best practices into one lightweight online self assessment guide that can be run as a self assessment or audit tool. HCYMAF supports the assessment of the maturity of each of the 15 specified domains to identify weak and strong practices and can be easily extensible in order to incorporate other domains, e.g., IoT, blockchain [32] etc.

3. Proposed Maturity Framework

An appraisal is an activity that helps identify the strengths and weaknesses of an organisation's processes and to examine how closely the processes relate to identified best practices. Appraisals are typically conducted to determine how well the organisation's processes are when compared to related identified security best practices and to identify areas where improvement can be made. Our proposed Maturity Assessment Framework (MAF) can be used in order to inform external customers and suppliers about how well the organisation's processes are when compared to related identified security best practices. The model can also be used as a gap-analysis and compliance-checking tool that any organisation can use in order to define how well contractual requirements are met. The MAF is established based on the following:

- A review of security requirements that HEIs must follow in order to demonstrate compliance with the General Data Protection Regulation (GDPR), Payment Card Industry Data Security Standard

(PCI DSS), Data Security and Protection Toolkit (DSPT), and any other regulation that may apply to them;

- A literature review of existing research on maturity models in cybersecurity as well as in other areas.

This framework, entitled "A Holistic Cybersecurity Maturity Assessment Framework (HCYMAF) for Higher Education Institutes (HEIs)", aims at designing a cybersecurity maturity assessment framework for all higher education institutes in the United Kingdom. The framework can be used as a self-assessment tool by the HEIs organisation in order to establish their security level and highlight the weaknesses and mitigation plans that need to be implemented. The framework is a mapping and codification tool for HEIs against all regulations that the HEIs must comply with, such as the GDPR, PCI DSS, DSPT, etc.

The framework uses six different levels of maturity against which the cybersecurity performance of each organisation can be measured. The framework will be validated through three pilot implementations, of which one has already been conducted with positive results and feedback obtained. This model is important and novel because HEIs, by using this framework, will be able to assess the security level of their organisation, conduct a gap analysis, and create appropriate mitigation plans. The model also informs whether the organisation is compliant with the expected regulations, thus helping them in self-assessment and improvement by producing relevant compliance reports.

It is necessary to design a maturity model that will be able to facilitate the organisations and the National Cyber Security Center (NCSC) of the UK. To achieve this, the model must have the following characteristics. It must

- Cover the full extent of the requirements of the different regulations;
- Be able to be used as a self-assessment tool;
- Be able to be used as a basis for an independent assessment;
- Provide clear results regarding the security posture of the organisations;
- Produce compliance reports;
- Be able to be used as guidance for implementation of a concrete security policy by the HEIs;
- Be measurable;
- Be easily extractable and reusable.

3.1. Security Requirements

As illustrated in Figure 1, the proposed maturity assessment model has 15 requirements. The 15 requirements followed are categorised as 'General Security Requirements'. The General Security Requirements, which are the foundation of the model, is based on cybersecurity best practices such as the CIS Controls, NIST Framework, etc. The 15 requirements were divided into 3 groups. IDENTIFY (I), PROTECT & DETECT (P), and RESPOND & RECOVER (R). It should be noted that the DETECT controls of NIST were merged into our protect & detect requirements in order to keep our model lightweight.

Requirements $I1$–$I4$ fall under Identify, Requirements $P5$–$P13$ fall under Protect & Detect, whilst Requirements $R14$–$R15$ fall under Respond & Recover. All the requirements of the category Identify are necessary for the facilitation of the understanding of the business and operational ecosystem of the organisation. All the requirements of Protect & Detect are necessary in order to detect incidents and protect all assets supporting the services of the organisation i.e., (people, procedures, and technologies). Lastly, all the requirements of Respond & Recover are necessary in order to respond and manage an information security incident that may have the ability to influence the provision of the services offered by the HEIs. Finally, it should be noted that some requirements do have sub-requirements (See Figure 2).

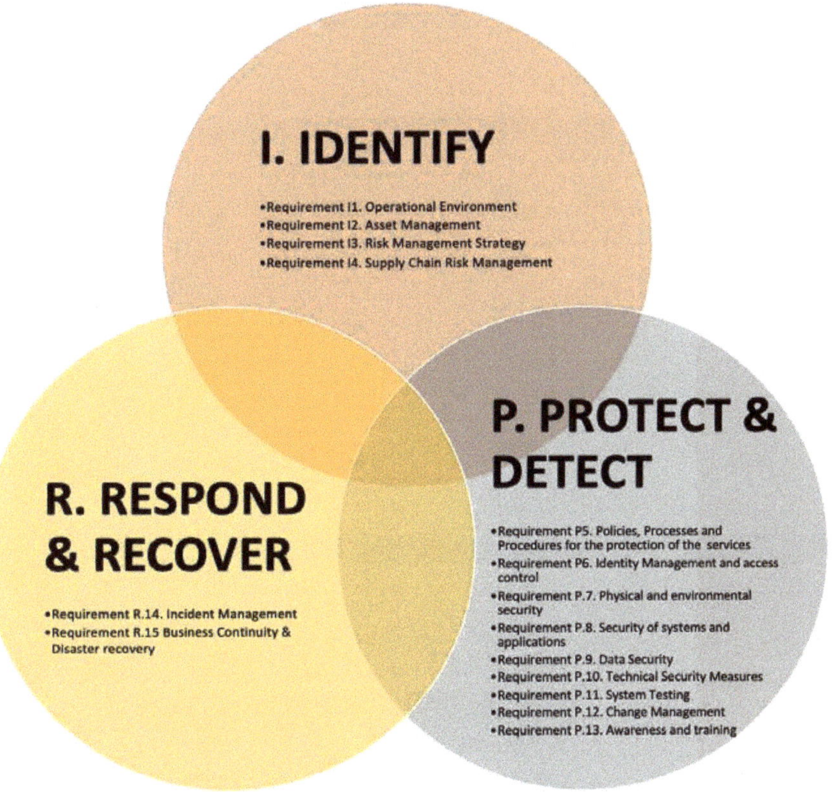

Figure 1. Holistic Cybersecurity Maturity Assessment Framework (HCYMAF) requirements are divided into three groups.

3.2. Mapping of Regulations

It is worth stating that we incorporated the regulation requirements of GDPR, PCI DSS, and DSPT into our General Security Requirements. The mapping of the different regulations into the HCYMAF is shown in the upcoming Figures 3–5. This was done by focusing on each individual regulation and mapping it into one of our requirements. For example, in terms of GDPR, we focused on the 7 principles of GDPR—as shown in Figure 3—and mapped each of the principles into one of our requirements. For example, the first principle of GDPR is lawfulness, fairness, and transparency. This was mapped into Sub-Requirement P5.1: GDPR Compliance. The second principle, which is purpose limitation, was mapped into Requirement P5: Policies, Processes, and Procedures for the protection of the services, and so on.

In terms of incorporating PCI DSS, we focused on the 6 principles of PCI DSS, as each of these principles had its requirements (See Figure 4). In terms of incorporating DSPT, we also focused on the 10 principles of the regulation and likewise each of those principles was mapped into one of our requirements (See Figure 5). Overall, all the aforementioned regulations were incorporated into our model and merged to form a solid maturity model, as illustrated in Figure 6.

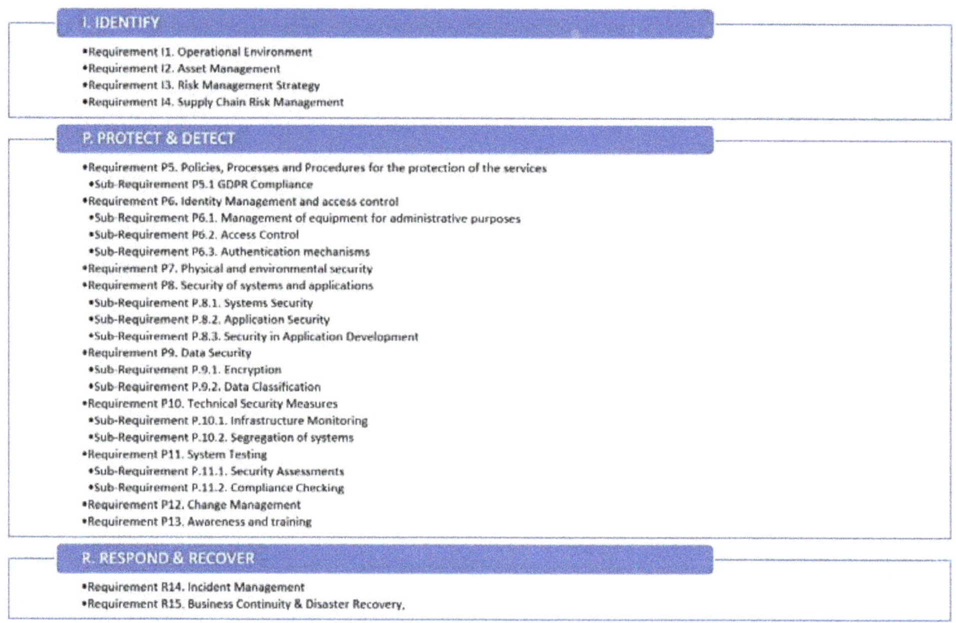

Figure 2. The proposed HCYMAF model in detail.

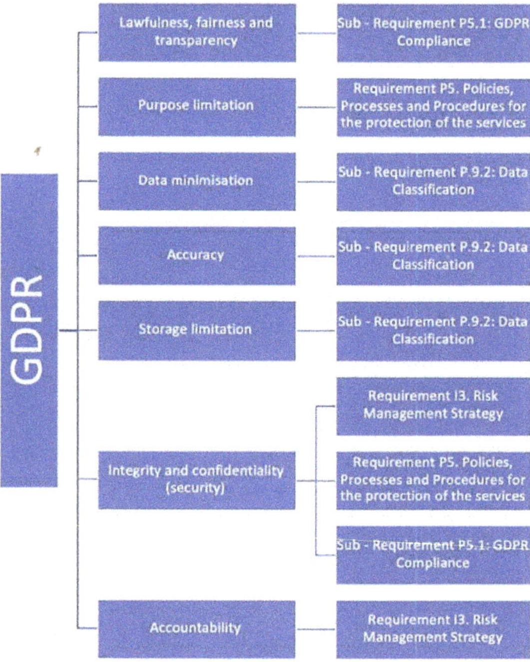

Figure 3. General Data Protection Regulation (GDPR) Mapping.

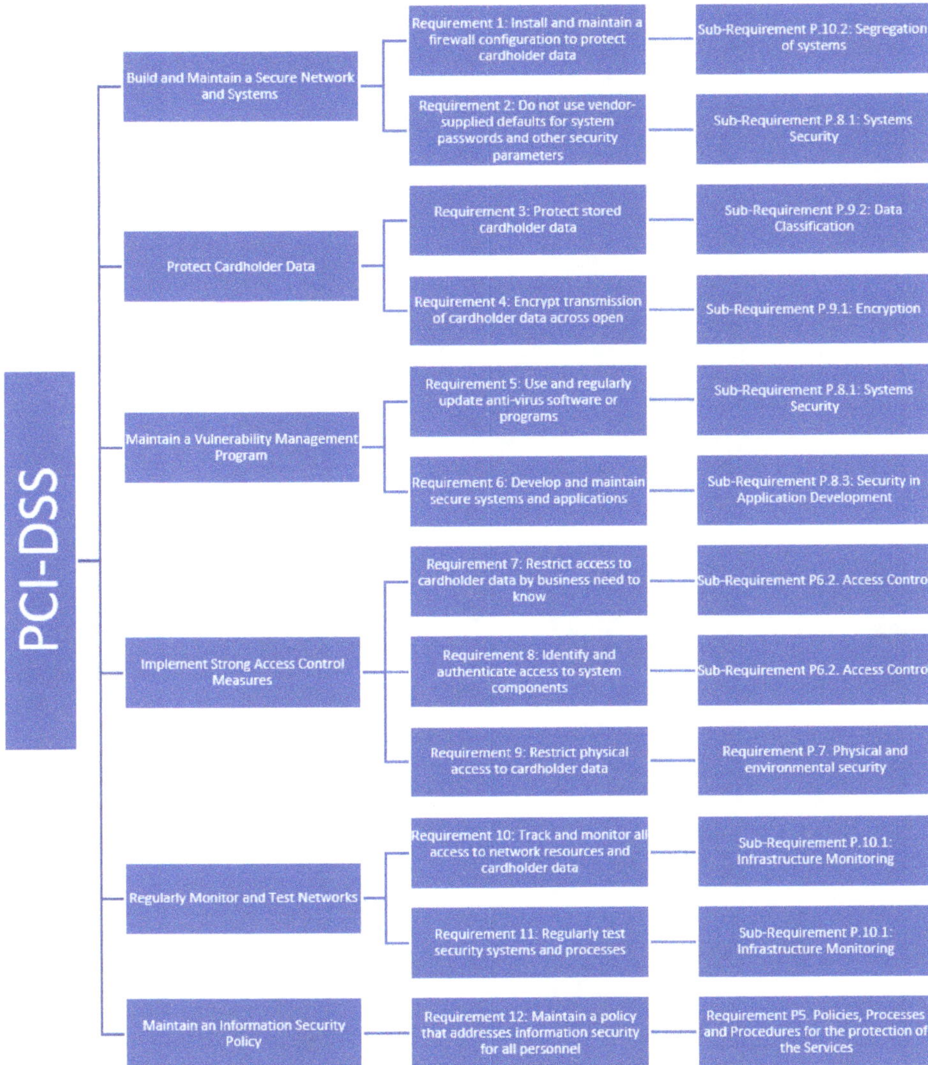

Figure 4. Payment Card Industry Data Security Standard (PCI DSS) Mapping.

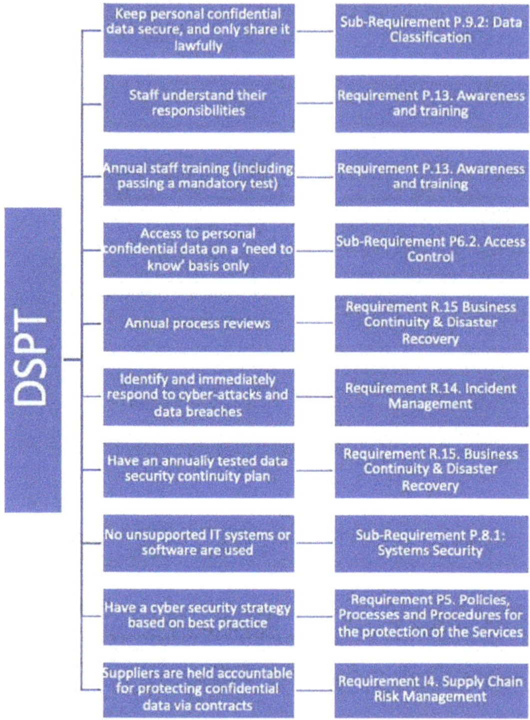

Figure 5. Data Security and Protection Toolkit (DSPT) Mapping.

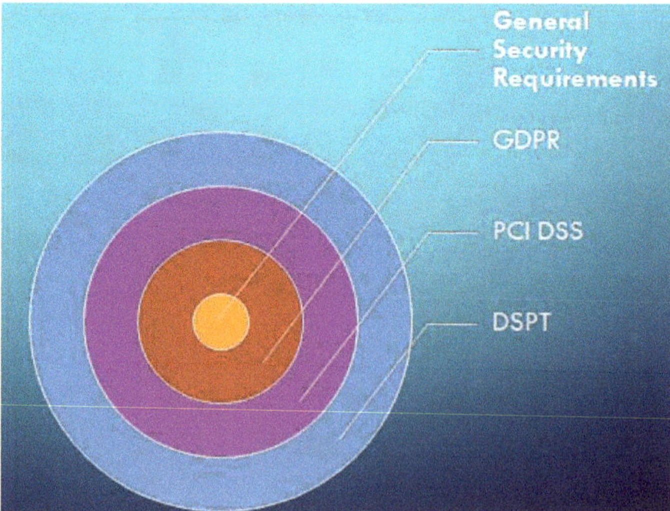

Figure 6. Merging of different requirements into the proposed HCYMAF.

3.3. Maturity Levels

The maturity model has its maturity levels. This means that each of the requirements and subrequirements has its own maturity levels. The maturity levels are 6 scores, from 0 to 5, with 0 being the lowest and 5 being the highest. Each of these maturity levels has a meaning, it represents a staged path for an organisation's performance and process improvement efforts based on predefined sets of practice areas. Each maturity level also builds on the previous maturity levels by adding new requirements. An example of such a scale is shown in Figure 7 below. A brief description of each level is presented:

- Level 0: Incomplete; Ad hoc and unknown. Work may or may not get completed.
- Level 1: Initial; Unpredictable and reactive. Work gets completed but is often delayed and over budget.
- Level 2: Managed; Projects are planned, performed, measured, and controlled.
- Level 3: Defined; the organisation is proactive, rather than reactive. There are organisation-wide standards that provide guidance across projects, programs, and portfolios.
- Level 4: Quantitatively Managed; the organisation is data-driven with quantitative performance improvement objectives that are predictable and align to meet the needs of internal and external stakeholders.
- Level 5: Optimising; the organisation is focused on continuous improvement and is built to pivot and respond to opportunity and change.

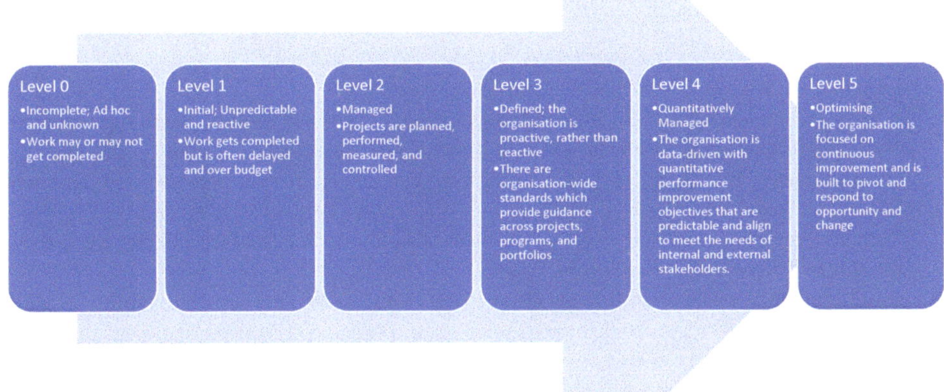

Figure 7. Maturity levels of the proposed model.

In terms of evaluation of the performance of an organisation against an individual requirement, the maturity notes should be read one at a time in ascending order (from 0 to 5). If all notes are fulfilled, then the next level should be read and examined. In order to assign a certain score, all of the lower levels must be completely fulfilled first. It should also be noted that some subrequirements have a Not Applicable (N/A) option, this is because not all subrequirements are applicable to every organisation.

3.4. Evaluation and Validation

The validation authenticates the contribution of the proposed maturity model, HCYMAF, as well as its usefulness, value, capability, and operational characteristics. A validation strategy was developed to provide a convincing argument for the model's effectiveness and demonstrate its function within its proposed and realistic environment. It included the following:

- Interview with experts in the field of security and data protection of HEIs (DPO or Cyber Security Officers) in order to identify the different regulations that the HEIs must be compliant with, the best practices that they follow, how do organisations manage the overlap between cybersecurity and data protection (GDPR), the integration of Risk Management and the Privacy Impact Assessment among others. Apart from structured interviews that were sent to the experts, members of the team that developed the framework had many discussions. Using their advice and suggestions, the final groups and requirements were developed.
- A case study: The objectives of the case study that was conducted include the validation of the proposed structure and categories by adding or removing them from the model, which is expected to make advances to the model, and collect information related to the processes used to manage security in HEIs
- Feedback from the scientific community through the submission and presentation of academic papers—A number of research outputs will be produced alongside the study course which further enhances the validation of the research outputs.
- A webinar that will be organised and will take place later this year, where HEIs in the UK will be invited. During the webinar, the representatives of the HEIs will be given an overview of the framework, the results of the conducted case studies and the option to run the HCYMAF either offline (through a dedicated excel file and a detailed guide that we have developed) or online through our website.

Structured Interviews with Experts

1. What are the regulations that universities have to be compliant with?
2. Are universities in the UK obliged to have a security officer?
3. How do you conduct the DPIA and Risk Assessment? Do you do it in parallel or one after the other? Is data protection impact assessment under Risk Assessment?
4. Can you please briefly tell us the procedures you follow in order to be compliant with those regulations?
5. We have some categories that might not be applicable to universities, for example, 'security for software development'—what is your opinion?
6. How are the roles and responsibilities between the DPO and the security officer split?
7. How do you actually merge security requirements and Data Protection requirements during the implementation of a new service?
8. What is the procedure that is followed when a security or data breach takes place?
9. What would be the added value of a cybersecurity assessment framework? What would you expect from such a model?
10. We have created an initial pool of sectors that our HCYMAF is going to investigate. Do you think that we may miss any important category?

Before the final model is released to the HEIs, it should be validated through several pilot implementations. The model should preferably be used by organisations of different sizes and regardless of the activities they have, e.g., provide health studies. The first pilot has already been conducted, and the team at De Montfort University (DMU) in cooperation with the NCSC has already planned to run the other two cases in the next period. In the meantime, the DMU team has released the first version of the website which HEIs will use in order to perform self-assessments and receive the results in a graphical model and a gap analysis that showcases the cybersecurity sectors of HEI's IT systems that need immediate actions. Additionally, compliance reports will be produced automatically from the HCYMAF, giving the opportunity for the organisation to react fast and avoid penalties.

Each HEI representative will need to register to the platform and then go through the guide. The process can be paused and continued at a later time since a lot of information and time are needed in order to conduct a full cybersecurity assessment. The results for each organisation are only visible to the organisation along with charts and reports that will help the security and data-protection-officers to take the appropriate measures. Aggregated results will be collected and used for analysis by the NCSC in order to prioritise future security plans.

4. Discussion

Our proposed framework defines a set of metrics for measuring organisational competency or maturity in terms of a set of recognised best practices, skills, or standards. It incorporates the General Data Protection Regulation (GDPR), Payment Card Industry Data Security Standard (PCI DSS), and Data Security and Protection Toolkit (DSPT), and can be used in order to conduct a gap analysis against 15 security requirements. The metrics are organised into categories and quantified on a performance scale. The measurable transitions between levels are based on empirical data that have been validated in practice, and each level in the model is more mature than the previous level.

By applying the proposed framework, organisations can achieve progressive improvements in their cybersecurity maturity by first achieving stability at the project level, then continuing to the most advanced-level, organisation-wide, continuous process improvement, using both quantitative and qualitative data to make decisions. For instance, at maturity level 2, the organisation has been elevated from ad hoc to managed by establishing sound security controls, procedures, and processes. As a university achieves the generic and specific goals at a maturity level, it is increasing its maturity and at the same time achieves compliance with relevant regulations and national laws.

Based on the experience we will gain out of this project, we will adapt the proposed HCYMAF for organisations in other sectors, e.g., water and power suppliers, in the future. We will incorporate the best practices, skills, or standards that are essential for different sectors. We also aim to create (working closely with the NCSC) a semiautomated self-assessment online framework. This online framework could be used by all critical organisations in the UK. The framework will include specific controls like IoT, SCADA, etc., where each organisation will fill the controls that are applicable to them. Finally, the information collected by this online tool will help the UK government to prioritise the mitigation plans related to security that need to be taken at a national level in terms of funding specific actions, launch new security tools, etc.

5. Conclusions

There have been a number of cyberattacks upon HEIs around the globe, and the recent JISC report reaffirmed that HEIs of the UK are not well prepared to defend against, or recover from, cyberattacks. Capability Maturity Models can enable organisations to benchmark current maturity levels against best practices. Although many maturity models have been already proposed in the literature, no model that integrates several regulations exists. Based on this finding, in this article we present a light, web-based model that can be used as a cybersecurity assessment tool for Higher Education Institutes (HEIs) of the UK that incorporates all security and privacy regulations and best practices that HEIs must be compliant with.

The proposed model consists of 15 security categories, six maturity levels, and is implemented on an online platform that can be used both as a self-assessment and audit tool, facilitating organisations to perform a gap analysis and to receive automated compliance reports and graphical representations of their security posture. Information that will be collected from the platform can be used, after proper aggregation and anonymisation processes from the NCSC, in order to identify current security problems and prioritise future security plans and funding actions.

Author Contributions: Conceptualization, A.A., Y.H., and L.M.; methodology, A.A., L.M., Y.H., and A.C.; software, A.A., I.Y., and L.M.; validation, H.J., E.B., A.C., and L.M.; formal analysis, A.A. and H.J., investigation, A.A., Y.H., and I.Y.; resources, A.A., Y.H., I.Y., and A.C.; data curation, A.A. and L.M.; writing—original draft preparation, A.A., Y.H., I.Y., and L.M.; writing—review and editing, E.B., H.J., and A.C.; visualization, A.A. and Y.H.; supervision, L.M. All authors have read and agreed to the published version of the manuscript.

Funding: We thankfully acknowledge the support of the NCSC, UK funded project (RFA: 20058).

Conflicts of Interest: The authors declare no conflicts of interest.

References

1. Chapman, J.; Francis, J. *Cyber Security Posture Survey Results 2019*; Joint Information Systems Committee (JISC): London, UK, 2019.
2. Katz, F.H. The effect of a university information security survey on instruction methods in information security. In *Proceedings of the 2nd Annual Conference on Information Security Curriculum Development*; Association for Computing Machinery: New York, NY, USA, 2005; pp. 43–48.
3. Kim, E.B. Recommendations for information security awareness training for college students. *Inf. Manag. Comput. Secur.* **2014**, *22*, 115–126. [CrossRef]
4. Kaspersky, G.C.I. *Global Corporate IT Security Risks: 2013*; Kaspersky Lab: Moscow, Russia, 2013.
5. Aloul, F.A. The need for effective information security awareness. *J. Adv. Inf. Technol.* **2012**, *3*, 176–183. [CrossRef]
6. Evans, M.; He, Y.; Maglaras, L.; Janicke, H. HEART-IS: A novel technique for evaluating human error-related information security incidents. *Comput. Secur.* **2019**, *80*, 74–89. [CrossRef]
7. Cook, A.; Smith, R.; Maglaras, L.; Janicke, H. *Using Gamification to Raise Awareness of Cyber Threats to Critical National Infrastructure*; BCS: Belfast, UK, 2016.
8. Rajewski, J. Cyber Security Awareness: Why Higher Education Institutions Need to Address Digital Threats. 2013. Available online: https://www.huffpost.com/entry/cyber-security-awareness-_b_4025200 (accessed on 22 May 2020).
9. Maglaras, L.; Ferrag, M.A.; Derhab, A.; Mukherjee, M.; Janicke, H.; Rallis, S. Threats, Protection and Attribution of Cyber Attacks on Critical Infrastructures. *arXiv* **2019**, arXiv:1901.03899.
10. Butkovic, M.J.; Caralli, R.A. Advancing Cybersecurity Capability Measurement Using the CERT-RMM Maturity Indicator Level Scale. 2013. Available online: https://resources.sei.cmu.edu/library/asset-view.cfm?assetid=69187 (accessed on 22 May 2020).
11. Humphrey, W. Characterizing the software process: A maturity framework. *IEEE Softw.* **1988**, *5*, 73–79. [CrossRef]
12. Caralli, R.; Knight, M.; Montgomery, A. *Maturity Models 101: A Primer for Applying Maturity Models to Smart Grid Security, Resilience, and Interoperability*; Technical Report; Carnegie-Mellon University, Software Engineering Institute: Pittsburgh, PA, USA, 2012.
13. Proença, D.; Borbinha, J. Information Security Management Systems—A Maturity Model Based on ISO/IEC 27001. In *Business Information Systems*; Abramowicz, W., Paschke, A., Eds.; Springer International Publishing: Cham, Switzerland, 2018; pp. 102–114.
14. Humphreys, E. *Implementing the ISO/IEC 27001: 2013 ISMS Standard*; Artech House: Washinton, DC, USA, 2016.
15. Brewer, D. *An Introduction to ISO/IEC 27001: 2013*; BSI Standard Limited: London, UK, 2013.
16. Barrett, M. *Framework for Improving Critical Infrastructure Cybersecurity*; Technical Report; National Institute of Standards and Technology: Gaithersburg, MD, USA, 2018.
17. Sabillon, R.; Serra-Ruiz, J.; Cavaller, V.; Cano, J. A comprehensive cybersecurity audit model to improve cybersecurity assurance: The cybersecurity audit model (CSAM). In Proceedings of the 2017 International Conference on Information Systems and Computer Science (INCISCOS), Quito, Ecuador, 23–25 November 2017; pp. 253–259.
18. Adler, R.M. A dynamic capability maturity model for improving cyber security. In Proceedings of the 2013 IEEE International Conference on Technologies for Homeland Security (HST), Waltham, MA, USA, 12–14 November 2013, pp. 230–235.

19. Almuhammadi, S.; Alsaleh, M. Information security maturity model for NIST cyber security framework. *Comput. Sci. Inf. Technol. CS IT* **2017**, *7*, 51–62.

20. Miron, W.; Muita, K. Cybersecurity capability maturity models for providers of critical infrastructure. *Technol. Innov. Manag. Rev.* **2014**, *4*, 33–39.. [CrossRef]

21. Akinsanya, O.O.; Papadaki, M.; Sun, L. *Current Cybersecurity Maturity Models: How Effective in Healthcare Cloud?* CERC: New Delhi, India, 2019; pp. 211–222.

22. Le, N.T.; Hoang, D.B. Can maturity models support cyber security? In Proceedings of the 2016 IEEE 35th International Performance Computing and Communications Conference (IPCCC), Las Vegas, NV, USA, 9–11 December 2016; pp. 1–7.

23. Akinsanya, O.O.; Papadaki, M.; Sun, L. Towards a maturity model for health-care cloud security (M2HCS). *Inf. Comput. Secur.* **2019**. [CrossRef]

24. Team, C.P. Capability maturity model® integration (CMMI SM), version 1.1. In *CMMI Product Team, "CMMI for Systems Engineering/Software Engineering/Integrated Product and Process Development/Supplier Sourcing, Version 1.1, Staged Representation (CMMI-SE/SW/IPPD/SS, V1.1, Staged)"*; Technical Report CMU/SEI-2002-TR-012; Software Engineering Institute, Carnegie Mellon University: Pittsburgh, PA, USA, 2002.

25. Keller, N. *CIS Controls Informative Reference Details*; NIST: Gaithersburg, MD, USA, 2019.

26. ENISA. *Guidelines on Assessing DSP Security and OES Compliance with the NISD Security Requirements*; European Union Agency For Network and Information Security: Heraklion, Greece, 2018.

27. Mbanaso, U.M.; Abrahams, L.; Apene, O.Z. Conceptual Design of a Cybersecurity Resilience Maturity Measurement (CRMM) Framework. *Afr. J. Inf. Commun.* **2019**, 1–26. [CrossRef]

28. Butkovic, M.; Caralli, R. *Advancing Cybersecurity Capability Measurement Using the CERT-RMM Maturity Indicator Level Scale*; Technical Report CMU/SEI-2013-TN-028; Software Engineering Institute, Carnegie Mellon University: Pittsburgh, PA, USA, 2013.

29. Markopoulou, D.; Papakonstantinou, V.; de Hert, P. The new EU cybersecurity framework: The NIS Directive, ENISA's role and the General Data Protection Regulation. *Comput. Law Secur. Rev.* **2019**, *35*, 105336. [CrossRef]

30. Lachaud, E. ISO/IEC 27701: Threats and Opportunities for GDPR Certification. 2020. Available online: https://research.tilburguniversity.edu/en/publications/isoiec-27701-threats-and-opportunities-for-gdpr-certification (accessed on 22 May 2020).

31. Hiller, J.S.; Russell, R.S. Privacy in crises: The NIST privacy framework. *J. Contingencies Crisis Manag.* **2017**, *25*, 31–38. [CrossRef]

32. Ferrag, M.A.; Maglaras, L.; Janicke, H. Blockchain and its role in the internet of things. In *Strategic Innovative Marketing and Tourism*; Springer: Berlin/Heidelberg, Germany, 2019; pp. 1029–1038.

Article

A Multi-Tier Streaming Analytics Model of 0-Day Ransomware Detection Using Machine Learning

Hiba Zuhair [1], Ali Selamat [2,3,4,*] and Ondrej Krejcar [4]

[1] Department of Systems Engineering, College of Information Engineering, Al-Nahrain University, Baghdad 64074, Iraq; hiba.zuhir@coie-nahrain.edu.iq
[2] School of Computing, Faculty of Engineering, UTM & Media and Games Center of Excellence (MagicX), Universiti Teknologi Malaysia (UTM), Johor Baharu, Johor 81310, Malaysia
[3] Malaysia Japan International Institute of Technology (MJIIT), Universiti Teknologi Malaysia, Jalan Sultan Yahya Petra, Kuala Lumpur 54100, Malaysia
[4] Center for Basic and Applied Research, Faculty of Informatics and Management, University of Hradec Kralove, Rokitanskeho 62, 500 03 Hradec Kralove, Czech Republic; ondrej.krejcar@uhk.cz
* Correspondence: aselamat@utm.my

Received: 25 March 2020; Accepted: 27 April 2020; Published: 4 May 2020

Abstract: Desktop and portable platform-based information systems become the most tempting target of crypto and locker ransomware attacks during the last decades. Hence, researchers have developed anti-ransomware tools to assist the Windows platform at thwarting ransomware attacks, protecting the information, preserving the users' privacy, and securing the inter-related information systems through the Internet. Furthermore, they utilized machine learning to devote useful anti-ransomware tools that detect sophisticated versions. However, such anti-ransomware tools remain sub-optimal in efficacy, partial to analyzing ransomware traits, inactive to learn significant and imbalanced data streams, limited to attributing the versions' ancestor families, and indecisive about fusing the multi-descent versions. In this paper, we propose a hybrid machine learner model, which is a multi-tiered streaming analytics model that classifies various ransomware versions of 14 families by learning 24 static and dynamic traits. The proposed model classifies ransomware versions to their ancestor families numerally and fuses those of multi-descent families statistically. Thus, it classifies ransomware versions among 40K corpora of ransomware, malware, and good-ware versions through both semi-realistic and realistic environments. The supremacy of this ransomware streaming analytics model among competitive anti-ransomware technologies is proven experimentally and justified critically with the average of 97% classification accuracy, 2.4% mistake rate, and 0.34% miss rate under comparative and realistic test.

Keywords: crypto-ransomware; locker-ransomware; static analysis; dynamic analysis; machine learning

---

1. Introduction

Motivated by fame and illegal profit, cyber-criminals have threatened users' privacy and information systems by ransomware attacks [1]. Thus, different anti-ransomware tools and anti-malware software have been developed to detect ransomware attacks of various ransomware families on desktop and portable platforms [2,3]. Along with them; are the anti-ransomware tools assisted by machine learning that learn ransomware data with a set of static and/or dynamic traits to examine ransomware attacks at their runtime successfully [3,4]. Although they perform better than their competitors, machine learning-based anti-ransomware tools still suffer from late ransomware tackle, somewhat incorrect categorization of a ransomware family among other malware families, variable performance outcomes versus various ransomware families, complex computations, and longtime reaction with heavy use of CPU and memory [4,5]. Accordingly, they still provide a chance of evasion to the cyber-criminals who advance their exploitations to evolve 0-Day versions of new ransomware families every day [6]. Then, more users' data loss, systems' data leakage, and users' money loss would be produced along with other tragic concerns to cyber-security [4–6]. Since data protection, systems' defense, and cyber-space survival is the superior aim of researchers in cyber-security [7]. A more proficient scheme to detect ransomware attacks in their runtime is required to overcome the previous issues of the existing anti-ransomware tools. The required solution should be efficacious with less performance overhead, and adaptive to operate on desktop and portable platforms of servers, PCs, tablets, and smartphones. Furthermore, it should identify the ransomware version among the generic versions of malware and good-ware apps as well as categorize its corresponding family decisively. In addition, it should analyze both *crypto* and *locker* ransomware families automatically to extract their static and dynamic traits that discriminating against them from other malicious families and good-ware apps with light use of CPU and memory.

To this end, we propose a ransomware streaming analytics model by integrating a compact set of 24 static and dynamic traits, a hybrid machine learner, a numeral measurement for ransomware's ancestor family attribution, and a statistic formula for a multi-descent ransomware version via a multi-tiered architecture. The proposed machine learner trains a set of 24 rich traits to characterize 14 ransomware families in a semi-realistic environment. Correspondingly, it identifies the ancestor family as well as the multi-descent ransomware versions among more than 40K of various ransomware, malware, and good-ware versions decisively. Overall, this is done through multiple and synchronous tiers, including ransomware characterization and family attribution tiers, multi-descent ransomware fusion tiers, and then ransomware classification tier. To affirm the efficacy and the supremacy of our proposed approach, an extensive study, test, and benchmarking are conducted across other machine learners and anti-ransomware tools that have been recently adopted in the anti-ransomware domain. Also, a critical qualification is achieved to distinguish and justify their limitations in terms of the type of machine learning algorithms, the employed traits of ransomware, the family attribution, the robustness against big datasets, and the performance outcomes. Hence, the addressed limitations demonstrate what is overlooked, which our proposed model boosts for the best detection performance? Precisely, this paper makes a six-fold contribution as follows:

- Revisiting state-of-the-art anti-ransomware technologies, particularly those assisted by machine learners to highlight the issues that they have overlooked in ransomware detection.
- Deploying an informative compact set of static and dynamic traits to holistically characterize the ransomware versions of both crypto and locker families.
- Enhancing numeral and statistic metrics to attribute a ransomware version to its ancestor family, and to fuse ransomware versions of multi- descent families, respectively.
- Proposing a machine learner that unearths the 0-Day versions of ransomware into a generalized class model.
- Designing a multi-tier streaming analytics model to implement in a realistic environment and operate on manifold platforms by integrating the compact set of traits, the numeral and statistic metrics, and the proposed machine learner.
- Significance of our proposals is manifested through testing, evaluating, and benchmarking on a big dataset consisting of 35,000 ransomware versions of 14 families, 500 versions of 10 malware, and 500 good-ware apps aggregated at a different time from different data archives.

For the aforesaid contribution, the remaining of this paper is organized as follows: Section 2 describes the ransomware life cycle, types, activities, and families. Section 3 revisits the state-of-the-art anti-ransomware tools and appraises them critically. In Section 4, the materials and methodology are elaborated, including the proposed machine learner, numeral, and statistic formulas. Results of the test, evaluation, and benchmarking are presented in Section 5. Then, Section 6 discusses the observations and addresses the pivot issues of the proposed model and machine learner. Finally, Section 7 concludes the overall facets investigated in this paper, along with future outlook.

2. Ransomware Versions and Ransomware Families

Even though, cyberspace provides useful communication media to the users with many services of e-commerce and e-government [1,2], it puts them at the risk of ransomware attacks that cause significant damage to their interconnected information systems and then money loss [2–4]. The risk of ransomware often occurs when the cyber-criminals exploit e-commerce services and e-government applications to access the victims' information systems fully and to gain a ransom from those victims [8]. Furthermore, the advanced electronic payment methods and electronic currencies like bitcoin and de facto payments enable the cyber-criminals to deploy their social engineering technologies to deceive the victims [8,9]. Usually, a ransomware attack is a malicious variant aiming at either locking the victim's information system or encrypting that system and its user files for a ransom acquirement to revive the system and regain the files access [1,4,10]. As we illustrate in Figure 1, cyber-criminals pursue a "recipe-to-success" strategy to deceive the users, intruding their information system on different platforms like Windows and Androids [4,10]. Through the "recipe-to-success" strategy, the victim catches the bait via an email attachment or spoofed link that is under the cyber-criminal's control. Then, the ransomware is disseminated to the victim's information system by exploiting that system's vulnerabilities to infect that system and encrypt particular system files and/or lock particular system's locations. Accordingly, the victim's access to his/her system or files is blocked, and the ransomware attack acquires a ransom from the victim to either decrypt the files or unlock the system [4,10].

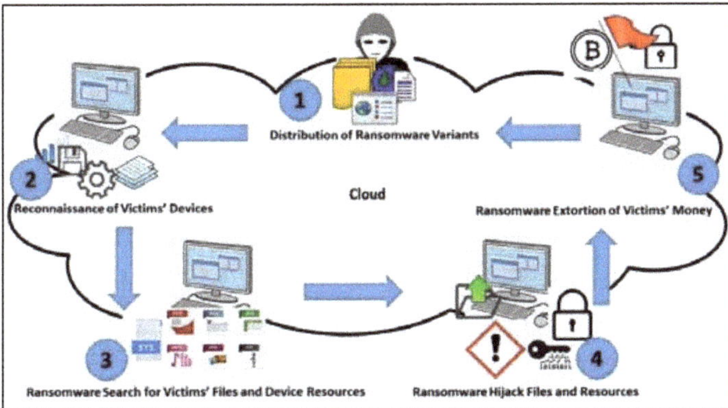

Figure 1. Ransomware routine.

Continually, cyber-criminals evolve ransomware families of either *crypto* or *locker* type, as described in Table 1. *Crypto*-ransomware leverages symmetric and asymmetric ciphering on the user data and system files, whereas *locker*-ransomware hijacks the hosted system's resources and apps to disable the user's access to them [3,4]. As they intend to bypass the existing system defense' settings to cause more potential damage and gain more profit; cyber-criminals used to create many patterns (i.e., versions) belonging to every evolving family [2,3,7]. Generally, ransomware versions run sophisticated intrusion actions and employ advanced exploits that may infect different or similar platforms [8,9]. Additionally, they act similarly to those of other malicious threats and/or they maneuver to those of good-ware apps [8–10]; for example, the ransomware families that are described in Table 1. Hence, ransomware versions and families require different analytics mechanisms assisted by different compact sets of traits to recognize them, among other malware and good-ware apps accurately [4,10].

Table 1. Time-line summary of ransomware families.

Family	Year	Type	Exploits and Actions	Damage (s)
AiDS [10]	1989	Crypto	It was delivered to computer-based information systems via floppy disks	• Leakage of Root Directories • Loss of System Files
GpCode [10,11]	2005	Crypto	It was developed with an asymmetric encryption algorithm to encrypt users' data files	• Holding Up the Banking Information Systems • Loss of System and User Files
Archiveus [10,11]	2006	Crypto	It was applied with the RSA algorithm to encrypt system files	• Ruining the Original Version of Windows Platform • Data Loss • Money Loss
WinLock [10,11]	2010	Locker	It locked the computer system and demanded ransom via sending SMS to the victim's phone number	• Holding Up The Operating System • Loss of Backup Data • Loss of Money
Reveton [4,10,11]	2012	Locker	It impersonated the law enforcement agencies to deceive users with rumor claims	• Holding Up The Operating System • Abusing the Prepaid Electronic Payment Platforms • Loss of Backup Data
Crypto-Locker [4,10,11]	2013	Crypto	It encrypted the file's contents by RSA algorithm with private and public keys	• The Halt of Targeting System • Loss of Backup Data • Loss of Money
Crypto-Wall [4,10,11]	2014	Crypto	It encrypted the system files and injected malicious codes which freezes the system's firewalls	• Leakage of Original System Files • Loss of User Files • Loss of Money in Bitcoins
Ransom as Service (RaaS) [6,10]	2015	Locker	It used the social engineering techniques to impersonate a good-ware website as a malicious website in the dark web	• The Halt of Targeting Systems • Loss of Backup Data • Loss of Money in Bitcoins
Cerber [10,11]	2016	Crypto	It injected the malicious instructions to overwrite an encrypted content onto the original system	• Loss of Original System Files • Deactivation of the system registry
Crysis [10]	2016	Crypto	It encrypted the system files, rewrote their contents, and renamed them	• Leakage of System Files • Loss of User Data • Loss of Backup Data
Locky [4,10,11]	2016	Locker	It used the social engineering techniques to intrude the system through vulnerabilities of system settings, deactivated the registry actions, and removed the backup data	• The Halt of the Targeting System • Loss of Backup Data • Money Loss
WannaCry [4,10]	2017	Crypto	It encrypted the contents of the system files and removed the original system files	• Data Loss • Holding Up the Operating System • Money Loss in Bitcoins
Sopra [12,13]	2017	Crypto	It encrypted the contents of the system files and removed the original system files	• Data Loss • Holding Up the Operating System • Money loss in Bitcoins
Zeus [12,13]	2018	Crypto	It encrypted the contents of systems files and created new files with extensions belonging to different ransomware versions	• The Halt of the Targeting System of Industrial Organization • Backup Data Loss • Money Loss

3. The State-of-The-Art of Anti-Ransomware Technology

To thwart ransomware families and their corresponding versions, researchers have developed various anti-ransomware technologies during the last decades, and they categorized them into misuse-based technology, anomaly-based technology, and machine learning-based technology [6,14–16], as described briefly in Table 2. For a clear description, we illustrate the categories above, along with their advantages, are presented in Figure 2.

Table 2. Brief description of the state-of-the-art anti-ransomware technology.

Table	Description	Demerits	Examples
Misuse-Based Technology [6,14–16]	They analyze ransomware versions to extract cryptographic primitives, suspicious scripts, built-in functions, infected files' paths, and extensions. They achieved moderate accuracy in lightweight performance	-User's data can be lost -They can be obfuscated -They can be defeated against 0-Day ransomware versions -A lot of false alarms -The database must be updated frequently	Bitdefender Kaspersky McAfee Avast
Anomaly-Based Technology [17–23]	They trace ransomware runtime activities, computer processes, CU and memory footprints, server actions and control to detect ransomware versions effectively and efficiently	-The infirm vs. scalable network traffic -They can be defeated vs. 0-Day ransomware versions -Fragile analysis vs. big and imbalanced dataset	R-Locker RansomFlare Poster UNIVEIL Talos
Machine Learning-Based Technology [11,24–32]	They apply machine learning algorithms to classify training and testing sets of ransomware and good-ware instances with a hybrid set of static and dynamic traits. Then, generated class models characterize generic and/or unseen ransomware versions with high detection accuracy, low false alarms, and misclassifications	-They can be evaded by adversarial ransomware traits and newly emerged ransomware families -Simi-real testbed and real-mode conditions are needed -Obsolescent detection vs. various ransomware families	EldeRan ShieldFS 2entFOX GAN NetConverse RansomWall RANDS DRTHIS

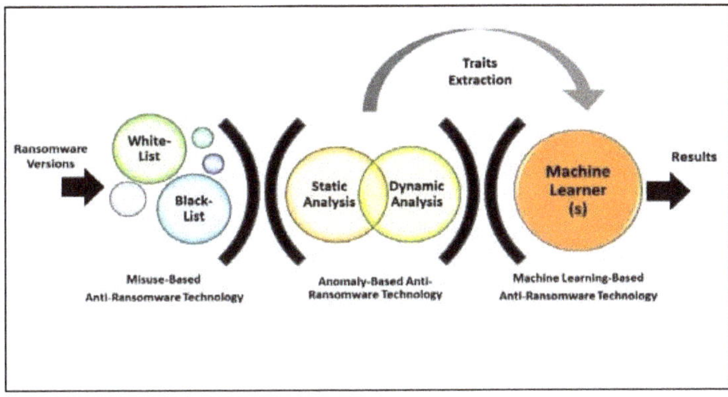

Figure 2. Essentials and appendages of the state-of-the-art anti-ransomware technologies.

The misuse-based anti-ransomware technology relies on either archiving the exploits and actions of ransomware families in a black-list or archiving exploits of good-ware apps in a white-list [6,14–16]. However, it can detect the generic versions of the prevalent ransomware families exclusively as if the built-in archives are not frequently updated by the data of 0-Day versions [6,16]. Furthermore, it consumes a longer time and more computer resources as well as human labor to trace ransomware exploits and actions [6,10,14]. Unlikely, the anomaly-based anti-ransomware technology can analyze ransomware' normal behaviors and generic processes statically and/or dynamically to deviate 0-Day versions [4,6] (see Figure 2). Although anomaly-based anti-ransomware technology outperforms the misuse-based technology against 0-Day versions, it is still bypassed by versions of more advanced crypto coding families [6,10,14].

As described in Table 2, machine learning-based anti-ransomware technology applies various machine learners to cope with the problems of the former anti-ransomware technologies at discriminating 0-Day versions of different families expertly [6,24]. Furthermore, the constructed machine learners take the advantages of static and/or dynamic analysis of ransomware traits to extract the vectors of traits and learn them with a compact set of traits [25,26]. Precisely, they rely on different decisive functions, induction parameters, design, and class attributes to promote their discriminating power; for example, naïve bayes (NB), support vector machine (SVM), decision tree (DT), logic regression (LR), bayesian network (BN), neural networks (k-NN), and random forest (RF), etc. as well as hybrid machine learners like (HMLC) that hybridizes multiple and complementary machine learners as presented in Table 3. Therefore, machine learning-based anti-ransomware technologies outperform their competitors [2–4]. However, they still fall short at characterizing ransomware families holistically, attributing ransomware versions to their ancestor families, fusing ransomware versions of multi-descents, learning big data stream, and running on manifold platforms [24], as presented in Table 3. Beyond the above, the next sections will emulate the milestones of our proposed solutions, justify its sufficiency versus the lack of state-of-the-art anti-ransomware technology, qualify its performance, and enumerate its future outlook.

Table 3. Prominent anti-ransomware tools assisted by machine learners.

Existing Tools	Method	Machine Learning Algorithm	Lacks
EldeRan [11]	It detected the updates of API calls, registry key and file system operations via a sandbox versus locker-ransomware families	DT	• Exclusive for locker ransomware • Heavyweight in use • Time-sensitive • High detection faults • Unaware of new ransomware families • Unaware of family attribution • Unaware of multiple datasets • Unaware of manifold platforms
ShieldFS [30]	It detected some crypto-ransomware families that exploit generic ransomware traits like I/O and low-level file system infections, encrypting file contents, and overwriting the original contents by monitoring file system activity and then updating the threat profile overtime on a realistic environment.	DT	• Exclusive for crypto-ransomware • Heavyweight in use • Time-sensitive • High detection faults • Unaware of new ransomware families • Unaware of multiple datasets • Unaware of manifold platforms • Unaware of family attribution
Net-Converse [26]	It used dynamic traits and six machine learning algorithms to detect 0-Day versions of locker-ransomware versions	LR, DT, BN	• Exclusive for locker-ransomware • Performance overhead • Unaware of new ransomware families • Unaware of runtime condition • Unaware of static analysis • Unaware of family attribution • Unaware of multiple datasets • Unaware of manifold platforms
2entFOX [31]	It detected static and dynamic traits of crypto-ransomware by using graph traversal network	BN	• Exclusive for crypto-ransomware • Performance overhead • Heavyweight in use • Unaware of new ransomware families • Unaware of family attribution • Unaware of manifold platforms
GAN [32]	It develops a generative adversarial network for detecting versions of locker-ransomware families with a generic set of dynamic traits	NN, NB, RF, SVM	• Exclusive for locker-ransomware • Performance overhead • Heavyweight in use • Unaware of multi-class decision • Unaware of new ransomware families • Unaware of runtime condition • Unaware of multiple datasets • Unaware of static analysis • Unaware of manifold platforms
Ransom-Wall [33]	A multi-layered tool detects 0-Day versions of crypto-ransomware families by developing a generalized model comprised of static and dynamic traits	LR, SVM, NN, RF	• Exclusive for crypto-ransomware • Time-sensitive • High detection faults • Unaware of multi-class decision • Unaware of family attribution • Unaware of multiple datasets

Table 3. *Cont.*

Existing Tools	Method	Machine Learning Algorithm	Lacks
RANDS [34,35]	A hybrid machine learning-based anti-ransomware tool that detects 0-Day versions of both crypto and locker ransomware families by using dynamic traits on Windows platform	HMLC	• Unaware of static analysis • Unaware of family attribution • Unaware of manifold platforms • Unaware of multiple datasets
DRTHIS [36]	A three-fold machine learner that identifies crypto and locker ransomware versions among malware and good-ware versions using dynamic traits.	NN	• Unaware of static analysis • Unaware of family attribution • Unaware of advanced and new ransomware versions • Time-sensitive • Unaware of manifold platforms

4. Materials and Methods

This section describes the design and emulates the milestones of the proposed ransomware streaming analytics model to justify its sufficiency versus the lacks of the state-of-the-art anti-ransomware technology in terms of the compactness of traits, hybrid machine learner, and decision margins for ransomware characterization and family attribution.

4.1. Ransomware Characterization

Like malware families, ransomware families expose static traits such as digital signatures, built-in scripts, fuzzy functions, and hashes. They exhibit the ten static traits described in Table 4 to intrude the targeting system, infect its firewalls, and disable its restore settings. Also, they utilize static traits to encrypt data, APIs, files' content, and files' paths as well as spoofing the links to particular directories. The ten static traits presented in Table 4, are usually exploited in the versions of 14 ransomware families that this work deploys to implement and testify the proposed model. In the extraction tier that of a semi-realistic environment, the existence of encrypts, packers, hashes, and suspicious scripts are traced to inspect. Furthermore, any infections like altered filenames and directories, file system locations, bootstraps, and registry keys that might be done by a ransomware version are tracked through created trap files. Whereas, the 14 dynamic actions those leveraged by the versions of the 14 ransomware families (as described in Table 4) are traced to state their exploits through the accessing queries of files and directories, read/write/delete operations, edit the system's digital certification, modify system files' headers as well as the entropies of buffering data. Then, the raw data of all traced traits and actions of an examined version are formulated into an input vector of traits along with the class of that version as either ransomware R or non-ransomware R' to be computationally readable by the machine learner. The values of the trait vector are normalized into "0" s and "1" s according to the existence of their traits. The class label R in the trait vector is represented by "1" and the class label R' is represented by "−1". In contrast, any suspicious version that does not belong to R and R' (an imperviously examined version that might be malicious) is remarked as "0".

4.2. Ancestor Family Attribution and Multi-Descent Fusion

Consequently, the attribute of the ransomware family that the examined version might belong to is assumedly represented by two digits ranging from "01" to "99" as the header of the version's trait vector. Since 14 ransomware families are investigated for this work; therefore, the headers would range from "01" to "14". Any additionally adopted ransomware families will be assigned with the remaining digits of the assumed range in the future work. It is noteworthy to mention that the assumption "00" is assigned to the headers of all good-ware trait vectors, and "99" is assumedly assigned to the headers of all malware apps in the dataset. Then, a lookup table of the assigned headers is created as a tracing file for the numeral computation of *Attribution Rate* $(AR_{t_{i,j}})$ of every trait in a trait vector across other trait vectors involved in ransomware vectors and non-ransomware vectors, as in Equation (1). By product, the header data *"h"* of all characterized trait vectors are exploited to identify their relevance to a particular ransomware family among other families.

$$AR_{t_{i,j}} = \sum_{j=1}^{n} \frac{h_{t_{i,j}\to R} - h_{t_{i,j}\to R^\sim}}{h_{t_{i,j}\to R} + h_{t_{i,j}\to R^\sim}} \qquad (1)$$

where R refers to the ransomware trait vectors and $R^\sim$ refers to all non-ransomware trait vectors in the batch of the dataset. Then, $AR_{t_{i,j}}$ is the frequency of each trait t_j belonging to a trait vector $t_{i,j}$ across all vectors encompassed in R and $R^\sim$. Unlikely, $h_{t_{i,j}\to R}$ is the frequency of that trait $t_{i,j}$ with respect to the headers of trait vectors in R. However, $h_{t_{i,j}\to R^\sim}$ is the frequency of the same trait $t_{i,j}$ with respect to all headers of trait vectors in $R^\sim$.

Table 4. Potential traits exploited by 0-Day versions of generic ransomware families.

Traits	Type	Crypto-Wall	Win-Lock	Reventon	Crypto-Locker	Archiveus	GpCode	AiDS	RaaS	Cerber	Locky	Crysis	Wanna-Cry	Sopra	Zeus
Windows API calls	Dynamic	✗	✓	✓	✗	✗	✗	✗	✗	✗	✓	✗	✗	✗	✗
Windows Cryptographic APIs	Dynamic	✓	✗	✓	✓	✓	✓	✗	✗	✓	✗	✓	✗	✓	✗
Registry Key	Dynamic	✓	✗	✓	✓	✗	✗	✗	✗	✗	✓	✗	✗	✗	✗
System File Process	Dynamic	✓	✗	✗	✓	✗	✗	✗	✓	✓	✓	✓	✗	✗	✗
Directory Actions	Dynamic	✓	✓	✗	✓	✗	✗	✗	✗	✗	✗	✓	✓	✓	✓
Application Folders	Dynamic	✓	✓	✗	✓	✗	✗	✗	✗	✗	✓	✓	✓	✓	✓
Control Panel Settings	Dynamic	✓	✗	✓	✓	✗	✗	✗	✗	✗	✓	✗	✓	✗	✓
System File Locations	Dynamic	✓	✗	✗	✓	✗	✗	✗	✓	✓	✓	✓	✗	✗	✗
Pay-loaders/Downloaders	Dynamic	✓	✗	✓	✓	✗	✗	✗	✗	✗	✓	✗	✓	✗	✓
Command and Control Server	Dynamic	✓	✓	✓	✓	✗	✗	✗	✓	✗	✓	✗	✓	✗	✓
Windows Volume Shadow (vssadmin.exe and WMIC.exe)	Dynamic	✗	✓	✓	✗	✗	✗	✗	✓	✗	✓	✗	✓	✗	✓
File Fingerprint	Dynamic	✓	✗	✗	✓	✗	✓	✗	✗	✓	✗	✓	✗	✗	✗
Directory Listing Queries	Dynamic	✓	✗	✗	✓	✗	✗	✗	✓	✓	✓	✓	✗	✗	✗
Windows Safe Mode Booting (bcdedit.exe)	Dynamic	✗	✓	✓	✗	✗	✗	✓	✓	✗	✓	✗	✓	✓	✓
File Extensions	Static	✓	✓	✓	✓	✓	✓	✗	✗	✓	✗	✓	✓	✓	✓
Files Names	Static	✗	✗	✗	✗	✓	✗	✓	✗	✓	✗	✓	✓	✓	✓
Portable Executable Header	Static	✗	✓	✓	✗	✗	✗	✓	✓	✗	✓	✗	✓	✓	✓
Embedded Resources	Static	✗	✓	✓	✗	✗	✗	✓	✓	✗	✓	✗	✓	✓	✓
Packers	Static	✗	✓	✓	✗	✗	✗	✓	✓	✗	✓	✗	✓	✓	✓
Shannon's Entropy	Static	✓	✗	✓	✓	✓	✓	✗	✗	✓	✗	✗	✗	✗	✗
Cryptors	Static	✓	✗	✓	✓	✓	✓	✗	✗	✓	✗	✗	✗	✗	✗
Portable Executable Signature	Static	✗	✓	✓	✗	✗	✗	✓	✓	✗	✓	✗	✓	✓	✓
Embedded Scripts	Static	✗	✓	✓	✗	✗	✗	✓	✓	✗	✓	✗	✓	✓	✓
Fuzzy Hashing	Static	✓	✗	✓	✓	✓	✓	✗	✗	✓	✗	✗	✓	✗	✗

On the other hand, the header data *"h"* of all characterized trait vector *"T_j"* would be checked-up across all extracted trait vectors from the batch of the dataset to categorize its mutual multi-descent to the other families of ransomware among malware and good-ware families as per Equations (2) and (3). This statistic probability is named *Multi − Descent Ratio* ($MDR(T_i)$), which prioritizes the ransomware versions that may belong to multi-descent families of ransomware with respect to their relative redundancy in malware and good-ware families on the learned trait vectors (T).

$$P_r(T_i \mid h_i) = \frac{N_{h_i \to R}}{N_{h_i \to R} + N_{h_i \to R^{\sim}}} \tag{2}$$

$$MDR(T_i) = \frac{\sum_{i=1}^{|K|} \Pr(T_i \mid h_i)}{|T_i|} \tag{3}$$

4.3. Ransomware Classification

As it is elaborated in Algorithm 1, two dominant machine learning algorithms, DT and NB, are synchronously hybridized in HML to optimize the adaptive categorization at the moment of tackling 0-Day versions of ransomware, malware, and good-ware. Since DT and NB are complements in their decisive functions and pruning margins, they are therefore hybridized to classify ransomware more accurately [28]. Conceptually, DT carries out a fast classification versus big training data throughout the tree structure such that the predictive classes of the input trait vectors can be arranged as the antecedent nodes, and their traits can be set as leaves of the tree. However, it might be ineffective in predicting the class of imperviously seen and relevant traits [27–29]. On the other hand, NB executes fast training of data; however, it is impractical against a big set of traits and heterogeneous trait values. However, it spends a short computation time in learning the training vectors of traits and predicting their actual classes by using Bayes' probabilistic theorem with the assumption that all the examined traits are independent of each other [27–29]. Thus, NB is applied by HML to trace the predictive classes of all overlooked traits in the indecipherable nodes of DT that optimizes the adaptive classification.

To do so, HML trains the fetching batch of trait vectors through cutting the decision edges of DT with NB pruning margins in an iterative splitting of the training trait vectors into sub-training vectors. Thus, the training trait matrix ($T = \{T_1, \ldots, T_K\}$) is given such that ($T_i = \{T_{i,j}\}_{i \in K, \ j \in |T_i|}$) with the predictive labels ($P_{class} = \{C_1, C_2\} : C_1 = 1, \ and \ C_2 = -1$). Each trait vector can be represented as ($T_i = \{C_m, T_{i, j}\}_{i \in |T_K|, \ m \in |C_M|}$. Then, the prior class probability $P(C_m)$ is computed as per Equation (4) to predict how often each class occurs over (T) relatively to the trait vector (T_i); whilst, the conditional probability of (T_i) is computed by Equation (5) to predict the relevance between the predictive class (C_m) and its corresponding trait ($T_{i, j}$) as it was indicated by ($P(T_{i,j}|C_m)$).

$$P(T_iC_m) = P(C_m) \prod_{e=1 \to p} (T_{i,j}|C_m) \tag{4}$$

$$C_m = C_i \to P_{ms}(T_i, C_m) \tag{5}$$

4.4. Structure of the Ransomware Multi-Tier Streaming Analytics Model

Figure 3 illustrates the structure of the proposed ransomware streaming analytics model. It consists of a trait extraction tier, an ancestor-family attribution tier, a multi-descent fusion tier, and a learning tier. The extraction tier works on the semi-realistic environment to analyze datasets (i.e., versions of ransomware, malware, and good-ware) into class-labeled trait space (i.e., a set of trait vectors with their class label). The semi-realistic environment is a virtual testbed used to create trap files (s) with realistic conditions, and it disseminates them into particular system directories. Once the trap (s) are exploited by ransomware, that ransomware runs its malfunctions and downloads its *crypto*-or *locker*-gadgets to either encrypt system data or halt the system. Correspondingly, the extracted trait vectors are attributed to their ancestor ransomware/non-ransomware families in the family attribution tier by normalizing the header data of each trait vector with respect to its family. Given that they assigned to their family attributes, all trait vectors are examined versus the case of a multi-descent family in the multi-descent fusion tier.

Algorithm 1

Definition of Semantic Codes

Let

S the stream of ransomware and non-ransomware versions such that $S = \{S_m\}_{m \in |S|}$

T the compact set of traits

T_{space} the generated space of extracted traits where $T_{space} = \{T_i\}_{i \in |T|}$

T_i a trait vector included in T_{space}

$TreeNode$ the decision tree of T_{space}

T_{sub} the splitting space of T_{space} such that $T_{space} = \{T_{sub}\}_{sub \in |sub|}$

C_m the class model of T_i, where $C = \{C_m\}_{m \in M}$, and M is the number of predictive classes

R the ransomware trait vectors (i.e., ransomware versions)

$R^\sim$ are the non-ransomware trait vectors (i.e., non-ransomware versions)

F the trace file

$AR_{t_{i,j}}$ the Attribution Rate of a trait vector

$MDR(T_i)$ the multi-descent ratio of a trait vector

Input: S and T

Output: R and R'

Begin

1. Generate T_{space} from S with headers
2. Repeat (3) to (c)
   a. Create $TreeNode$
3.   b. IF (all $\{T_i\}_{i \in |T|}$ in T_{space} have similar class C_m) THEN $TreeNode \leftarrow LeafNode$
   c. Until $T = \{\}$ THEN attach $TreeNode$ to the majority class model C_m
4. For each T_i in T_{space}
   a. find prior probability C_j over T_{space} by Equation (4)
   b. Find the conditional probability of $t_{i, j}$ about C_j over T_{space} by Equation (5)
   c. Update T_i in T_{space} with the maximal $P(t_{i,j}|C_m)$ such that $P(C_{jm}|t_{i,j})$; $C_m \rightarrow P_{ml}(C_m|t_{i,j})$
   d. Partition T_{space} into $T_{space} = \{T_{sub}\}_{sub \in |T|}$ and $T_{space} \leftarrow \{T_{sub}\}_{sub \in |T|}$
5. Repeat (6) Until $(T_{space} \neq \{\})$ AND $(T \neq \{\})$
6. Keep all computed probabilities in R and $R^\sim$ for classification decision

   For each T_i in R and $R^\sim$
   a. find $AR_{t_{i,j}}$
7.   b. find $MDR(T_i)$
   c. keep $AR_{t_{i,j}}$ and $MDR(T_i)$ in the trace file F_i

End

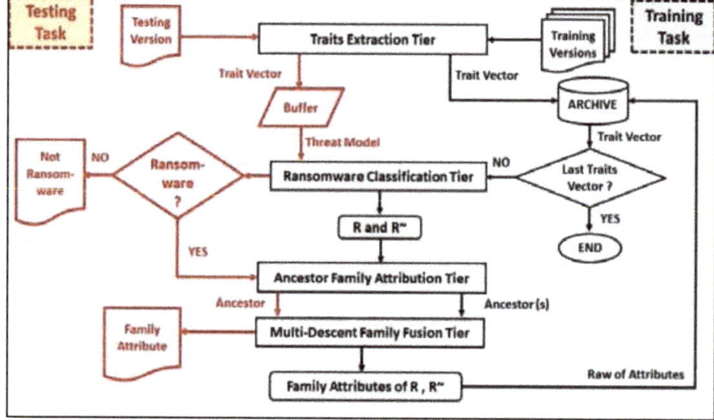

Figure 3. Structure of multi-tier ransomware streaming analytics model.

The diagnoses of ancestor family and multi-descent families that are pursued by family attribution and multi-descent fusion tiers are implemented synchronously along with learning done by the hybrid machine learner in the learning tier. So far, the proposed ransomware streaming analytics model carries out the aforesaid multi-tiers in a multi-disciplinary manner across data, trait space, header data, and predictive classes during both the training and testing tasks, as shown in Figure 3. Thus, the input (unknown) ransomware version is analyzed on its runtime during the testing task of the proposed multi-tiered streaming analytics architecture as the same as the data of the training corpus but in a different order. The training task can be traced by the black path, whereas the testing task is tracked throughout the red path, as it can be seen in Figure 3.

5. Experiments and Results

This section elaborates the experimental workflow that is conducted for the purpose of performance assessment along with the description of data corpora and evaluation metrics as well as the discussion of obtained results. As illustrated in Figure 4, the experimental workflow is conducted through several tasks, including a collection of data, construction of data into training corpus and testing corpus, implementation of the benchmarking machine learners as well as the proposed machine learner (HML). In addition, the experimental workflow pursues the implementation task of the benchmarking anti-ransomware tools and the proposed ransomware multi-tiered streaming analytics model. Then, the evaluation task is carried out to qualify the overall experimental results and to justify the asserted findings.

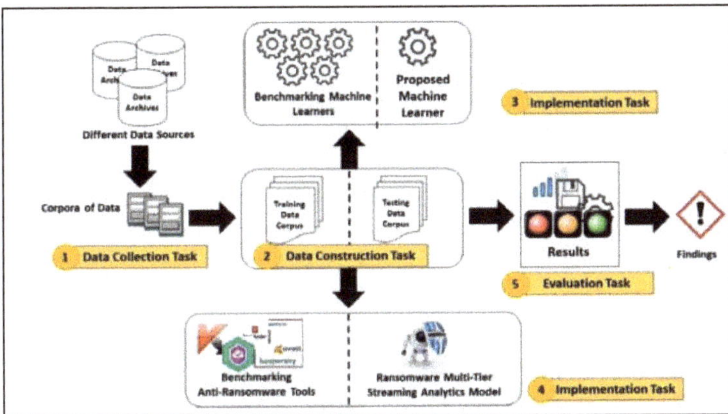

Figure 4. Experimental workflow.

5.1. Runtime Test Routine

Figure 5 shows how the proposed multi-tier streaming analytics model runs its run-time test routine to detect a ransomware version in a realistic environment. In Figure 5, a suspicious version downloads itself onto the targeting computer system when the user browses the web. To do so, the suspicious attack uses many toolkits for this purpose, such as Trojans, spoofing links, and downloading software and apps. Meanwhile, the suspicious version tackles the trap files created by the proposed model during the extraction tier. It activates its static and dynamic traits to infect the trap files. Then, a scan disk is implemented to check-up the actions and infections on the computer system. Whenever infections are observed, they are analyzed statically and characterized dynamically. Accordingly, the ancestor families are attributed, and multi-descent ransomware versions are fused to be ready for further learning by the hybrid machine learner. Consequently, the class label of the fetched version is predicted throughout the learning tier. By tracking and analyzing the observed actions, the proposed model attempts to determine to what ancestor family this version does belong? Thus, it affirms that the suspicious version is either ransomware, or malware, or good-ware. Finally, it either warns the user by a popped up message or; it pops-up a safe acknowledgment to the user's screen.

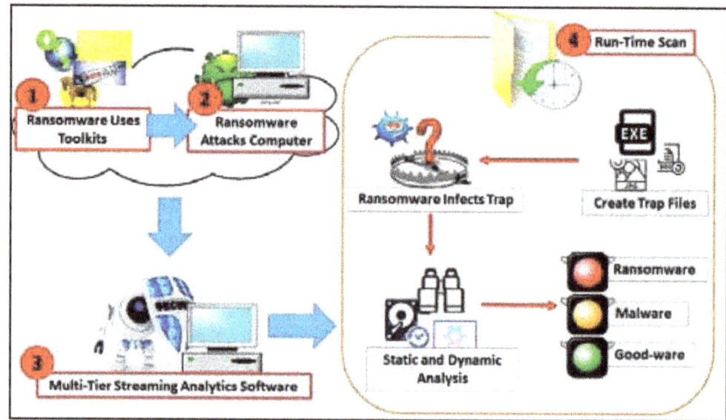

Figure 5. The run-time test routine of the multi-tier streaming analytics model.

5.2. Data Collection and Evaluation Metrics

By searching publically used archives like the Virus Total Intelligence Platform and Virus Share, the executable and portable ransomware and malware versions are aggregated to construct data corpora. Exclusively, our data aggregation targets those versions that are submitted at least three times to the aforesaid data archives between 1 January 2019 and 1 September 2019, and they are still undetected by any existing antivirus software and/or anti-ransomware tools. Similarly, the good-ware instances are aggregated and merged randomly to the aforesaid data corpora based on particular heuristics of benign software apps. As described in Table 5, the aggregated data corpora are divided randomly into $\frac{2}{3}^{nd}$ and $\frac{1}{3}^{rd}$ as training data corpus and testing data corpus to be used in training and testing tasks, respectively. On the other hand, the rates of true positive (TPR), false positive (FPR), false negative (FNR), and classification accuracy as well as mistake rate, miss rate, and elapsed time; all are used as standard performance evaluation metrics for the evaluation task in recently published works [11,27–29,34,35]. TPR, FPR, and FNR are derived from the confusion matrix calculations as follows:

$$TPR = \frac{TP}{TP + FN} \tag{6}$$

$$FPR = \frac{FP}{TN + FP} \tag{7}$$

$$FNR = \frac{FN}{TP + FN} \tag{8}$$

where TPR refers to the rate of correctly classified ransomware data, FPR indicates the rate of wrongly classified good-ware data as ransomware, and FNR indicates the rate of wrongly labeled ransomware data as good-ware data, respectively. While TP is the number of good-ware samples classified as ransomware, FN is the number of ransomware samples classified as good-ware, TN is the number of ransomware samples that correctly classified, and FP is the number of good-ware samples that classified as ransomware; respectively.

Based on the above mentioned standard metrics, the classification accuracy rate (ACCR) is calculated to validate the effectiveness of the applied machine learner and/or anti-ransomware tool at detecting valid ransomware (TP) and valid good-ware samples (TN) relatively to the whole data corpora as follows:

$$ACCR = \frac{TP + TN}{TP + FP + TN + FN} \tag{9}$$

Accordingly, mistake rate is computed to qualify the abilities of the comparable anti-ransomware tools as well as the proposed ransomware multi-tier streaming analytics model on how they rationally detect the valid ransomware versions with least false classifications as per Equation (10); whereas, the miss rate is computed to qualify their abilities to rationally detect valid ransomware versions with least misclassification cost as per Equation (11).

$$\text{Mistake Rate} = \frac{FP}{N_G} \tag{10}$$

$$\text{Miss Rate} = \frac{FN}{N_R} \tag{11}$$

where, N_G and N_R are the number of good-ware data and ransomware data, respectively.

In addition, the Elapsed Time plays an important to assert how long the comparable machine learners and anti-ransomware tools spend to execute and respond by examining a batch of data with a nominal cost of computations [35–37]. On the other hand, the proposed HML is assessed against its competitor machine learners in terms of AUC against the up-to-date ransomware, malware, and good-ware data during the real test. AUC calculation is widely used to justify the performance of a machine learning algorithm on experimental data by devoting the scalar value of the receiver operating characteristic curve (ROC), which is a plot of TPR versus FPR [35,37,38]. If the AUC score closes to 0.9, then it signifies an excellent performance in the realistic practice; while score values of 0.8, 0.7, and less could signify good, moderate, and then poor performance [38,39].

Table 5. The corpora of data.

Description		Corpus
Number of Valid Ransomware Versions		35,000
Number of Malware Versions		500
Number of Good-ware Versions		500
Data Archives		[40–42]
Aggregation Time		1/1/2019–1/9/2019
Training Data Corpus		17,332
Testing Data Corpus		17,666
Ransomware Families	AiDS	4000
	GpCode	8000
	Archiveus	1500
	WinLock	3620
	Reveton	2400
	CryptoLocker	1720
	CryptoWall	3250
	RaaS	1300
	Cerber	1535
	Locky	2000
	Crysis	1320
	WannaCry	1300
	Sopra	1570
	Zeus	1500

Thus, overall performance metrics are used during performance evaluation task that involves comparative experiments, and realistic experiments. Such experiments quantify how often and how long the proposed machine learner and the multi-tier ransomware streaming analytics model takes for ransomware detection and ransomware families' categorization versus the dominant machine learners and the benchmarking anti-ransomware tools. Both comparative and real-time experiments are evaluated by using the corpora of training and testing data. It is worthy of mention that simulated machine learners in Weka, as well as soft computing by Python, are used for the conducted experiments.

5.3. Comparative Experiment

Three comparative experiments are conducted to address several problematic issues of machine learners-based anti-ransomware technology like the rich set of traits to classify ransomware, ransomware family attribution, and learn big and imbalanced corpora of data. The first comparative experiment is conducted in between to validate the efficacy of the proposed hybrid machine learner (HML) against state-of-the-art machine learners, including LR, SVM, RF, DT, and NB. On the other hand, the second experiment is conducted to compare the proposed ransomware streaming analytics model versus the most salient signature-based anti-ransomware tools like *BitDefender*, misuse-based anti-ransomware tools like *R-Locker*, and the machine learner-based anti-ransomware tools such as *EldeRan* and *RANDS*. As shown in Figures 6 and 7, HML outperforms significant TPRs, FPRs, and FNRs versus all ransomware versions of the 14 families.

Consequently, HML contributes the proposed ransomware streaming model progressively to classify ransomware versions effectively, among other examined anti-ransomware tools. Plots of Figures 6 and 7 demonstrate how the previous questionable issues can be improved to enrich ransomware classification throughout (i) handling variety, heterogeneity, and quantity of the employed set of traits (static and dynamic traits), (ii) leveraging the commonness among ransomware families, (iii) learning a scalable and variable corpus of data adaptively, and (iv) affecting by different emerging and aggregation time of the input versions. Furthermore, plots of Figures 6d and 7d demonstrate the efficiency of the proposed HML and then the proposed ransomware streaming model within its multi-tiered design in time, and light-weight use of computer resources. The attitudes above could be decisive factors to escalate the detection accuracy and lessen the false detections against ransomware versions.

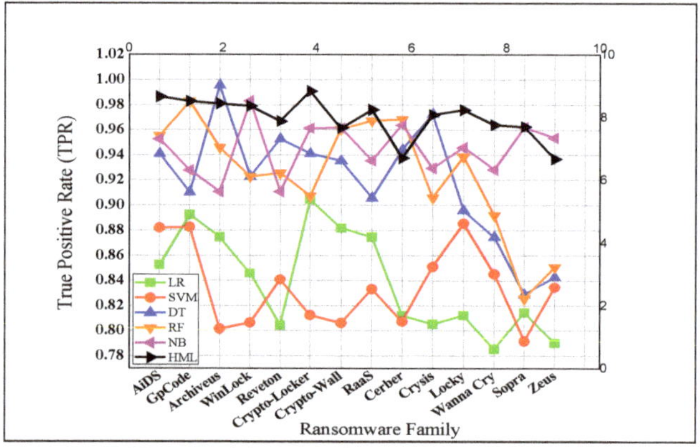

(a) True positive rate (TRR) vs. ransomware families

Figure 6. *Cont.*

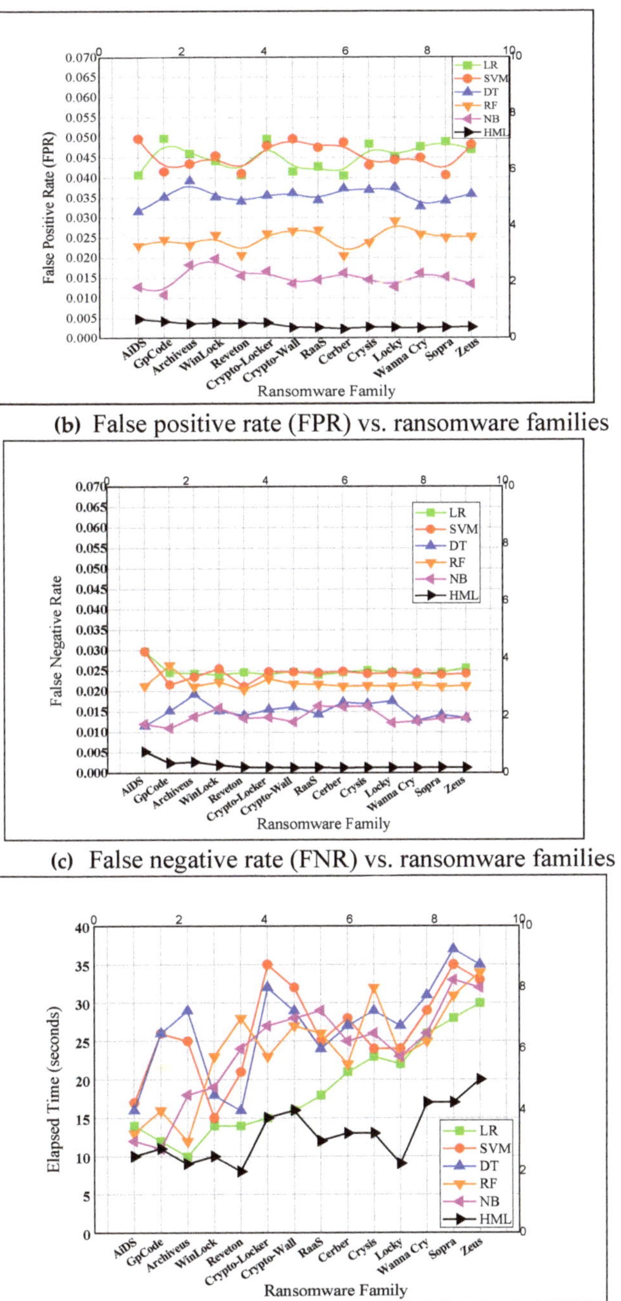

(b) False positive rate (FPR) vs. ransomware families

(c) False negative rate (FNR) vs. ransomware families

(d) Cost of elapsed time vs. ransomware families

Figure 6. Evaluation of the proposed machine learner against state-of-the-art machine learners.

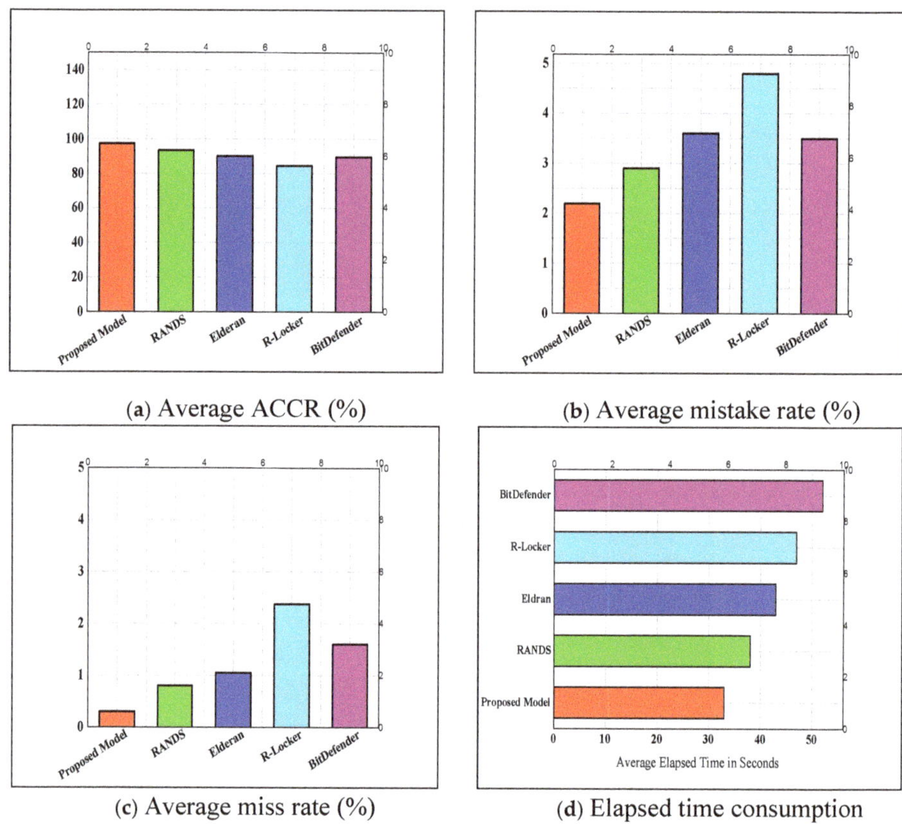

Figure 7. Evaluation of the proposed ransomware streaming analytics model versus different examples of anti-ransomware tools.

So far, we attempt to investigate how do state-of-the-art machine learners, as well as the proposed HML, can leverage the issues above in the realistic environment against 0-Day ransomware versions. Such investigation points out the effects of real-life data corpus that might contain many different and daily emerged versions of cyber-attacks, including ransomware, malware, and good-ware. Almost cyber-attacks are of a multi-descent family; for example, malicious and ransomware attacks. The existence of cyber-attacks causes an abundance of attacks' versions, imbalance in those versions' classes, variety in versions' ancestor families, and versions' multi-descent families as well as their commonness in their latent traits and dynamic behaviors that are suggested by this work. Thus, a third comparative experiment is conducted in the realistic mode during one month (specifically from 1 November to 30 November 2019), and its outcomes are evaluated by the scores of AUC as they are plotted in Figure 8.

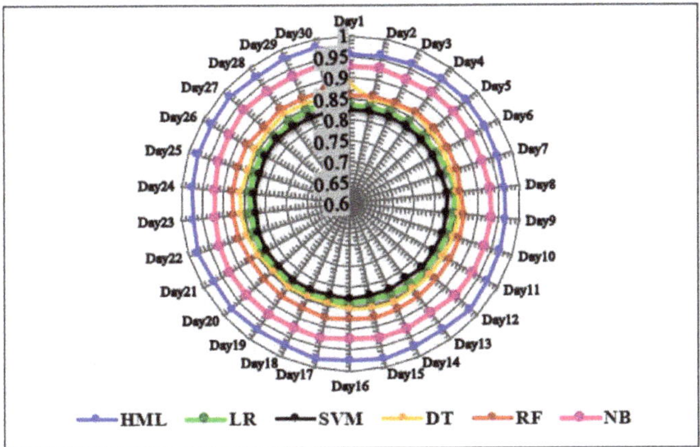

Figure 8. Outcomes of the third comparative experiment in a realistic environment with respect to AUC scores.

Unlike the proposed HML, the state-of-the-art machine learners achieve approximately high to moderate scores of AUC at classifying 0-Day ransomware versions among other cyber-versions during the month of the test (see Figure 8). AUC scores state that the examined machine learners require more crucial functions with probably regulating margins to adapt 0-Day versions involved in such real-life data corpus. That is due to some cases like that any suspicious version might be classified as invalid ransomware, or invalid good-ware, or invalid malware, or a new version of ransomware inaccurately. Furthermore, the examined machine learners show their consumption of changeable periods of time and complex computations to charactserize every version as well as recognizing its identity and its probability of a multi-descent family.

5.4. Realistic Experiment

Besides the previous comparative experiments, a daily based test is conducted in a realistic environment to evaluate how the proposed ransomware streaming analytics model manifests its efficacy against different and 0-Day ransomware versions of those of different ancestor-families and multi-descent family. The plotted charts of Figure 9 demonstrate the detection ability and holistic characterization of the proposed ransomware streaming analytics model with minimal performance overhead through its multi-tiered design versus daily escalating/deescalating and/or imbalanced corpus of data during one month. Unlike the anti-ransomware tool *RANDS*, which is devoted to our previous work [34,35], a minor escalation or de-escalation of efficacy's trend line is reported by our proposed ransomware streaming analytics model at certain days (see Figure 9). This is due to its attitude in solving the problems of the multi-descent family case as well as ancestor family attribution by using numeral and statistic metrics that are pursued by the proposed HML. Furthermore, the proposed ransomware streaming analytics model achieves shorter elapsed time than that of *RANDS* in [34], as shown in Figure 9d, due to its minor consumption of CPU and memory.

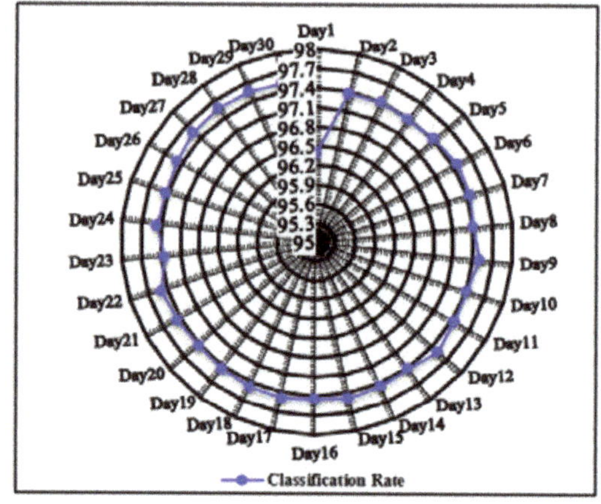

(a) Classification Accuracy Rate

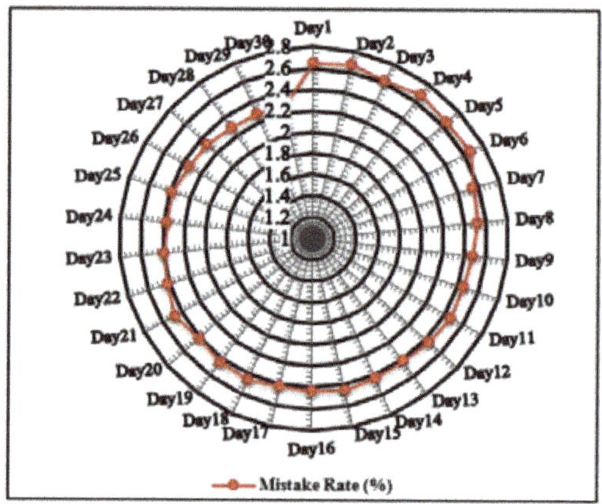

(b) Mistake Rate

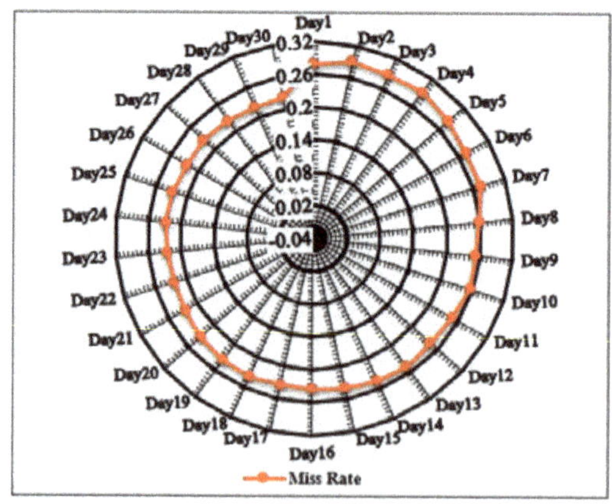

(c) Miss Rate

Figure 9. *Cont.*

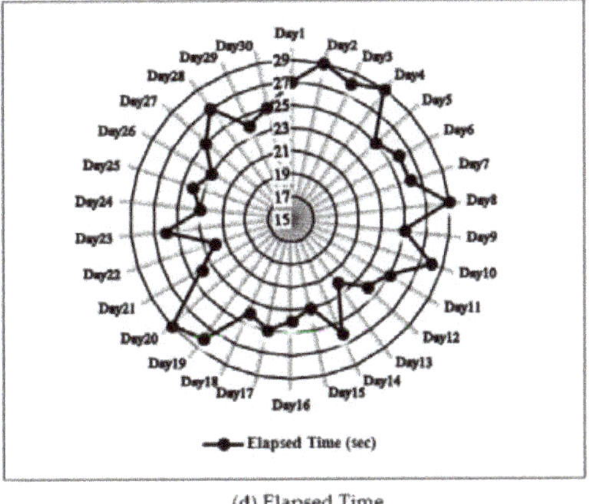

(d) Elapsed Time

Figure 9. Evaluation outcomes of the proposed ransomware streaming analytics model in the realistic test.

6. Discussion

The previous experiments have pointed out "why the state-of-the-art anti-ransomware tools those assisted by machine learners are still insufficient to detect ransomware versions of different ancestor families and multi-descent families in the realistic environment?" The answers to this question can be summarized as follows:

- Shortage in static analysis. Not all the state-of-the-art machine learners leverage static traits to characterize versions of all-inclusive ransomware families. Indeed, many ransomware families might exploit more adversarial scripts to intrude on their target operating systems [6,24]. Thus, the examined machine learners showed their defeatism against static exploits of ransomware, and they achieved low to sensible TPRs along with nontrivial FPRs and FNRs, as shown in Figures 6 and 8. In addition, some of the examined anti-ransomware tools revealed partial characterization of such static exploits on data corpora, as shown in Figure 7.
- Shortage in dynamic analysis. Tracking the behavioral traits, executed server controls, I/O resources, buffers, system activity, file events, CPU processes, and memory usage; all together, these are more distinctive to classify ransomware versions than the static exploitations [13,15,21]. However, their analysis and extraction require a virtual testbed which must be designed under realistic conditions to trigger ransomware payloads, linking libraries, and access permissions. As shown in Figure 6, the state-of-the-art machine learners fall short in tracing a ransomware's infection chain, and they produced trivial FPRs because they are mostly devoted to semi-realistic environments and lack the necessary triggers. Then, they are rendered vulnerable against any 0-Day ransomware versions during the second comparative experiment in realistic practice (see Figure 7).
- Elapsed time-accuracy trade-off. The trade-off between elapsed time and classification accuracy is a double-edges sword in machine learning [37]. Almost all, the examined machine learners and anti-ransomware tools achieved acceptable accuracy rates but at unacceptable elapsed time for detecting an incremental stream of training and testing corpora (see Figures 6 and 7). Elapsed time is a crucial factor to consider in ransomware detection since its computation depends on the traits' relevance and redundancy to the data corpus, the number of traits to be extracted from data corpus, the dimensionality of data corpus to learn, the extensive utilization of device's resources like CPU and memory and the computation complexity of the machine learning algorithm during both learning and testing tasks. Thus, some machine learners showcase high accuracy rates, but they spend a long elapsed time to obtain the detection results.

- Unaware of multi-descent versions of ransomware. The training and testing corpora of versions might involve versions of other malware and scareware families that probably share similar and mutually inter-related static exploits and dynamic behaviors and functionalities [6,24]. This may confuse machine learners, which leverage binary-class inductive function rather than a multi-class inductive function [36,37]. As a result, almost examined machine learners achieved high to moderate false classifications versus valid ransomware versions; those are descents of multiple ransomware and/or malware families, as shown in Figures 6–8. Hence a manifold induction could be attained by hybridizing multiple machine learners and/or statistic pruning formularization as it is adopted in HML

- Imbalanced corpora of data. Experimentally, the state-of-the-art machine learners produced their own predictions by learning the employed set of traits on the corpus of training data that was (i) various in the versions of other cyber-attacks like scareware and malware, etc. (ii) sub-optimally representative of ransomware detection, (iii) imbalanced in ransomware family population and ransomware-type distribution (i.e., crypto and locker), and (iv) varied in aggregation time. Altogether, it caused substantial FPRs, FNRs, and insignificant AUCs, as shown in Figures 6 and 7, respectively. Similarly, the examined anti-ransomware tools achieved incompetent outcomes versus the aforesaid corpus of data, as shown in Figure 8.

- Unreliable source of data. Most of the popular archives of ransomware fall short in providing reliable and unique ransomware datasets that is a complementary factor of boosting the implicit ransomware classification against 0-Day ransomware versions and imperviously detected ransomware families [4,6,7,12]. It is observed in Figures 6 and 7 that predicting valid ransomware versions across unreliable training corpus of data could be crucial to adjust the decision margins of the examined machine learners. Hence, resembling a reliable training corpus by re-learning the default predictions would solve the problem of substantial FPRs [37,38].

- Unaware of family attribution. Escalated ransomware streams from 2005 to 2019 yield more than 30 different ransomware families of both locker and crypto types [21,24,34,35]. This is attributed to the availability of e-services, social engineering, and primitive ciphering technologies that enable cyber-criminals to advance their ransomware without sophisticated knowledge [21,24]. However, the big stream of ransomware versions might share mutually inter-related traits that cause overfitting to decisive margins of machine learners and then limited detection against some ransomware families with high rates of FNR, as shown in Figure 6.

- Realistic and semi-realistic environment. The comparable machine learners, except HMLC that is devoted by [34,35], are still inactive to learn the misclassified versions of ransomware, and they achieve sub-optimal AUCs (see Figure 9). This is attributed to their decisive default settings, which fall short in minimizing the future cases of their predictions, which rely on their high rates of falsely detected versions versus their low rates of the truly detected versions. By product, the examined machine learners are still unaware of imperviously classified ransomware families on chronologically increasing ransomware stream.

The aforesaid observations raise another question to answer: "Why and how did our proposed HML and then our proposed multi-tier ransomware streaming analytics model outperform the state-of-the-art machine learners and anti-ransomware technologies?" The answer is summarized as follows:

- Set of traits. Great care was put on exploring and utilizing the most generic and evolutionary traits of ransomware families. Twenty-four different traits of both static-type and dynamic-type are used to provide a holistic characterization of ransomware versions among the versions of other cyber-attacks. Thus, our proposed work was privileged at classifying ransomware in the semi-realistic and realistic modes as it is observed from the experimental outcomes in Figures 6–9.

- Hybrid machine learner. The evaluation outcomes in Figures 6–9 restated that the proposed HML was more competent among its competitors. This was attributed to (i) its dual inductive function that is a hybrid of NB and DT functions, and (ii) its adjustable decisive boundary that is complementary of numeral and statistic metrics, *AR* and *MDR*. Accordingly, it achieved higher detection rates than others against the versions of ever-seen and/or never-seen ransomware families on big and different corpora of data.

- Training and testing corpora of data. The corpora of data were collected during a one-year period of time at different runtimes $\mathcal{T}$. Due to this periodic aggregation of data corpora every $\mathcal{T}$, the training, and testing tasks would be less biased versus the implicit and explicit class distribution problem in the examined data stream. Moreover, the data corpora were privileged in terms of quantity and variety of versions 45 K of ransomware, malware, and good-ware versions), quantity and variety of families (14 ransomware families), the difference of aggregation time, the difference of data archives, and imbalance of versions belong to common ancestors as well as the multi-descent versions (see Figures 6–9).
- Realistic conditions. To customize the overall performance, the proposed ransomware streaming analytics model with the presence of the proposed HML was run with the assumption of learning a corpus of data at its actual runtime $\mathcal{T}$. Hence, a suspicious version R that was inspected at time $\mathcal{T}$ could be classified as a ransomware version in the future runtime iteration of ($\mathcal{T} + \Delta$). That, in turn, would control the trade-off between the elapsed time, which relied on both CPU and memory consumption of the hosted system and the rate of detection accuracy in the realistic environment (see Figures 8 and 9).

7. Conclusions and Future Work

By studying the performance trade-offs across the state-of-the-art machine learners-based anti-ransomware tools experimentally, this paper affirms that they were computationally insufficient to classify 0-Day versions of different ransomware families on 40K corpora of data throughout in semi-realistic and realistic environments. They were still limited in static and dynamic analyses, set of traits to extract, type of ransomware to examine, the number of versions to learn, variety of ancestor ransomware families to attribute, multi-descent versions to recognize, the significant and imbalanced data stream to analyze, semi-realistic and realistic testbeds, decisive margins to prune, and inductive functions to adjust. Thus, this paper devotes a multi-tier ransomware streaming analytics model, which is empowered by a rich set of 24 static and dynamic traits, a novel machine learner, statistic, and numeral rates of family attribution and multi-descent family fusion those overlooked by the state-of-the-art anti-ransomware technologies. The proposed solution pursues four tiers of traits extraction, ransomware classification, ancestor family attribution, and multi-descent family fusion to discriminate ransomware versions from malware and good-ware versions. Experiments have qualified how the proposed solution enriches the accuracy, reduces the mistakes and misclassifications, and shortens the elapsed time versus escalating, big, lifelike, and imbalanced corpora of data. Overall results that averaged by 97% of accuracy rate, 2.4% of mistake rate, and 0.34% of miss rate; justify its maximal efficacy and cost-efficiency among its competitors.

For future improvement, it is recommended to explore more distinctive traits to classify the imperviously seen ransomware families by investigating versions of other ransomware families. Herewith, 14 *crypto*-type, and *locker*-type ransomware families are investigated with 24 traits solely. Furthermore, a hybrid machine learner of other base machine learners rather than NB and DT could be designed as either a single-based learner or an ensemble-based learner for pivoting an ideal decision that could satisfy the aforesaid cases.

Author Contributions: Conceptualization, H.Z.; methodology, H.Z.; software, H.Z.; validation, H.Z. and A.S.; formal analysis, A.S., and O.K.; investigation, A.S. and O.K.; resources, H.Z.; data curation, H.Z.; writing—original draft preparation, H.Z.; writing—review and editing, A.S. and O.K.; visualization, H.Z.; supervision, A.S. and O.K.; project administration, A.S. and O.K.; funding acquisition, A.S. and O.K. All authors have read and agreed to the published version of the manuscript.

Funding: The Article Processing Charge was funded by the SPEV project 2102/2020, Faculty of Informatics and Management, University of Hradec Kralove.

Acknowledgments: This research was partially funded by Universiti Teknologi Malaysia (UTM) under Research University Grant Vot-20H04, Malaysia Research University Network (MRUN) Vot 4L876, and the Fundamental Research Grant Scheme (FRGS) Vot 5F073 supported under the Ministry of Education Malaysia. The work is partially supported by the SPEV project, (ID: 2102-2020), "Smart Solutions in Ubiquitous Computing Environments" Faculty of Informatics and Management, University of Hradec Kralove. We are also grateful for the support of Ph.D. students Michal Dobrovolny and Ayca Kirimtat in consultations regarding application aspects from Hradec Kralove University, Czech Republic.

Conflicts of Interest: The authors declare no conflict of interest.

References

1. Bhardwaj, A.; Avasthi, V.; Sastry, H.; Subrahmanyam, G.V.B. Ransomware digital extortion: A rising new age threat. *Indian J. Sci. Technol.* **2016**, *9*, 1–5. [CrossRef]
2. Richardson, R.; North, M. Ransomware: Evolution, mitigation and prevention. *Int. Manag. Rev.* **2017**, *13*, 10–21.
3. Azmoodeh, A.; Dehghantanha, A.; Conti, M.; Choo, K.K.R. Detecting crypto-ransomware in IoT networks based on energy consumption footprint. *J. Ambient Intell. Humaniz. Comput.* **2018**, *9*, 1141–1152. [CrossRef]
4. Al-rimy, B.A.S.; Maarof, M.A.; Shaid, S.Z.M. Ransomware threat success factors, taxonomy, and countermeasures: A survey and research directions. *Comput. Secur.* **2018**, *74*, 144–166. [CrossRef]
5. Tailor, J.P.; Patel, A.D. A comprehensive survey: Ransomware attacks prevention, monitoring and damage control. *Int. J. Res. Sci. Innov.* **2017**, *4*, 2321–2705.
6. Kok, S.; Abdullah, A.; Jhanjhi, N.; Supramaniam, M. Ransomware, threat and detection techniques: A review. *Int. J. Comput. Sci. Netw. Secur.* **2019**, *19*, 136.
7. Yaqoob, I.; Ahmed, E.; Rehman, M.H.; Ahmed, A.I.A.; Al-Garadi, M.A.; Imran, M.; Guizani, M. The rise of ransomware and emerging security challenges in the Internet of Things. *Comput. Netw.* **2017**, *129*, 444–458. [CrossRef]
8. Pathak, P.B.; Nanded, Y.M. A dangerous trend of cybercrime: Ransomware growing challenge. *Int. J. Adv. Res. Comput. Eng. Technol.* **2016**, *5*, 371–373.
9. Herrera Silva, J.A.; Barona López, L.I.; Valdivieso Caraguay, Á.L.; Hernández-Álvarez, M. A survey on situational awareness of ransomware attacks—detection and prevention parameters. *Remote Sens.* **2019**, *11*, 1168. [CrossRef]
10. Zavarsky, P.; Lindskog, D. Experimental analysis of ransomware on windows and android platforms: Evolution and characterization. *Proced. Comput. Sci.* **2016**, *94*, 465–472.
11. Sgandurra, D.; Muñoz-González, L.; Mohsen, R.; Lupu, E.C. Automated dynamic analysis of ransomware: Benefits, limitations and use for detection. *arXiv* **2016**, arXiv:1609.03020.
12. Kok, S.H.; Abdullah, A.; Jhanjhi, N.Z.; Supramaniam, M. Prevention of crypto-ransomware using a pre-encryption detection algorithm. *Computers* **2019**, *8*, 79. [CrossRef]
13. Morato, D.; Berrueta, E.; Magaña, E.; Izal, M. Ransomware early detection by the analysis of file-sharing traffic. *J. Netw. Comput. Appl.* **2018**, *124*, 14–32. [CrossRef]
14. Sihwail, R.; Omar, K.; Ariffin, K.A.Z. A survey on malware analysis techniques: Static, dynamic, hybrid and memory analysis. *Int. J. Adv. Sci. Eng. Inf. Technol.* **2018**, *8*, 1662. [CrossRef]
15. Stiborek, J.; Pevný, T.; Rehák, M. Probabilistic analysis of dynamic malware traces. *Comput. Secur.* **2018**, *74*, 221–239. [CrossRef]
16. Cybersecurity, K.E. The Protection Technologies of Kaspersky Endpoint Security. Available online: https://mediacircle.de/pdf/Protection_Technologies_Whitepaper.pdf (accessed on 3 March 2020).
17. Kharraz, A.; Kirda, E. Redemption: Real-time protection against ransomware at end-hosts. In *International Symposium on Research in Attacks, Intrusions, and Defenses*; Springer: Cham, Switzerland, 2017; pp. 98–119.
18. Kharaz, A.; Arshad, S.; Mulliner, C.; Roberson, W.K.; Krida, E. UNVEIL: A large scale, automated approach to detecting ransomware. In Proceedings of the 2017 IEEE 24th International Conference on Software Analysis, Evolution and Reengineering (SANER), Klagenfurt, Austria, 20–24 February 2017; pp. 757–772.
19. Gómez-Hernández, J.A.; Álvarez-González, L.; García-Teodoro, P. R-Locker: Thwarting ransomware action through a honeyfile-based approach. *Comput. Secur.* **2018**, *73*, 389–398. [CrossRef]
20. Cabaj, K.; Gregorczyk, M.; Mazurczyk, W. Software-defined networking-based crypto ransomware detection using HTTP traffic characteristics. *Comput. Electr. Eng.* **2018**, *66*, 353–368. [CrossRef]

21. Hampton, N.; Baig, Z.; Zeadally, S. Ransomware behavioural analysis on windows platforms. *J. Inf. Secur. Appl.* **2018**, *40*, 44–51. [CrossRef]
22. Feng, Y.; Liu, C.; Liu, B. Poster: A new approach to detecting ransomware with deception. In Proceedings of the 38th IEEE Symposium on Security and Privacy, San Jose, CA, USA, 22–24 May 2017.
23. Cimitile, A.; Mercaldo, F.; Nardone, V.; Santone, A.; Visaggio, C.A. Talos: No more ransomware victims with formal methods. *Int. J. Inf. Secur.* **2018**, *17*, 719–738. [CrossRef]
24. Zhang, H.; Xiao, X.; Mercaldo, F.; Ni, S.; Martinelli, F.; Sangaiah, A.K. Classification of ransomware families with machine learning based on N-gram of opcodes. *Future Gener. Comput. Syst.* **2019**, *90*, 211–221. [CrossRef]
25. Alhawi, O.M.; Baldwin, J.; Dehghantanha, A. Leveraging machine learning techniques for windows ransomware network traffic detection. In *Cyber Threat Intelligence*; Advances in Information Security (ADIS, Volume 70); Springer: Cham, Switzerland, 2018; pp. 93–106.
26. Bae, S.I.; Lee, G.B.; Im, E.G. Ransomware detection using machine learning algorithms. *Concurr. Comput. Special Issue* **2016**. [CrossRef]
27. Aburomman, A.A.; Reaz, M.B.I. A survey of intrusion detection systems based on ensemble and hybrid classifiers. *Comput. Secur.* **2017**, *65*, 135–152. [CrossRef]
28. Tsai, C.-F.; Hsu, Y.-F.; Lin, C.-Y.; Lin, W.-Y. Intrusion detection by machine learning: A review. *Expert Syst. Appl.* **2009**, *36*, 11994–12000. [CrossRef]
29. Shabtai, A.; Moskovitch, R.; Elovici, Y.; Glezer, C. Detection of malicious code by applying machine learners on static features: A state-of-the-art survey. *Inf. Secur. Tech. Rep.* **2009**, *14*, 16–29. [CrossRef]
30. Continella, A.; Guagnelli, A.; Zingaro, G.; De Pasquale, G.; Barenghi, A.; Zanero, S.; Maggi, F. ShieldFS: A self-healing, ransomware-aware filesystem. In Proceedings of the 32nd Annual Conference on Computer Security Applications, ACM, New York, NY, USA, 5–8 December 2016; pp. 336–347.
31. Ahmadian, M.M.; Shahriari, H.R. 2entFOX: A framework for high survivable ransomwares detection. In Proceedings of the 13th International Iranian Society of Cryptology Conference on Information Security and Cryptology (ISCISC), Tehran, Iran, 7–8 September 2016; pp. 79–84.
32. Zimba, A. Malware-free Intrusion: A novel approach to Ransomware infection vectors. *Int. J. Comput. Sci. Inf. Secur.* **2017**, *15*, 317.
33. Shaukat, S.K.; Ribeiro, V.J. RansomWall: A layered defense system against cryptographic ransomware attacks using machine learning. In Proceedings of the 10th International Conference on Communication Systems & Networks (COMSNETS), Bengaluru, India, 3–7 January 2018; pp. 356–363.
34. Zuhair, H.; Selamat, A. RANDS: A machine learning-based anti-ransomware tool. In *Advancing Technology Industrialization through Intelligent Software Methodologies, Tools and Techniques, In Proceedings of the 18th International Conference on New Trends in Intelligent Software Methodologies, Tools and Techniques (SoMeT2019), Kuching, Sarawak, Malaysia, 23–25 September 2019*; IOS Press: Amsterdam, The Netherlands, 2019; pp. 573–587. [CrossRef]
35. Zuhair, H.; Selamat, A. An Intelligent and Real-Time Ransomware Detection Tool Using Machine Learning Algorithm. *J. Theor. Appl. Inf. Technol.* **2019**, *97*, 3448–3461.
36. Homayoun, S.; Dehghantanha, A.; Ahmadzadeh, M.; Hashemi, S.; Khayami, R.; Choo, K.K.R.; Newton, D.E. DRTHIS: Deep ransomware threat hunting and intelligence system at the fog layer. *Future Gener. Comput. Syst.* **2019**, *90*, 94–104. [CrossRef]
37. Krawczyk, B.; Minku, L.L.; Gama, J.; Stefanowski, J.; Woźniak, M. Ensemble learning for data stream analysis: A survey. *Inf. Fusion* **2017**, *37*, 132–156. [CrossRef]
38. Huang, G.; Huang, G.B.; Song, S.; You, K. Trends in extreme learning machines: A review. *Neural Netw.* **2015**, *61*, 32–48. [CrossRef]
39. Kwon, O.; Sim, J.M. Effects of data set features on the performances of classification algorithms. *Expert Syst. Appl.* **2013**, *40*, 1847–1857. [CrossRef]
40. Benign Software. Available online: http://software.informer.com/software/ (accessed on 4 April 2019).
41. *Virus Share*, "Malware Repository". Available online: https://virusshare.com (accessed on 13 January 2019).
42. Virus Total-Intelligence Search Engine, "Free Online Virus, Malware URL Scanner". Available online: https://www.virustotal.com (accessed on 21 August 2019).

Article

Sisyfos: A Modular and Extendable Open Malware Analysis Platform

Dimitrios Serpanos [1,2,*], Panagiotis Michalopoulos [2], Georgios Xenos [1,2] and Vasilios Ieronymakis [2]

1   ATHENA Research Center, Industrial Systems Institute, GR-26504 Patras, Greece; gxenos@upnet.gr
2   Department of Electrical and Computer Engineering, University of Patras, GR-26504 Patras, Greece; ece8125@upnet.gr (P.M.); bma2476@upnet.gr (V.I.)
*   Correspondence: serpanos@ece.upatras.gr

Abstract: Sisyfos is a modular and extensible platform for malware analysis; it addresses multiple operating systems, including critical infrastructure ones. Its purpose is to enable the development and evaluation of new tools as well as the evaluation of malware classifiers. Sisyfos has been developed based on open software for feature extraction and is available as a stand-alone tool with a web interface but can be integrated into an operational environment with a continuous sample feed. We present the structure and implementation of Sisyfos, which accommodates analysis for Windows, Linux and Android malware.

Keywords: malware analysis; static malware analysis; dynamic malware analysis; malware classification; machine learning; random forest; support vector machines

Citation: Serpanos, D.; Michalopoulos, P.; Xenos, G.; Ieronymakis, V. Sisyfos: A Modular and Extendable Open Malware Analysis Platform. *Appl. Sci.* **2021**, *11*, 2980. https://doi.org/10.3390/app11072980

Academic Editor: Leandros Maglaras

Received: 22 February 2021
Accepted: 23 March 2021
Published: 26 March 2021

Publisher's Note: MDPI stays neutral with regard to jurisdictional claims in published maps and institutional affiliations.

1. Introduction

Malicious software undoubtedly poses a serious threat to the security of computer systems. In the last decade, malware has been widely used by threatening actors to target private corporations, public organizations and individuals. Ransomware, for example, is increasingly used to attack major companies, organizations and persons; recent cases include Advantech, Canon and Cognizant, whose computer systems were encrypted by the attackers [1]. Recently, attackers also used malware to target government organizations and critical infrastructure with novel attacks, e.g., SolarWinds attacks [2]. Such attacks not only lead to serious data leakage but also have significant financial impact due to damages. It is estimated that the yearly cost of malware will surpass $ 6 trillion by 2021 [3]; the AV-TEST institute detects over 350,000 novel malware samples and potentially unwanted applications every day [4].

To defend against such attacks, a great deal of effort has been made to design effective malware detection and analysis systems. Researchers often use features derived from static analysis, in which information is extracted from a binary file without executing it [5–7]. As malware is usually heavily obfuscated and static analysis has limits [8], another approach is the use of dynamic analysis, where a suspicious program gets executed in a virtual environment and certain measurements are made [9–14]. Then, to differentiate between malware and benign files, classification techniques are employed, including machine learning algorithms. Common machine learning approaches include algorithms like random forests or support vector machines [15–18], while the use of deep learning neural networks is becoming increasingly prevalent [19–22].

Several efforts have been made to provide complete end-to-end solutions for malware analysis and detection; these include static analysis, dynamic analysis and learning models, which prove to be quite efficient at detecting specific pieces of malware on specific systems. Existing tools and platforms of wide use include VirusTotal [23], Joe Sandbox [24], Hybrid Analysis [25] and ANY.RUN [26], where a user uploads a suspicious sample file and receives a complete report that contains information gathered from both static and dynamic

141

analysis. Alternatives include systems like the VMRay Analyzer and Detector [27], where a malware analysis platform is either installed on-premises or is deployed in the Cloud. While these services are effective at analyzing and detecting malware, they typically include proprietary tool chains and require paid subscriptions.

Importantly, since malware classifiers adopt learning systems, they require significant amounts of reliable data for effective and efficient training. Up to date, no public data are available for malware classifier training and, actually, there is a significant lack of common data for training and comparison of alternative analyzers and classifiers. The inherent security and privacy concerns make it very difficult for researchers and private companies alike to publicize and share their data, thwarting collaboration among researchers.

To address these issues, we have developed and employed Sisyfos, a novel extensible and openly available malware analysis platform. Sisyfos has been developed based on a set of open software tools for feature extraction (from static and dynamic analysis) and integrates our own classifiers; the tools have been integrated through an orchestrator that has been designed to provide a robust and modular environment for malware analysis and classification. Although Sisyfos is a platform suitable for educational and research environments, for training and experimentation with new tools and classifiers, promoting collaboration within the research community, it also integrates into operational environments effectively and efficiently, providing invaluable feedback for the performance and requirements of operational environments.

The paper is organized as follows: Section 2 briefly introduces the general architecture of Sisyfos and provides a description of its functionality; Section 3 describes the implementation of Sisyfos; and Section 4 presents our classifiers, which employ appropriate machine learning models that are effective in malware detection.

2. The Architecture of Sisyfos

Sisyfos is a general-purpose platform for malware analysis and is organized in three stages: static analysis, dynamic analysis and classification. Sisyfos accepts as input a software sample and extracts its classification category, currently as malware or benign software. Figure 1 depicts the structure of Sisyfos, showing a pipeline of analysis and classification steps.

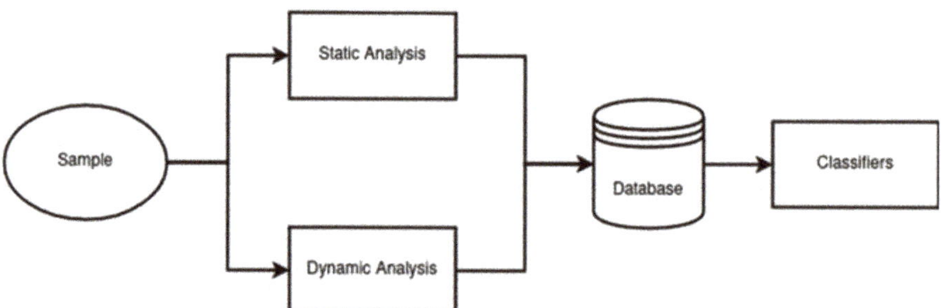

Figure 1. Sisyfos platform architecture.

Static analysis constitutes the first stage of analysis of a processed sample. Static analysis extracts features and measures parameters of the software sample through tools that analyze the sample code without executing it. This kind of analysis allows for quick and computationally inexpensive feature extraction; such typical features include information about the different sections of a file and the resources the executable is using as well as the calls to external libraries or to the underlying operating system. Static analysis includes scanning the binary sample to detect possible YARA rule matches. Sisyfos also submits the sample's hash to VirusTotal in order to obtain the existing analysis reports in the case that the sample has been analyzed in the past by VirusTotal; VirusTotal is a publicly available service integrating a multitude of well-known antivirus systems [23].

After the features are extracted and measured, analysis continues with the second stage, dynamic analysis. Dynamic analysis is the process by which the sample is executed inside a safe and isolated environment (a sandbox) and various behavioral characteristics related to its execution are collected. These features can be broadly categorized in the following general categories: file system, registry, network, and dynamic signatures. The first category contains features that describe the interaction of the sample with the file system, such as the number of files it created or opened. The second relates to Windows PE executables and contains information related to the Windows Registry, such as the keys accessed by the sample. The third category summarizes the network activity of the sample. It contains the number of TCP and UDP connections that were established, the number of unique IPs the sample connected to, the number of DNS queries performed, etc. Finally, the last category contains the names of the dynamic signatures created by the sandbox, which can be used to summarize the overall execution of the sample. These signatures are created by parsing the memory dump and the execution trace of the sample and searching for suspicious patterns, such as the allocation of read-write-execute memory areas (a possible indication of payload unpacking) or the injection of code into child processes. The raw results of the dynamic analysis, along with the extracted features, are stored in a central database, from which they can be accessed at later stages of analysis.

Finally, the features and measurements that result from the static and dynamic analysis are fed to the classification stage, which classifies the sample according to a category of the used classification scheme.

Sisyfos is built mainly with open software tools that are combined, which process and extract features of the analyzed software sample (the suspected malware), in order to provide accurate metrics and enable effective detection and classification of the sample. The platform has been designed to be highly modular and easily expandable. Its modularity and scalability are key to its success, since new tools can be easily and quickly integrated; Sisyfos' easy setup and highly scalable design enable implementation with various sizes of organizations, from small private enterprises to large public organizations.

3. The Implementation of Sisyfos

Sisyfos is implemented by exploiting open software. Various open software tools for static and dynamic analysis have been employed with the platform, whose implementation is shown in Figure 2. Dynamic analysis is performed using sandboxes that are implemented with VirtualBox through the Cuckoo Sandbox [28] for Windows and Linux; currently, work is under way to integrate the Mobile Software Framework (MobSF) for Android [29] to Sisyfos.

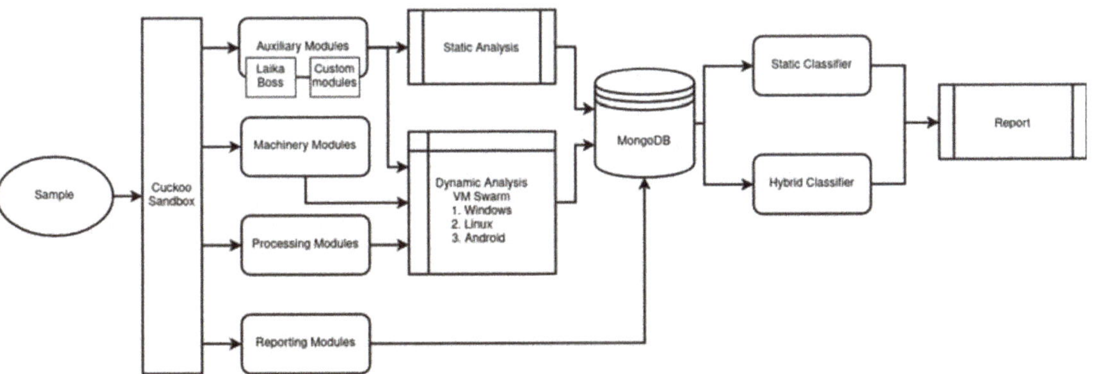

Figure 2. Sisyfos implementation diagram.

The tools have been integrated using a non-trivial orchestrator which achieves two main goals, resilience and modularity. For the first goal, the orchestrator spawns a new process for every submitted sample, which manages the analysis (e.g., interacts with Cuckoo, stores the results in the database and passes the sample to the classifier). This strengthens the resilience of Sisyfos, because if an analysis fails, the rest of the platform will continue normal operation; at the same time, Sisyfos' throughput capabilities increase. For the second goal, the orchestrator is constructed in a tool-agnostic way, enabling the easy replacement of one tool (e.g., Cuckoo) with another, increasing the modularity of Sisyfos. The orchestrator has been built in Python.

Importantly, Sisyfos is built so that it can be used either as a stand-alone Web service, where a user uploads a software sample for analysis, or as a service that can be integrated and automated for a continuous feed of samples in an operational setting.

In the following, we present the main implementation characteristics of Sisyfos for: (i) the analyses stages (static and dynamic) and (ii) the classification stage. The presentation focuses on Windows and Linux malware analysis.

Static analysis is implemented using two different open software tools, Cuckoo [28] and LaikaBOSS [30]. First, the executable sample is sent to Cuckoo, which processes it with its own static analysis tools. Cuckoo collects the hash of the file along with the names, addresses and sizes of its different sections. The names, offsets and sizes of the resources that are used by the binary sample are also extracted. Furthermore, Cuckoo extracts the printable strings contained in the binary; Cuckoo specifies these printable strings as UTF-8/16 strings with lengths between 6 and 1024 characters and limits the number of extracted strings to 2048. Finally, it searches for whether the sample matches any sample on VirusTotal [23] and fetches the associated signatures extracted by the detecting antivirus systems.

Cuckoo also allows for auxiliary modules to be loaded. One such module measures the entropy of the file and its different sections. Using entropy measurements, the system can detect if the file is using obfuscation techniques, e.g., packers, to avoid detection.

In parallel with Cuckoo static analysis, the sample is sent to LaikaBOSS, which is integrated as an auxiliary module in Sisyfos. LaikaBOSS, an open software tool, is an object scanner and intrusion detection system that recursively extracts child objects from a file. Such objects can be archives, wrappers or obfuscators. Finally, Laika scans the binary against a repository of YARA rules and creates a detection flag for each rule that gets matched.

Dynamic analysis is implemented using Cuckoo, taking special steps to conceal the virtual analysis environment and make it appear as a normal workstation. Three elements contribute to this. First, the Disguise module of Cuckoo and the VMCloak tool [31] change specific environmental values that could otherwise reveal the virtualization environment. Second, we install commonly found software such as office suites and web browsers and ensure the creation of dummy files in the user's personal folders, along with creating browser history. Finally, we allocate typical resources to the virtual machines (hard disk larger than 60 GB, RAM > 1 GB, and CPU cores more than 2). We have evaluated these modifications with the use of the Paranoid Fish tool [32] which tries to detect if it is running inside a virtualized environment using a wide array of techniques commonly used by malware.

During dynamic analysis, Cuckoo identifies the requirements to execute each sample and invokes the appropriate analysis module. The sandbox includes special software that may be required by the sample, e.g., PDF reader. Then, the sample is uploaded to a virtual machine and through a monitoring process, Cuckoo intercepts the system calls that are made and the network traffic that is initiated by the sample. Cuckoo is highly modular and configurable and is capable of outputting a large volume of behavioral information and execution statistics for each sample it analyses. Extracted measurements and statistics by the dynamic analysis stage include:

- File System: number of files opened, read, deleted, created, recreated, written and checked for existence;
- Registry: number of records opened and read;
- Network: number of TCP sessions, UDP sessions, unique IPs, DNS A/CNAME/PTR/ MX queries, HTTP GET/HEAD/POST requests and HTTP 200/300/400/500 response status codes.

Importantly, Sisyfos implements a sophisticated interface, enabling users to upload samples to Sisyfos in two ways. The first one is through a web UI, shown in Figure 3, which has been set up as an easy-to-use web application; anyone can upload a sample that will be delivered to our platform. As Figure 3 shows, the submission page of the web interface consists of a modified version of the corresponding Cuckoo page. The backend interface of the platform, which provides real-time information for each submitted sample, is shown in Figure 4. The second way to transfer a sample to Sisyfos is through a transfer.sh service. Using this transfer service, one can send a sample to Sisyfos, at any time, through a simple curl command over the terminal; this enables connectivity of the platform to multiple network computers or many sample sources.

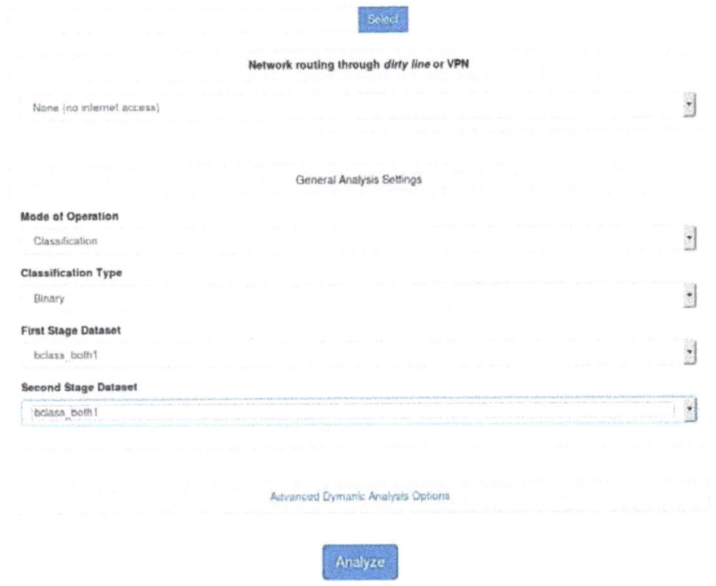

Figure 3. Sisyfos web interface.

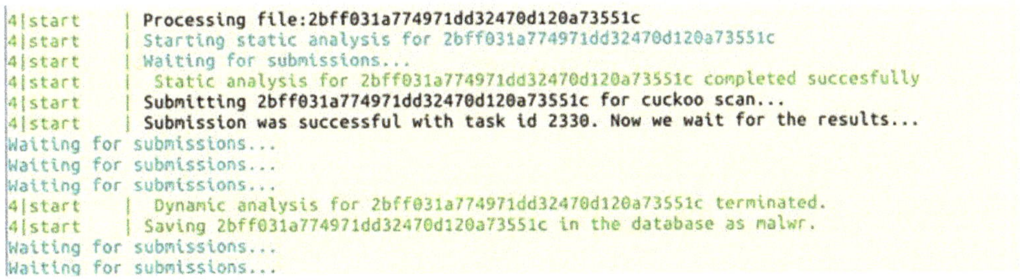

Figure 4. Real time information we collect from each sample.

Considering privacy issues in Sisyfos, in addition to the typical disclaimer that raises the privacy risks and ascertains the corresponding liabilities to uploaders of samples in the main page, Sisyfos offers a mechanism to users to erase the uploaded samples and the analysis features from the systems, if they wish. Furthermore, Sisyfos does not have an agent that automatically collects samples from users and, thus, it does not create any privacy risks as do systems that collect client data automatically through agents [33].

4. Classification

The final step of Sisyfos operation is to classify the software sample, based on the classification scheme that is employed by Sisyfos. The platform is independent of the classification scheme and can accommodate any scheme that is associated with the features and measurements of the Sisyfos tools. Currently, we use binary classification for all samples, i.e., we classify them either as malicious or benign; this decision is made based on the information extracted during both static and dynamic analysis phases.

4.1. Static and Dynamic Classifiers

Sisyfos currently includes classifiers that exploit machine learning techniques. We have developed two different classifiers, one based on static features and one based on dynamic ones. We evaluate the effectiveness of these classifiers through a small dataset that contains 6000 Windows Portable Executable samples which include both malware and benign samples. The samples originate from VirusShare, a popular web repository of malware files. The VirusShare malware samples are augmented with benign installation files of well-known benign programs, in order to construct a balanced dataset of samples. We verified the labels of our dataset using the VirusTotal API [23]. The samples are then processed by the platform to produce feature matrices. We divide our dataset into two subsets, one containing 5000 samples used for training and one containing 1000 samples used for testing. Figure 5 shows the distribution of our dataset. Our current classification method uses approximately 500 features derived from static analysis and approximately 150 derived from dynamic analysis. Examples of such features are data extracted from the different sections of the executable, different matching YARA rules, DNS and HTTP requests, files dropped, downloaded files or detected PowerShell commands that are suspicious.

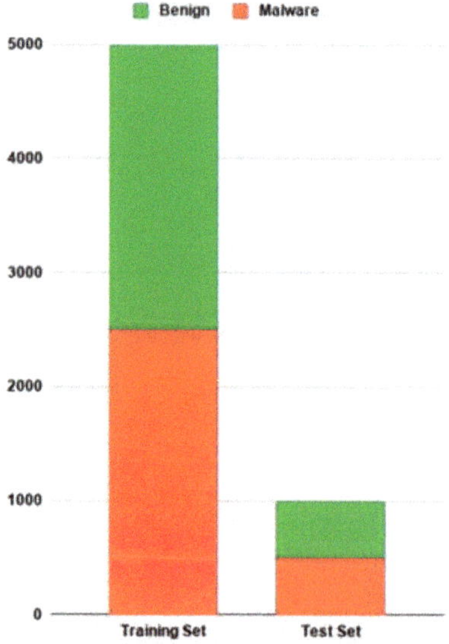

Figure 5. Sample distribution in our training and test subsets.

In order to identify the most efficient machine learning algorithm for our dataset, we use TPOT [34], an open source tool, which executes different algorithms and tries different values for the hyper-parameters and calculates the optimal combination. For our specific dataset, the optimal algorithm calculated by TPOT is the gradient boosting classifier. Table 1 presents the optimal hyper-parameters for our case and Table 2 presents the results of our static and dynamic model.

Table 1. Optimal hyper-parameters of the gradient boosting classifier.

Hyper-Parameter	Value
Number of trees	5000
Minimum samples in leaf	1
Number of features per tree	Sqrt (m) [1]
Learning rate	0.1
Max depth	9

[1] Where m is the number of all features.

Table 2. Accuracy, precision and recall of the static and dynamic models.

Model	Accuracy	Precision	Recall
Static Model	99.21%	98.78%	99.86%
Dynamic Model	96.53%	96.99%	96.99%

4.2. A Random Forest Static Classifier

Sisyfos enables sample classification exploiting static and dynamic features extracted through static and dynamic analysis as demonstrated in Section 4.1. We exploit Sisyfos to evaluate classifiers using various data sets of software samples, focusing on developing a reliable static classifier with machine learning techniques. However, the unavailability of standard datasets raises issues for the objective comparison of research efforts by different research groups. There is significant and well-known lack of standardized sample data sets in the research community; such datasets should include both malware and benign samples. Reproducible and comparable experiments require common samples and features. Although many online malware repositories exist today, such as VirusTotal [23], one would need specific parameters (such as the MD5 hash of each file) in order to extract specific sample subsets. Most of the researchers though, do not publish the specific parameters for selecting samples from such online repositories. To address these issues in our static classifier, we use one of the very few publicly available datasets, the EMBER dataset [35], which contains both malware and benign files. EMBER unfortunately does not distribute the binary files, but instead provides a feature set for each sample. Due to Sisyfos' modularity, we can train the model on those features and then we can easily integrate it into Sisyfos, as we describe below.

4.2.1. Model and Training

Our classifier adopts a model based on random forests, an algorithm that uses bagging, where many smaller noisy decision trees are averaged to produce the final model. This algorithm is selected because it can capture complex interactions between the features, and due to its inherent parallelism, can construct multiple decision trees at the same time.

The EMBER dataset [35], which was used to train the model, contains 2351 features extracted with static analysis from 1.1 million malicious and benign portable executable samples. EMBER comes with a baseline pre-trained model along with a distinct test dataset, allowing us to directly compare and evaluate the efficiency of our algorithm.

During the training phase we used features from 600,000 samples labeled as either malware (300,000 samples) or benign (300,000 samples), as shown in Figure 6.

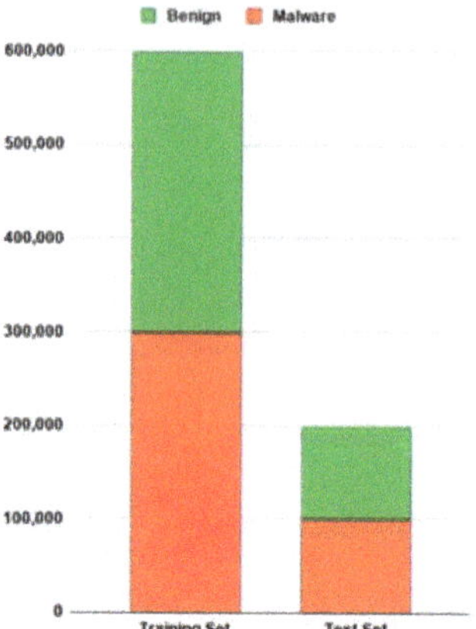

Figure 6. EMBER dataset distribution of samples in the training set (labeled data only).

In order to boost the accuracy and reduce the training time of our model, we performed a light hyper-parameter tuning, searching only in a small subspace of commonly used values [36]. To achieve this, we used an open software tool called TPOT [34] that allowed us to choose a specific range of possible values for each hyper-parameter that it then used to find the optimal combination of values. To expedite the hyper-parameter finding process, we used only a subset of the whole dataset, specifically features from 100,000 samples: 50,000 malware and 50,000 benign, maintaining the original balance of the dataset. In order to validate the results on a separate dataset, we did an additional 80/20 training/validation split of our data; specifically, we shuffled the dataset and picked at random 20% of the samples, while maintaining again the balance of the dataset as shown in Figure 7.

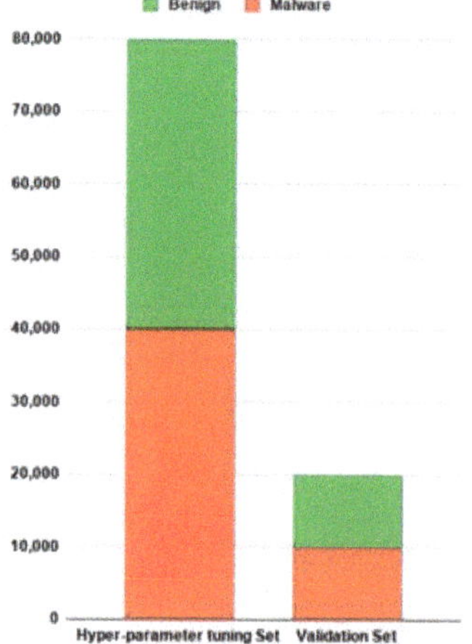

Figure 7. Sample distribution we used during the hyper-parameter tuning process.

Table 3 presents the optimal major hyper-parameters of the final model.

4.2.2. Model Integration into Sisyfos

The modular architecture of the platform then allows us to easily integrate our model into the system. The feature extraction logic that is not already included in our tools can be described in a single module that gets executed whenever a user submits a file. Additionally, the platform supports quick swapping between different machine learning models, enabling us to easily load and use the trained model. This, combined with the fact that the platform saves every sample in a database, allows us to update the model often with new data, in order to have an up-to-date prediction system at all times.

Table 3. Optimal hyper-parameters of the model.

Hyper-Parameter	Value
Number of trees	1000
Minimum samples in leaf	1
Number of features per tree	Sqrt (m) [1]

[1] Where m is the number of all features.

4.2.3. Results

We measured the performance of the model through its accuracy at different false positive rates (FPR) and we compared it with the baseline model that came with EMBER.

The model was trained with 600,000 samples using all the 2351 features at about 36 h, on a single 8-core processor, and achieved 99% accuracy at 1% FPR and 98.8% accuracy at 0.1% FPR. Table 4 provides a comparison of our results with the EMBER baseline model, demonstrating that the model is capable of very accurately classifying both malware and benign executables at very low FPRs, performing better than the baseline model.

Table 4. Accuracy, precision and recall of the model for different false positive rates (FPRs) in comparison to the baseline model.

Recall		Precision		Accuracy		
Random Forest	**Ember**	**Random Forest**	**Ember**	**Random Forest**	**Ember**	**False Positive Rate**
99%	98%	99%	98.98%	99%	98.50%	1%
97.7%	92.9%	99.89%	99.89%	98.80%	96.40%	0.10%

Another important metric we used to evaluate the performance of our model was the receiver operating characteristic (ROC) curve. The ROC curve is a plot that illustrates the detection ability of a classifier at various false positive (FP) rates. To plot it, we measured the false positives while varying the discrimination threshold of the algorithm. Figure 8 plots the ROC curves of our model and the EMBER baseline model, demonstrating that our model clearly outperforms the EMBER baseline model in the FPR range of $[10^{-5}, 10^{-2}]$ with improvements reaching 3% in true positive rate.

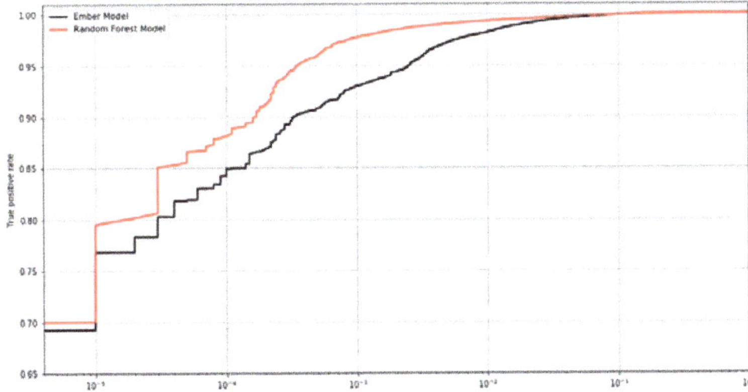

Figure 8. Receiver operating characteristic (ROC) curve of our model, in comparison with the EMBER baseline.

5. Conclusions and Future Work

Malware analysis platforms constitute fundamental infrastructure for the detection and mitigation of malware. Malware platforms not only classify and detect malware but enable the analysis of their features and enhance understanding of their structure, enabling efficient and effective detection. Machine learning techniques are quite promising in malware detection and classification but require significant sizes of reliable data for effective classifier training as well as for the selection of the appropriate machine learning techniques in variable operational environments.

Sisyfos constitutes a significant step in the development of open, modular and extensible malware platforms that support operational environments, including critical infrastructures, and enable the academic community to develop new tools and experiment with novel classifiers. However, significant effort needs to be spent in the specification, collection and public sharing of appropriate datasets, which will enable objective comparison of malware analysis methods and platforms and will lead to effective solutions.

In this direction, our future work includes exploring and improving classification methods by training and evaluating more machine learning algorithms with the adoption of larger datasets such as the newly released dataset, SoReL-20M [37]; this includes designing binary as well as multi-class classification methods in order to differentiate among different malware types and families. Furthermore, in addition to the next version of Sisyfos, which will include the ability to analyze Android files by integrating open-source tools such as MobSF [29], we will work toward improving Sisyfos' robustness and fault tolerance, especially in boundary cases where samples may lead to platform failures.

Author Contributions: All authors have contributed equally to this work. All authors have read and agreed to the published version of the manuscript.

Funding: Part of the work that was performed at ISI/ATHENA was supported by the project "I3T—Innovative Application of Industrial Internet of Things (IIoT) in Smart Environments" (MIS 5002434) which is implemented under the "Action for the Strategic Development on the Research and Technological Sector", funded by the Operational Programme "Competitiveness, Entrepreneurship and Innovation" (NSRF 2014-2020) and co-financed by Greece and the European Union (European Regional Development Fund. Part of the work that was performed at ISI/ATHENA was supported by the European project "CONCORDIA".

Institutional Review Board Statement: Not applicable.

Informed Consent Statement: Not applicable.

Data Availability Statement: Openly available dataset. Data was obtained from EMBER: An Open Dataset for Training Static PE Malware Machine Learning Models and can be found here: https://github.com/elastic/ember (accessed on 20 February 2021).

Acknowledgments: Not applicable.

Conflicts of Interest: The authors declare no conflict of interest. The funders had no role in the design of the study; in the collection, analyses, or interpretation of data; in the writing of the manuscript, or in the decision to publish the results.

References

1. Choudhury, A. Top 8 Ransomware Attacks of 2020 that Shook the Internet. *Anal. India Mag.* Available online: https://analyticsindiamag.com/top-8-ransomware-attacks-of-2020-that-shook-the-internet/ (accessed on 20 February 2021).
2. Halpern, S. After the SolarWinds Hack, We Have No Idea What Cyber Dangers We Face. Available online: https://www.newyorker.com/news/daily-comment/after-the-solarwinds-hack-we-have-no-idea-what-cyber-dangers-we-face (accessed on 19 February 2021).
3. Morgan, S. Report: Cyberwarfare in the C-Suite. 2021. Available online: https://1c7fab3im83f5gqiow2qqs2k-wpengine.netdna-ssl.com/wp-content/uploads/2021/01/Cyberwarfare-2021-Report.pdf (accessed on 20 February 2021).
4. Malware Statistics & Trends Report | AV-TEST. Available online: /en/statistics/malware/ (accessed on 19 February 2021).
5. Shalaginov, A.; Banin, S.; Dehghantanha, A.; Franke, K. Machine Learning Aided Static Malware Analysis: A Survey and Tutorial. *arXiv* **2018**, arXiv:1808.01201.

6. Baldangombo, U.; Jambaljav, N.; Horng, S.-J. A Static Malware Detection System Using Data Mining Methods. *arXiv* **2013**, arXiv:1308.2831. [CrossRef]

7. Amin, M.; Tanveer, T.A.; Tehseen, M.; Khan, M.; Khan, F.A.; Anwar, S. Static Malware Detection and Attribution in Android Byte-Code through an End-to-End Deep System. *Future Gener. Comput. Syst.* **2020**, *102*, 112–126. [CrossRef]

8. Moser, A.; Kruegel, C.; Kirda, E. Limits of Static Analysis for Malware Detection. In Proceedings of the 23rd Annual Computer Security Applications Conference (ACSAC 2007), Miami Beach, FL, USA, 10–14 December 2007; pp. 421–430.

9. Anderson, B.; Quist, D.; Neil, J.; Storlie, C.; Lane, T. Graph-Based Malware Detection Using Dynamic Analysis. *J. Comput. Virol.* **2011**, *7*, 247–258. [CrossRef]

10. Wong, M.Y.; Lie, D. IntelliDroid: A Targeted Input Generator for the Dynamic Analysis of Android Malware. In *Proceedings of the 2016 Network and Distributed System Security Symposium*; Internet Society: San Diego, CA, USA, 2016.

11. Sgandurra, D.; Muñoz-González, L.; Mohsen, R.; Lupu, E.C. Automated Dynamic Analysis of Ransomware: Benefits, Limitations and Use for Detection. *arXiv* **2016**, arXiv:1609.03020.

12. Mohaisen, A.; Alrawi, O.; Mohaisen, K. AMAL: High-fidelity, behavior-based automated malware analysis and classification. *Comput. Secur.* **2015**, *52*. [CrossRef]

13. Park, Y.; Reeves, D.; Mulukutla, V.; Sundaravel, B. Fast Malware Classification by Automated Behavioral Graph Matching. In Proceedings of the 6th Annual Workshop on Cyber Security and Information Intelligence Research (CSIIRW '10), Oak Ridge, TN, USA, 21–23 April 2010; pp. 45:1–45:4. [CrossRef]

14. Shijo, P.V.; Salim, A. Integrated Static and Dynamic Analysis for Malware Detection. *Procedia Comput. Sci.* **2015**, *46*, 804–811. [CrossRef]

15. Li, W.; Ge, J.; Dai, G. Detecting Malware for Android Platform: An SVM-Based Approach. In Proceedings of the 2015 IEEE 2nd International Conference on Cyber Security and Cloud Computing, New York, NY, USA, 3–5 November 2015; pp. 464–469.

16. Zhao, M.; Ge, F.; Zhang, T.; Yuan, Z. AntiMalDroid: An Efficient SVM-Based Malware Detection Framework for Android. In *Information Computing and Applications*; Liu, C., Chang, J., Yang, A., Eds.; Springer: Berlin, Germany, 2011; Volume 243, pp. 158–166. ISBN 9783642275029.

17. Zhu, H.-J.; .; Cheng, L. HEMD: A Highly Efficient Random Forest-Based Malware Detection Framework for Android. *Neural. Comput. Applic.* **2018**, *30*, 3353–3361. [CrossRef]

18. Garcia, F.C.C.; Muga, F.P., II. Random Forest for Malware Classification. *arXiv* **2016**, arXiv:1609.07770.

19. Yuan, Z.; Lu, Y.; Wang, Z.; Xue, Y. Droid-Sec: Deep Learning in Android Malware Detection. In Proceedings of the 2014 ACM conference on SIGCOMM, Chicago, IL, USA, 17 August 2014; pp. 371–372.

20. Vinayakumar, R.; Alazab, M.; Soman, K.P.; Poornachandran, P.; Venkatraman, S. Robust Intelligent Malware Detection Using Deep Learning. *IEEE Access* **2019**, *7*, 46717–46738. [CrossRef]

21. Raff, E.; Barker, J.; Sylvester, J.; Brandon, R.; Catanzaro, B.; Nicholas, C. Malware Detection by Eating a Whole EXE. *arXiv* **2017**, arXiv:1710.09435.

22. Fleshman, W.; Raff, E.; Sylvester, J.; Forsyth, S.; McLean, M. Non-Negative Networks Against Adversarial Attacks. *arXiv* **2019**, arXiv:1806.06108.

23. VirusTotal. Available online: https://www.virustotal.com/gui/ (accessed on 22 February 2021).

24. Automated Malware Analysis—Joe Sandbox Cloud Basic. Available online: https://www.joesandbox.com/ (accessed on 22 February 2021).

25. Free Automated Malware Analysis Service—Powered by Falcon Sandbox. Available online: https://www.hybrid-analysis.com/ (accessed on 22 February 2021).

26. ANY.RUN—Interactive Online Malware Sandbox. Available online: https://any.run/ (accessed on 22 February 2021).

27. Malware Analysis Sandbox & Malware Detection Software. Available online: https://www.vmray.com/products/analyzer-malware-sandbox/ (accessed on 22 February 2021).

28. Cuckoo Sandbox—Automated Malware Analysis. Available online: https://cuckoosandbox.org/ (accessed on 20 February 2021).

29. OpenSecurity Mobile Security Framework (MobSF). Available online: https://github.com/MobSF/Mobile-Security-Framework-MobSF (accessed on 22 February 2021).

30. Lockheed Martin. Laika BOSS: Object Scanning System. Available online: https://github.com/lmco/laikaboss (accessed on 22 February 2021).

31. Hatching VMCloak. Available online: https://github.com/hatching/vmcloak (accessed on 22 February 2021).

32. Ortega, A. Paranoid Fish. Available online: https://github.com/a0rtega/pafish (accessed on 22 February 2021).

33. Goodin, D. Kaspersky: Yes, We Obtained NSA Secrets. No, We Didn't Help Steal Them. ArsTechnica, 16 November 2017. Available online: https://arstechnica.com/information-technology/2017/11/kaspersky-yes-we-obtained-nsa-secrets-no-we-didnt-help-steal-them/ (accessed on 15 March 2021).

34. Olson, R.S.; Bartley, N.; Urbanowicz, R.J.; Moore, J.H. Evaluation of a Tree-Based Pipeline Optimization Tool for Automating Data Science. In Proceedings of the Genetic and Evolutionary Computation Conference 2016, Denver, CO, USA, 20 July 2016; pp. 485–492.

35. Anderson, H.S.; Roth, P. EMBER: An Open Dataset for Training Static PE Malware Machine Learning Models. *arXiv* **2018**, arXiv:1804.04637.

36. Hastie, T.; Tibshirani, R.; Friedman, J.H. *The Elements of Statistical Learning: Data Mining, Inference, and Prediction*, 2nd ed.; Springer series in statistics; Springer: New York, NY, USA, 2009; ISBN 9780387848570.
37. Harang, R.; Rudd, E.M. SOREL-20M: A Large Scale Benchmark Dataset for Malicious PE Detection. *arXiv* **2020**, arXiv:2012.07634.

Article

A Study on the Concept of Using Efficient Lightweight Hash Chain to Improve Authentication in VMF Military Standard

Dohoon Kim [1], Sang Seo [1], Heesang Kim [1], Won Gi Lim [2] and Youn Kyu Lee [3],*

[1] Department of Computer Science, Kyonggi University, Suwon-si, Gyeonggi-do 16227, Korea; karmy01@kgu.ac.kr (D.K.); tjtkd8271@kgu.ac.kr (S.S.); victoriousian@kgu.ac.kr (H.K.)
[2] Agency for Defense Development (ADD), Seoul 05661, Korea; wklim@add.re.kr
[3] Department of Information Security, Seoul Women's University, Seoul 01797, Korea
* Correspondence: younkyul@swu.ac.kr

Received: 23 November 2020; Accepted: 14 December 2020; Published: 16 December 2020

Abstract: Authentication algorithms in the form of cryptographic schemes, such as the Secure Hash Algorithm 1 (SHA-1) and the digital signature algorithm (DSA), specified in the current variable message format (VMF) military standard have numerous reliability-related limitations when applied to tactical data link (TDL) and multi-TDL networks (MTN). This is because TDL and MTN require maximum tactical security, communication integrity, and low network overhead based on many protocol header bits for rapid communication with limited network resources. The application of such authentication algorithms to TDL and MTN in a rapidly changing battlefield environment without reinforcement measures will lead to functional weaknesses and vulnerabilities when high-level digital-covert activities and deception tactics are implemented. Consequently, the existing VMF authentication scheme must be improved to secure transmission integrity, lower network transaction, and receive authentication tactical information in VMF-based combat network radio (CNR) networks. Therefore, in this study, a tactical wireless ad hoc network topology, similar to that of the existing CNRs, is considered, and a lightweight multi-factor hash chain-based authentication scheme that includes a time-based one-time password (T-OTP) for network overhead reduction and terminal authentication is proposed, coupled with exception handling. The proposed method enhances the confidentiality of tactical message exchanges and reduces unnecessary network transactions and transmission bits for authentication flows between real-time military terminals owned by squads, while ensuring robustness in limited battlefields. Based on these approaches, in the future, we intend to increase the authentication reliability between wireless terminals in the Korean variable message format (KVMF)-based CNR networks based on the Korean Army Corps network scenarios.

Keywords: military; VMF; hash chain; T-OTP; lightweight secure hash (LSH); CNR

---

1. Introduction

The variable message format (VMF) [1] is a military digital information exchange standard established by the Joint Interoperability of Tactical Command and Control Systems (JINTACCS) program under the United States Joint Chiefs of Staff. It provides common interoperability standards, including command data elements and protocol standards for information transfer in command, control, communications, computer and intelligence (C4I) systems over limited battlefield networks. Further, it enables the real-time exchange of digital tactical information in surveillance systems, command and control (C2) systems, and striking systems in a limited resource-based network-centric operational environment (NCOE) using both wired and wireless military communications. These VMF-based tactical information transmissions involve the transmission of only the required command data to minimize the message size. This technique enables faster message transmission compared to that of previous systems and presents the minimum essential technical parameters in the form of a mandatory system standard and optional design objectives for interoperability and compatibility among digital message transfer devices.

1.1. Limitations of the Authentication Process within the Existing VMF Military Standard and Importance

However, because the VMF-based tactical communication uses the Secure Hash Algorithm 1 (SHA-1)-based digital signature algorithm (DSA) as the primary authentication process without reinforcement measures, this can lead to potential limitations owing to major challenges, such as practically proven hash collision vulnerability [2–5], network overhead-based protocol format, and low transmission integrity. This will be problematic within the combat network radio (CNR) in tactical data link (TDL) networks that require maximum tactical security, communication integrity, and low network overhead based on several protocol header bits. In addition, such vulnerabilities will become limitations when high-level adaptive digital-covert activities and deception tactics are successfully implemented in hierarchical corps communication-based network-centric warfare (NCW) because of the unnecessary network transactions required for authentication. Furthermore, RSA encryption, which is primarily used for VMF authentication, will cause a slow processing speed and computational overhead when transferring secure tactical information in rapidly changing unstable battlefield networks with limited resources. Therefore, the existing VMF authentication must be improved to support the reliable transmission and reception of authentication information in limited CNRs, satisfying the requirements for military tactics, strategies, and interoperability.

To improve the overall terminal authentication scheme in VMF-based CNR networks, it is vital to introduce simplified approaches to enable rapid authentication and realize an authorization network connecting the terminals of the CNRs. In addition, a lightweight modification detection and authentication scheme with multiple exception handling processes to quickly generate small robust hash values and reduce authentication flows must be applied. To maintain communication robustness and realize network migration in a limited network environment, the following are vital: reducing unnecessary network transactions for authentication between military terminals owned by squad members; introducing advanced methods to enhance the confidentiality of tactical information exchange; and securing network transmission integrity.

However, in most VMF-based studies and related documents, the efficiency of VMF message transfer for application in the Force XXI Battle Command Brigade and Below (FBCB2)-based embedded ground weapon systems, aerospace trace tracking systems, and tactical mobile platooning communication systems for infantry position reporting has been analyzed only in terms of specialization in obtaining the interoperability and adaptability to alert propagation, fire control, tracking and striking, detection, identification, and defense systems. Differentiated research supporting authentication and data integrity between terminals in a VMF-based operating environment has not been officially reported. Therefore, with increasing cybersecurity threats and a changing NCW system, it is essential to secure military authentication based on the VMF standard.

1.2. Security Requirements for Enhancing VMF Authentication and Research Contribution

As such, the following approach to improving the VMF authentication scheme in a military CNR network based on a wireless terminal device is required.

- The feasibility of implementing a lightweight authentication and integration scheme to improve the existing VMF terminal authentication, such as a hash chain, must be analyzed. The network configuration in CNRs with hash chain-based authentication schemes must be similar to that of the dynamic military ad hoc network-based mobile ad hoc network (MANET) and vehicle ad hoc network (VANET) in the cell planning area.
- Cryptographic hash chain-based authentication and transmission integrity approaches must be considered. Interoperability within the next-generation NCW system, and technical adaptability

and continuity in CNR networks with TDL, should also be ensured, and there should not be any deviations from the specifications of the existing VMF military standard.

- Rather than simply assembling the conceptual technology suggested by previous hash chain-based studies, a comparative analysis based on the pros and cons, differences, and limitations of each study should be considered to suit VMF-based CNR networks.
- Active security requirements that consider various military network scenarios must be established. Examples include a hash chain-based authentication, including a time-based one-time password (T-OTP) keyless point, low latency and transaction, robustness to external and internal auth attacks, a specialization of various FBCB2-based All-IP CNR network types, the introduction of re-authentication and revocation steps in the authentication scheme, and deep interoperability. Functional support and availability for rapidly changing battlefield environments and secure lightweight communication for covert activities are also required.

Accordingly, in this study, an authentication scheme to be employed in rapidly changing VMF-based environments is proposed. It is based on the cryptographic hash chain-based authentication technology that includes T-OTP. In particular, assuming a flexible CNR network environment, the hash chain structure is applied to various military ad hoc networks in dynamic cell planning areas, and its effectiveness is analyzed. Based on the final established authentication scheme within the military ad hoc network and the multi-factor hash chain structure, an improved lightweight authentication scheme is proposed to satisfy the demands of a rapidly changing battlefield network and any additional security requirements based on VMF standards.

1.3. Paper Organization

This hash chain-based lightweight authentication method with T-OTP, including comparative analysis of related studies and tactical network scenarios in CNR networks, is organized as follows. In Section 2, the research and standards related to the hash chain, ad hoc network, and VMF standard are introduced. In Section 3, the unique features and limitations, the advantages and disadvantages, and the applicability of the most relevant prior hash chain-based studies to VMF-based CNR networks are compared and analyzed. In addition, ad hoc network studies on utilizing and assembling such processes in the construction of a VMF-based lightweight authentication scheme are also described. In Section 4, a lightweight authentication scheme applicable to CNRs based on flowcharts and related simulation parameters is proposed. Finally, in Section 5, concluding remarks are provided. Further, the tactical background and technical limitations of the lightweight authentication scheme proposed in this study, as well as future military authentication research directions based on the Korean variable message format (KVMF) in the Korean Army Corps network, are described.

2. Background and Related Studies

2.1. VMF Authentication Improvements Based on MIL-STD-2045-47001

In the existing VMF standard applied to CNRs with TDL, the MIL-STD-2045-47001 standard defines the VMF network protocol and application header, security parameter information (SPI), and group parameter (GP). The purpose of the MIL-STD-2045-47001 standard is to stipulate the technical parameters and procedures, such as application header generation, message packaging and unpackaging, data segmentation, and assembly, which are essential for the message exchange and communication of digital tactical information between the digital message transfer (DMT) equipment and C4I systems in All-IP-based CNRs with limited bandwidth, resources, and device energy. It also presents the VMF message header, RSA-based encryption, and SHA-1-based DSA authentication rules. This determines the DSA-based authentication according to the SPI value in a VMF binary message header, which falls within the scope of this study.

The existing protocols and authentication schemes in the VMF standard have several unique characteristics.

- **Variability**—Variability minimizes message transmission and processing time by dynamically selecting only required real-time tactical messages and information when performing various operations in a limited resources-based network;
- **Bit-coded transmission**—Bit-coded transmission subdivides the transmission unit of a tactical message containing military operation information into bit and octet. It can maximize the transmission efficiency of the operational information and tactical data and ensure integrity;
- **Applicability of multiple layers**—The VMF standard is independent of arbitrary network structures applied in various types of military tactical networks. The MIL-STD-188-220 standard was applied for the physical and link layers, and the MIL-STD-6017 and MIL-STD-2045-47001 standards were applied in the application layer based on CNR networks. Therefore, the multi-layer applicability enables a configuration for the independence of the form of military tactical network and the handling of each squad's platooning behavior and related networks;
- **Integration and interoperability**—The integration of VMF is an essential feature in the design, development, testing, certification, and continuous operation of the automated tactical data system (that is, it satisfies the necessary requirements for promptly transmitting command and control information through joint boundaries). In addition, an integration is required to exchange tactical data and situation awareness information between heterogeneous weapon systems, and to exchange the command system within the allied operation in real-time through an interlocking of the joint tactical data link system (JTDLS). Currently, this is being applied, with a focus on FBCB2-based platooning networks in cell planning areas and mosaic warfare.

However, these protocols and authentication systems used in the VMF standard can cause several issues and limitations.

- **Weak authentication process by SHA-1 cryptographic hash function:** In the existing VMF military standard, the SHA-1-based DSA authentication and modification detection process is performed according to the SPI setting in a dynamic limited CNR network without additional encryption. However, SHA-1 currently has theoretical collision vulnerability and exploits a practical proof of concept (PoC). Thus, to protect confidential military information from leakage and theft in a tactical CNR network under the VMF standard, cryptographic hash-based authentication algorithms (e.g., SHA-2, SHA-3, and LSH of Korean cryptographic hash standard) that are more robust than SHA-1 are urgently required. In addition, although the SPI setting is immediately applied, it is only assumed that authentication in the current VMF-based CNR networks is safe from the perspective of the closed military network. Therefore, it is difficult to protect consistent security in CNR networks from parameter tampering because no separate encryption process in the modification detection and authentication scheme is applied (RSA public encryption may be applied in a separate protocol header area for tactical message encryption, but this may cause additional overhead on network performance and calculation).
- **Lack of factor robustness:** The existing VMF standard includes variables from a rapidly changing battlefield environment (e.g., information loss from environmental changes, latency, jitter, and authentication delay), enemy types (e.g., active and passive internal and external attackers and masqueraders), and operational environments (e.g., bandwidth, energy, frequency, squad, and resources by limited CNRs) that are not well defined. Therefore, it is necessary to introduce additional requirements and establish parameter-based measures.
- **Undefined re-authentication and revocation scheme:** The existing VMF standard does not suggest a continuous authentication scheme when it is necessary to re-establish the reliability of the participating nodes because it takes a certain amount of time after the initial authentication for the operation of the squad to be completed. In addition, processes related to authentication rejection and exclusion as well as the elimination of hostilities are not considered. Therefore, to support a consistent authentication system in a rapidly changing VMF-based CNR network, definitions of re-authentication and revocation schemes for all nodes are urgently required.

- **Non-existent exception process:** The existing VMF standard does not provide any method for the detection, tolerance, and attenuation of node malfunctioning caused by dynamic changes in the network or enemy nodes corrupting operational data and stealing confidential information. Therefore, it is necessary to establish a variety of exception handling methods to satisfy the unique information exchange demand in the battlefield, and support both functional and structural stability. This can also be related to the following challenges: low speed and low bandwidth in real-time VMF-based CNR, the deployment of military base stations for constructing the cell planning area, guaranteeing the availability of authentication for low-spec networks, and the network delay and tolerance related to the establishment of the initial FBCB2-based CNR in mosaic warfare networks, which requires rapidity.

As described above, SHA-1 was proposed by the National Security Agency (NSA) based on message digest 4 (MD4) and adopted in DSA. However, concerns regarding the collision attack vulnerabilities of the SHA-1-based cryptographic hash function were raised in 2005. In 2017, the PoC for generating collision pairs for the SHA-1 hash function was fully released, thereby verifying its theoretical and practical vulnerabilities. As such, owing to the existing vulnerabilities of SHA-1, the VMF standard for executing the authentication scheme between terminals through the SHA-1-based DSA in CNR networks also has limitations. Applying this method to the VMF authentication scheme without overcoming these vulnerabilities can result in the exposure of operational information and the exchange of tactical data to unauthorized nodes in real-time. Even during normal confidential operations, indiscriminate intrusions by enemy forces related to authentication in a tactical network can occur without preventative and detective measures. Thus, the cryptographic hash function must be transitioned into a function with guaranteed robustness, such as SHA-2, SHA-3, or LSH, the Korean cryptographic hash standard [6].

As indicated in Table 1, when the SPI is 0, the data in the VMF-based communication channel do not go through separate one-way encryption types in the authentication and transmission processes, but do undergo one-way encryption based on the weak cryptographic hash function standard, SHA-1. This aspect can also be observed in Tables 2 and 3.

Table 1. Security parameter information type codes.

Code	Reference
0000 (0)	Authentication (using SHA-1 and DSA)/No Encryption
0001–1111 (1–15)	Undefined

Table 2. Typical SPI field sizes.

Field Name	Size (Bits)
Keying Material ID	0–64
Cryptographic Initialization	0–128
Key Token	0–512
Authentication Data (A)	320–1024
Authentication Data (B)	320–1024
Message Security Padding	0–128

Table 3. Digital signature.

Octet	Field Identification	Value
1	Block Number: Identifies specific data block.	15
2	Length: Indicates the length of the Address Designation Parameters block in octets.	Variable length: 13 + size of Digital Signature
3	Hash Algorithm: Used to produce the hash.	0 = MD5 1 = SHA-1
4	Crypto Algorithm: Identification of the crypto algorithm used to encrypt the hash to produce the digital signature.	0 = Not encrypted 1 = RSA 2–255 User defined
5–13	Key ID: signer's public key.	8 octet binary field
14-Length field	Digital Signature: Authentication of the sender of the message.	

2.2. Related Research and Improvements for Introduction of Hash Chain-Based Authentication in VMF

To overcome the limitations related to the authentication and integrity of VMF-based CNR networks, a hash chain-based lightweight authentication scheme including T-OTP should be introduced. The primary considerations of hash chain-based authentication are the security extension protocol (SEP)-related security requirements, computation and network transmission overhead, latency issues, maximum tactical security, and communication integrity. Moreover, to establish a hypothesis for these issues and provide a clear basis for further studies, the potential possibilities of applying a hash chain-based authentication structure, including T-OTP steps, proactive re-initialization for efficient network transactions, specialization of existing military ad hoc networks in cell planning areas, and the configuration of processes related to misbehaving node detection and anomaly auditing, are considered. Therefore, the unique features, advantages, and limitations described in previous related studies based on hash chain and ad hoc must be compared and analyzed for the construction of lightweight authentication in VMF-based CNR networks.

A hash chain-based one-time password, as proposed by Lamport, uses the hash values of a hash chain by applying the same cryptographic hash function in reverse order, thereby preventing the calculation of the hash value used in the next authentication session and addressing the vulnerability of a simple password [7–11]. The S/Key standard proposed by Haller et al. addresses the difficulty of reducing the calculation weight when generating and re-registering the root hash values in a hash chain structure. In addition to enhancing the efficiency in a limited network environment, this standard also prevents reuse attacks [12–14].

The timed efficient stream loss-tolerant authentication (TESLA) protocol proposed by Perrig et al. used a multi-factor method, involving message authentication code chaining and the concepts of time-delayed key disclosure and loose time synchronization. This would mitigate any vulnerabilities to theft or abuse that might arise from a non-combination with other authentication schemes, as well as overhead problems related to the hash chain initialization that were not addressed by S/Key [15–18]. Zhu et al. proposed a lightweight hop-by-hop authentication protocol (LHAP) specialized for MANET networks and related authentication based on TESLA [19,20]. Akbani et al. proposed the hop-by-hop efficient authentication protocol (HEAP) to be employed in wireless networks [21,22], thereby ensuring scalability.

In addition, Zhang et al. proposed a self-renewable hash chain (SRHC) [23], Hamdy et al. proposed an OTP-based two-factor authentication [24], and Bittl et al. proposed an efficient construction of infinite-length hash chains with perfect forward secrecy using two independent cryptographic hash functions [25], further alleviating problems related to initialization, root re-registration, and overhead in existing hash chains. Subsequently, in a T/Key study, a two-factor authentication based on T-OTP and S/Key structure was proposed that further reduced the computational overhead, compared to those in previous studies, while avoiding the storage of the client's secret key in the server [26]. Yin et al. proposed a binary hash tree-based Merkle tree structure to solve potential issues, such as the management of the finite hash chain length, the complexity of computation according to the hash chain length, the lack of a self-reinitialization scheme, and security problems dependent on the hash chain length, related to the efficiency of T-OTP generation and verification in T/Key. Consequently, it was possible to use less storage than T/Key, lower network transactions, and maximize the efficiency of OTP generation and verification time [27].

However, despite these unique features and advantages, previous studies related to hash chain-based authentication have not presented specific exception handling concepts at an algorithmic level for functional problems that potentially arise in a rapidly changing limited battlefield network, such as VMF-based CNRs. In particular, in a low-resource wireless CNR network, where the bandwidth varies according to the layers because of dynamic operational environments, a method for specialized authentication has not been clearly described. An existing hash chain authentication approach in poor communication environments causes several restrictions to network-centric operations, and the performance of the entire network potentially deteriorates when the traffic increases rapidly, such as during wartime. In addition, these naïve hash chain-based authentications have a limitation in that their parameters do not consider the structure or functional aspects under such limited tactical circumstances. In particular, when topology update messages are periodically transmitted to prevent VMF-based CNR networks from disconnecting when the squad commander and related members perform combined platooning operations, poor communication environments may increase the corruption of the authentication status of each squad member, causing network overhead owing to transactions that are required to establish authentication.

Therefore, rather than utilizing the same existing hash chain-related authentication scheme in this military ad hoc network that focuses on cell planning areas and mosaic warfare networks, an alternative approach is necessary to identify and analyze unique features that can be optimized for All-IP or non-All-IP-based CNRs. Accordingly, a comparative analysis of the pros and cons of existing hash chain-based approaches is required.

3. Comparative Analysis of the Existing Hash Chain-Based Authentication Approaches

As described in Section 2, improvements to the DSA-based VMF message authentication processes of the existing SHA-1 are necessary, and their integrity and availability must also be secured for various limited battlefield networks. Therefore, additional requirements must be satisfied, including authentication-related elements provided in the SPI of the existing VMF authentication header; that is, data origin authentication (whether the data transmitted by the sender have been forged along the path), connectionless integrity (limited connectionless configuration for detecting any modifications or retransmissions of the tactical data while preventing hostile nodes from analyzing VMF messages, thereby preventing the identification of the sender's or receiver's tasks), and non-repudiation with proof of origin and proof of delivery. In addition, adaptability to other constraints, such as operational security (covertness), intermittent connectivity, risk of capture and compromise, and limited resources based on low bandwidth, frequency, and device energy, is required in order to operate in rapidly changing tactical network environments [28].

Accordingly, additional security requirements specific to VMF-based CNR networks for a hash chain-based authentication scheme including T-OTP are preemptively defined so as to achieve functional authentication stability over a certain level and high tactical and strategic diversity. Based on these requirements, this approach contains a comparative analysis of prior studies related to hash chains in terms of their differences, unique features, advantages, and limitations, as well as their potential for being applied to VMF-based limited tactical networks.

3.1. Definitions of Additional Tactical Security Requirements in VMF

The following additional security requirements are predefined and analyzed based on general requirements in the existing VMF standards and various national defense documents [29,30]:

- **Configuration of keyless point (CKP) (①)**—Because the hash chain-based authentication with a T-OTP value is used in a rapidly changing battlefield environment, authentication in a VMF-based CNR must be able to verify friendly tactical ad hoc networks in reliable cell planning areas and warfare networks, thereby necessitating robustness against authentication data loss. In addition, to reduce the regular authentication flow, related network overhead and the security cost of server storage after adjusting the initial tactical network, the system is configured to exclude the key ownership stage in one of the two terminals using the authentication algorithm and scheme side. Interference from a variety of environmental variables and wireless jitter must also be considered [31,32];

- **Low latency (LL) (②)**—To rapidly conduct tactical operations and insert this scheme in wireless static terminals, the entire initial authentication process of VMF must be quickly completed. Moreover, to secure the robustness of military-based adaptive strategies [33] in line with developing network technology trends, components of latency-related requirements, such as time-to-transmit, time-to-preprocessing, time-to-retransmission, and bias or noise in battlefield [34,35], based on general parameters in VMF, are reflected and must be defined [36–39];

- **Robustness to authentication attack (RA) (③)**—Security against active and passive internal and external authentication attacks on random nodes in VMF-based CNRs is necessary, which can be achieved using various defensive mechanisms, such as preventing the reuse of keys in hash chain- and state-based exception handling. In particular, the hash chain-based authentication system including a T-OTP value must be configured to minimize the impact of various ad hoc network-based wireless authentication attacks, such as rushing attacks [40] wormhole, blackhole and sinkhole attacks, jellyfish, flooding and fragmentation attacks, man-in-the-middle (MITM), eavesdropping and sniffing [41–45]. Specifically, considering the military security requirements, the resistance capabilities should focus on replay, MITM, and Byzantine authentication attacks.

- **Low authentication overhead in limited networks (LAO) (④)**—To secure interoperability across hierarchical limited-resource CNRs and connected tactical terminals with low specifications, it is necessary to minimize the authentication overhead of the computation and network transactions based on the purpose of VMF standards. In addition, unnecessary transmission and energy consumption must be reduced through exception handling, punishment, and the monitoring of participating nodes in VMF-based closed CNRs, thereby achieving higher efficiency in limited battlefields. Furthermore, it is necessary to enable rapid decision-making and provide high-quality real-time combat environment features [46];

- **Re-authentication and revocation for operations (RR) (⑤)**—Re-authentication and topology updation processes must be provided to ensure that new nodes continue to participate as legitimate nodes even after the initial authentication is completed, re-establish the reliability of internally authenticated nodes, and prevent authentication attacks in an environment where intrusions from enemy forces can lead to theft and damage. It is also necessary to develop immediate authentication rejection and routing-based removal processes for the detected hostile nodes in hash chain-based authentication, including a T-OTP value, and identify internal malfunctioning nodes through an online-based culprit recognition and detection mechanism. Furthermore, more realistic military scenarios related to authentication in VMF must be established to ensure consistent authentication capabilities in actual CNRs. These scenarios must be standardized based on previously proposed military features, such as heterogeneous velocity, tactical areas, optimal paths, obstacles, and unit join and leave scenarios [47,48];

- **Deep interoperability in authentication scheme (DI) (⑥)**—The US Army plans to employ the VMF standard for data exchange in most of its TDL- and CNR-based systems, and the US Navy uses VMF-based TDL to satisfy the tactical requirements of information exchange between ground and maritime operations. According to the unit-specific operability described in US military materials [49] published in 2008, in an actual combat environment in which 64 vehicles and one unmanned aerial vehicle are active, it is possible to observe the movement of nodes constituting squads without disconnections. When one squad moves safely through a specific operational area, other squads often follow, confirming the presence of mobility through group units. On this basis, if multiple nodes forming a squad receive similar command data while moving as a squad, platooning at the same speed and in the same direction, it is necessary to maintain the authentication robustness of the squad CNRs at a high level to ensure the security of not only a given CNR, but also of the overall tactical environment. In addition to forming a smooth C4I between squads within the same country, it is necessary to establish combined operations with the militaries of other nations. As such, any hash chain-based authentication applied in CNRs must adhere to the interoperability demands based on the purpose of VMF standards.

3.2. Applicability Analysis of VMF-Based CNRs in Previous Studies on Hash Chain-Based Authentication

A hash chain is a cryptographic hash function-based one-way chain structure originally devised by the mathematician Lamport. By continuously calculating the hash value using a one-time password and a random value as the initial seed determined by the client, the cryptographic hash function has a preimage resistance. Therefore, it is impossible to recover the original message from the arbitrary hash value of a given message. Because one hash-based password contains both the actual data value and the hash value for the next hash-based password, it is used in the authentication process, which utilizes the continuity and sequentiality of the passwords formed in the chain structure. Therefore, by the comparative analysis of various prior authentication studies using this hash chain structure with the security requirements presented in Section 3.1, its applicability in a VMF-based CNR network in a pre-built cell planning area is determined as follows. The functional satisfaction scores in each study, presented in Table 4, are the key points required in VMF standards. In particular, to logically distinguish the satisfaction of these qualitative requirements, it is necessary to express them as relative indexes, such as weak, slightly weak, slightly strong, and strong, compared to the proposed research method.

Table 4. Taxonomy of hash chain research applicable to VMF authentication process.

	CKP (①)	LL (②)	RA (③)	LAO (④)	RR (⑤)	DI (⑥)
Haller et al. [12]	X	X	△	△	△	△
Perrig et al. [15]	X	△	▲	▲	▲	△
Zhu et al. [19]	X	△	▲	▲	▲	▲
Zhang et al. [23]	X	△	△	△	▲	△
Hamdy et al. [24]	X	△	X	▲	▲	△
Bittl et al. [25]	X	△	O	X	△	△
Kogan et al. [26]	O	△	O	△	▲	O
Yin et al. [27]	O	▲	O	▲	▲	▲

[Functional Satisfaction Score] (X = weak, △ = slightly weak, ▲ = slightly strong, O = strong).

First, Haller et al. [12] proposed "S/Key," an authentication scheme that uses an exclusive OR and a one-way chain structure based on a cryptographic hash function. This technique makes it difficult for attackers to use old message exchange information, despite the existence of all previous communications and related messages between the client and server. Moreover, because this scheme does not allow for duplicate hash values in the chain and establishes overall security based on the preimage resistance, it is suitable for lightweight authentication environments. In the S/Key standard, however, because all hash values in the one-way chain are sequential, unless a random one-time hash value is used, the hash value is valid for an indefinite amount of time and is also vulnerable to theft, abuse, race condition, loop, MITM, and small-n attacks. Therefore, the widespread use of other cryptographic protocols that can secure an entire session, and not only the password, can render S/Key insignificant if mainly used by itself. Despite these limitations, applying such features of the S/Key standard to the authentication process in VMF-based CNR networks will achieve the following:

- The basic verification and re-authentication processes can be standardized using only one S/Key structure-based hash chain without needing multiple independent keys to periodically identify normal nodes in a resource-limited tactical network;
- S/Key has an extensible structure that is easiest to employ in multi-factor authentication without duplication for a randomly exposed key or hash value;
- To transit to a joint operation front with other squads during a combined operation, interoperability is achieved by applying the S/Key-based one-way chain structure used in the previous squad unit to the new squad unit, and based on the sequential hashing process, a rapidly lightweight tactical message transmission and authentication scheme can be secured according to the VMF-based information control. Thus, the existing authentication schemes and related parameters presented in VMF standards can be satisfied and further strengthened based on S/Key as an initial interface for the construction of a lightweight authentication scheme.

Next, Perrig et al. [15] proposed "TESLA", based on the S/Key standard. Despite a recipient knowing the specific signature, the signature for the next sequence cannot be computed in advance until it is disclosed in the next time interval; therefore, malicious nodes in the network are prevented from sending packets after they have been declared false through packet theft and abuse. Thus, by applying the loosely time synchronization concept, this process handles situations in which a specific hash value for authentication is valid for an indefinite period, while reducing the overhead in generating and verifying authentication information, as compared to S/Key (aspects of network transactions and computations), expanding the recipient nodes, and preparing for a packet loss [16]. However, the TESLA authentication has limitations, including the non-guaranteed problem of non-repudiation, and the problem of storage space if packets transmitted during a specific time interval before the release of a random secret key must be continuously buffered in the receiver. Other limitations include the delayed authentication of buffered packets until the secret key is disclosed, the use of the same cryptographic hash function as that in the S/Key, and the inadequate exception handling of malicious nodes. Despite these limitations, applying the features of the TESLA standard to the authentication process in VMF-based CNR networks will achieve the following:

- TESLA is an authentication method for adding the calculated message authentication code (MAC), including a secret key generated through a hash function, to a packet based on a one-way chain. Therefore, based on the act of revealing the hash value owing to the loose time synchronization between the sender and the receiver, it is possible to minimize the exposure of information to the attacker and block false packet attacks. In addition, when there are multiple recipients in a multicast environment, quick and accurate individual authentication for each recipient can also be performed;
- The TESLA standard that was first proposed did not support non-repudiation, but TESLA++, an improved version, can secure a higher non-repudiation function than the elliptic curve digital signature algorithm (ECDSA). Through the TESLA-based VAST combined with the existing ECDSA, it is possible to derive specialization for authentication in military ad hoc networks such as VMF-based CNRs, including MANET and VANET [50];
- In addition to protection from valid authentication values that have not been used for long periods of time in VMF-based CNRs, a system can be established to monitor and identify spoofed nodes using such valid authentication keys. Moreover, hash chain-based authentication schemes can be constructed for an update and expansion of the squad member authorization list for long-term operations.

Zhu et al. [19] proposed "LHAP", which uses a TESLA-based bootstrap trust structure and a one-way chain for traffic and device authentication. As a protocol concept specialized in ad hoc networks based on MANET, this study implemented a network access control scheme while preventing unauthorized nodes from injecting traffic into a given MANET. To implement this process, each node in the network applied token-based hop-by-hop authentication before specific neighboring nodes one hop away transmitted a random packet, thereby deleting abnormal packets. To achieve this, a structure was adopted that combined one-way chain-based authentication and trust management between nodes. Based on this lightweight structure, issues with a buffering-based authentication delay and a large-capacity storage demand owing to delayed key disclosure and loose time synchronization in the existing TESLA could be addressed. Moreover, issues related to securing independence and transparency for multiple routing protocols, increasing the authentication and computational efficiencies compared to those of TESLA, and abnormal node exception handling that was not well defined in TESLA could be addressed through a standardization based on a trust relationship. However, the limitations of the LHAP must be considered, including a central management problem in which all nodes in the network had to share the secret key in advance, a limited key maintenance package, and the increased complexity of the authentication process owing to a larger number of keys, latency caused by one node performing duplicate hashing for verification of abnormal packets and nodes, and authentication delays that could not be fully resolved owing to vulnerabilities to basic intrusion attacks such as wormhole and MITM [20,21]. Despite these limitations, applying the features of the LHAP to the authentication process in VMF-based CNR networks will achieve the following:

- LHAP was proposed to overcome the vulnerability of basic TESLA standards applied to multi-hop ad hoc networks such as MANET. Therefore, the network and computational overhead can be reduced by reducing the number and capacity of requests for nonce and hash values for initial authentication and re-initialization, and removing authentication information based on each node participating in or leaving the VMF-based ad hoc network.
- The concept of hop-by-hop access control can be realized for the behavior of malicious nodes and collaborative hostile nodes that have personal channels. In addition, the issues of exception processing for authentication delay and disconnection according to changes in the ad hoc network can be alleviated by generating a separate control packet. Furthermore, an additional lightweight authentication for distributed VMF-based WSNs can be calculated.
- Beyond filtering invalid authentication keys and packets at the authentication algorithm level, by establishing a system to authenticate and identify legitimate squad nodes and punish enemy nodes, a small-scale VMF-based CNR network specialized for more realistic battlefield environments can be constructed. Establishing trust relationships through specific keys will mitigate issues with authentication delays and network disconnections that arise in limited tactical networks and low-spec military wireless terminals.

Various studies supplementing the weaknesses of the hash chain-based authentication, including the T-OTP value, must also be analyzed, including the problems of the server-side re-registration and initialization after the hash values in the hash chain are exhausted [22], and the client-side overhead based on multiple hash calculations when the hash values are stored beforehand.

Zhang et al. [23] proposed an "SRHC" that does not require a separate re-registration, despite all hash chains being exhausted after registering the root value of the first hash chain in the Lamport's OTP. Here, during the authentication process, the bits constituting the root value of the new hash chain are transmitted to the server in advance, whereas a random hash value is transmitted through the OTP. A one-time signature (OTS) based on a cryptographic hash function standard is also applied as a secondary modification detection and authentication technique for protection against abnormal nodes and attackers during this transmission process. Consequently, all hash values in the current hash chain are exhausted, and the root value of the new hash chain is automatically registered based on a four-way protocol. However, further strategies to improve the SRHC are required owing to other limitations, such as a lack of exception handling in the case of mismatches between specific hash values of the client and server, and the computational and network overhead caused by the increasing length of the transmitted messages. Despite these limitations, applying the features and advantages of the SRHC structure to the authentication in VMF-based CNR networks will achieve the following:

- The central system will automatically mitigate the problem of separate re-initialization owing to the exhaustion of the root keys used for node confirmation, identification, and updation in VMF-based closed CNRs when conducting independent operations for long periods of time [51];
- When applying this hash chain-based authentication in CNR networks, issues such as the potentially limited operational time and resources, unnecessary network transactions associated with checking the remaining hash values, and securing independence owing to specificity in the military environment can be overcome.

Hamdy et al. [24] proposed "OTP-based two-factor authentication," an infinitely superimposed hash chain structure that applies a two-factor authentication-based multi-hash function to address the cost, time, and performance limitations of the authentication process. In this study, the hash chain generation and usage processes are applied in the forward direction, as opposed to the existing Lamport method. In addition, when the server side sends a specific index value to a client based on the hash chain generated in the forward direction, the client responds with an OTP. As such, because the cryptographic hash function operates in the forward direction rather than in the reverse direction, as in other hash chain structures, a re-registration owing to the exhaustion of the hash values is not required. However, specific strategies to improve the two-factor authentication-based infinitely superimposed forward hash chain structure are required owing to various limitations, such as the existing simple-password problem owing to the seed value being secret information immediately shared by the client and server, as well as the inability to realize the advantages of the OTP. Despite these limitations, applying the features of a two-factor authentication-based infinitely superimposed forward hash chain structure to the authentication in VMF-based CNRs will achieve the following:

- The re-initialization problem occurring from the exhaustion of keys used for authentication and updation when performing independent operations over a long period of time will be simply resolved. The performance cost owing to the participation of multiple new squad and member nodes and the establishment of a joint operation with other squads will also be minimized.

Similar to other advanced studies, to solve the problem of hash chain re-registration, an infinite hash chain technique based on the double independent hash function proposed by Bittl et al. [25] was designed such that a client had to apply a specific cryptographic hash function to a random seed value and another hash function shared by only the client and server to the output value, which was used as an OTP. Compared to the techniques used in previous studies, this technique enabled the realization of robust security at higher levels of OTP. However, this technique had a limitation in that the client and server had to share secret information, such as seed, in advance. Thus, prior studies on improving the hash chain-based OTP assumed that the secret key, cryptographic hash function, and digital signature, among other factors, were shared between the client and the server in advance. Owing to this limitation, they could not be efficiently established compared to other cryptographic primitives, which were fundamental advantages of a hash chain structure. Specific strategies to address this issue are necessary. Despite these limitations, applying the features of a double independent hash function-based infinite hash chain to the authentication process in VMF-based CNR networks will achieve the following:

- The security of the authentication in each closed CNR network will be further strengthened, and the scheme will be more robust than those in other methods against the issues of limited key maintenance, authentication delays from duplicate hashing, wormhole attacks, and MITM;
- Application of this approach also enables the conversion of the authentication scheme into one that is specialized for changing battlefields in which enemy forces employ active confusion tactics based on the network.

Kogan et al. [26] proposed a "T/Key" based on a multi-authentication scheme that combines the S/Key and T-OTP and does not store the secret key on the server. T/Key uses independent hash functions to address any potential security instabilities that arise from deriving hash values through the same cryptographic hash function in the S/Key. In addition, regarding the vulnerability to phishing attacks owing to the indefinite validity of the hash value by applying the concept of the time interval length, the validity of a specific hash value can be limited up to only a predetermined period of time. Furthermore, to alleviate the ripple effects of preprocessing attacks, an independent salt value is assigned to each cryptographic hash function in the hash chain, thereby verifying the lower limit of the hash value for each function and establishing proper security. However, because the hash values in the T/Key structure are time-limited, the structure is significantly longer than the existing S/Key structure, thereby leading to potential structural vulnerabilities. Moreover, other potential issues, such as the management of finite hash chain length, the complexity of computation according to hash chain length, the lack of a self-reinitialization scheme, the security problems dependent on hash chain length, and the efficiency of T-OTP generation and verification, remain. Therefore, possible designs to be employed in limited-resource CNRs as well as a specific authentication formulation based on T/Key are required. Despite these limitations, applying these features of the T/Key to the authentication process in VMF-based CNR networks will achieve the following:

- Limiting the lifetime of valid authentication keys that have not been used for long periods based on a predefined time span will enable the preparation of functional countermeasures to prevent, identify, and attenuate the continuous collection of information by specialized enemy nodes and conspirator nodes;
- Laying the foundation for military cyber agility [52,53] in a limited-resource CNR, based on independent cryptographic hash functions that can be deployed at any given time for each node in a tactical network, will facilitate the development of an authentication scheme that proactively prevents intrusions from specialized hostile nodes in a battlefield environment;
- Establishing a multi-factor hash chain-based authentication process including T-OTP will secure high levels of robustness and responsiveness against any initial reconnaissance attempts by enemy nodes, maintain tactical covert activities, and maximize the covert activities of friendly forces based on a scheme specialized for low-spec and low-speed closed squad networks;
- Furthermore, the ability to easily establish joint operation systems with other squad members and foreign militaries by synchronizing the hash functions will help satisfy various interoperability-related scenarios.

Based on these potential possibilities, introducing the T/Key concept is vital for schemes related to the multi-factor hash chain-based authentication scheme including T-OTP in VMF-based CNRs.

4. Proposed Lightweight Authentication Based on Multi-Factor Hash Chain with T-OTP in VMF

This section describes the design principle, initialization and registration, authentication and verification, re-authentication and revocation, exception handling flow processes based on tactical scenarios, and multi-factor hash chain-based authentication with T-OTP that satisfy the predefined requirements of the VMF-based CNRs. In addition, the conditions and additions satisfied by the final proposed scheme are compared with those in the methods in previous studies.

4.1. Design Principle

The DSA authentication used with the existing VMF standard faces severe challenges, such as collision vulnerabilities of the SHA-1-based cryptographic hash function. In the CNR networks, any confidential tactical messages, such as situation reports, location reports, and reconnaissance reports, are exchanged in real-time according to the VMF standards. Therefore, VMF standards should be robust against cryptographic problems using strong cryptographic hash functions such as SHA-2, SHA-3, and LSH. In addition, it is also vital to secure inter-message operability for low-spec connected wireless terminals, such as position reporting equipment based on infantry communication devices, armored terminals in tank and helicopter types, and limited networks used in NCW-based tactical operations that change in real-time as well as adding a re-authentication procedure based on sudden unexpected military operational scenarios. The concepts of our proposed processes were improved based on the multi-factor hash chain and T-OTP proposed by Kogan and Yin. The related authentication scheme, with multi-factor hash chain and T-OTP values that satisfy the overall requirements of the existing VMF standards, is illustrated in Figure 1.

①　Initialization and registration phases: The OTP seed hash generation in an LSH-based multi-factor hash chain between the administrator (the military authority or the commander of squad) and the user (the connected wireless device or squad member) in CNRs.

②　Authentication and verification phases: Rapid authentication and verification with time-stamp-based clock synchronization values for T-OTP, and the limitation of the lifetime of arbitrary hash values in a hash chain that have not been used for a long time.

③　Re-authentication and revocation phases: Tactical scenario-based network regularization of authentication status and exception handling for CNRs, including wireless terminals.

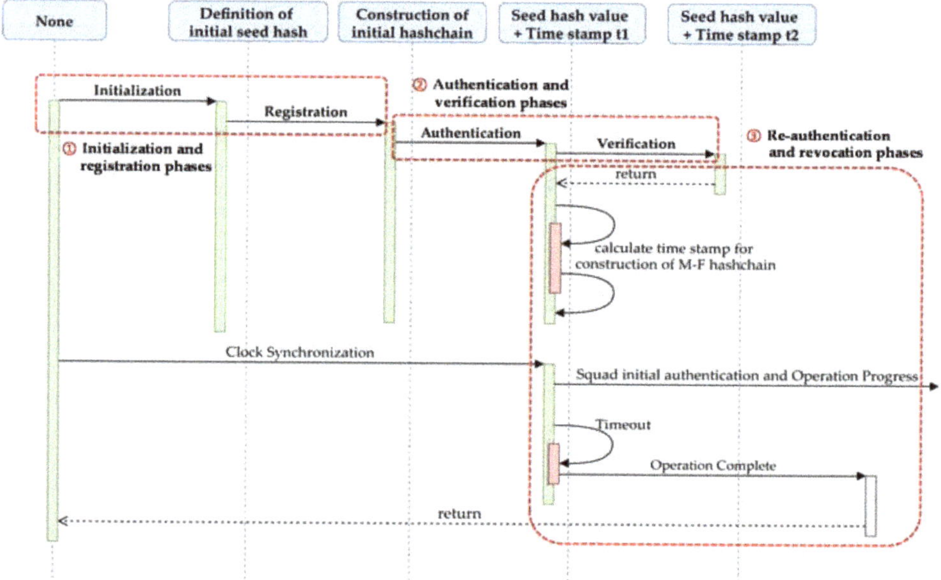

Figure 1. Proposed overall authentication scheme with multi-factor hash chain and T-OTP in VMF.

The proposed processes included three specific phases: initialization and registration, authentication and verification, and re-authentication and revocation. The scheme proposed in this study used the hash value of the LSH-based hash chain with T-OTP in the CNR networks, including wireless military devices and internal terminals. After the initial authentication, the status of the current authentication progress was reflected in a real-time value, such as the serialization of a QR code, and the participating tactical equipment nodes were updated based on static transmission rules in CNR networks in cell-planning areas. Subsequently, if an internal or external malfunctioning node or hostile node was detected and re-authentication was required, an additional authentication scheme could be easily executed at any time through clock synchronization of the hash functions in the LSH-based multi-factor hash chain of combined operation.

4.2. Lightweight Authentication Processes Based on Hash Chain

4.2.1. Initialization and Registration Phases

The processes proposed in this study were improved by applying the existing multi-factor hash chain and T-OTP authentication proposed by Kogan and Yin. The one-time initialization and registration phases similar to the hash chain-based OTP method are depicted in Figure 2.

① Fetch user information: Execution of the initialization and registration phases.

② Challenge/Reject: Initial settings of the range of the time slot length for the configuration of T-OTP values in the multi-factor hash chain structure by the administrator.

③ Key matching: Clock synchronization (periodic linkage of time between the user and administrator in CNR networks within a predetermined time-error range).

④ Establish session: Delivery of the seed hash value on T-OTP by the administrator to the user.

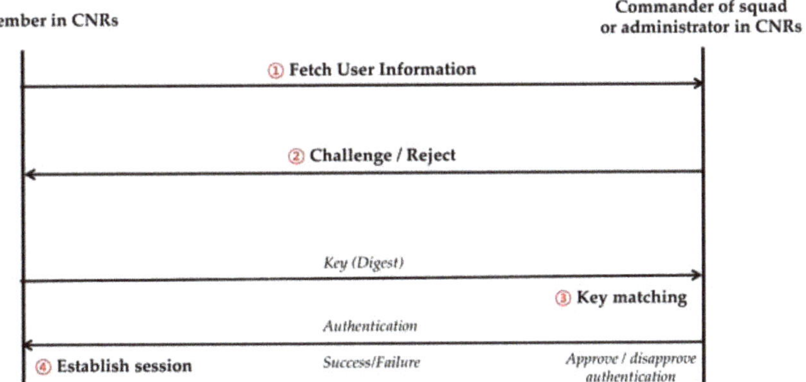

Figure 2. Initialization and registration phases.

The administrator was defined as the operation control authority or commander of the squad in any military operation, and the user was defined as a connected device or squad members that executed the operations under the received tactical messages. During the initialization and registration phase, the hash value of the hash chain was used as the T-OTP to execute the authentication phase. A user shared a bit-length seed to calculate the hash chain used as the OTP, after which the hash chain $h(x)$ was continued up to length i of the repetitive hash chain. The hash chain length was defined as 2×10^7 to 2×10^8 owing to the 2^{114} or 2^{224} based minimum security strengths and open multi-factor standard, and the length of the key used for certification was dynamically composed of values within the range of 130–224 bits because of protocol header storage limitations in VMF standards. In contrast to the Kogan method, a periodic clock synchronization process existed between the user and the administrator within a predetermined time-error range. In addition, for the authentication and verification phases that directly followed the initialization and registration phases, the user and administrator did not have to share secret keys, cryptographic hash functions, or digital signatures during the T-OTP authentication process based on the hash value of the hash chain. They had to share only pre-built abstract hash chain structures in wireless military terminals at the time of creation of the VMF-based CNR networks.

4.2.2. Authentication and Verification Phases

After the initialization and registration phases, the user utilized the T-OTP in an LSH-based multi-factor hash chain structure that used the hash values generated by applying the same cryptographic hash function in reverse order to prevent the calculation of the hash value used in the next authentication session, thereby addressing the vulnerability of a simple password.

The objective of this phase was to deliver to the administrator the seed hash value of the LSH-based hash chain that the user would apply as the T-OTP. During the previous phase, authentication was performed using the time-based multi-factor method, utilizing the seed during the clock synchronization process. In the S/Key, because all hash values in the hash chain were sequential, they were indefinitely valid and vulnerable to theft and abuse. To mitigate this vulnerability, prior studies on improving a hash chain-based OTP assumed that a secret key, cryptographic hash function, and digital signature, among other factors, were shared, in advance, between the client and the server, resulting in limitations compared to other cryptographic primitives. However, through time-stamp-based clock synchronization, our proposed authentication process utilized hash values and related T-OTP values by applying a predefined time slot length between nodes participating in the VMF message exchange in the operation execution in CNRs. As such, the lifetimes of the hash values and T-OTPs of valid authentication keys that have not been used for long periods of time were limited to prevent, identify, and attenuate the continuous collection of information by specialized hostile nodes. The related phases are illustrated in Figure 3.

①   Definition of initial seed hash: The initial seed hash of T-OTP, in which clock synchronization is executed and defined based on static transmission rules in military corps scenarios.

②   Clock synchronization: The real-time authentication-based squad interoperability between the user and administrator within a predetermined time-error range is confirmed from the perspective of the operation start time in CNRs in cell-planning areas.

③   Sharing and calculating based on time slot length: The calculated hash values between the administrator and user within the range of the dynamic time slot length is verified. Next, authentication is performed based on T-OTP in the LSH-based hash chain structure.

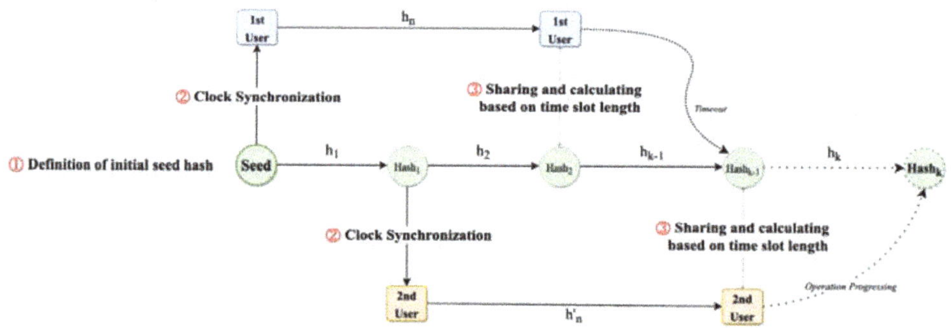

Figure 3. Authentication and verification phases.

The significant advantage of the LSH-based hash chain with multi-factor seed management was the independence of the T-OTP table between the administrator and the user in CNRs. Security was ensured because the administrator did not know the user's secret key. In terms of the feature securing availability in a mobile environment, such as ad hoc networks, by applying a finite dynamic time slot length, the validity of a specific hash value was limited up to the moment a predetermined period of time had passed. Thus, the proposed hash chain-based T-OTP yielded better computational and utilization efficiency compared to other cryptographic primitives. In addition, an independent salt value was assigned to each cryptographic hash function in the hash chain to verify the lower limit of the hash value of each function, thereby establishing the level of security. Here, the administrator could change the predetermined period for each operation group; as such, the clock synchronization process did not cause any issues owing to a low influence on real-time operation.

4.2.3. Re-Authentication and Revocation Phases

This phase included security considerations for the initial operation of various VMF-based closed CNRs and the rapid additional authentication of new squad nodes based on an extension of secure communication and authentication sessions. In addition, it was necessary to maintain an authenticated connection between squad members in the command system who were conducting squad operations and refresh the new authentication status with legitimated squad members after a certain amount of operation time had elapsed. The related phases are shown in Figure 4.

①   Identify the state: The current authentication state of each squad member node in VMF-based CNR networks and a tactical system is identified.

②   Check if a failure session occurs within this arbitrary time: The re-authentication variable-based military scenario in dynamic battlefields is recognized.

③   Whether a user has already achieved authentication within this arbitrary time is checked: The squad node authentication and re-authentication are executed in CNR networks for refreshing.

④   Setting of a time scheduler for clock synchronization: The re-authentication time scheduler for a configuration of the dynamic time stamp is set.

⑤   Re-configuration of time slot length: The slot length for clock synchronization is configured.

⑥   Phase execution: The re-authentication and execution of this overall phase is awaited.

⑦   Phase repetition: The authentication and verification phases with the re-authentication scheduler (if necessary) are repeated.

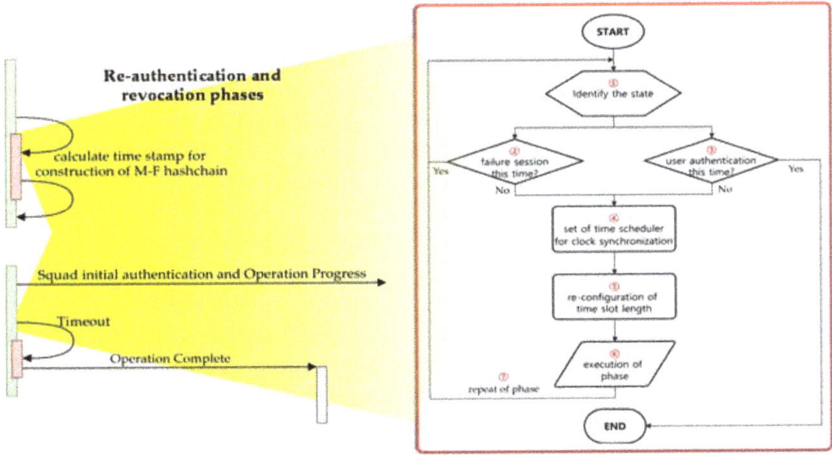

Figure 4. Re-authentication and revocation phases.

During the re-authentication and revocation phases, a scenario-based theoretical scheme was proposed to improve the robustness of each operation phase, the responsiveness and speed of each squad, the robustness and agility of the authentication for rapidly changing battlefield environments, the maximization of allied nodes and minimization of potential enemy nodes in CNRs related to digital-covert activities, and the independence of seed hash re-initialization based on securing transmission integrity, lower network transaction, and the reception of tactical information.

4.2.4. Testing of Exception Handling Flows in Specialized Military Scenarios in VMF-Based CNRs

This phase involved testing to prove the reliability of the LSH-based multi-factor hash chain with T-OTP value schemes in CNRs, including specialized military scenarios. Certain example scenarios to test exception handling flows are as follows.

- **Scenario 1**: Masquerading hostile nodes for exfiltration of tactical information (Figure 5).
- **Scenario 2**: Internal friendly nodes deliberately or accidentally conspiring with malicious outsiders based on deceptive collusion attacks (Figure 6).
- **Scenario 3**: Selfish friendly nodes monopolizing limited network resources with an enemy node that has been compromised (Figure 7).

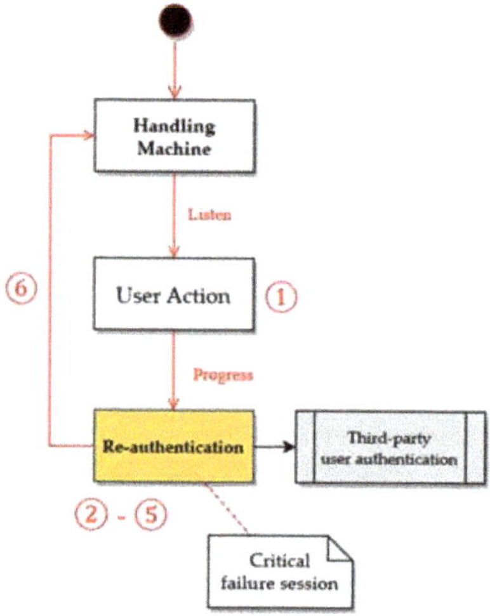

[Scenario 1]

When an enemy node deceives a friendly node.

Example: Enemy nodes disguised as allies.

① Detection of abnormal behavior of unspecified nodes during military operation in a dynamic battlefield.

② Situational awareness of critical failure session-based state machine.

③ Instant node authentication based on hash chain with T-OTP and each node authentication in third-party environments, such as central control systems.

④ Early execution of identification and authorization.

⑤ Elimination of a node from the network when it is identified as an enemy node.

⑥ Reflection of the variation state value and related time slot length for clock synchronization in the hash chain.

Figure 5. Scenario 1—based exception handling flows.

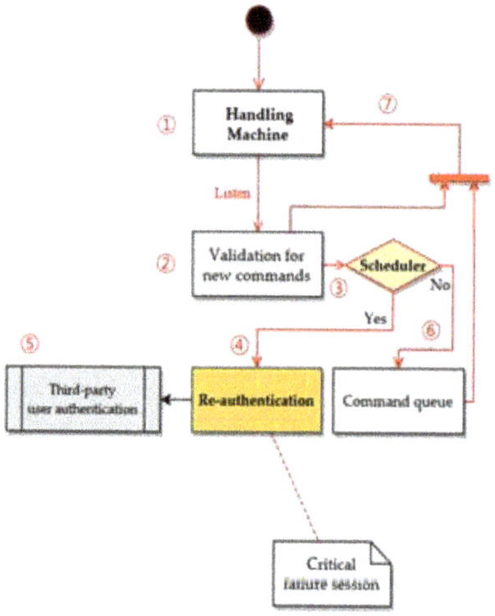

[Scenario 2]

When an enemy conspires maliciously with friendly nodes.

Example: Friendly node intentionally or unintentionally conspiring with outsiders.

① Validation if a node in VMF-based CNRs is online.

② Pass when no exception occurs with tactical inspection.

③ Parameterization of clock cycles in re-authentication.

④ Re-authentication according to scheduler parameters during communication; elimination of a node from the network if it is identified as an enemy node when a critical failure session is recognized during the re-authentication process.

⑤ Early execution of identification and authorization.

⑥ Waiting for a message to be received when the scheduler is not required.

⑦ Reflection of the variation state and the related time slot length for clock synchronization in the hash chain.

Figure 6. Scenario 2—based exception handling flows.

[Scenario 3]

When a friendly node is terminated and removed under the administrator's authorization owing to a problem occurring during a tactical operation.

Example: Termination of existing nodes for efficient processing of restricted resources.

① Awaiting new commands and user authentication of each squad in third-party environments.

② Activation of operation activity termination, such as operation end command.

③ Elimination of nodes that are no longer used for a tactical combat network.

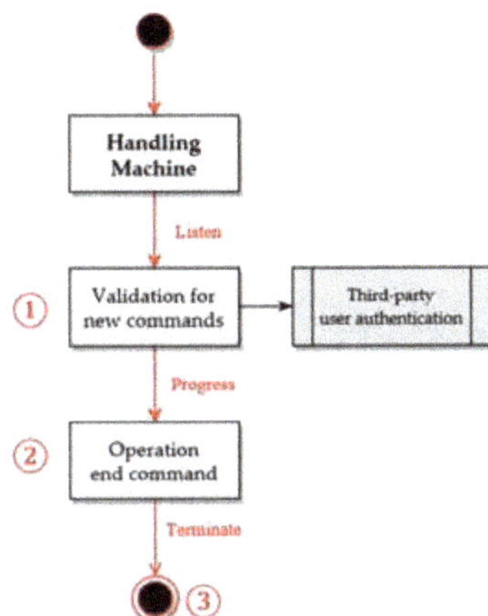

Figure 7. Scenario 3—based exception handling flows.

Scenario 1 checked the maintenance power of tactical confidentiality and fraud detection of an occupied hostile node or an enemy node disguised as a friendly node (that is, when abnormal behavior is detected and recognized as a critical failure session, the LSH-based hash chain including T-OTP was used to conduct an emergent authentication between the commander of the squad and the squad members in CNR networks). In addition, when a specific abnormal node was identified as an enemy, it was removed from the tactical squad status, and the change was reflected in the CNR. Scenario 2 checked the detection and prevention of internal nodes that might conspire with unspecified hostile nodes. While the inter-connection was maintained, re-authentication was performed according to the re-authentication scheduler (periodic). Depending on the resilience of the operation, the clock cycle for the re-authentication scheme was dynamically configured according to the purpose of the tactical operation (non-periodic). When a critical failure session was recognized during the re-authentication process, the corresponding enemy node was removed from the network, and the state of the belonging squad nodes was reflected in the VMF-based CNRs. Scenario 3 checked the robustness of the network structure when removing abnormal nodes that occupied limited resources, such as low transaction for lightweight authentication or nodes that were currently not required for additional allied operations. In the event that member nodes in the squad network were changed to conduct additional tactical operations, unwanted nodes were safely removed.

In addition to the aforementioned three scenario-based exception handling flows, attack scenarios that can be considered in limited military ad hoc-based CNR networks should reflect the characteristics of military communication environments in small-scale combat situations, for example, hierarchical high-capacity transmitter radio-net (HCTR) communication for independent battalions above brigade levels, low capacity transmitter radio-net (LCTR) communication below battalion levels, tactical mobile communication system (TMCS), and master–slave-based swarm communication. These include malicious radiowave interference in wideband network waveforms (WNW), spoofing and meaconing-based GPS jamming, advanced DoS-based operation availability violations, deceptions, and other cyber threats that can significantly impact limited tactical networks.

4.3. Configuration of Experimental Environments and Related Variables Based on VMF-Based CNR Networks

The configuration of related experimental environments and variables focused on comparative analysis combined with compound military metrics and wireless tactical operation scenarios based on Korean Army Corps networks in VMF-based CNRs. An overview of the NS-3- and MATLAB-based co-testbed with limited resources in military ad hoc, cell-planning areas, and mosaic warfare networks is depicted in Figure 8.

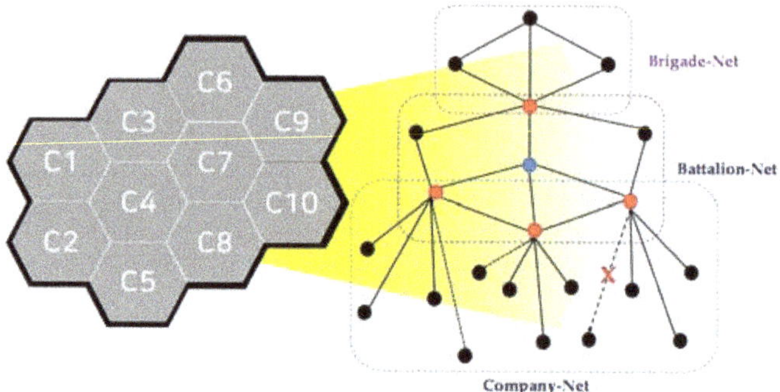

Figure 8. Overview of VMF-based CNR testbed with ad hoc and static cell-planning concepts.

Here, the red dot is the commander node in the CNRs and is the only one involved in establishing a communication channel between the upper and lower military ad hoc networks based on VMF. The blue dot is a communication forwarding node and exists to ensure the level of communication tolerance according to regional changes in CNRs and cell planning areas. In general, for VMF-based CNRs and related military ad hoc networks, communication environments and related configurations and structures are determined, in advance, before the operation is performed. However, because the communication structure of the VMF-based military ad hoc networks may change according to changes in operations, such as performing combined operations with other squads, the situation, such as the participation and withdrawal of nodes, is also reflected.

In addition, as shown in Figure 9, VMF-based CNR networks have different ad hoc structures for each wireless communication device, such as FM- or All-IP-based radio. The green dot is the transmitter node of the upper network when configuring the cluster swarming communication network types related to the ground and aerial unmanned systems, such as drones, Unmanned Aerial Vehicle (UAV) and Unmanned Ground Vehicle (UGV). The orange dot is a master or leader node constituting a cluster swarming communication network.

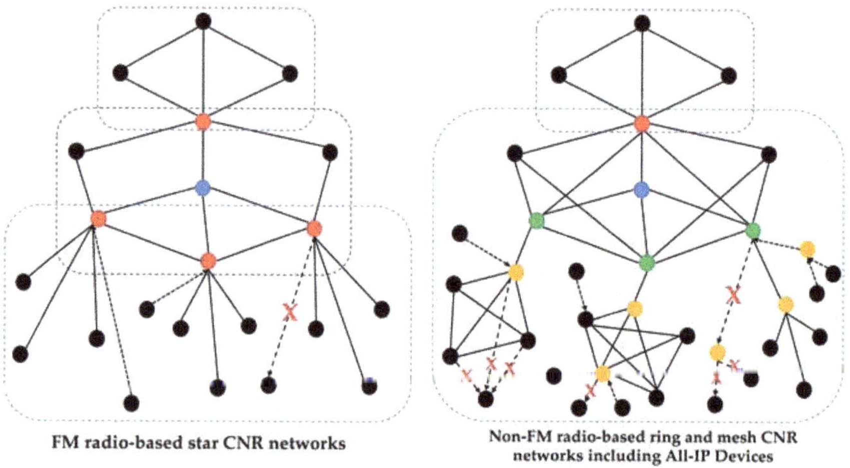

Figure 9. VMF–CNR network architecture based on FM radio and All-IP next-generation radio.

The overall experimental variables related to the VMF-based CNR network co-testbeds are presented in Tables 5–8, and Figure 10. In addition, when configuring related experimental variables, the following major tactical perspectives are considered.

①   Currently, the VMF-based CNR networks, to which the LSH-based multi-factor hash chain with a T-OTP authentication system is applied, are considered to support a maximum bandwidth of approximately 2 Mbps and BPSK modulation. The primary goal is to transfer messages as quickly and with as low an overhead as possible, while ensuring low-transaction-based authentication and integrity within the CNRs. Therefore, preferentially, based on the extreme configuration from a minimum of 2.4 kbps bandwidth to a maximum of 192 kbps bandwidth, additional variables such as transmission rate, transmission and reception electric power, modulation and demodulation, maximum transmit time (MTT), and jitter should be variably defined.

②   There are differences in the communication power specifications of wireless devices for each squad class and military specification in the VMF-based CNRs. In addition, there is a difference in the equipment used for mixed voice and multi-media transfers. In other words, the wireless communication for infantry in a small squad network does not have an amplifier; as such, the bandwidth is low owing to small energy resources. However, the pieces of wireless communication terminal equipment mounted on tanks, corps helicopters, and corps unmanned vehicles have an amplifier, resulting in high radio energy emissions; as such, the bandwidth is higher. Therefore, all these aspects should be applied within the VMF-based military ad hoc network for each communication environment such as MANET and VANET, and the concepts of mobility variations and ground device specifications to simulate rapidly changing battlefields and limited network resources should be defined and calculated.

③   The VMF-based CNR networks are largely classified into an FM radio-based communication type and an All-IP-oriented non-FM radio-based communication type. The FM radio environment includes a hierarchical star cluster network such as "brigade–battalion–company–platoon," and each commander-specific communication uses different frequency bands. The All-IP-based non-FM radio environment includes a ring-type network, determines upper, lower, and unmanned swarming communication based on WAN-LAN, and uses different frequency bands for each operation or set of military occupational skills (MOS). Therefore, by applying all these aspects, isolated closed radio networks within VMF-based CNRs and related ad hoc networks should be constructed.

④   The communication nodes are physically separated from the network, and only commander nodes, such as the brigade commander, battalion commander, company commander and platoon commander, can perform hierarchical sequential communication. In other words, in the FM radio environment, commander nodes other than the rank of platoon commander use two static radio channels instead of one, and squad members can communicate only to commanders in the same squad networks. Commander nodes in an All-IP-based non-FM radio environment have two IP classes themselves, such that they can simultaneously configure closed or open radio channels. All radio channels or IP addresses in FM and non-FM radio environments are not dynamically assigned, but are statically injected and maintained until the end of the operation. In addition, when an arbitrary commander node belonging to the lower cluster network requests communication from the upper commander, the lower node must directly join the upper network and change to the frequency of the upper cluster network. Therefore, all these aspects should be considered and applied.

⑤   The authentication variable values related to the LSH-based multi-factor hash chain with the T-OTP system are obtained in a form that is injected before the start of the tactical operation. In addition, wireless communication authentication in the cell-planning areas based on corps ground and air control stations is similarly considered. Therefore, all these aspects should be considered and applied.

Table 5. Examples of message variables in CNR networks based on infantry networks.

Message	Generation Cycle (s)	Generation Length (Bit)	Importance	Allocation
Location reporting	10	200, 400, 600	Routine	Broadcast
Reconnaissance reporting	1000–2000	2000, 4000, 12,000	Priority	Multicast
Volley ordering	1000–2000	150	Urgent	Unicast

Table 6. Examples of communication parameters in physical layer-related VMF standards.

Parameter	Value	Description
Number of static cells	1–7	Number of cell clusters with center
Number of nodes	8–1200	Number of participating nodes in CNRs
Execution time (s)	120–259,200 (3 days)	Operation time in VMF-based CNRs
Join nodes	0–200	Number of join nodes in ad hoc
Withdrawal nodes	0–30	Number of withdrawal nodes in ad hoc
Bandwidth (kbps)	2.4–192	Frequency band size
Frequency	Declared according to bandwidth	Frequency band
Power (w)	0.01–0.2	Transmit and receiver power
Coverage	Declared according to power	Device communication coverage
Distance (m)	1–3000	Distance of transmission
Number of channels	1–3	Number of sessions for communication
Modulation	BPSK, QPSK	Modulation method
Data rate (bps)	2400–7200	Data transfer rate
MTT	0.0–0.02	Maximum transmit time
PER	0.0–0.1	Packet error rate
BER	0.0–0.1	Bit error rate

Table 7. Configuration of parameters in LSH-based multi-factor hash chain with T-OTP scheme.

Parameter	Value
Hash algorithm	LSH-256, LSH-512
Size of nonce (bit)	224, 256, 384, 512
Minimum security length (bit)	112 (NIST), 224 (BSA), 256 (NSA)
Execution time based on hash chain (s)	128–1024
Time slot length for T-OTP (s)	5, 10, 30, 60, 300, 600
Message digest (bit)	150, 200, 400, 600, 2000, 4000, 12,000
Crypto period (year)	1, 2, 3

Table 8. Configuration of authentication testbed environment (not military spec).

Element	Value
Processor 1 for central administration	Intel Xeon E series, Intel i7-10700
Processor 2 for commander	Intel i7-10875H, Samsung Exynos 8890 (virtual)
Processor 3 for member	Qualcomm Snapdragon 805 (virtual)
Type of communication nodes	Producer, Interpreter, Forwarder, Consumer, and Noise node (Declared according to number of nodes)
Enabling jitter nodes for causing overhead such as PDV, PER, and BER	True (0–20)
Type of communication packets	Voice, Data, Image, and Video
Type of ACK	TWOACK and S-ACK
Re-transmission of max count-based ACK	0–3
Type of wireless radio-based SDR	HF (AM), VHF (AM), and UHF (AM/FM)
Enabling cell-planning node	True (0–10)/False
Enabling terrain information	False

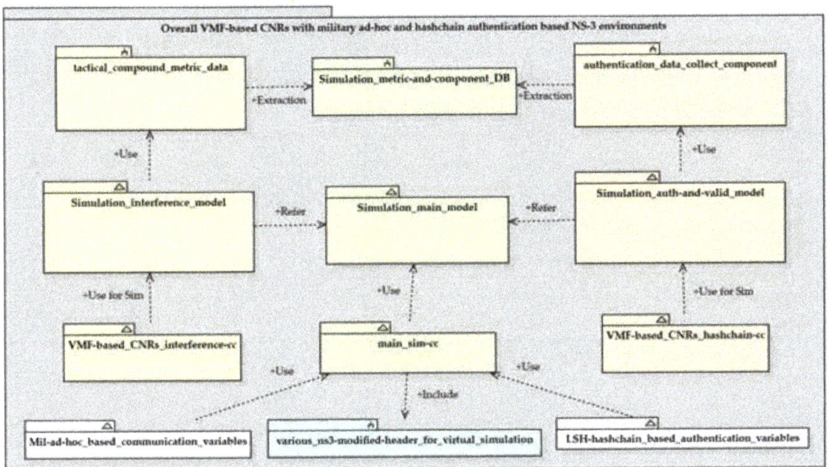

Figure 10. Overview of overall VMF-based CNR testbed architecture with hash chain authentication.

4.4. Comparison between Overall Proposed Model and Previous Studies

The existing hash chain-based studies required improvements owing to network operation constraints as a result of the poor communication environment in rapidly changing VMF-based wireless CNR networks. The proposed authentication model addressed these limitations through a four-phase process of initialization and registration, authentication and verification, re-authentication and revocation, and various exception handling flows. The related comparison results and the taxonomy of the 112 bit-based minimum security strengths are presented in Figure 11 and Table 9, respectively.

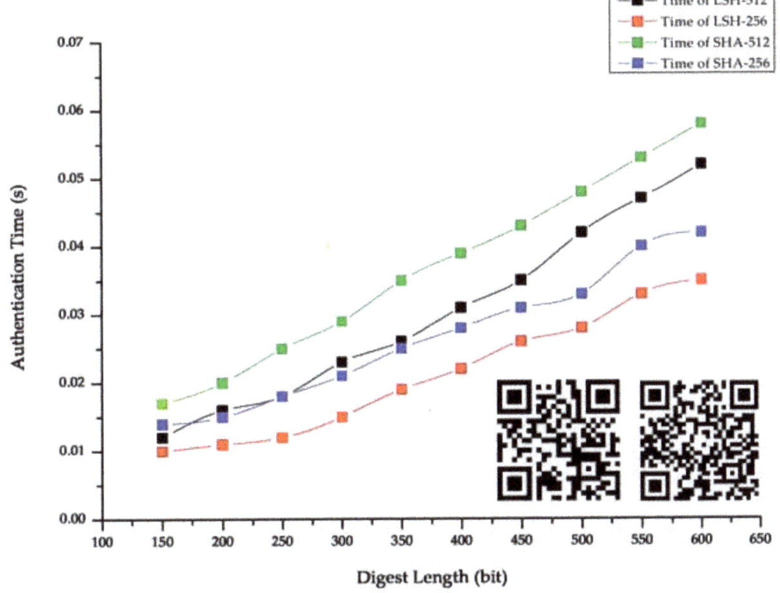

Figure 11. Brief comparative result of LSH and SHA hash functions based on each QR code.

Table 9. Taxonomy of existing LSH hash chain-based authentication and proposed model.

	CKP (①)	LL (②)	RA (③)	LAO (④)	RR (⑤)	DI (⑥)
VMF [1]	O	O	O	O	X	X
Haller et al. [12]	X	X	△	△	△	△
Perrig et al. [15]	X	△	▲	▲	▲	△
Zhu et al. [19]	X	△	▲	▲	▲	▲
Zhang et al. [23]	X	△	△	△	▲	△
Hamdy et al. [24]	X	△	X	▲	▲	△
Bittl et al. [25]	X	△	O	X	△	△
Kogan et al. [26]	O	△	O	△	▲	▲
Yin et al. [27]	O	▲	O	▲	▲	▲
Our proposed authentication model	O	▲	O	▲	O	O

[Functional Satisfaction Score] (X = weak, △ = slightly weak, ▲ = slightly strong, O = strong).

During the initialization and registration phases, for the conditions of LL (②), the entire initial authentication process of the VMF-based tactical CNR networks based on wireless terminals and squad devices was completed within a short period of time, allowing the operations to be rapidly conducted and the multiple squads to be quickly input into the battlefield environment. In terms of LAO (④), smooth communication and message transmission were ensured for hierarchical low-speed, low-bandwidth networks and the low-spec connected equipment with limited resources. Because the authentication process was an improvement over existing processes in VMF standards, it also secured interoperability, satisfying the DI (⑥) conditions based on military ad hoc networks, such as MANET and VANET, with pre-built cell planning areas. During the authentication phase, by satisfying the CKP (①) conditions, the authorization of friendly squad nodes in arbitrary tactical networks could be verified at a reliable level, even in a rapidly changing limited network. In addition, to maintaining a regular authentication process and reducing the overhead, the system was configured to exclude the key ownership stage in one of the two terminals from the authentication algorithm, while satisfying the RA (③) conditions. Thus, security against overall wireless attacks in a CNR was achieved by activating various defensive mechanisms, such as preventing the reuse of random keys and exception handling. The RR (⑤) conditions were satisfied during the verification phase, which ensured that any new nodes continued to participate as legitimate nodes even after the initial authentication was completed. An environment in which theft and damage might occur owing to intrusions from unforeseen hostile enemy forces during combined operations was considered when designing this scheme.

5. Conclusions

In this study, for the first time, a multi-factor hash chain-based lightweight authentication scheme including T-OTP values applicable to VMF-based CNR networks was proposed. The existing military tactical message system using the VMF standard has potential vulnerabilities in terms of message integrity and authentication, which arise from the use of a digital signature algorithm based on SHA-1- and RSA-based encryption. Accordingly, through comparative analysis of previous studies based on pros and cons, in this study, a lightweight authentication scheme was proposed. This model could enhance the integrity of tactical message exchanges and reduce unnecessary network transactions and transmission bits for authentication flow in VMF-based CNR networks, while ensuring robustness with limited resources. In addition, by considering the actual limited military communication environment, such as HCTR, LCTR, TMCS, and unmanned ground and aerial system-based swarm communication, and presenting functional exception handling flows that are not implemented in the existing VMF military standard as a basis for possible cyber threats, military authentication processes applicable to small-scale combat situations can be practically constructed.

In the future, we intend to increase the reliability between wireless devices in the KVMF, and apply it in-depth as lightweight authentication in Korean Army network scenarios. Finally, the limitations of this study and plans for future research are as follows.

- Because this hash chain-based study focused on the design and analysis of a lightweight authentication for a VMF standard, future studies will assess its quantitative performance through network load tests in TDL based on the Korean Army Corps network scenarios with related equations, pseudo codes, and compound All-IP- or non-All-IP-based wireless topologies.
- During the re-authentication and re-validation phases, aiming to take advantage of the dedicated hash chain configured using the Byzantine fault tolerance property among other failure models in a distributed system, specific scenario-based schemes and attack graphs will be presented. In addition, they will be judged on their ability to be classified into specific tactical scenarios, including state machines for critical failure between actual operations.
- To obtain a dynamic adaptation of the proposed authentication process for a rapidly changing battlefield environment, the responsiveness of each squad must be considered for the rapid deployment of various nodes and the successful execution of various tactics and strategies. Therefore, by further devising a process for applying concrete state-based conditional exception handling modeling, a military communication system-based VMF message transfer standard will be produced that can also consider real-time authentication situations.

Author Contributions: Conceptualization, D.K., S.S., and H.K.; methodology, S.S. and H.K.; software, D.K., Y.K.L., and W.G.L.; validation, D.K., S.S., and Y.K.L.; formal analysis, S.S. and H.K., investigation, D.K., Y.K.L., and H.K.; resources, D.K., W.G.L., and S.S.; data curation, D.K., S.S., H.K. and Y.K.L.; writing—original draft preparation, Y.K.L., D.K., and W.G.L.; writing—review and editing, S.S. and D.K.; visualization, D.K. and S.S.; supervision, W.G.L. All authors have read and agreed to the published version of the manuscript.

Funding: This research was funded by the ADD project, the Defense Acquisition Program Administration and the Agency for Defense Development under the contract UD200023ED.

Acknowledgments: The authors gratefully acknowledge the financial support provided by the Defense Acquisition Program Administration and the Agency for Defense Development under the contract UD200023ED.

Conflicts of Interest: The authors declare no conflict of interest.

References

1. MIL-STD-2045/47001D (w/CHANGE 1), Department of Defense Interface Standard: Connectionless Data Transfer Application Layer Standard. June 2008. Available online: http://everyspec.com/MIL-STD/MIL-STD-2000-2999/MIL-STD-2045_47001D_CHANGE-1_25098 (accessed on 1 November 2020).
2. The FIPS 180-1 Secure Hash Standard. April 1995. Available online: https://nvlpubs.nist.gov/nistpubs/Legacy/FIPS/fipspub180-1.pdf (accessed on 1 November 2020).
3. Wang, X.; Yin, Y.; Yu, H. Finding collisions in the full SHA-1. In *Advances in Cryptology—CRYPTO 2005, Proceedings of the Annual International Cryptology Conference, Santa Barbara, CA, USA, 14–18 August 2005*; Springer: Berlin/Heidelberg, Germany, 2005; Volume 3621, pp. 17–36.
4. Wang, X.; Yin, Y.L.; Yu, H. Finding Collisions in the Full SHA-1. In *Advances in Cryptology–CRYPTO 2005*; CRYPTO 2005; Lecture Notes in Computer Science; Springer: Berlin/Heidelberg, Germany, 2005; Volume 3621. [CrossRef]
5. Stevens, M.; Bursztein, E.; Karpman, P.; Albertini, A.; Markov, Y. The first collision for full SHA-1. In *Proceedings of the 37th Annual International Cryptology Conference, Santa Barbara, CA, USA, 20–24 August 2017*; Springer: Berlin/Heidelberg, Germany, 2005; Volume 10401, pp. 570–596.
6. Lee, J.; Kim, J. LSH: A new fast secure hash function family. In *Proceedings of the 17th International Conference on Information Security and Cryptology, Seoul, Korea, 3–5 December 2014*; Springer: Berlin, Germany, 2005; Volume 8949, pp. 286–313.
7. Lamport, L. Password authentication with insecure communication. *Commun. ACM* **1981**, *24*, 770–772. [CrossRef]

8. Eldefrawy, M.H.; Khan, M.K.; Alghathbar, K. One-time password system with infinite nested hash chains. In *Security Technology, Disaster Recovery and Business Continuity*; Communications in Computer and Information Science: Berlin, Germany, 2010; Volume 122, pp. 161–170.

9. Hu, Y.C.; Perrig, A. A survey of secure wireless ad hoc routing. *IEEE Secur. Priv.* **2004**, *2*, 28–39.

10. Hyun, S.; Ning, P.; Liu, A.; Du, W. Seluge: Secure and DoS-resistant code dissemination in wireless sensor networks. In Proceedings of the International Conference on Information Processing in Sensor Networks, St. Louis, MO, USA, 22–24 April 2008; pp. 445–456.

11. Deng, J.; Han, R.; Mishra, S. Secure code distribution in dynamically programmable wireless sensor networks. In Proceedings of the 5th International Conference on Information Processing in Sensor Networks, Nashville, TN, USA, 19–21 April 2006; pp. 292–300.

12. Haller, N.M. The S/Key™ one-time password system. In Proceedings of the Internet Society Symposium on Network and Distributed System, San Diego, CA, USA, 1 January 1994; pp. 151–157.

13. Zhang, Y.; Chen, Y.; Sun, Y.; Chen, M. Training demand analysis for airlines safety manager based on improved OTP model. In *Proceedings of the International Conference on Human-Computer Interaction, as Vegas, NV, USA, 15–20 July 2018*; Springer: Berlin/Heidelberg, Germany, 2018; pp. 334–342.

14. Mitchell, C.J.; Chen, L. Comments on the S/KEY user authentication scheme. *ACM SIGOPS Oper. Syst. Rev.* **1996**, *30*, 12–16. [CrossRef]

15. Perrig, A.; Canetti, R.; Tygar, J.D.; Song, D. The TESLA broadcast authentication protocol. *RSA CryptoBytes Tech. Newsl.* **2002**, *5*, 2–13.

16. Perrig, A.; Song, D.; Canetti, R.; Tygar, J.; Briscoe, B. Timed Efficient Stream Loss-Tolerant Authentication (TESLA): Multicast Source Authentication Transform Introduction. *Req. Comments* **2005**, RFC-4082. Available online: https://tools.ietf.org/html/rfc4082 (accessed on 12 December 2020).

17. Jakobsson, M. Fractal hash sequence representation and traversal. In Proceedings of the IEEE International Symposium on Information Theory, Lausanne, Switzerland, 30 June–5 July 2002; p. 437.

18. Hu, Y.-C.; Jakobsson, M.; Perrig, A. Efficient constructions for one-way hash chains. In Proceedings of the International Conference on Applied Cryptography and Network Security, New York, NY, USA, 7–10 June 2005; pp. 423–441.

19. Zhu, S.; Xu, S.; Setia, S.; Jajodia, S. LHAP: A lightweight hop-by-hop authentication protocol for ad-hoc networks. In Proceedings of the 23rd International Conference on Distributed Computing Systems Workshops, Providence, RI, USA, 19–22 May 2003; pp. 749–755.

20. Lu, B.; Pooch, U.W. A lightweight authentication protocol for mobile ad hoc networks. In Proceedings of the International Conference on Information Technology: Coding and Computing (ITCC'05), Las Vegas, NV, USA, 4–6 April 2005; pp. 546–551.

21. Akbani, R.; Korkmaz, T.; Raju, G.V.S. HEAP: Hop-by-hop efficient authentication protocol for mobile ad-hoc networks. In Proceedings of the 2007 Spring Simulaiton Multiconference, Norfolk, VA, USA, 25–29 March 2007; pp. 157–165.

22. Goyal, V. How to re-initialize a hash chain. *IACR Cryptol. ePrint Arch.* **2004**, *97*, 1–9.

23. Zhang, H.; Li, X.; Ren, R. A novel self-renewal hash chain and its implementation. In Proceedings of the IEEE/IFIP International Conference on Embedded and Ubiquitous Computing, Shanghai, China, 17–20 December 2008; pp. 144–149.

24. Eldefrawy, M.H.; Alghathbar, K.; Khan, M.K. OTP-based two-factor authentication using mobile phones. In Proceedings of the Eighth International Conference on Information Technology: New Generations, Las Vegas, NV, USA, 11–13 April 2011; pp. 327–331.

25. Bittl, S. Efficient construction of infinite length hash chains with perfect forward secrecy using two independent hash functions. In Proceedings of the 11th International Conference on Security and Cryptography (SECRYPT), Vienna, Austria, 28–30 August 2014; pp. 213–220.

26. Kogan, D.; Manohar, N.; Boneh, D. T/Key: Second-factor authentication from secure hash chains. In Proceedings of the ACM SIGSAC Conference on Computer and Communications, Dallas, TX, USA, 30 October–3 November 2017; pp. 983–999.

27. Yin, X.; He, J.; Guo, Y.; Han, D.; Li, K.C.; Castiglione, A. An efficient two-factor authentication scheme based on the Merkle tree. *Sensors* **2020**, *20*, 5735. [CrossRef] [PubMed]

28. Burbank, J.L.; Chimento, P.F.; Haberman, B.K.; Kasch, W.T. Key challenges of military tactical networking and the elusive promise of MANET technology. *IEEE Commun. Mag.* **2006**, *44*, 39–45. [CrossRef]

29. Tang, H.; Salmanian, M.; Chang, C. *Strong Authentication for Tactical Mobile Ad Hoc Networks*; Defence R & D Canada: Ottawa, ON, Canada, 2007.

30. Lee, K.; Lee, S.; Kim, Y.; Kwon, K.; Lim, W. A study for hop count on the ad-hoc of wireless communication. In Proceedings of the 14th International Conference on Advanced Communication Technology (ICACT), PyeongChang, Korea, 19–22 February 2012; Volume 176, pp. 931–935.

31. Tu, W.; Lai, L. Keyless authentication and authenticated capacity. *IEEE Trans. Inf. Theory* **2018**, *64*, 3696–3714. [CrossRef]

32. Jiang, S. Keyless authentication in a noisy model. *IEEE Trans. Inf. Forensics Secur.* **2014**, *9*, 1024–1033. [CrossRef]

33. Kong, J.; Luo, H.; Xu, K.; Gu, D.L.; Gerla, M.; Lu, S. Adaptive security for multilevel ad hoc networks. *Wirel. Commun. Mob. Comput.* **2002**, *2*, 533–547. [CrossRef]

34. Ji, H.; Park, S.; Yeo, J.; Kim, Y.; Lee, J.; Shim, B. Ultra-reliable and low-latency communications in 5G dwnlink: Physical layer aspects. *IEEE Wirel. Commun.* **2018**, *25*, 124–130. [CrossRef]

35. Ji, H.; Park, S.; Shim, B. Sparse vector coding for ultra reliable and low latency communications. *IEEE Trans. Wirel. Commun.* **2018**, *17*, 6693–6706. [CrossRef]

36. Parvez, I.; Rahmati, A.; Guvenc, I.; Sarwat, A.I.; Dai, H. A survey on low latency towards 5G: RAN, core network and caching solutions. *IEEE Commun. Surv. Tutorials.* **2018**, *20*, 3098–3130. [CrossRef]

37. Gao, P.X.; Narayan, A.; Karandikar, S.; Carreira, J.; Han, S.; Agarwal, R.; Ratnasamy, S.; Osdi, I.; Gao, P.X.; Narayan, A.; et al. Network Requirements for Resource Disaggregation. In Proceedings of the 12th USENIX Symposium on Operating Systems Design and Implementation, Savannah, GA, USA, 2–4 November 2016; pp. 249–264.

38. Seok, B.; Sicato, J.C.S.; Erzhena, T.; Xuan, C.; Pan, Y.; Park, J.H. Secure D2D communication for 5G IoT network based on lightweight cryptography. *Appl. Sci.* **2019**, *10*, 217. [CrossRef]

39. Vladyko, A.; Khakimov, A.; Muthanna, A.; Ateya, A.A.; Koucheryavy, A. Distributed edge computing to assist ultra-low-latency VANET applications. *Future Internet* **2019**, *11*, 128. [CrossRef]

40. Hu, Y.C.; Perrig, A.; Johnson, D.B. Rushing attacks and defense in wireless ad hoc network routing protocols. In Proceedings of the 2nd ACM Workshop on Wireless Security, San Diego, CA, USA, 19 September 2003; pp. 30–40.

41. Wu, B.; Chen, J.; Wu, J.; Cardei, M. A survey of attacks and countermeasures in mobile ad hoc networks. In *Wireless Network Security*; Springer: Berlin/Heidelberg, Germany, 2007; pp. 103–135.

42. Aad, I.; Hubaux, J.P.; Knightly, E.W. Impact of denial of service attacks on ad hoc networks. *IEEE/ACM Trans. Netw.* **2008**, *16*, 791–802. [CrossRef]

43. Abdul-Ghani, H.A.; Konstantas, D.; Mahyoub, M. A comprehensive IoT attacks survey based on a building-blocked reference model. *Int. J. Adv. Comput. Sci. Appl.* **2018**, *9*, 355–373.

44. Ponsam, J.G.; Srinivasan, R. A survey on MANET security challenges, attacks and its countermeasures. *Int. J. Emerg. Trends Technol. Comput. Sci.* **2014**, *3*, 274–279.

45. Alani, M.M. MANET security: A survey. In Proceedings of the IEEE International Conference on Control System, Computing and Engineering (ICCSCE), Batu Ferringhi, Malaysia, 28–30 November 2014; pp. 559–564.

46. Bar-Noy, A.; Cirincione, G.; Govindan, R.; Krishnamurthy, S.; Laporta, T.F.; Mohapatra, P.; Neely, M.; Yener, A. Quality-of-information aware networking for tactical military networks. In Proceedings of the IEEE International Conference on Pervasive Computing and Communications Workshops (PERCOM Workshops), Seattle, WA, USA, 21–25 March 2011; pp. 2–7.

47. Aschenbruck, N.; Gerhards-Padilla, E. A survey on mobility models for performance analysis in tactical mobile networks. *J. Telecommun. Inf. Technol.* **2008**, *2*, 54–61.

48. Suri, N.; Benincasa, G.; Lenzi, R.; Tortonesi, M.; Stefanelli, C.; Sadler, L. Exploring value-of-information-based approaches to support effective communications in tactical networks. *IEEE Commun. Mag.* **2015**, *53*, 39–45. [CrossRef]

49. Lu, X.; Chen, Y.-C.; Leung, I.; Xiong, Z.; Liò, P. A novel mobility model from a heterogeneous military MANET trace. In Proceedings of the International Conference on Ad-Hoc Networks and Wireless, Sophia-Antipolis, France, 10–12 September 2008; Volume 5198, pp. 463–474.

50. Studer, A.; Bai, F.; Bellur, B.; Perrig, A. Flexible, extensible, and efficient VANET authentication. *J. Commun. Netw.* **2009**, *11*, 574–588. [CrossRef]

51. Chefranov, A.G. One-time password authentication with infinite hash chains. In *Novel Algorithms and Techniques in Telecommunications, Automation and Industrial Electronics*; Springer: Berlin/Heidelberg, Germany, 2008; pp. 283–286.
52. Mitchell, W. *Three C2 Models for Military Agility in the 21st Century*; Royal Danish Defence College Press: Copenhagen, Denmark, 2012.
53. Alexeev, A.; Henshel, D.S.; Levitt, K.; McDaniel, P.; Rivera, B.; Templeton, S.; Weisman, M. Constructing a science of cyber-resilience for military systems. In Proceedings of the NATO IST-153 Workshop on Cyber Resilience, Munich, Germany, 23–25 October 2017; pp. 23–25.

Publisher's Note: MDPI stays neutral with regard to jurisdictional claims in published maps and institutional affiliations.

Article

Post Quantum Cryptographic Keys Generated with Physical Unclonable Functions

Bertrand Cambou *, Michael Gowanlock, Bahattin Yildiz, Dina Ghanaimiandoab, Kaitlyn Lee, Stefan Nelson, Christopher Philabaum, Alyssa Stenberg and Jordan Wright

College of Engineering Informatics and Applied Sciences (CEIAS), Northern Arizona University (NAU), Flagstaff, AZ 86011, USA; michael.gowanlock@nau.edu (M.G.); bahattin.yildiz@nau.edu (B.Y.); dg856@nau.edu (D.G.); kdl222@nau.edu (K.L.); swn34@nau.edu (S.N.); cp723@nau.edu (C.P.); ajs937@nau.edu (A.S.); jaw566@nau.edu (J.W.)
* Correspondence: bertrand.cambou@nau.edu; Tel.: +1-928-523-7824

Featured Application: Using physical unclonable functions (PUFs), in support of networks secured with a public key infrastructure, to generate, on demand, the key pairs needed for lattice and code PQC algorithms.

Abstract: Lattice and code cryptography can replace existing schemes such as elliptic curve cryptography because of their resistance to quantum computers. In support of public key infrastructures, the distribution, validation and storage of the cryptographic keys is then more complex for handling longer keys. This paper describes practical ways to generate keys from physical unclonable functions, for both lattice and code-based cryptography. Handshakes between client devices containing the physical unclonable functions (PUFs) and a server are used to select sets of addressable positions in the PUFs, from which streams of bits called seeds are generated on demand. The public and private cryptographic key pairs are computed from these seeds together with additional streams of random numbers. The method allows the server to independently validate the public key generated by the PUF, and act as a certificate authority in the network. Technologies such as high performance computing, and graphic processing units can further enhance security by preventing attackers from making this independent validation when only equipped with less powerful computers.

Keywords: lattice cryptography; code cryptography; post quantum cryptography; physical unclonable function; public key infrastructure; high performance computing

Citation: Cambou, B.; Gowanlock, M.; Yildiz, B.; Ghanaimiandoab, D.; Lee, K.; Nelson, S.; Philabaum, C.; Stenberg, A.; Wright, J. Post Quantum Cryptographic Keys Generated with Physical Unclonable Functions. *Appl. Sci.* **2021**, *11*, 2801. https://doi.org/10.3390/app11062801

Academic Editors: Leandros Maglaras, Ioanna Kantzavelou and Mohamed Amine Ferrag

Received: 13 February 2021
Accepted: 18 March 2021
Published: 21 March 2021

Publisher's Note: MDPI stays neutral with regard to jurisdictional claims in published maps and institutional affiliations.

1. Introduction

In most public key infrastructure (PKI) schemes for applications such as cryptographic currencies, financial transactions, secure mail and wireless communications, the public keys are generated by private keys with Rivest–Shamir–Adleman (RSA) and elliptic curve cryptography (ECC). These private keys are natural numbers, typically 3000-bit long for RSA and 256-bits long for ECC. For example, in the case of ECC, the primitive element of the elliptic curve cyclic group is multiplied by the private key to find the public key. It is now anticipated that quantum computers (QC) will be able to break both RSA and ECC when the technology to manufacture enough quantum nodes becomes available. The paper entitled "A Riddle Wrapped in an Enigma" by N. Koblitz and A. J. Menezes suggested that the ban of RSA and ECC by the National Security Agency is unavoidable, and that the risk of QC is only one element of the problem [1]. Plans to develop post quantum cryptographic (PQC) schemes have been proposed to secure blockchains by Kiktenko et al. [2], and for cryptocurrency security by Semmouni et al. [3], even if the timeline for the availability of powerful QC is highly speculative. Recently, Campbell et al. [4], and Kampanakisy et al. [5], proposed distributed ledger cryptography and digital signatures with PQC. In 2015, the National Institute of Standards and Technology (NIST) initiated a large-scale program to

Appl. Sci. **2021**, *11*, 2801. https://doi.org/10.3390/app11062801

standardize PQC algorithms. One possible implementation of PQC algorithms for a PKI is the one in which each client device, or designate, generates the key pairs, and sends the public keys to a certificate authority (CA). This assumes a separate authentication process, and that each client device can securely store the key pairs.

The research question that is the subject of this paper is the feasibility of using physical unclonable functions (PUFs), together with a handshake process with the CA that generate new key pairs from the PUF at each transaction, thereby eliminating the need to store the key pair. Attempts to retrieve the secret keys are not relevant anymore as they are only used once. Such a configuration is raising several structural and technical questions. A secure enrollment process of each PUF needs to be established, and the CA has to store the challenges and reference values of each PUF. Such an infrastructure is already known when the security is based on secure hardware elements and tokens and requires special protections against opponents. From a technical standpoint, it is questionable that the long key pairs necessary for PQC algorithms can be generated from physical elements. While a single bit mismatch is not acceptable for PQC algorithms, the natural drifts of PUFs over environmental conditions and aging are real concerns that need to be addressed. This paper is structured in the following way:

[**Section** 2]: The lattice and code-based cryptographic algorithms under consideration for standardization by NIST are presented. These algorithms are well documented, and the software stack written in C can be downloaded for IoT implementation. The schemes are based on the generation of random numbers, and the computation of public–private key pairs. The digital signature algorithms (DSA), and key encapsulation mechanisms (KEM) are not more complex to implement with PQC than with the existing asymmetrical cryptographic schemes.

[**Section** 3]: We present some of the challenges associated with the use of PUF technology to secure PKI architectures. The proposed methods are based on existing cryptographic schemes, and commercially available PUFs. We present how the response based cryptographic (RBC) scheme can overcome the bit error rates (BER) that occur when keys are generated from physical elements. Finally, we present some hardware considerations in the implementation of PQC for PKI.

[**Section** 4]: In this section, we propose schemes that use PUFs to generate the public-private key pairs for lattice and code-based cryptography. We show how the combination of random number generators, combined with the streams generated by the PUF can generate key pairs with relatively low error rates. We show how the error in these streams can be corrected using a search engine.

[**Section** 5]: Finally, in the implementation and experimental section, we compare cryptographic schemes and algorithms. We analyze the experimental results comparing the efficiency of RBC operating with various PQC schemes, ECC, and advanced encryption standard (AES). As expected, asymmetrical schemes are slower than AES; however, the performance of the selected PQC algorithms is encouraging for the implementation of PUF-based architecture, using the RBC to handle the expected BER.

2. Lattice and Code-Based Post Quantum Cryptography

In 2019, the number of candidates of the NIST standardization effort was narrowed to 26, as part of phase two of the program [6]. In July 2020, NIST announced the selection of seven likely finalists for phase three of the program [7]: CRYSTALS-Kyber, CRYSTALS-Dilithium, SABER, NTRU, and FALCON with lattice cryptography [8–12]; RAINBOW with multivariate cryptography [13], and Classic McEliece with code-based cryptography [14,15]. The software developed is mainly targeting DSA applications, as well as KEM. Lattice cryptography is relatively mature, well documented, and is most likely to become mainstream for cybersecurity. Lattice-based algorithms exploit hardness to resolve problems such as the closest vector problem (CVP), learning with error (LWE), and learning with rounding (LWR) algorithms, and share similarities with the knapsack cryptographic problem.

2.1. Learning with Error Cryptography

The LWE of the CVP problem was first introduced by Regev [16]. The knowledge of integer-based vector t, and matrix A with $t = A.s_1$ cannot hide the vector s_1; however, the addition of a "small" vector of error s_2 with $t = A.s_1 + s_2$, makes it hard to distinguish the vectors s_1 and s_2 from t. The vector s_2 needs to be small enough for the encryption/decryption cycles, but large enough to prevent a third party from uncovering the private key $(s_1; s_2)$ from the public information $(t; A)$. The public–private cryptographic key pair generation for client device i can be based on polynomial computations in a lattice ring, and is described in Figure 1:

1. The generation of a first data stream called seed $a_{(i)}$ that is used for the key generation; in the case of LWE, the seed $a_{(i)}$ is shared openly in the network.
2. The generation of a second data stream called seed $b_{(i)}$ that is used to compute a second data stream for the private key $Sk_{(i)}$; the seed $b_{(i)}$ is kept secret.
3. The public key $Pk_{(i)}$ is computed from both data streams and is openly shared.
4. The matrix $A_{(i)}$ is generated from seed $a_{(i)}$.
5. The two vectors $s_{1(i)}$ and $s_{2(i)}$ are generated from seed $b_{(i)}$.
6. The vector $t_{(i)}$ is computed: $t_{(i)} \leftarrow A_{(i)}\, s_{1(i)} + s_{2(i)}$.
7. Both seed $a_{(i)}$ and $t_{(i)}$ become the public key $Pk_{(i)}$.
8. Both $s_{1(i)}$ and $s_{2(i)}$ become the private key $Sk_{(i)}$.

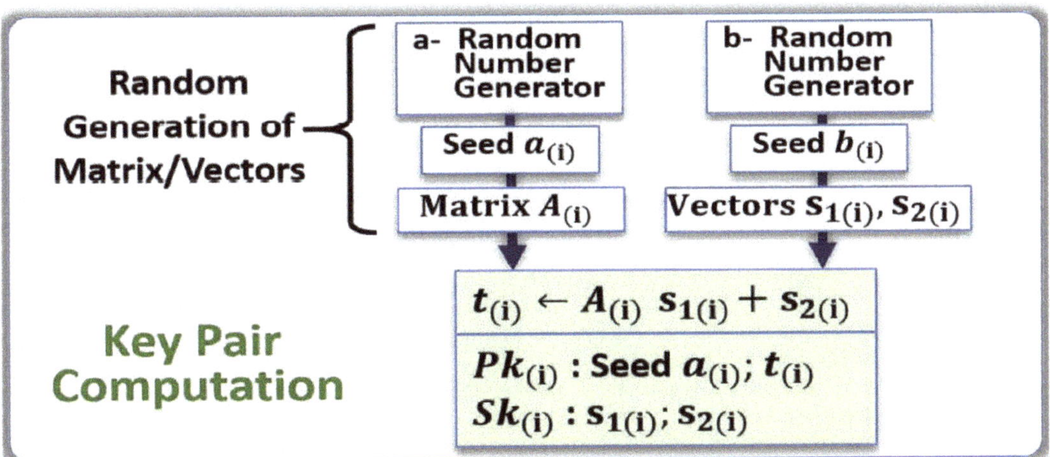

Figure 1. Example of public–private key generation for LWE based cryptography. The matrix $A_{(i)}$ and vectors $S_{1(i)}$ and $S_{2(i)}$ are generated from random number generators. The vector $t_{(i)}$ is computed from these elements for the generation of the key pair.

A digital signature algorithm (DSA) can be realized from the LWE instance by first generating a public–private key pair as in Figure 1. The secret key is then used to sign a message, and the public key is used to verify this signed message. In CRYSTALS-Dilithium [8], the authors use a Fiat-Shamir with Aborts approach [17] for their signing and verification procedure. The outline of the signing procedure is as follows:

1. Generate a masking vector of polynomials y.
2. Compute vector $A.y$ and set w_1 to be the high-order bits of the coefficients in this vector.
3. Create the challenge c, as the hash of the message and w_1.
4. Compute intermediate signature $z = y + c.s_1$.
5. Set parameter β to be the maximum coefficient of $c.s_1$.
6. If any coefficient of z is larger than $\gamma_1 - \beta$, then reject and restart at step 1.

7. If any coefficient of the low-order bits of $A.z - c.t$ is greater than $\gamma_2 - \beta$, then reject and restart at step 1.

Note: γ_1, γ_2, and β are set such that the expected number of repetitions is between 4 and 7.

The general outline of the verification procedure is given by the following:

Compute w_1' to be the high-order bits of $A.z - c.t$ and accept if all coefficients of z are less than $\gamma_1 - \beta$ and if c is the hash of the message and w_1'.

Encapsulation allows for two parties to securely share a symmetric key by encapsulating the key in ciphertext. When both parties have the symmetric key, they are then able to use a symmetric-key encryption algorithm to communicate (e.g., AES). These algorithms are known as key encapsulation mechanisms (KEM) and a few examples from NIST are SABER [18], Classic McEliece [14,15], CRYSTALS-Kyber [19], and NTRU [20]. The process of using encapsulation with LWE/LWR is described below:

- The public and private keys of both parties are constructed as described in Figure 1.
- Person A sends Person B their public key.
- Person B randomly generates a symmetric key and encapsulates it in a ciphertext with the public key of person A.
- Person B sends the ciphertext to person A.
- Person A decapsulates the ciphertext with their private key.
- Both parties now have the symmetric key in their possession.

In summary, LWE schemes are now relatively mature, and very well documented. The methods selected by the NIST standardization program, presented here, are straightforward to use. The codes are widely available online for download, and we successfully deployed them in our research environment to study the use of PUFs for key generation.

2.2. Learning with Rounding Cryptography

The learning with rounding problem was first introduced by Banerjee [21]. It is the derandomized version of learning with error, which deterministically generates the noise in the LWE by rounding coefficients. This will eliminate the noise sampling, and significantly reduce the bandwidth [22]. The LWR is proved to be as hard as LWE to solve; hence, it remains secure to be used in cryptographic applications. In schemes such as "Saber", a constant h is added as a constant vector to simulate the rounding operation by bit shifting, therefore playing a similar protecting role than the error vectors of LWE [18]. Saber, which is one of the NIST's finalists in the key encapsulation category, uses LWR for key generation in public key encryption and key encapsulation. Below all three steps of PKE and KEM are described:

Saber PKE Key Generation

1. Similar to LWE, seed $a_{(i)}$ is used to generate matrix $A_{(i)}$.
2. Seed $b_{(i)}$ is used to generate vector $s_{(i)}$.
3. The vector $t_{(i)}$ is computed: $t_{(i)} \leftarrow A_{(i)}. s_{(i)} + h_{(i)}$.
4. Both seed $a_{(i)}$ and $t_{(i)}$ become the public key $Pk_{(i)}$.
5. $s_{(i)}$ becomes the private key $Sk_{(i)}$.

Saber PKE Encryption

1. The seed $a_{(i)}$ and $t_{(i)}$ is extracted from public key to encrypt the message m.
2. Matrix $A_{(i)}$ and vector $s'_{(i)}$ are generated.
3. The vector $t'_{(i)}$ is computed by rounding the product of $A_{(i)}. \; s'_{(i)}$: $t'_{(i)} \leftarrow A'_{(i)}. s'_{(i)} + h_{(i)}$
4. Polynomial $v'_{(i)}$ is calculated as: $v'_{(i)} = t_{(i)}. s'_{(i)}$.
5. $v'_{(i)}$ is used to encrypt the message m which denoted as c_m.
6. Ciphertext consists of c_m and $t'_{(i)}$.

Saber PKE Decryption

1. $v_{(i)}$ is calculated as: $v_{(i)} = t'_{(i)}. s_{(i)}$.

2. The message m' is decrypted by reversing computations with $v_{(i)}$ and c_m.

The Saber key encapsulation mechanism has three steps: Saber KEM Key Generation, Saber KEM Encapsulation, and Saber KEM Decapsulation:

Saber KEM Key Generation

1. Saber PKE key generation is used to return seed $a_{(i)}$, $t_{(i)}$ and $s_{(i)}$.
2. Both seed $a_{(i)}$ and $t_{(i)}$ become the Saber KEM public key $Pk_{(i)}$.
3. The hashed public key $Pkh_{(i)}$ is generated using SHA3-256.
4. Parameter z is randomly sampled.
5. z, $Pkh_{(i)}$ and $s_{(i)}$ become the Saber KEM secret key.

Saber KEM Encapsulation

1. Message m and public key $Pk_{(i)}$ are hashed using SHA3-256.
2. Saber PKE encryption is used to generate ciphertext.
3. Hash of the $Pk_{(i)}$ and ciphertext are concatenated, then hashed to encapsulate the key.

Saber KEM Decapsulation

1. Message m' is decrypted by using Saber PKE Decryption.
2. The decrypted message m' and hashed public key $Pkh_{(i)}$ are hashed to generate K'.
3. Ciphertext c'_m is generated from saber PKE Encryption for message m'.
4. If $c_m = c'_m$ then the K = Hash(K',c), if not, K = Hash(z,c).

The level of documentation available on LWR is not quite as complete as what is available for LWE. However, the proposed implementation of LWR for NIST's PQC program is solid. The use of PUFs to secure PKIs based on LWR is not more challenging than the one based on LWE.

2.3. NTRU Cryptography

Cryptographic algorithms such as FALCON, which uses NTRU (*Nth* degree of TRUncated polynomial ring) arithmetic, are also based on lattice cryptography. The parameters of the scheme include a large prime number N, a large number **q** and a small number **p** that are both used for modulo arithmetic. Two numbers **df** and **dg** are used to truncate the polynomials $f_{(i)}$ and $g_{(i)}$. The key generation cycle for client device (i), as shown in Figure 2, is the following:

1. Generation of the two truncated polynomials $f_{(i)}$ and $g_{(i)}$ from seed $a_{(i)}$ and seed $b_{(i)}$.
2. Computation of $Fq_{(i)}$, which is the inverse of polynomial $f_{(i)}$ modulo **q**.
3. Computation of $Fp_{(i)}$, which is the inverse of polynomial $f_{(i)}$ modulo **p**.
4. Computation of polynomial $h_{(i)}$: $h_{(i)} \leftarrow$ p. $Fq_{(i)}$. $g_{(i)}$.
5. The private key $Sk_{(i)}$ is $\{f_{(i)}; Fp_{(i)}\}$.
6. The public key $Pk_{(i)}$ is $h_{(i)}$.

As the polynomials $f_{(i)}$ and $g_{(i)}$ are not always usable, they are subject to some preconditions such as invertible modulo p and q. The client device needs to try several possible random numbers, and select the ones giving acceptable private keys. Once sufficient public and private keys are available, the encryption of the plaintext message m, $m \in \{-1, 0, 1\}^N$ is done by finding the random polynomial r, $r \in \{-1, 0, 1\}^N$, which uses a corresponding parameter d_r, and calculating the ciphertext with the equation $e \equiv r.h + m \pmod{q}$. To retrieve m from e, we first calculate $a \equiv f.e \pmod{q}$ and lift the coefficients of a to be between $\pm q/2$. Then, $a \pmod{p}$ is equal to m. [23].

NTRU lattices can also be applied to DSA. This was originally introduced in NTRUSign, but NIST submissions such as Falcon expand on these algorithms [9]. Falcon utilizes the GPV framework applied to NTRU lattices; that is, the public key is a long basis for an NTRU lattice while the private key is a short basis. From here, the message m is sent a non-lattice point C, utilizing a random value salt and hash function H. Using the short basis, a user signs by finding the closest vector v to c. The signature is (salt, $s = c - v$), verified by checking if s is short and H (msg || salt) $- s$ is a point on the lattice (verified using the long basis [24]).

The cryptography based on NTRU is well known, and extremely well documented. The polynomial arithmetic truncating the N-th element is elegant and effective. Like other lattice algorithms under consideration by NIST, we are considering the NTRU as a strong candidate, both for DSA, and KEM.

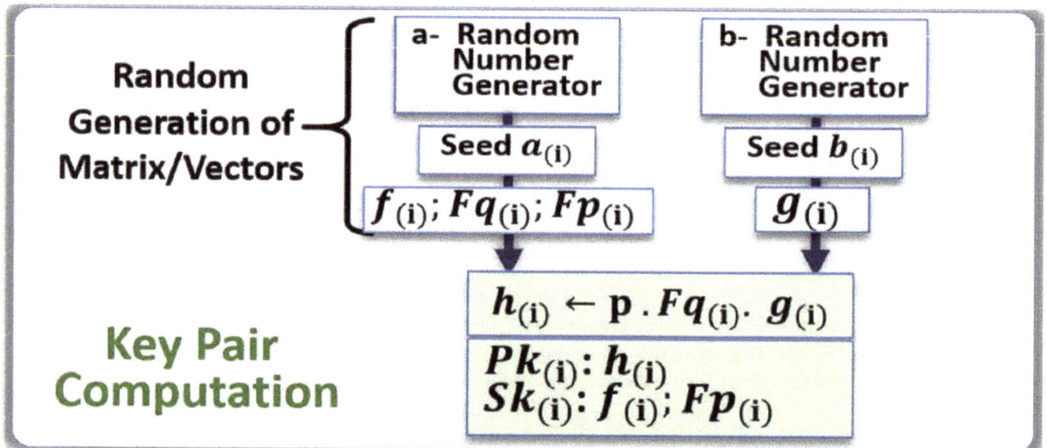

Figure 2. Example of key generation for NTRU cryptography. The polynomials $f_{(i)}$ and $g_{(i)}$ and are generated from random number generators, from which the inverses $Fq_{(i)}$ and $Fp_{(i)}$ are computed. The public key $h(i)$ is also computed from these polynomials.

2.4. Code-Based Cryptography

Code-based algorithms such as Classic McEliece are implemented with binary Goppa codes, that is, Goppa codes with underlying computations in finite Galois fields **GF(2^m)** The parameters are an irreducible polynomial of degree **t**, field exponent **m**, and code length **n**. The resulting code has error-correction capability of **t** errors, the information-containing part of the code word has a size of **k = n − m × t** and has generator matrix **G** with a size of **k × n** [14,15].

The block diagram of Figure 3 is showing an example of public–private key generation for code-based cryptography, and client device i.

1. Seed $a_{(i)}$ is used to create a random invertible binary **k × k** scrambling matrix $S_{(i)}$.
2. Seed $b_{(i)}$ is used to create a random **n × n** permutation matrix $P_{(i)}$.
3. The public key $Pk_{(i)} = \hat{G}_{(i)}$ is computed with the generator matrix G: $\hat{G}_{(i)} \leftarrow S_{(i)} \cdot G \cdot P_{(i)}$
4. The private key $Sk_{(i)}$ is $\{G; S_{(i)}{}^{-1}, P_{(i)}{}^{-1}\}$.

Given a generator matrix of a binary Goppa code G, an irreducible polynomial of degree t, the field exponent m, and the code length n, the encryption process involves the following steps:

1. Create the public key, $\hat{G}_{(i)}$ as described above.
2. Multiply the message m by $\hat{G}_{(i)}$, creating the ciphertext message $\hat{m}$.
3. Add a random error vector e of Hamming weight t to $\hat{m}$ to obtain the ciphertext **c**.

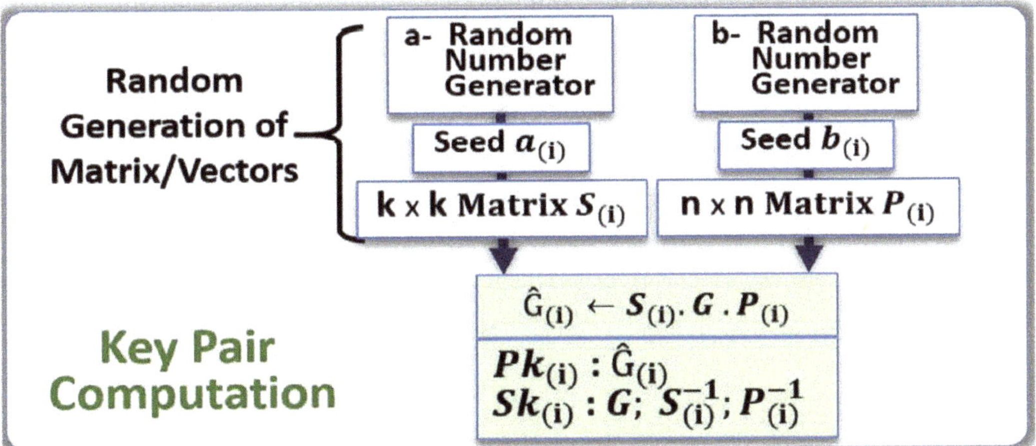

Figure 3. Example of key generation for code-based cryptography. The two matrixes $S_{(i)}$ and $P_{(i)}$ are generated from random number generators. The key pairs are computed from the matrixes, and the generator matrix G.

Given a ciphertext c, a decoding algorithm, and the private key $\{G; S_{(i)}^{-1}, P_{(i)}^{-1}\}$, decryption involves the following steps:

1. Compute $\hat{c} = c \, P_{(i)}^{-1}$.
2. Use the decoding algorithm to correct the errors to obtain $\hat{m}$.
3. Obtain the original message by computing $m = \hat{m}.S_{(i)}^{-1}$.

One example of a decoding algorithm is Patterson's algorithm. This algorithm calculates the error-locator polynomial which has roots corresponding with the locations of the error bits added to the encrypted message. This algorithm can be implemented as follows [25]:

Input: Syndrome polynomial s, Goppa polynomial g of degree t
Patterson (s, g):

1. $t = s^{-1} \bmod g$.
2. $t = \sqrt{t + x}$.
3. Find polynomials a, b such that $b.t \equiv \bmod g$ with deg(a) $\leq |t/2|$ and deg(b) $\leq |(t\text{-}1)/2|$ using the extended Euclidean algorithm.
4. Calculate and return the error locator polynomial, $e = a^2 + x.b^2$.

Once the error locator polynomial is found, the Berlekamp Trace Algorithm can be used to find the roots of the polynomial via factorization. These correspond to the locations of the error bits added to the message. The Berlekamp Trace Algorithm can be implemented as follows [26]:

Input: Polynomial to factor p, trace polynomial t, basis index i
Berlekamp Trace (p, t, i):

1. if deg(p) ≤ 1
2. return the root of p.
3. $p_0 = gcd(p, t(B_i. x))$.
4. $p_1 = gcd(p, 1 + t(B_i. x))$.

return berlekampTrace $(p_0, i + 1)$, berlekampTrace $(p_1, i + 1)$.

Code-based is probably the most mature, and well documented PQC algorithm currently under consideration. The new implementations are highly effective for KEM; the extended output functions allow the quick generation of the two matrixes needed for key generation.

3. Public Key Infrastructure

3.1. Public-Private Key Pairs

As part of a PKI, the public–private key pairs can be used to securely transmit shared secret keys though KEM and to digitally sign messages with DSA (see Figure 4). The public key $Pk_{(2)}$ of Client 2 encapsulates the shared secret key of Client 1, that can only be viewed by the client (2), thanks to their private key $Sk_{(2)}$ that reverses the encapsulation. Client 1 uses their private key $Sk_{(1)}$ to digitally sign a message that is verified with the public key $Pk_{(1)}$, providing non-alteration and non-repudiation in the transaction. The trust and integrity of such an architecture relies on the following:

i. The secure generation and distribution of the public–private key pairs to the client devices that are participating in the PKI.

ii. The identification of the client devices, and trust in their public keys.

iii. The sharing of the public keys among participants.

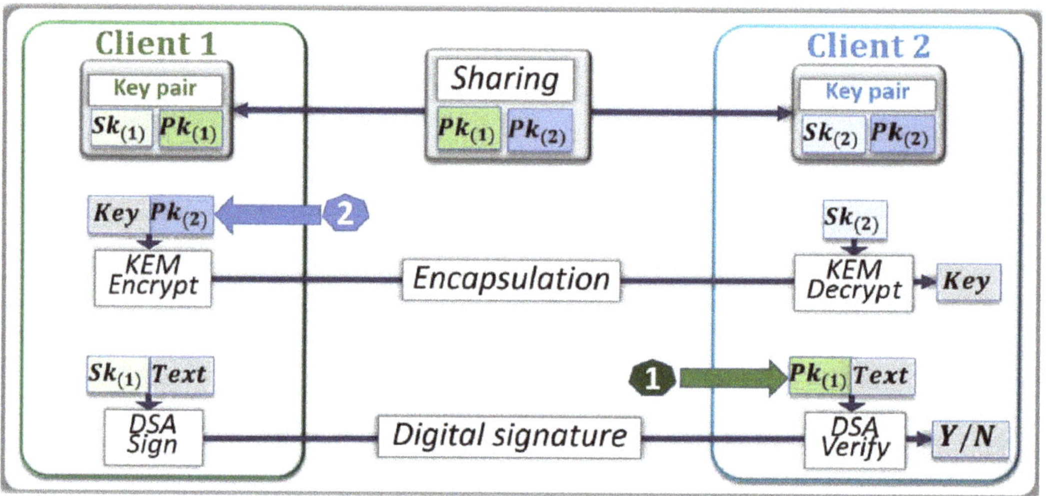

Figure 4. Communication protocol between two client devices with shared public keys. Each client device is set with its key pair. For KEM, Client 1 uses the public key of Client 2 to encrypt a shared key; Client 2 retrieves the key with their private key. For DSA, Client 1 signs a message with their private key, Client 2 verifies the message with the public key of Client 1.

Most PKIs rely on certificate authorities (CA) and registration authorities (RA) to offer such an environment of trust and integrity. The architecture is vulnerable to several threats, including loss of identity, man-in-the-middle attacks, and side channel attacks in which the private keys are exposed during KEM, and DSA.

3.2. PKI with Network of PUFs

The use of networks of PUFs can mitigate the vulnerabilities of PKIs. PUF technology exploits the variations created during fabrication to differentiate each device from all other devices, acting as a hardware "fingerprint" [27–29]. Solutions based on PUFs embedded in the hardware of each node can mitigate the risk of an opponent reading the keys stored in non-volatile memories. The keys for the PKI can be generated on demand with a one-time use; stealing a key becomes useless as new keys are needed at each transaction. During enrollment cycles, the images of the PUFs are stored in look-up tables in the CA (see Figure 5); enrollment has to be done only once in a secure environment. Handshake protocols [30] can select a portion of the PUFs—and their image is stored in the CA—to extract a data stream that generates the key pairs. The PUFs can be erratic, therefore the generation of cryptographic keys, the focus of this work, is challenging. A single-bit mismatch in a cryptographic key is not acceptable for most encryption protocols. Therefore, the use of error correcting code (also use the acronym ECC, not to be confused with "elliptic curve cryptography) methods, helper data, and fuzzy extractors can minimize the levels of errors [31–33]. The alternate method is one where the CA has search engines, such as response-based-cryptography (RBC), that can handle the validation of erratic keys [34–37].

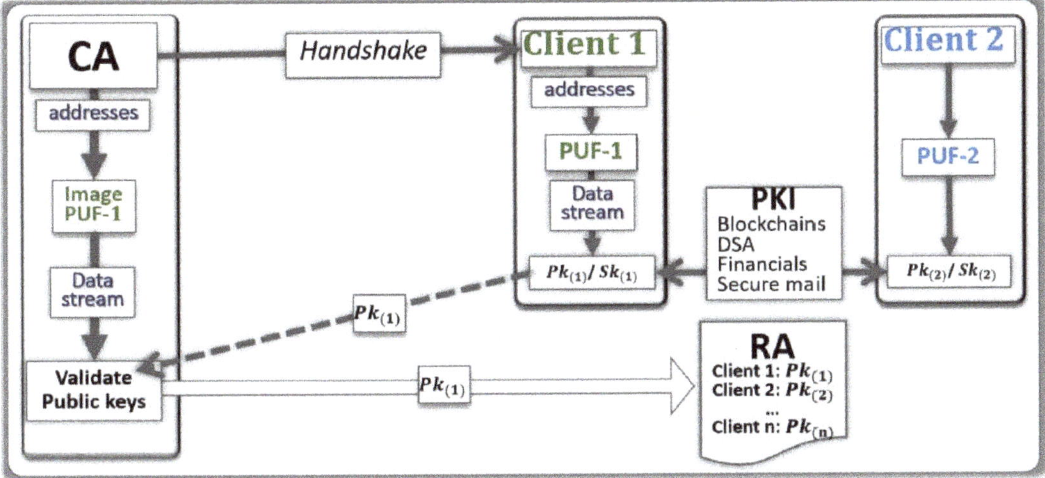

Figure 5. Physical unclonable function (PUF)-based public key infrastructure (PKI). The Certificate Authority (CA) verify the validity of the public keys and transfer them to the Register Authority (RA). The key pairs are generated from the PUFs embedded in each client device. The architecture enables cloud-based peer-to-peer secure transactions protected by asymmetrical cryptography.

The RBC engine that validates public keys, shown in Figure 5, generates public/private key pairs until the public key matches the client's provided key. The server searches over a seed (e.g., a 256 bit seed), and uses that seed for key generation. If the generated public key matches the client's public key, then the client is authenticated. If the public keys do not match, then the server flips one bit of the seed at a time (increasing the Hamming distance) until the public keys match. Thus, the search is carried out by generating the public/private key pairs by iterating over the seed and increasing the Hamming distance until the seed is found that matches the client's public key. The search space for a 256 bit key is 2^{256} and it would be nearly impossible to authenticate a user in a fixed time without the use of parallel computing. High-performance computing HPC and graphics processing unit (GPU) technologies are valuable to enhance the ability of the CA to validate the public key generated by the client devices. For instance, graphics processing units (GPUs) can be employed to parallelize and accelerate the authentication process. By using a GPU, the server can search over multiple keys in parallel.

3.3. Implementation of PQC Algorithms for PKI

The CRYSTALS-Dilithium digital signature algorithm consists of the following procedures: key generation, signing, and verification. These procedures are computationally bound by two operations: multiplication in the polynomial ring noted $\mathbb{Z}_q[X]/(X_n+1)$, and matrix/vector expansion via an extendable output function (XOF). Therefore, any attempt to optimize Dilithium should target these operations. We describe below the literature that focuses on such optimizations.

The operation of polynomial multiplication has a quasi-linear time complexity bound by the Number Theoretic Transform (NTT) implementation, and the operation of expansion via XOF is bound by the SHAKE-128 implementation. Using the AVX2 instruction set, the matrix and vector expansion is optimized by using a vectorized SHAKE-128 implementation that operates on four sponges that can absorb and squeeze blocks in parallel Additionally, Ducas et al. [8] use the AVX2 instruction set to optimize the NTT thus speeding up the polynomial ring multiplication by about a factor of two. This optimization is achieved by interleaving the vector multiplications and Montgomery reductions so that parts of the multiplication latencies are hidden.

Nejatollahi et al. [38] outline two different works that optimize the NTT using an Nvidia GPU. The first reports higher throughput polynomial multiplication [39] and the second is a performance evaluation between several versions of the NTT, including iterative NTT, parallel NTT, and CUDA-based FFT (cuFFT) for different polynomial sizes [40] Strictly algorithmic optimizations of the NTT are presented in other works [41,42]. Longa et al. [41] show that limiting the coefficient length in polynomials to 32 bits yields an efficient modular reduction technique. By employing this new technique in NTT, reduction is only required after multiplication, and significant performance gains are achieved when compared to a baseline implementation. Additionally, the authors use signed integer arithmetic which decreases the number of add operations necessary in both sampling and polynomial multiplication. Greconici et al. [42] use signed integer arithmetic to decrease the number of add operations, which leads to performance gains in several functions including NTT and SHAKE-128. The authors also employ a merging layers technique in NTT that reduces the number of loads and stores by about a factor of two.

The SABER KEM algorithm is similarly computationally bound by polynomial multiplication and hashing functions. As mentioned by D'Anvers et al. [18], since SABER uses power-of-2 moduli, this eliminates the need for rejection sampling and makes modular reduction fast by using bit shift operations. However, one drawback of using power-of-2 moduli is the inability to take advantage of faster NTT multiplication since the moduli are not prime. As described above, Akleylek et al. [40] examines the performance of different multiplication techniques. By implementing a version of cuFFT in a similar fashion for SABER, we may observe a speedup in polynomial multiplication. In addition, SABER is computationally bound by hashing and extendible functions. SABER uses SHA3_256 and SHA3_512 functions for hashing and SHAKE128 as an XOF. Roy et al. [43] demonstrate parallelizing SHAKE128 using AVX2 and batching four operations, thus achieving a 38% increase in throughput for SABER's key generation. Additionally, optimizing the hashing functions and SHAKE128 in a different way, the SABER technical documentation describes replacing the SHA3 functions with SHA2 and replacing SHAKE128 with AES in counter mode [18].

Focusing on three PQC algorithms, SABER, CRYSTALS-Dilithium, and NTRU, a breakdown of the fraction of time spent (as a percentage) in the hashing/XOR and polynomial multiplication components of the algorithms is reported in Table 1. NTRU spends the majority of its time doing polynomial multiplication first, then hashing second [44], but no benchmarks have been calculated thus far. The times spent for the hashing and polynomial multiplication components of CRYSTALS-Dilithium, and SABER are reported as percentages of the total execution time for the key pair generation procedure where the percentages are an average of 10 time trials.

Table 1. Breakdown of the fraction of the time hashing and extendable output function XOF compared with the time performing polynomial multiplication. Both SABER and CRYSTALS are constrained by the polynomial multiplication. The light versions considered in the post quantum cryptographic (PQC) implementations of hashing and XOF functions such as SHA3 with SHAKE are extremely effective.

	Hashing and XOF	Polynomial Multiplication	Reference
SABER	~30%	~60%	Our benchmarks
CRYSTALS-Dilithium	~42%	~33%	Our benchmarks
NTRU	second bottleneck	first bottleneck	[40,41]

4. PUF-Based Key Distribution for PQC

4.1. PUF-Based Key Distribution for LWE Lattice Cryptography

The proposed generic protocol to generate public–private key pairs with PUFs for LWE lattice cryptography is shown in Figure 6. The random number generator (a) is used for the generation of seed $a_{(i)}$, which is public information. However, Seed k that is needed for the generation of the private key $Sk_{(i)}$ is generated from the PUF. The outline of a protocol generating a key pair for LWE cryptography is the following:

1. The CA uses a random numbers generator and hash function to be able to point at a set of addresses in the image of the PUF-i.
2. From these addresses, a stream of bits called Seed K′ is generated by the CA.
3. The CA communicates to the Client (i), through a handshake, the instructions needed to find the same set of addresses in the PUF.
4. Client (i) uses the PUF to generate the stream of bits called Seed K. The two data streams Seed K and Seed K′ are similar, however slightly differ from each other due to natural physical variations and drifts occurring in the PUFs.
5. [If needed, Client (i) applies error correcting codes to reduce the difference between Seed K and Seed K′; the corrected, or partially corrected, data stream is used to generate the vectors $s_{1(i)}$ and $s_{2(i)}$]
6. Client (i) independently uses a random numbers generator (a) to generate a second data stream Seed $a_{(i)}$, which is used for the computation of the matrix $A_{(i)}$.
7. The vector $t_{(i)}$ is computed: $t_{(i)} \leftarrow A_{(i)} \, s_{1(i)} + s_{2(i)}$.
8. The private key $Sk_{(i)}$ is $\{s_{(1(i))}; s_{2(i)}\}$.
9. The public key $Pk_{(i)}$ is $\{a_{(i)}; t_{(i)}\}$.
10. Client (i) communicates to the CA, through the network, the public key $Pk_{(i)}$;
11. The CA uses a search engine to verify that $Pk_{(i)}$ is correct. The search engine initiates the validation by generating a public key from Seed $a_{(i)}$ and Seed K′ with lattice cryptography codes. If the resulting public key is not $Pk_{(i)}$, an iteration process gradually injects errors into Seed K′ and computes the corresponding public keys. The search converges when a match in the resulting public key is found, or when the CA concludes that the public key should be bad.
12. If the validation is positive, the public key $Pk_{(i)}$ is posted online by the RA.

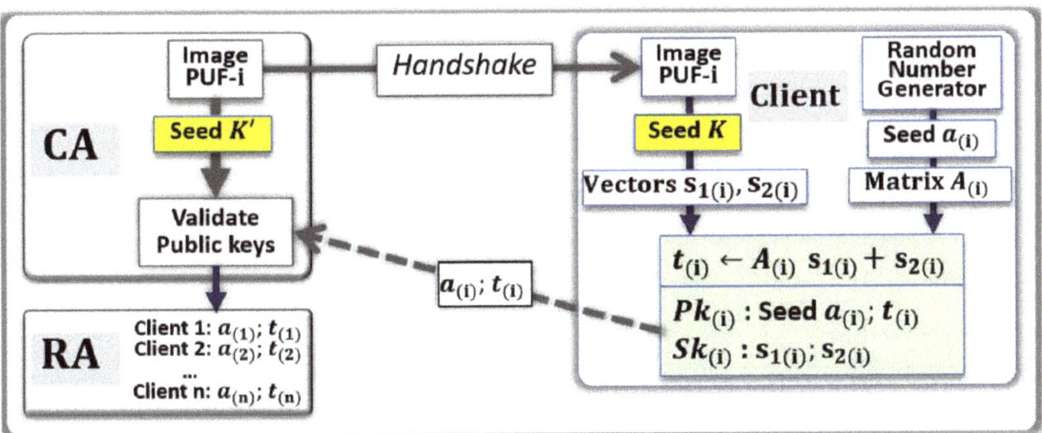

Figure 6. PUF-based key pair generation for learning with error (LWE). The private keys, i.e., vectors $S_{1(i)}$ and $S_{2(i)}$, are generated from the seed K that is extracted from the PUF. The matrix $A_{(i)}$ continues to be generated from a random number. The search engine of the CA has access to an image of the PUF, and can independently validate the validity of public key, which is posted by the RA for cloud-based transactions. A new key pair can be generated, and validated by the CA, at each handshake cycle.

This protocol is applicable for single use key pairs that are generated for each transaction. The random number generators of the first step of the protocol can generate new data streams, which point at different portions of the PUFs, thereby triggering the generation of new key pairs. The search engine described above can benefit from noise injection and high-performance computing. The injection of noise in Seed K will make the search too difficult for CA, unless equipped with HPCs, or GPUs. This can preclude hostile CAs from participating.

4.2. PKI Architecture with PUF-Based Key Distribution and LWE

The PUF-based key pair generation scheme with LWE cryptography, as presented in the previous section, can be integrated in a PKI securing a network of i clients. Figure 7 shows two client devices communicating directly, either by exchanging secret keys through KEM or DSA. The client devices independently generate the seed $a_{(i)}$, while the PUFs and their images are used for the independent generation of the vectors $s_{1(i)}$ and $s_{2(i)}$. The role of the CA is to check the validity of the vectors $t_{(i)}$, and to transmit both the seeds $a_{(i)}$ the vectors $t_{(i)}$ to the RA, which maintain a ledger with valid public keys. Such an architecture is secured assuming the following conditions:

i. The enrollment process in which the PUFs are characterized to generate their image is accurate and not compromised by the opponent.

ii. The database stored in the CA that contains the image of the PUFs for the i client devices is protected from external and internal attacks.

iii. The PUFs embedded in each client device are reliable, unclonable, and tamper resistant.

iv. The key generation process, KEM, and DSA are protected from side channel analysis.

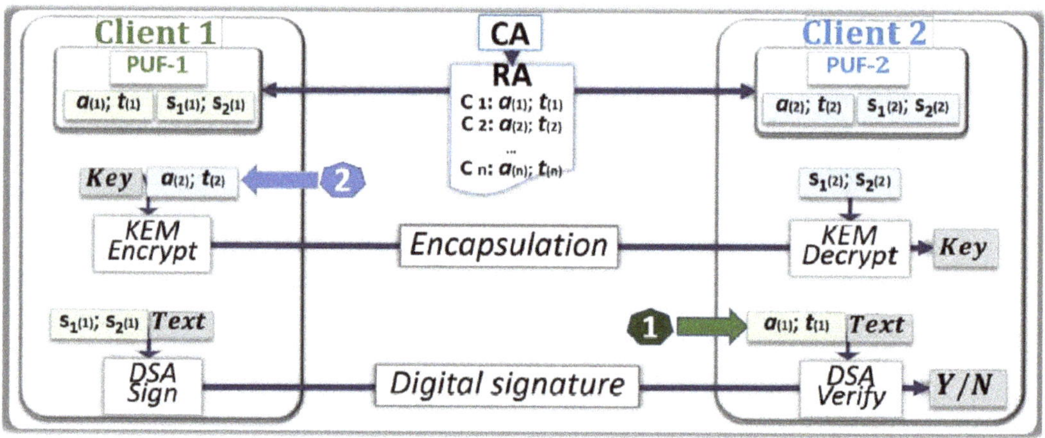

Figure 7. PUF-based public key infrastructure (PKI) with LWE cryptography. The use of PUFs in the PKI network does not impact the user experience during peer-to-peer secure communication. The ability to generate one time use key pairs, and authenticate each client device at each transaction, enhances the root of trust. Such PUF based architectures are only valid if the latencies are kept below a few seconds.

As we experimentally verified that the latencies of the key generation process from the PUFs are low enough, such a protocol can be used to change the key pairs after each encryption cycle. Therefore, the potential loss of the secret keys during an encryption/decryption cycle has minimum impact as different keys will be used during the subsequent cycles.

4.3. PUF-Based Key Distribution for LWR Lattice Cryptography

There are some similarities between LWE and LWR implementations. The seed k of the PUF is only used to generate one vectors $s_{1(i)}$, while a constant vector $h_{(i)}$ can be generated independently. The public vector $t_{(i)}$ is computed in a similar way: $t_{(i)} \leftarrow A_{(i)} \cdot s_{1(i)} + h_{(i)}$.

4.4. PUF-Based Key Distribution for NTRU Lattice Cryptography

The protocol to generate key pairs from PUFs for NTRU cryptography is similar than the one presented above in Section 4.1 for LWE, see Figure 8. We are suggesting a method where the only source of randomness is the PUF, Seed **K**, to compute both the public key $Pk_{(i)}$, and the private key $Sk_{(i)}$. In our implementation, Seed **K** feeds the hash functions SHA-3, and SHAKE, to generate a long stream of bits, then compute the two polynomials $f_{(i)}$ and $g_{(i)}$.

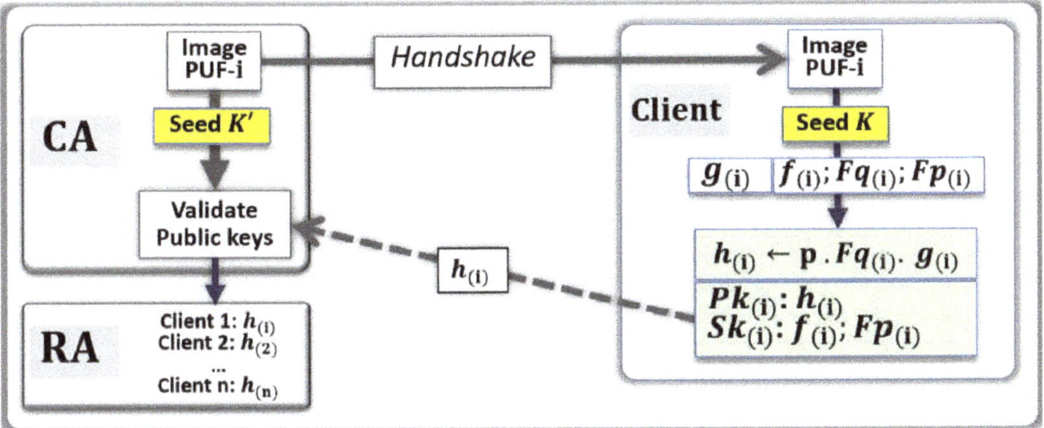

Figure 8. PUF-based key pair generation for NTRU. After each handshake, the polynomials $g_{(i)}$ and $f_{(i)}$ are generated from the Seed K that is extracted from the PUF. The public and private key pairs are computed from these two polynomials. The search engine of the CA can independently validate the validity of public key with the image of the PUF.

As previously discussed in Section 2.3, the polynomials $f_{(i)}$ and $g_{(i)}$ are not always usable due to pre-conditions, therefore a scheme to try several possible ways of addressing the PUF has to be developed. One way is to implement a deterministic method that is known by both the client device and the CA, which can have a negative impact on the latencies. We preferred the solution driven by the client device that asks the CA to initiate new handshakes. The summary of the method used to generate the key pairs for NTRU cryptography is the following:

1. The CA uses random numbers to point at a set of addresses in the image of the **PUF-i**.
2. From these addresses, a stream of bits called Seed **K'** is generated by the CA.
3. The CA sends the handshake to the client (i) to find the same addresses.
4. Client (i) uses the PUF to generate Seed **K**.
5. Client (i) applies error correction to Seed **K** and generates the truncated polynomials $f_{(i)}$ and $g_{(i)}$.
6. Computation of $Fp_{(i)}$ and $Fq_{(i)}$ and verify that the pre-conditions are fulfilled.
7. If needed, ask for a new handshake and iterate.
8. The polynomial $h_{(i)}$ is computed: $h_{(i)} \leftarrow p \cdot Fq_{(i)} \cdot g_{(i)}$.
9. The private key $Sk_{(i)}$ is $\{f_{(i)}; Fp_{(i)}\}$.
10. The public key $Pk_{(i)}$ is $h_{(i)}$.
11. Client (i) communicates to the CA, through the network, the public key $h_{(i)}$.

12. The CA uses a search engine to verify that $h_{(i)}$ is correct.
13. If the validation is positive, the public key $h_{(i)}$ is posted online by the RA.

It is important to notice that steps six and seven of the proposed method, "Computation of $Fp(i)$ and $Fq(i)$ and verify that the pre-conditions are fulfilled; if needed ask for a new handshake and iterate", could be handled differently to minimize backward and forward communications cycles between the CA and the client device. One example of implementation is to have a pre-arranged way to modify the seed generated by the PUF and its image. When the CA fails to validate the public key, several pre-arranged modifications of Seed K' will be tested.

4.5. PUF-Based Key Distribution for Code-Based Cryptography

An example of a protocol to generate the key pairs with PUFs for code-based cryptography is shown in Figure 9. The overall protocol is similar to the one presented above for lattice cryptography. Much like NTRU, the only source of randomness is Seed **K** that is generated from the PUF to compute the two matrixes $S_{(i)}$ and $P_{(i)}$. The brief outline of the protocol for generating key pairs for code-based cryptography is the following:

1. The CA uses random numbers to point at a set of addresses in the image of the **PUF-i**.
2. From these addresses, a stream of bits called Seed **K'** is generated by the CA.
3. The CA sent the handshake to the client (i) to find the same addresses.
4. Client (i) uses the PUF to generate Seed **K**.
5. Client (i) applies ECC) on Seed **K** and generates the matrixes $S_{(i)}$ and $P_{(i)}$.
6. Computation of $S_{(i)}^{-1}$ and $P_{(i)}^{-1}$.
7. The public key $Pk_{(i)} = \hat{G}_{(i)}$ is computed with the generator matrix G: $\hat{G}_{(i)} \leftarrow S_{(i)} \cdot G \cdot P_{(i)}$.
8. The private key $Sk_{(i)}$ is $\{G; S_{(i)}^{-1}, P_{(i)}^{-1}\}$.
9. Client (i) communicates to the CA, through the network, the public key $\hat{G}_{(i)}$.
10. The CA uses a search engine to verify that $\hat{G}_{(i)}$ is correct.
11. If the validation is positive, the public key $\hat{G}_{(i)}$ is posted online by the RA.

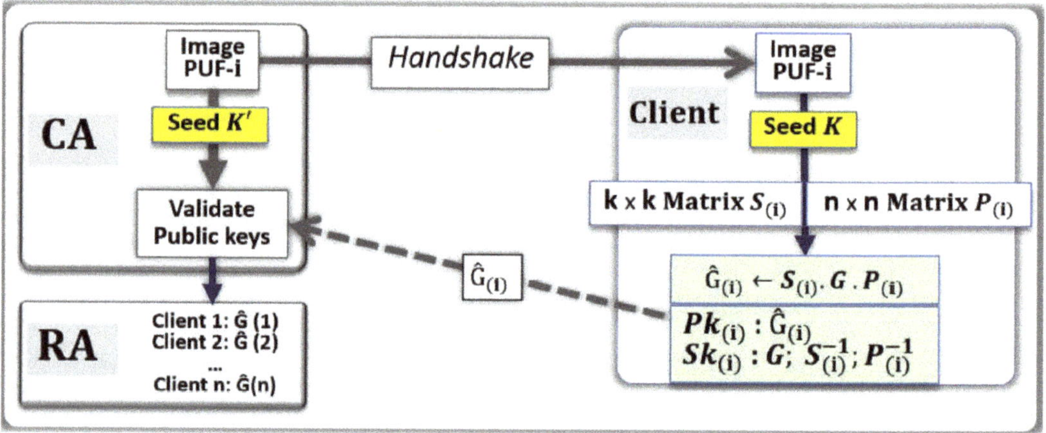

Figure 9. PUF-based key pair generation for code-based cryptography. The two matrixes $S_{(i)}$ and $P_{(i)}$ are generated from the Seed K that is extracted from the PUF. As was done in the previously presented PQC schemes, the search engine of the CA has access to an image of the PUF and can independently verify the validity of the public key.

Step six, the computing of inverses, does not work for all matrixes: request for new handshakes, or shared ways to find invertible matrixes and iterate.

5. Experimental Evaluation

The purpose of this evaluation is to demonstrate the practicality of the proposed protocols and understand the potential limitations. The replacement of the random number generators by the PUFs follow a similar path for various PQC algorithms; therefore we reduced the scope of this evaluation to LWE (qTESLA, CRYSTALS-Dilithium), and LWR (LightSABER). The generation of the Seed K from the PUF is done using known methods, and the computation of the key pairs is based on the PQC codes made available by NIST, which are also considered known. The unknown in the practicality of the protocol is the sensitivity to bit error rates (BER) of the search engine of the CA for verifying the public keys. The RBC method [34–37] uses the Seed K' as a starting point to generate an initial public key, then iterates by incrementally adding errors, eventually finding the public key computed from Seed K by the client device. At high BER, it is desirable to use cryptographic algorithms that have the ability to generate the key pairs at high throughput.

The RBC itself is an interesting simulation platform for this evaluation, because of the possibility to directly measure the throughput in term of the number of key pairs generated by second. We selected the RBC to experimentally demonstrate that the PQC protocols are fast enough. We designed an experiment to benchmark three algorithms, (qTESLA, CRYSTALS-Dilithium, and LightSABER) with two known cryptosystems (AES and ECC). Each of these cryptosystems have a list of parameter sets as a part of their specifications. We chose parameter sets that were inherently compatible with a 256-bit output from a hypothetical PUF as well as these that were best optimized between performance, size, and security for IoT devices. For these reasons, the parameter sets AES256, ECC Secp256r1, qTESLA-p-I, CRYSTALS-Dilithium 2, and LightSABER were used for the performance comparison. In this analysis, the comparison with ECC is the most relevant one, because the PQC codes under consideration and ECC are similar in their objective to generate public–private keys pairs for PKI. Therefore, ECC and the three PQC algorithms are tested here with their software versions. The comparison with AES was included as a benchmark of excellence; we used the hardware implementation of AES, natively available in Intel processors. One of the objectives of the PQC standardization program driven by the NIST is to encourage private industry to eventually design hardware implementations of the selected codes.

We summarize the parameter sets for each algorithm and our motivation for their selection in Table 2, as shown below.

Table 2. Selected cryptosystems, parameter sets, whether the PQC algorithms are NIST round 3 candidates, specialized instructions employed in the implementations, and our motivation for selecting the algorithm and its configuration.

Algorithm	Parameter Set	NIST Candidate	Instructions	Selection Reason
AES	AES256	N/A	AES-NI, SSE	Benchmark: HW implementation Lacks DSA and KEM capabilities
ECC	Secp256r1	N/A	AVX, SSE	Benchmark: mainstream for PKI Uses 256-bit long keys for DSA
qTESLA	p-I	Dropped	AVX, SSE	PQC dropped by NIST: too slow DSA uses relatively small keys
CRYSTALS-Dilithium	2	Phase 3	AVX, SSE	Active LWE PQC algorithm One of preferred DSA scheme
SABER	LightSABER	Phase 3	AVX, SSE	Active LWR PQC algorithm One of the preferred KEM scheme

5.1. Experimental Methodology

As of the time of writing, there are few implementations of RBC engines proposed. In this paper, we focus on executing the RBC protocol on a single machine equipped with multi-core CPUs. Our implementations are parallelized using OpenMP. To terminate the search when a thread finds the correct key, we use a flag in shared memory that is atomically updated. All implementations utilized the same overall structure and key iteration mechanism. We also use AVX instructions in all cryptosystems where applicable; however, further optimizations can be made by taking advantage of AVX2 or other wide vector technologies. The AES256 implementation takes advantage of the AES-NI instruction set, whereas all other cryptosystems tested do not use any additional vectorized instructions except AVX and SSE.

RBC engines targeting purely CPU platforms were only considered for demonstrative purposes. The purpose of this experiment is to compare the relativistic performance between all five chosen cryptosystems. The ease of porting one cryptosystem to another all on the CPU influenced the scope of experiments. Future experimental evaluations exploring GPU focus will require more dedicated, specialized programming for each cryptosystem. The CPU used for the experiments was a $2\times$ Xeon Gold 6132 (Skylake) CPU with 28 total physical cores. Experiments were executed on a dedicated platform. All codes were written in C/C++, compiled using O3 optimization flag.

The experiments were executed by randomly selecting a target thread and using the 256-bit permutation that is the middle of that given thread's workload. This guarantees that each run accurately reflects the average case where execution stops halfway through the 256 choose k search space, for any Hamming distance k. We decided to use this approach to reduce the need for a high number of iterations to reach a statistical central point. Thus, 10 iterations were performed for each cryptosystem, and the median response time was selected from the set of 10 time trials.

The major key performance index (KPI) of the RBC search is key search throughput. Therefore, our performance evaluation uses the "effective key throughput" performance metric. Since the search increases exponentially with Hamming distance, the fraction of time spent initializing/dismantling our procedure will dominate the response time on small workloads (small Hamming distances). Consequently, to measure the effective key throughput, we use a sufficiently large Hamming distance in each algorithm such that we observe peak throughput, indicating that the initialization and dismantling procedures (freeing memory, deconstructing objects, etc.) constitute a negligible fraction of the total response time.

For AES256, the minimum Hamming distance is 4, while for ECC Secp256r1, qTESLA-p-I, CRYSTALS-Dilithium 2, and LightSABER the minimum Hamming distance is 3. Unfortunately, due to the intractable nature of the problem, the single bit error jump from a Hamming distance of 3 to 4 makes it impractical to run a statistically sufficient number of runs for ECC and qTESLA-p-I. For this reason, the AES256 benchmarks ran at a Hamming distance of 4, and the remaining cryptosystems ran at a distance of 3.

5.2. Evaluation of the Effective Key Throughput

In this section, we evaluate the effective (peak) key throughput. Figure 10 plots the median of each RBC cryptosystem's effective key throughput on a logarithmic scale. The AES256 implementation, aided by AES-NI, runs several orders of magnitude more efficiently than the public key cryptography variants at 2.17×10^8 keys per second. ECC Secp256r1 performed the second slowest at 4.77×10^4 keys per second. The post-quantum algorithms largely performed better than ECC with 1.97×10^5 and 6.83×10^5 keys per second for CRYSTALS-Dilithium 2 and LightSABER respectively. qTESLA-p-I was the worst performing PQC and overall cryptosystem out of all five at 2.24×10^4 keys per second.

To get a better sense of the relativistic scaling, we set ECC Secp256r1's effective key throughput results as the reference point since we are interested in how the PQC algorithms perform when replacing it in future PKI cryptosystems. This is plotted in Figure 11, where now the response variable is displayed in a percentage of the throughout relative to ECC Secp256r1's. Shown here, AES256 is roughly 4550 times more performant than ECC Secp256r1. CRYSTALS-Dilithium 2 is over 4.14 times more efficient than ECC Secp256r1. The most efficient PQC was LightSABER at 14.3 times faster, and the worst overall cryptosystem was qTESLA-p-I at 0.469 times slower.

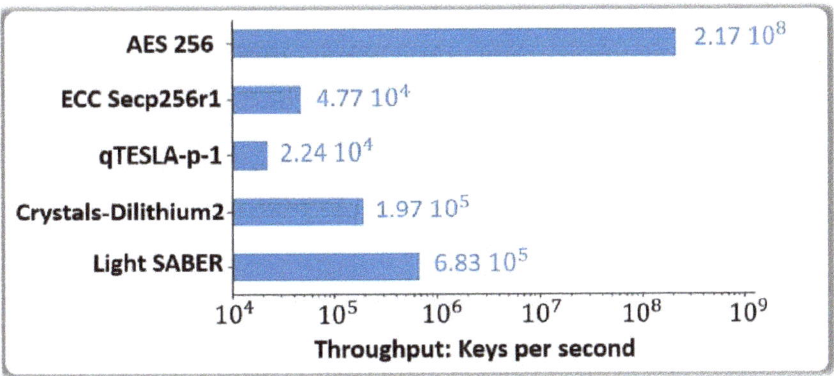

Figure 10. Key performance index (KPI) and the effective throughput in keys per second achieved by the response based cryptographic (RBC) search engine powered with AMD Ryzen 9 3900X. Benchmark of the post quantum cryptographic (PQC) algorithms are compared to the reference codes AES 256 and ECC.

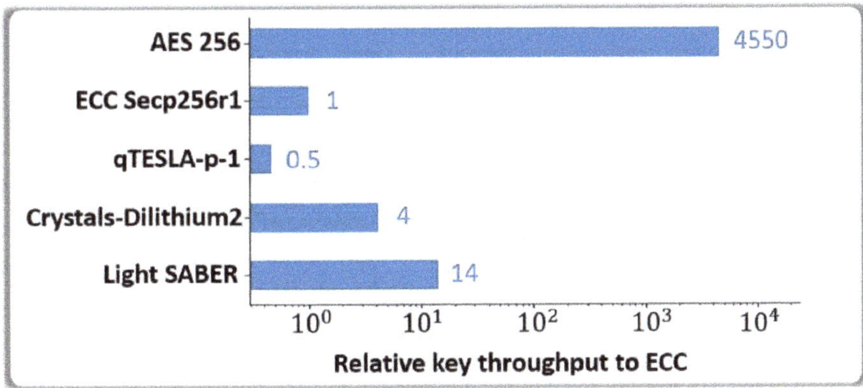

Figure 11. Maximum effective throughput relative to the performance of ECC Secp256r1 achieved for each response based cryptographic (RBC) cryptosystem implementation powered by AMD Ryzen 9 3900X. Light SABER performance is approximately 14 times faster than elliptic curve cryptography (ECC).

From these results, we confirm NIST's position that qTESLA is slower than the algorithms selected in round three. Out of what was tested, this leaves CRYSTALS-Dilithium as the strongest candidate for DSA in a PQC environment. For key encapsulation, our results show that SABER is a strong candidate for its relatively fast key generation. Future testing might consider comparing FALCON against CRYSTALS-Dilithium for DSA, and CRYSTALS-Kyber, NTRU, and Classic McEliece against SABER for KEM.

6. Conclusions and Future Work

The PQC algorithms under standardization are encouraging and the latencies are reasonable, making the protocols suitable for PKIs securing networks of client devices and IoTs. The generation, distribution, and storage of the public–private key pairs for PQC can be complex because the keys are usually very long. This paper proposes to generate the public–private key pairs by replacing the random number generators with data streams generated from addressable PUFs to get the seeds needed in the PQC algorithms. Unlike the key pairs computed by PQC algorithms, the seeds are relatively short, typically 256-bits long. The use of PUFs as a source of randomness is applicable to all five lattice-based codes under consideration in the phase III investigation of NIST, and to the code-based Classic McEliece scheme. In order to simultaneously generate key pairs from a server acting as the certificate authority, and the client device with access to its PUF, it is critical to handle the bit error rates (BERs) that are frequent with physical elements. We verified in the experimental section that the RBC can find the erratic seeds by testing an excess of 10^5 seeds per second with CRYSTALS-Dilithium 2 and LightSABER, which is faster than what we measured with mature algorithms such as the ones with elliptic curves. The experimental evaluation conducted in this work, with the RBC, also lets us conclude that the pre-selection by NIST of CRYSTALS-Dilithium for DSA and SABER for KEM are promising from a performance standpoint. Our results show that the key generation performance is at least comparable to that of ECC. The PQC algorithms under consideration are excellent in an environment targeting PUF-based key exchange. The AES hardware-accelerated AES-NI implementation yields roughly 220 million keys per second throughput on a single machine, which serves a practical real world upper bound for future hardware-accelerated PQC implementations

In this work we have not yet studied the multivariate-based RAINBOW code, which is also an important scheme under consideration for standardization; we are currently studying ways to use PUFs for key generation. The task needed to deploy PUF-based PQC solutions is not underestimated by the authors of this paper. This will include the use of highly reliable PUFs, and the optimization of the cryptographic protocol pointing simultaneously at the same set of addresses in the PUF, and in the look up table capturing the challenge–response pairs stored in the server. Further optimizing the PUF's protocols and the RBC for PQC algorithms is seen as an opportunity. The use of noises, nonces, errors, and rounding vectors can exploit the stochasticity of PUFs, and the ability to handle erratic streams of the RBC. The PQC algorithms analyzed in this paper can also benefit from the use of distributed memories, high performance computing, and parallel computing, which have the potential to further reduce the latencies of the RBC.

Author Contributions: Conceptualization, B.C.; methodology, B.C., M.G., B.Y.; software, M.G., D.G., K.L., S.N., C.P., A.S., J.W.; validation, C.P.; formal analysis, M.G., B.Y., D.G., K.L., S.N., C.P., A.S., B.C.; investigation, M.G., B.Y., D.G., K.L., S.N., C.P., A.S., B.C.; resources, B.C., M.G., B.Y.; data curation, M.G., B.Y., D.G., K.L., S.N., C.P., A.S., B.C.; writing—original draft preparation, B.C., D.G., K.L., S.N., C.P., J.W.; writing—review and editing, B.C., C.P., M.G.; visualization, M.G., B.Y., D.G., K.L., S.N., C.P., A.S., B.C.; supervision, B.C., M.G., B.Y.; project administration, B.C.; funding acquisition, B.C., M.G. All authors have read and agreed to the published version of the manuscript.

Funding: This material is based upon the work funded by the Information Directorate under AFRL award number FA8750-19-2-0503.

Institutional Review Board Statement: Not applicable.

Informed Consent Statement: Not applicable.

Data Availability Statement: Not applicable.

Acknowledgments: The authors thank the staff, students, and faculty from Northern Arizona University (NAU) in particular, Brandon Salter who is a software engineer in NAU's cybersecurity lab. We also thank the professionals of the Information Directorate of the Air Force Research laboratory (AFRL) of Rome, New York (US), who supported this effort.

Conflicts of Interest: The authors declare no conflict of interest. The funders had no role in the design of the study, collection, analyses, interpretation of data, writing of the manuscript, or decision to publish the results.

Disclaimer: (a) Contractor acknowledges Government's support in the publication of this paper. This material is partially based upon the work funded by the Information Directorate, under the Air Force Research Laboratory (AFRL); (b) any opinions, findings and conclusions or recommendations expressed in this material are those of the author(s) and do not necessarily reflect the views of AFRL

References

1. Koblitz, N.; Menezes, A. A Riddle Wrapped in an Enigma. 2015. Available online: http://eprint.iacr.org/2015/1018 (accessed on 18 May 2015).
2. Kiktenko, E.; Pozhar, N.; Anufriev, M.; Trushechkin, A.; Yunusov, R.; Kurochkin, Y.; Lvovsky, A.; Fedorov, A. Quantum Secured Blockchains. Open Source. *arXiv* **2018**, arXiv:1705.09258v3.
3. Semmouni, M.; Nitaj, A.; Belkasmi, M. Bitcoin Security with Post Quantum Cryptography. 2019. Available online: https://hal-normandie-univ.archives-ouvertes.fr/hal-02320898 (accessed on 19 October 2019).
4. Campbell, R. Evaluation of Post-Quantum Distributed Ledger Cryptography. Open Access, JBBA. 2019; Volume 2. Available online: https://doi.org/10.31585/jbba-2-1-(4)2019 (accessed on 16 March 2019).
5. Kampanakisy, P.; Sikeridisz, D. Two Post-Quantum Signature Use-Cases: Non-issues, Challenges and Potential Solutions. In Proceedings of the 7th ETSI/IQC Quantum Safe Cryptography Workshop, Seattle, WA, USA, 5–7 November 2019.
6. Ding, J.; Chen, M.-S.; Petzoldt, A.; Schmidt, D.; Yang, B.-Y. *Rainbow*; NIST PQC Project Round 2, Documentation. In Proceedings of the 2nd NIST Standardization Conference for Post-Quantum Cryptosystems, Santa Barbara, CA, USA, 22–24 August 2019.
7. NIST Status Report of Phase 3 of PQC Program, NISTIR.8309. Available online: https://www.nist.gov/publications/status-report-second-round-nist-post-quantum-cryptography-standardization-process (accessed on 22 July 2020).
8. Ducas, L.; Kiltz, E.; Lepoint, T.; Lyubashevsky, V.; Schwabe, P.; Seiler, G.; Stehlé, D. CRYSTALS-Dilithium Algorithm Specifications and Supporting Documentation. Part of the Round 3 Submission Package to NIST. Available online: https://pq-crystals.org/dilithium (accessed on 19 February 2021).
9. Fouque, P.-A.; Hoffstein, J.; Kirchner, P.; Lyubashevsky, V.; Pornin, T.; Prest, T.; Ricosset, T.; Seiler, G.; Whyte, W.; Zhang, Z. Falcon: Fast-Fourier Lattice-Based Compact Signatures over NTRU, Specification v1.2. Available online: https://falcon-sign.info/falcon.pdf (accessed on 1 October 2020).
10. Peikert, C.; Pepin, Z. *Algebraically Structured LWE Revisited*; Springer: Berlin/Heidelberg, Germany, 2019; pp. 1–23.
11. IEEE Computing Society. *IEEE Standard 1363.1-2008—Specification for Public Key Cryptographic Techniques Based on Hard Problems over Lattices*; IEEE: Piscataway, NJ, USA, 2009.
12. Regev, O. New lattice-based cryptographic constructions. *J. ACM* **2004**, *51*, 899–942. [CrossRef]
13. Casanova, A.; Faugere, J.-C.; Macario-Rat, G.; Patarin, J.; Perret, L.; Ryckeghem, J. GeMSS: A Great Multivariate Short Signature; NIST PQC Project Round 2, Documentation. Available online: https://csrc.nist.gov/Projects/post-quantum-cryptography/round-2-submissions (accessed on 3 January 2017).
14. McEliece, R.J. *A Public-Key Cryptosystem Based on Algebraic Coding Theory*; California Institute of Technology: Pasadena, CA, USA, 1978; pp. 114–116.
15. Biswas, B.; Sendrier, N. McEliece Cryptosystem Implementation: Theory and Practice. In *Post-Quantum Cryptography. PQCrypto. Lecture Notes in Computer Science*; Buchmann, J., Ding, J., Eds.; Springer: Berlin/Heidelberg, Germany, 2008; Volume 5299.
16. Regev, O. On Lattices, Learning with Errors, Random Linear Codes, and Cryptography. In Proceedings of the Thirty-Seventh Annual ACM Symposium on Theory of Computing—STOC'05, Baltimore, MD, USA, 22–24 May 2005; ACM: New York, NY, USA, 2005; pp. 84–93.
17. Lyubashevsky, V. Fiat-Shamir with Aborts: Applications to Lattice and Factoring-Based Signatures. Available online: https://www.iacr.org/archive/asiacrypt2009/59120596/59120596.pdf (accessed on 31 December 2009).
18. D'Anvers, J.-P.; Karmakar, A.; Roy, S.; Vercauteren, F. Saber: Module-LWR Based Key Exchange, CPA-Secure Encryption and CCA-Secure KEM. Cryptology ePrint Archive, Report 2018/230. Available online: https://eprint.iacr.org/2018/230 (accessed on 7 May 2018).
19. Bos, J.; Ducas, L.; Kiltz, E.; Lepoint, T.; Lyubashevsky, V.; Schanck, J.; Schwabe, P.; Seiler, G.; Stehle, D. CRYSTALS—Kyber: A CCA-Secure Module-Lattice-Based KEM. In Proceedings of the 2018 IEEE European Symposium on Security and Privacy (EuroS&P), London, UK, 24–26 April 2018; pp. 353–367.
20. Hülsing, A.; Rijneveld, J.; Schanck, J.; Schwabe, P. High-Speed Key Encapsulation from NTRU. IACR Cryptol. Available online: https://www.iacr.org/archive/ches2017/10529225/10529225.pdf (accessed on 28 August 2017).
21. Banerjee, A.; Peikert, C.; Rosen, A. Pseudorandom functions and lattices. In Proceedings of the Annual International Conference on the Theory and Applications of Cryptographic Techniques, Cambridge, UK, 15–19 April 2012; Springer: Berlin/Heidelberg, Germany, 2012; pp. 719–737.
22. Alwen, J.; Stephan, K.; Krzysztof, P.; Daniel, W. Learning with rounding, revisited. In Proceedings of the Annual Cryptology Conference, Athens, Greece, 26–30 May 2013; Springer: Berlin/Heidelberg, Germany, 2013; pp. 57–74.
23. Nurshamimi, S.; Kamarulhaili, H. NTRU Public-Key Cryptosystem and Its Variants. *Int. J. Cryptol. Res.* **2020**, *10*, 21.
24. Gentry, C.; Peikert, C.; Vaikuntanathan, V. Trapdoors for Hard Lattices and New Cryptographic Constructions. In Proceedings of the 14th Annual ACM Symposium on Theory of Computing. pp. 197–206. Available online: https://doi.org/10.1145/1374376.1374407 (accessed on 25 August 2008).
25. Heyse, S. Post-Quantum Cryptography: Implementing Alternative Public Key Schemes on Embedded Devices. Ph.D. Thesis, For the Degree of Doktor-Ingenieur of the Faculty of Electrical Engineering and Information Technology at the Ruhr-University Bochum, Bochum, Germany, 2013.
26. Menezes, A.; van Oorschot, P.; Vanstone, S. *Some Computational Aspects of Root Finding in GF(q^m)*; Lecture Notes in Computer Science; Springer: Berlin/Heidelberg, Germany, 1989; Volume 358.

27. Papakonstantinou, I.; Sklavos, N. *Physical Unclonable Function Design Technologies: Advantages & Tradeoffs*; Daimi, K., Ed.; Computer and Network Security; Spinger: Berlin/Heidelberg, Germany, 2018; ISBN 978-3-319-58423-2.

28. Herder, C.; Yu, M.; Koushanfar, F. Physical Unclonable Functions and Applications: A Tutorial. *Proc. IEEE* **2014**, *102*, 1126–1141. [CrossRef]

29. Cambou, B.; Orlowski, M. Design of Physical Unclonable Functions with ReRAM and Ternary states. In Proceedings of the Cyber and Information Security Research Conference, CISR-2016, Oak Ridge, TN, USA, 5–7 April 2016.

30. Cambou, B.; Telesca, D. Ternary Computing to Strengthen Information Assurance, Development of Ternary State based public key exchange. In Proceedings of the SAI-2018, Computing Conference, London, UK, 10–12 July 2018.

31. Taniguchi, M.; Shiozaki, M.; Kubo, H.; Fujino, T. A Stable Key Generation from PUF Responses with A Fuzzy Extractor for Cryptographic Authentications. In Proceedings of the IEEE 2nd Global Conference on Cons Electronics (GCCE), Tokyo, Japan, 1–4 October 2013.

32. Kang, H.; Hori, Y.; Katashita, T.; Hagiwara, M.; Iwamura, K. Cryptography Key Generation from PUF Data Using Efficient Fuzzy Extractors. In Proceedings of the 16th International Conference on Advanced Communication Technology, Pyeongchang, Korea, 16–19 February 2014.

33. Delvaux, J.; Gu, D.; Schellekens, D.; Verbauwhede, I. Helper Data Algorithms for PUF-Based Key Generation: Overview and Analysis. *IEEE Trans. Comput. Des. Integr. Circuits Syst.* **2014**, *34*, 889–902. [CrossRef]

34. Cambou, C.; Philabaum, D.; Booher, D. Telesca; Response-Based Cryptographic Methods with Ternary Physical Unclonable Functions. In Proceedings of the 2019 SAI FICC Conference, San Francisco, CA, USA, 14–15 March 2019.

35. Cambou, B. Unequally powered Cryptograpgy with PUFs for networks of IoTs. In Proceedings of the IEEE Spring Simulation Conference, Tucson, AZ, USA, 29 April–2 May 2019.

36. Cambou, B.; Philabaum, C.; Booher, D. Replacing error correction by key fragmentation and search engines to generate error-free cryptographic keys from PUFs. *CryptArchi* **2019**. Available online: https://in.nau.edu/wp-content/uploads/sites/223/2019/11/Replacing-Error-Correction-by-Key-Fragmentation-and-Search-Engines-to-Generate-Error-Free-Cryptographic-Keys-from-PUFs.pdf (accessed on 21 March 2021).

37. Cambou, B.; Mohammadi, M.; Philabaum, C.; Booher, D. Statistical Analysis to Optimize the Generation of Cryptographic Keys from Physical Unclonable Functions. Available online: https://link.springer.com/chapter/10.1007/978-3-030-52243-8_22 (accessed on 16 July 2020).

38. Nejatollahi, H.; Dutt, N.; Ray, S.; Regazzoni, F.; Banerjee, I.; Cammarota, R. Post-Quantum Lattice-Based Cryptography Implementations. *ACM Comput. Surv.* **2019**, *51*, 1–41. [CrossRef]

39. Emeliyanenko, P. Efficient Multiplication of Polynomials on Graphics Hardware. In Proceedings of the 8th International Symposium on Advanced Parallel Processing Technologies, Rapperswil, Switzerland, 24–25 August 2009; pp. 134–149.

40. Akleylek, S.; Dağdelen, Ö.; Tok, Y. On The Efficiency of Polynomial Multiplication for Lattice-Based Cryptography on Gpus Using Cuda. In *Cryptography and Information Security in the Balkans*; Springer: Berlin/Heidelberg, Germany, 2016.

41. Longa, P.; Naehrig, M. Speeding up the Number Theoretic Transform for Faster Ideal Lattice-Based Cryptography. *Comp. Sci. Math. IACR* **2016**, 124–139. [CrossRef]

42. Greconici, D.; Kannwischer, M.; Sprenkels, D. Compact Dilithium Implementations on Cortex-M3 and Cortex-M4. *IACR Cryptol. ePrint Arch.* **2020**, *2021*, 1–24.

43. Roy, S. SaberX4: High-Throughput Software Implementation of Saber Key Encapsulation Mechanism. In Proceedings of the 37th IEEE International Conference on Computer Design, ICCD 2019, Abu Dhabi, United Arab Emirates, 17–20 November 2019; pp. 321–324.

44. Farahmand, F.; Sharif, M.; Briggs, K.; Gaj, K. A High-Speed Constant-Time Hardware Implementation of NTRUEncrypt SVES. In Proceedings of the International Conference on Field-Programmable Technology (FPT), Naha, Okinawa, Japan, 10–14 December 2018; pp. 190–197. [CrossRef]

*applied
sciences*

Article

Threat Modelling and Beyond-Novel Approaches to Cyber Secure the Smart Energy System

Heribert Vallant *, Branka Stojanović, Josip Božić and Katharina Hofer-Schmitz

Joanneum Research, DIGITAL—Institute for Information and Communication Technologies,
A-8010 Graz, Austria; branka.stojanovic@joanneum.at (B.S.); josip.bozic@joanneum.at (J.B.);
katharina.hofer-schmitz@joanneum.at (K.H.-S.)
* Correspondence: heribert.vallant@joanneum.at

Abstract: Smart Grids (SGs) represent electrical power systems that incorporate increased information processing and efficient technological solutions. The integration of local prosumers, demand response systems and storage allows novel possibilities with regard to energy balancing and optimization of grid operations. Unfortunately, the dependence on IT leaves the SG exposed to security violations. In this paper, we contribute to this challenge and provide a methodology for systematic risk assessment of cyber attacks in SG systems. We propose a threat model and identify possible vulnerabilities in low-voltage distribution grids. Then, we calculate exploitation probabilities from realistic attack scenarios. Lastly, we apply formal verification to check the stochastic model against attack properties. The obtained results provide insight into potential threats and the likeliness of successful attacks. We elaborate on the effects of a security violation with regard to security and privacy of energy clients. In the aftermath, we discuss future considerations for improving security in the critical energy sector.

Keywords: smart grid; risk assessment; threat modeling; formal verification; probabilistic model checking

Citation: Vallant, H.; Stojanović, B.; Božić, J.; Hofer-Schmitz, K. Threat Modelling and Beyond-Novel Approaches to Cyber Secure the Smart Energy System. *Appl. Sci.* **2021**, *11*, 5149. https://doi.org/10.3390/app11115149

Academic Editors: Leandros Maglaras, Ioanna Kantzavelou and Mohamed Amine Ferrag

Received: 7 May 2021
Accepted: 26 May 2021
Published: 1 June 2021

Publisher's Note: MDPI stays neutral with regard to jurisdictional claims in published maps and institutional affiliations.

1. Introduction

A Smart Grid represents an enhanced energy supply network that relies on information and communication technologies for enhanced energy supply services. It offers greater efficiency than traditional grids, where the latter's centralized one-way flow direction is replaced with two-way communication and energy flows [1]. The initial concept of an SG was defined by the National Institute of Standards and Technology (NIST) [2] and, in fact, it is still under development [3]. Devices from the operational technologies (OT) side of this critical infrastructure, which were physically segregated in the past, are now more and more connected to the internet in a series of highly-distributed hierarchical network systems. By integrating distributed renewable energy sources, this next generation electric power system offer enhanced efficiency and reliability. In general, SGs are revolutionizing the energy supply sector and this trend is expected to rise [4,5].

Unfortunately, these facts make SGs a target of cyber attacks as well. In fact, a short time after its foundation, exploitation attempts by adversaries have been reported [6]. One of the first large-scale attacks on power systems was carried out against the Iranian nuclear program. A distributed malware caused severe malfunction and self-destruction of the system [7]. In 2016, a series of cyber attacks was successfully carried out against the power systems in Ukraine. The incident affected 225,000 clients by disconnecting them from the grid system for three hours. The subsequent investigation found out that the attackers possessed sophisticated hacking skills and a broad knowledge about the functionality of the power system [8]. Furthermore, several attacks were committed against the energy infrastructure and high-profile US organizations and private companies over the years (e.g., [9,10]). Apparently, the TRITON attack in 2017 targeted the Triconex safety instrumented system (SIS) of Schneider Electrics. SIS is responsible for safety from a higher level and takes immediate action in case of process control failure [11]. In addition to that, known attacks like phishing, Denial-of-Service (DoS), malware and eavesdropping are carried out against SGs [1]. Furthermore, more destructive attacks like BlackEnergy [12] or WannaCry [13,14] are common. Furthermore, another issue for security in the grid domain represents the human factor [15]. Therefore, fake honey pots can be inserted into the system in order to increase the uncertainty of the attacker [16]. For this matter, ensuring the security of smart grids represents a critical issue in the domain of power infrastructure

This paper proposes a formal risk assessment of cyber attacks in SG systems, based on the threat modeling and probabilistic model checking. An extensive literature survey, as one of the contributions of this paper, is provided at the beginning of the paper, in order to determine the state of the art in the related fields. Afterwards, a methodology that systematically describes involved steps and processes is given.

A smart grid demonstration case proposed in this paper serves as the common basis on which further security analysis are carried out. It takes into account the customer, the prosumer and the grid operator perspective, and encompasses three different attack surfaces in a form of three use cases: (i) smart home and HVAC hijacking attack, (ii) smart home and smart meter hijacking attack and (iii) smart grid and black-out attack. Threat modeling is applied on the whole demonstration case in order to identify threats and vulnerabilities within the system architecture. The resulting threat list is used as an input for vulnerability detection and for defining potential attack scenarios. An important parameter in the risk assessment is the exploitation probability–the likelihood that one particular vulnerability will be successfully exploited. While this parameter, in similar approaches, is most commonly determined through extensive literature survey, our approach includes calculation of exploitation probabilities using Common Vulnerability Scoring System (CVSS) [17]. As the last step, we apply formal verification on defined use cases, in order to perform formal risk analysis and to obtain an indication on how safety and security requirements can be fulfilled within the given environment. Subsequently, a model checker is used to check the probabilistic model against attack properties. In the conclusion, we elaborate on the effects of a security violation with regard to security and privacy of energy clients, and we discuss future considerations for improving security in the critical energy sector.

The remainder of the paper is structured as follows: Section 2 provides an overview about existing literature in the domain. Section 3 describes methodology used. Then, Section 4 describes demonstration case, including three use cases, and included components. Section 6 includes modeled attack scenarios and results of exploitation probability assessment. Section 7 describes available probabilistic formal verification tools and includes a description of modeled use cases and scenarios, and resulting attack probabilities. Section 8 elaborates on implications on security risks in SGs and concludes the paper.

2. Related Work

Ensuring the security of SG systems against cyber attacks represents a great challenge for public infrastructure. Therefore, an important step in this task is the identification of potential security issues. Subsequently, the classification of these attacks serves as a starting point in ensuring effective defense mechanisms. Therefore, several approaches exist that tackle the problem from different angles. In this paper, the existing literature is divided into three distinctive topics. The first topic discusses stochastic methods, including risk assessment, for the estimation of cyber threats in SGs. The second section discusses works on real security exploits against SGs to date. The final topic provides an overview on formal verification methods for the same challenge.

2.1. Stochastic Modeling of Cyber Attacks in Smart Grid Systems

In the context of cyber security, stochastic models are used to estimate the effect of the behavior of an attacker. For this sake, they analyze probability distributions of specific variables for a vulnerability. On the other hand, risk assessment is applied in order to identify existing threats, vulnerabilities or potential risks from potential consequences, as given by Paté-Cornell et al. [10].

Langer et al. [18] performs a cyber security risk assessment for attacks in SGs. For this matter, they methodically analyze the impact and likelihood of cyber attacks against such systems. Therefore, they implemented the toolbox Smart Grid Information Security (SGIS), which is applied for the estimation of risks for information assets. The toolbox is applied to two use cases, namely, voltage control and power flow optimization. The authors possible consequences for SGs due to exploitation of information assets.

The paper by Jauhar et al. [19] introduces a generalized model-based approach to assess realistic security risks for failure scenarios in SGs. The authors apply an assessment tool to analyze several of such scenarios. Each scenario is formalized in a structured model, which contains information about concrete vulnerabilities and attacker characteristics. These models are used in order to reason about failures in every scenario. The authors emphasize the advantage of their tool, namely the re-usability of generated models for different attacker profiles and systems.

Lee [20] provides a thorough theoretical analysis on risk assessment and failure scenario ranking in the electric sector. The work analyzes vulnerabilities as well as the resulting consequences and possible mitigation mechanisms. These vulnerabilities are listed and suggestions are given to define their naming conventions and classification. In fact, this work couples failure scenarios with security controls from [21].

Salehi Dobakhshari and Ranjbar [22] proposes a cloud-based solution for fault location estimation in grid systems. For this matter, the approach relies on a weighted least-squares (WLS) method, which is used to model a fault location estimation problem as a non-linear estimation problem. The authors observe that the probability of inaccuracy estimation increases as a consequence of multiple measurements. However, since the approach detects erroneous measurements, the overall estimation error remains below 1% at different points in the SG system.

The work from Rao et al. [23] introduces game-theoretic models for the estimation of attack probabilities. For this sake, the authors analyze the interactions between attacker and defender in a grid network. Then, they apply game theory in order to implement a probabilistic model. This represents a systematic Boolean attack–defense model, thereby analyzing the overall infrastructure in a top-down manner. In their example, in authors assume that the attacker obtained information about the target system prior to the attack. In this way, the probabilistic model returns success rates for both the attacker and defender. In the attack scenario, both parties assume certain action after calculating the opponent's success rate. However, in order to reproduce this approach, broad knowledge about the capabilities of the infrastructure is necessary, which is provided by facility users. For both sides to draft a probabilistic strategy, knowledge about incidental degradation, incurred costs, defender strategies and defense priorities is mandatory.

The approach Gao et al. [24] applies a Monte Carlo method, which is based on Markov chains for operation reliability in power systems. For this sake, the authors focus on time-dependent state probabilities for state prediction. In this method, a system state is defined in terms of a specific situation of its composite states. In this case, the probability of each state is depicted with corresponding state models. Subsequently, the resulting multi-state Markov model relies on a state transition matrix in order to derive failure and repair rates of a specific state in the system. In this way, security issues in states can be addressed and evaluated in advance.

Hao et al. [25] focuses on a false-data injection against SG distribution systems. In this approach, false-data injection is used in order to exploit a voltage regulation mechanism. This stealthy attack sends false voltage data to the system in order to cause load distur-bances in voltage regulators. Eventually, the attack causes a malfunction in the system. In order counter face the attack, the authors compute the best response strategy by relying on a concept from game theory, namely the adaptive Markov strategy (AMS). This method is especially suitable in situations where a system is attacked by unpredictable attackers.

Therefore, the AMS comprehends online learning mechanisms so optimal defense strategies are computed against an estimated behavior of the attacker.

In Leszczyna [26], the author provides a review on standards for cyber security assessments in Smart Grids. He identifies six relevant standards that provide a general guidance on this matter. However, they do not present a technical specification so far.

2.2. Cyber Security in Smart Grid Systems

As already mentioned, cyber attacks against the SG infrastructure can lead to severe consequences. Due to the complexity of this emerging technology, the behavior of attacker is difficult to predict. Several types of attacks like injection and malware target multiple systems across the network. Until now, several successful exploitations are executed in theory and practice.

Khan et al. [12] provides a detailed overview on one of the most destructive malwares in the energy sector, namely BlackEnergy. This Trojan horse has a long history with critical infrastructures so its capabilities evolved in time. In general, it targets real-time grid monitoring and control and causes coordinated DoS, eavesdropping, information theft, remote access, etc. Due to its evolving nature, providing effective defense mechanisms represents a difficult challenge. Therefore, the authors explain existing protection strategies and warn about the persistence of the attack in the future.

The work from Sun et al. [27] presents a description on the impact of coordinated cyber attacks against SG systems. Distributed attacks are usually well-organized and contain an elaborated attack plan before the execution. The latter constitutes an attack pattern, where each step is related to other steps that constitute an attack plan.

On the other hand, Soltan et al. [28] proposes a new class of potential attacks against SGs, which they call Manipulation of demand via IoT (MadIoT). For this matter, they address inherited vulnerabilities from the technology of IoT in the domain of power grids. In this scenario, attackers compromise less protected IoT devices and disrupt the functionality of an SG. Such attacks cause consequences like frequency instability, line and cascading failures and subsequently, increase in operational costs. Additionally, the authors utilize an IoT botnet that is meant to execute an attack against a power system. They perform the MadIoT attacks against an SG simulator and analyze different scenarios with regard to produced impact.

Security considerations for SCADA systems are discussed in detail by Pliatsios et al. [29]. First, a broad overview on the technology of SCADA and the underlying communication protocols is given. Then, security incidents are discussed as well as their impacts with regard to public health and safety. In the aftermath, the authors discuss the implementation of defense mechanisms. Thorough surveys of cybersecurity threats and issues in SCADA networks are given by Nazir et al. [30], Irmak and Erkek [31], Ghosh and Sampalli [32] and Antón et al. [33], respectively.

Furthermore, a broad discussion about cybersecurity issues in Industrial Control Systems is given by McLaughlin et al. [34]. In their review of theoretical frameworks, the authors assume worst case scenarios, i.e., that attackers are highly competent and in possession of complete knowledge about the system. Security measures are discussed that fit such attacker's profile for this matter.

A slightly different approach is given by Garcia et al. [35], which introduces a physics-aware cyber attack against power systems. The analyzed attacks target the system's underlying embedded programmable logic controller (PLC) by manipulating existing control commands. Thus, the malware is constructed in a way to cause severe impact to the physical power equipment. In order to make the malware functional and more efficient, the authors reverse engineer the central PLC to extract information about its control instructions. In this way, an adversary model is drafted in order to carry out attacks against the control system. Similar to other attack scenarios, the approach assumes that knowledge about the target domain is available prior to the attack. In fact, the attack is very effective because it is executed in a stealthy fashion. In this way, its existence is not obvious and causes a more enduring impact than strong short attacks.

Other attacks against PLCs in power systems are presented by Brüggemann and Spenneberg [36] and Klick et al. [37].

The work by Amini et al. [38] analyzes dynamic load altering attacks (LAAs) in power grids. Such cyber attacks are meant to control and destabilize a system by targeting its unsecured loads. The paper provides a model about the system and an estimation on the consequence of attacks against the system under a specific configuration. For this matter, the authors take into consideration the exposure of parameters to fluctuation and discuss mitigation mechanisms.

Security-related aspects of connected SGs in IoT infrastructures are discussed in the work from Dvorkin and Garg [39]. In these scenarios, distributed electric loads are interconnected and simultaneously controlled by a supervisory IoT system. For this sake, cybersecurity issues must be considered specifically for IoT-operated loads in order to guarantee undisrupted grid operation. Therefore, the authors provide a hypothetical scenario where an SG is targeted by distributed cyber attacks. Hence, the proposed modeling framework reveals the propagation and impact of cyber attacks on IoT-controlled electrical loads. In the aftermath, an attacker is able to alter the power consumption and cause severe damage to the system. In addition to that, the compromised IoT devices can be exploited further to launch additional attacks, like DoS, against other services. In addition to that, the authors discuss several attack strategies and the resulting consequences.

Li et al. [40] focuses on challenges for producing efficient defense mechanisms in SGs. For this reason, the authors introduce a sequential detector that is based on an adaptive sampling technique. In order to test their approach, malicious data is submitted to the monitoring systems of an SG. The efficiency of attack detection is compared in centralized and distributed attack detections. The authors claim their method is efficient against a broad variety of attacking strategies and unexpected situations. Furthermore, the proposed method considerably reduces overheads in the system without having a negative impact on the robustness and average detection performance.

Another discussion on coordinated data injection attacks against a power network and appropriate detection methods is given by Cui et al. [41]. First, the authors analyze an attack model for a stealth attack against an SG network. Therefore, they introduce a system model for state estimation on the example of a linearized model. In this method, the goal is to estimate possible electrical states by dynamically analyzing real-time redundant meter measurement data in a system. Subsequently, the authors provide a survey on common approaches against data injection attacks and discuss future research directions.

A more practical approach is given by Marksteiner et al. [42]. The paper provides a smart grid-specific methodology based on risk assessment and threat modeling for the development of security requirements. The given approach takes into account cyber threats on different levels, namely the architectural, the protocol and the device level.

Major security challenges and evolving cyber attacks in SGs are described by Li et al. [43], Rawat and Bajracharya Chandra [44], Shapsough et al. [45] and Gunduz and Das [46]. Yan et al. [47] summarizes cyber security requirements and existing solutions in communication systems and protocols. Security challenges with regard to technical foundations of SGs are described in detail by El Mrabet et al. [7].

2.3. Formal Verification in Smart Grids

The application of formal methods for the prevention of cyber attacks in the SG domain is addressed from different perspectives. These include quantitative as well as qualitative properties. On the protocol level, two relevant publications exist for this matter. Rashid et al. [48] addresses the performance and efficiency of Smart Grids by checking a ZigBee-based routing protocol for the communication network. The focus of this work is the functionality of the protocol. This is confirmed by the conducted verification of collision avoidance and liveness properties. On the other hand, Odelu et al. [49] focuses on the secure authentication key agreement in the Canetti–Krawczyk adversary model. The authors emphasize the importance of several software components for the security and privacy in the SG domain. Therefore, the authors propose a provable secure authentication key agreement scheme.

In Naseem et al. [50], PRISM is used for a reliability analysis and investigation of accuracy, stability and efficiency. The analysis was committed in cases where faults were detected in the transmission line system of a Smart Grid. In addition to that, the approach is applicable to calculate failure probabilities in network components. The reliability of a G3-Power Line Communication (PLC) networks is considered by Uddin et al. [51]. The work puts focus on the accuracy and reliability of the information flow. For this sake, it specially emphasizes the Fault Detection, Isolation and Supply Restoration behavior Hamman et al. [52] uses the model checker SPIN in order to check the wide-area backup protection system (WABPS). This is specifically designed to offer a high degree of reliability in different failure scenarios. However, due to the advanced design, the calculation of the fault tolerance proves difficult. Therefore, the authors applied SPIN to check the limits of a failure. Garlapati [53] applies the model checker UPPAAL for verification of the Agent Aided Distance Relaying Protection Scheme. This represents a hierarchically distributed and non-intrusive concept for decreasing blackout probability. In their work, this scenario occurred due to hidden failures of distance relays in the Advanced Metering Infrastructure

Bashar et al. [54] provides a security risk management solution for the protection of Smart Grids in case of cyber attacks. The authors use probabilistic model checking to model the attacker's behavior in form of a Markov Decision Process. For this sake, they rely on the model checker PRISM. Two types of attacker behaviors are considred, namely non-persistent and persistent behaviors, as well as their effect on critical components in an SG. On the other hand, Diovu and Agee [55] studies the effects of distributed DoS attacks on the Advanced Metering infrastructure. The authors estimate best and worst case analyses in the context of a cloud-based openflow firewall. In the aftermath, they study DoS attacks with respect to different detection probabilities of the firewall.

Similar to our proposed work, Krivokuća et al. [56] conduct a risk analysis for Smart Water Distribution use case. The corresponding authors focus on two main scenarios, namely water contamination and water tank overflow. Therefore, risk exposure scores are calculated for both scenarios. Additionally, these scores provide recommendations for system configurations to improve safety and security.

3. Methodology

The methodology for the research presented in this paper is shown on Figure 1. The steps in the process are sequentially marked in Figure 1. The first step is the definition of the system architecture, its components and communication channels. The system architecture is used for the threat modeling step, which as a result provides a list of threats. The vulnerabilities of the system and the attack possibilities are identified based on the threat list. Next, the exploitation probabilities for the identified vulnerabilities are calculated and the attacker's behavior is modeled. The formal system model is created based on the system architecture, the identified vulnerabilities with exploitation probabilities, and the modeled attacks. The formal properties of the attacks are identified next, and the model is checked against the identified properties using the model checker. This finally results in risk exposure scores.

A significant part of this research was the development of a non-deterministic system model as Markov Decision Process (MDP). The Markov model incorporates the system architecture, the attacker's behaviour and the existing vulnerabilities of the system with identified exploitation probabilities. The required input elements for formal verification are the system model, together with the identified attack properties. The chosen tool for formal verification is the PRISM model checker.

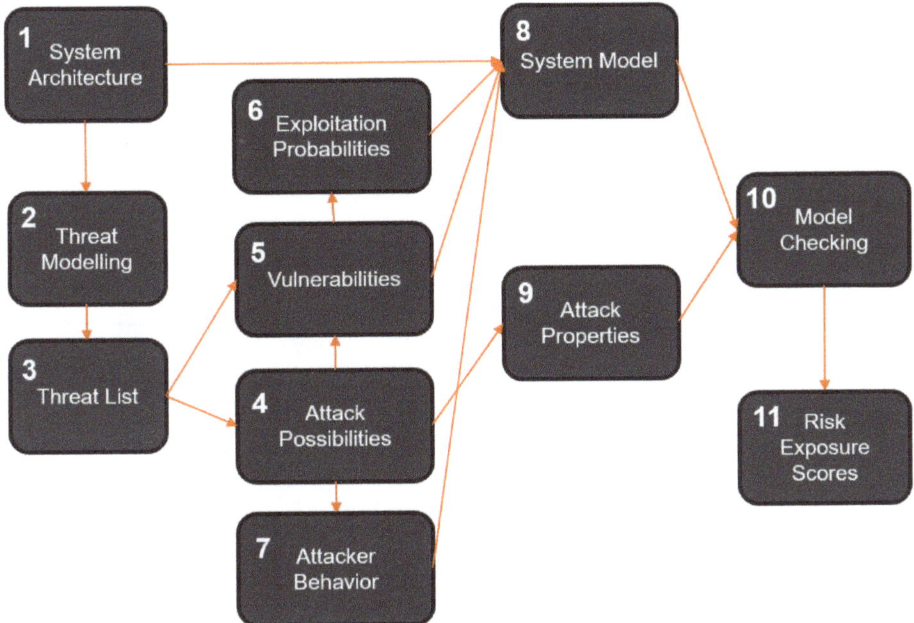

Figure 1. Risk analysis based on threat modeling and formal methods methodology.

4. Demonstration Use Case

The digitized bi-directional power infrastructure connecting production, distribution and prosumer assets offers the attackers different ways to penetrate the LV Grid via the ICT environment. The so called attack vectors, describing the attack route and the attack technique, can be very diverse depending on the hardware, software, communication channels and physical access. This large attack surface built up by different attack vectors and assets under different ownership has to be carefully addressed. Each new component added to such a configuration can introduce new unforeseen risks, both from HW and SW side.

The use case architecture described in this paper (Figure 2) simplifies the redesign of the energy network, which will be common in the future, towards a decentralized distributed generation and independent islanding. It serves as a starting point for our approach to a differentiated cyber risk assessment and takes into account the customer, the prosumer and the grid operator perspective.

The described architecture consists of various devices and services both on consumer and production/distribution side and, as such, it is analyzed from the cyber security perspective, taking into account three attack surfaces:

- UC1: Smart home and HVAC hijacking attack
- UC2: Smart home and smart meter hijacking attack
- UC3: Smart grid and black-out attack

The components included in each use case are described below. The components are carefully considered in the different steps of the proposed research, including threat modeling, attack assessment and model checking.

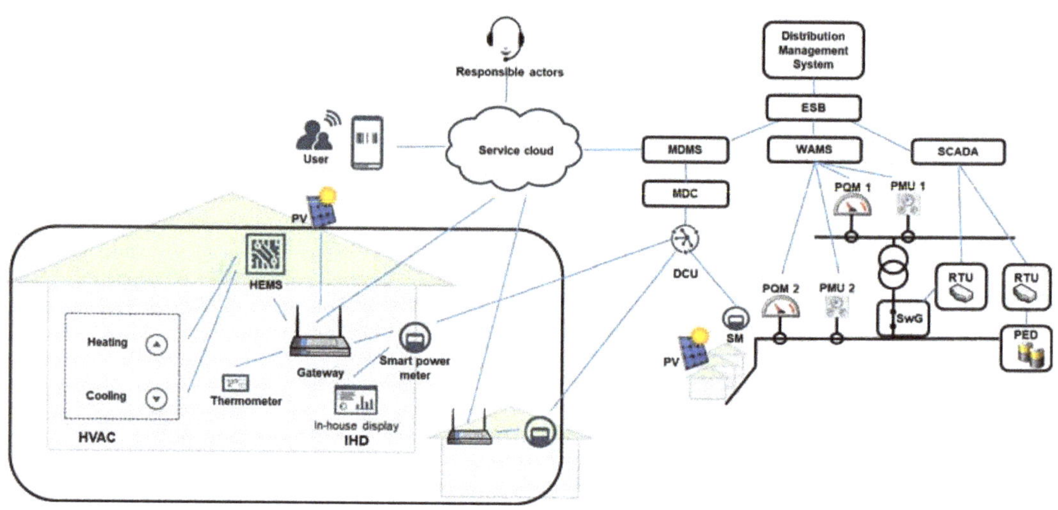

Figure 2. Smart energy system cyber security use case architecture.

4.1. Smart Home and HVAC Hijacking Use Case Components

Humidity, Ventilation and Air Condition (HVAC) system, including actuators controlling heating and cooling, and *thermometer*, is a technology used to control indoor thermal comfort and air quality. It is used in indoor applications, including for example residential, industrial and professional, environments, as well as in vehicular environments.

Home Energy Management System (HEMS) is a technology platform that introduces and connects sensors within smart home devices, via home networks. It consists of hardware and software components, and it is used to carry out home automation and control processes.

Gateway (G), or a residential gateway, is a consumer-grade system which provides network and internet access to various smart systems connected to a local area network (LAN).

4.2. Smart Home and Smart Meter Hijacking Use Case Components

Smart Meter (SM) is a digital device used in smart home applications, with the main function to collect and record information such as electric energy consumption, current, voltage levels, etc.

In-house display (IHD) is a user interface, in a form of a small monitor, directly connected to a smart meter. It usually shows the current system parameters, like current energy consumption.

This use case also includes distributor side components such as the Data Concentrator Unit (DCU), the Meter Data Collector (MDC) and the Meter Data Management System (MDMS), that are described bellow, in the third use case.

4.3. Smart Grid and Black-Out Attack Use Case Components

Power Electronic Device (PED) consists of an Intelligent Local Energy Manager (ILEM), which is responsible for the management of the device components, a Power Conversion System and a Battery Management System with hybrid energy storage features, based on predefined voltage support actions configured by the operator.

Phasor Measurement Unit (PMU) is a metering and control device for measuring phasor data (voltage and current synchronization), voltage and current waveforms.

Power Quality Monitor (PQM) is a lightweight metering device with a measurement and control unit and enables the calculation of all power quality parameters required according to EN 50160.

Wide Area Monitoring System (WAMS) collects processes and monitors geo- referenced spatial temporal field data coming from distributed PMUs and PQMs located both on low voltage and on the medium voltage side.

The *Remote Terminal Unit (RTU)* provides communications features to field devices to exchange telemetry data and control messages with the capability of processing data.

Distribution Management System (DMS) provides sophisticated functionality for advanced monitoring and control of the distribution grid to the DSO, which is important to handle increased customer generation and energy storage facilities as well as the management of demand response capabilities.

Supervisory Control and Data Acquisition System (SCADA) is in charge of real-time monitoring and control of the distribution grid and manages diverse remote control systems via a telecommunication network and provides data analytics and data storage functionality.

Enterprize Service Bus (ESB) is a middleware that enables the messaging and routing capabilities as well as the integration of distributed services and applications.

Data Concentrator Unit (DCU) is a unit responsible for the gathering of measurement data from multiple metering devices installed at the substation.

Meter Data Management System (MDMS) stores large quantities of smart meter data performs their validation and analytical processing.

Meter Data Collector (MDC) collects and manages the measurement data received by the data-concentrating units (DCUs) which is then forwarded to MDMS.

Photovoltaic system (PV) is a standalone solar PV with a power conditioning unit, a DC/AC converter and which is connected via the gateway to delivers data.

Switchgear (SwG) is a motorized switchgear which enables coupling and decoupling of the grid sector.

5. Threat Modeling

In general, threat modeling aims to identify threats and vulnerabilities within IT-related system architectures [57]. Furthermore, it helps to put security and privacy by design into practice. In this paper, a threat modelling approach is meant to secure a project setup with a systematic security analysis. In [42], a feasible list of requirements obtained by risk assessment for different components. In our paper, however, risk assessment is performed for each individual component, without analyzing their interaction within the overall system.

Figure 3 depicts the threat model based, which is based on the demonstration use case from Section 4. The model was created by using the Microsoft Threat Modeling Tool [58], which works on data flow diagrams that describes data stores, processes and communication lines and provides threats based on the STRIDE model [59]. STRIDE divides threats into the following six categories:

- **Spoofing:** refers to the illegally access and usage of foreign authentication information to obtain illegitimate access.
- **Tampering:** is associated with the malicious modification of data.
- **Repudiation:** denies an action that was performed by an entity without having a possibility to uncover this malicious action.
- **Information disclosure:** denotes the exposure of information to users who are not authorized to have access to this information.
- **Denial of service:** affects the availability of services so that they are not accessible for authorized users.
- **Elevation of privilege:** denotes obtaining privileged access by an unprivileged user and thus ultimately the possibility of compromising the entire system.

The Microsoft Threat Modeling Tool is not limited to a set of threats but offers the possibility to create individual templates for a given domain. Furthermore, we rely on the Azure cloud and IoT templates from Microsoft for the smart home area. We combine these templates with our own, which are based on our research in the smart energy domain. In the model itself, different trust zones were identified according to our use case. First, the trust zone of components is identified within the smart home, then in the outsourced cloud area and the immediate personal area. In the SG environment, the trust zones at the control center include DMS, SCADA and a monitoring system. At the substation level, two zones exist, namely, one that covers the PED and SwG, and a second dealing with metering components. Finally, measurement components that cover additional metering points in the network constitute a separate trust zone.

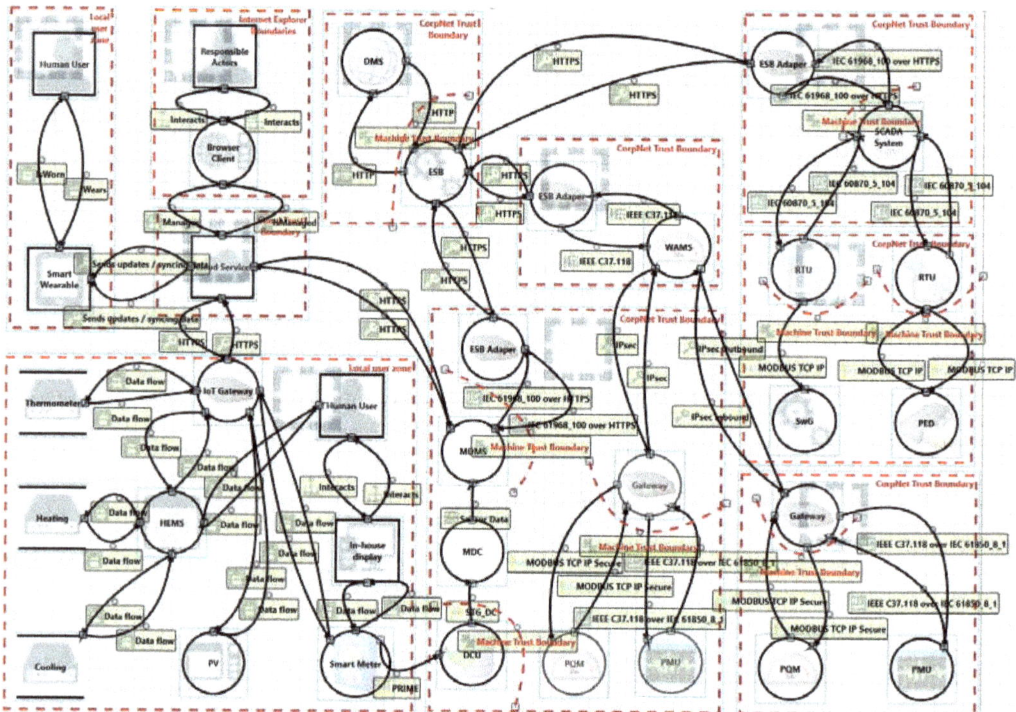

Figure 3. Threat Model.

In total, the modelling approach resulted in the identification of 1137 threats. These are classified according to STRIDE, with additional one that describes threats in smart home components. The latter also cover physical threats, which contribute to the overall number of threats. A list of all categories is given in Table 1.

However, for the conducted assessment we rely just on the threats from Table 2.

Table 1. Threat asignment to category.

	Category	Amount
S	Spoofing identity	51
T	Tampering with data	15
R	Repudiation	38
I	Information disclosure	33
D	Denial of Service	85
E	Elevation of privilege	116
-	Smart energy components related	798

Table 2. List of threats.

Id	Title	Category	Description
51	Elevation by Changing the Execution Flow in WAMS	Elevation Of Privilege	Gateway may be able to remotely execute code for WAMS.
187	PQM May be Subject to Elevation of Privilege Using Remote Code Execution	Elevation Of Privilege	Gateway may be able to remotely execute code for PQM.
230	Data Flow MODBUS TCP IP Secure Is Potentially Interrupted	Denial Of Service	An external agent interrupts data flowing across a trust boundary in either direction.
353	Data Flow IEEE C37.118 over IEC 61850-8-1 Is Potentially Interrupted	Denial Of Service	An external agent interrupts data flowing across a trust boundary in either direction.
478	Data Flow IEEE C37.118 over IEC 61850-8-1 Is Potentially Interrupted	Denial Of Service	An external agent interrupts data flowing across a trust boundary in either direction.
593	Data Flow MODBUS TCP IP Secure Is Potentially Interrupted	Denial Of Service	An external agent interrupts data flowing across a trust boundary in either direction.
597	Spoofing the ESB Adaper Processd	Spoofing	ESB Adaper may be spoofed by an attacker and this may lead to unauthorized access to ESB.
618	Data Flow IEC 60870-5-104 Is Potentially Interrupted	Denial Of Service	An external agent interrupts data flowing across a trust boundary in either direction.
621	Elevation by Changing the Execution Flow in RTU	Elevation Of Privilege	An attacker may pass data into RTU in order to change the flow of program execution within RTU to the attacker's choosing.
640	Data flowing sniffing across MODBUS TCP IP	Information Disclosure	Data flowing across MODBUS TCP IP may be sniffed by an attacker. Information may may be used to attack other parts of the system or simply be a disclosure of information leading to compliance violations.
647	Elevation by Changing the Execution Flow in PED	Elevation Of Privilege	RTU may be able to remotely execute code for PED.
661	Elevation by Changing the Execution Flow in SCADA System	Elevation Of Privilege	An attacker may pass data into SCADA System in order to change the flow of program execution within SCADA System to the attacker's choosing.
672	Data Flow IEC 60870-5-104 Is Potentially Interrupted	Denial Of Service	An external agent interrupts data flowing across a trust boundary in either direction.
686	Elevation by Changing the Execution Flow in RTU	Elevation Of Privilege	An attacker may pass data into RTU in order to change the flow of program execution within RTU to the attacker's choosing.
857	Elevation by Changing the Execution Flow in DCU	Elevation Of Privilege	An attacker may pass data into DCU in order to change the flow of program execution within DCU to the attacker's choosing.
880	Spoofing the In-house display External Entity	Spoofing	In-house display may be spoofed by an attacker and this may lead to unauthorized access to Smart Meter.
903	An adversary may block access to the application or API hosted on In-house display through a denial of service attack	Denial Of Service	An adversary may block access to the application or API hosted on In-house display through a denial of service attack
1005	An adversary may gain elevated privileges and execute malicious code on HEMS host	Elevation Of Privilege	If an application runs under a high-privileged account, it may provide an opportunity for an adversary to gain elevated privileges and execute malicious code on host machines
1013	An adversary may execute unknown code on Heating/Cooling	Tampering	An adversary may launch malicious code into Heating and execute it
1028	Potential Excessive Resource Consumption for IoT Gateway or Thermometer	Denial Of Service	Resource consumption attacks can be hard to deal with, and there are times that it makes sense to let the OS do the job.

Table 2. *Cont.*

Id	Title	Category	Description
1052	Elevation by Changing the Execution Flow in PMU	Elevation Of Privilege	An attacker may pass data into PMU in order to change the flow of program execution within PMU to the attacker's choosing.
1057	Data Flow Sniffing	Information Disclosure	Data flowing across IEEE C37.118 over IEC 61850 8 1 may be sniffed by an attacker. Information may be used to attack other parts of the system or simply be a disclosure of information leading to compliance violations
1111	Elevation by Changing the Execution Flow in IoT Gateway	Elevation Of Privilege	An attacker may pass data into IoT Gateway in order to change the flow of program execution within IoT Gateway to the attacker's choosing.

6. Attack Scenarios and Exploitation Probability Assessment

In this section, three different exploitation attempts against the system architecture are described (Figure 2). Each scenario describes a sequence of attack steps, which lead from an attacker to the exploitation of the system. In general, the SG system represents an interconnection of devices that are linked in an online infrastructure. Usually, every connection is implemented using different technology and communication protocols. Because of this, each device in the network represents a distinctive security challenge for the attacker. From the outside, the system can be accessed either physically or via a web interface. In the latter case, exploiting such interface represents the first security challenge for an attacker. In fact, connecting to the system and gaining access to an online device remains a challenge in every attack scenario. This paper omits the technological details about the target devices and focuses on the ways to exploit their vulnerabilities. However, it should be noted that advanced hacking skills and domain knowledge are mandatory in order to carry out the described attacks.

6.1. Smart Home and HVAC Hijacking Attack Scenarios

In this use case (UC1), the attacker successfully exploits the vulnerabilities in temperature regulation systems and their sensors. The ultimate target represents the HVAC system, which represents the last link in a chain of devices. As already mentioned, the attacker must gain control over the internet system of the smart home. Then, the attacker must find a way to compromise the gateway of a single smart home. Afterwards, she injects a payload to obtain operational access in the corresponding HEMS system. Finally, by controlling HEMS, instructions can be sent to the targeted HVAC system. The overall attack steps are depicted in Figure 4.

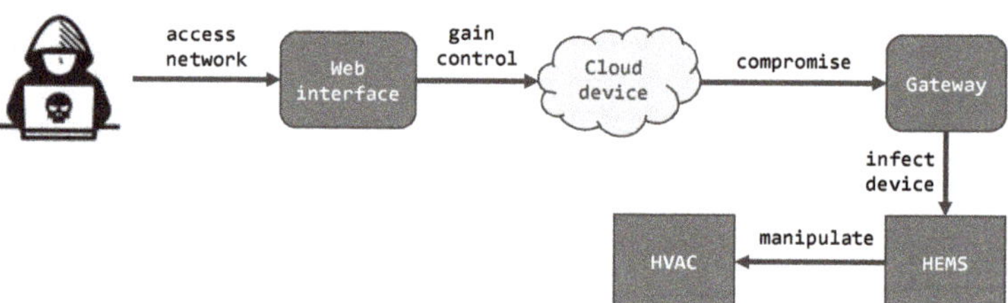

Figure 4. Attack scenario for hijacking of heating and the HVAC system.

6.1.1. Attack Scenario 1.1—Hijack Heating

In case the attacker gains control over the heating, she can manipulate the internal thermal sensors at will. The heating can be switched on and off, which eventually leads to discomfort and might cause health issues for inhabitants. Furthermore, extreme overheating can cause irreparable damage to the smart home system or surrounding devices.

6.1.2. Attack Scenario 1.2—Hijack HVAC

In addition to manipulating the heating, a remote attacker can affect ventilation and air conditioning in smart homes as well. By assuming control over the HVAC system, she is able to cause a temperature overload in the system. In this way, the system can be damaged and increases the power consumption. A known intrusion against the HVAC system, called HVACKer [60], relies on an already installed malware in order to manipulate temperature settings. This attack exploits a vulnerability that is caused by the lack of insights about thermal communication protocols.

6.2. Smart Home and Smart Meter Hijacking Attack Scenarios

The attack in this use case (UC2) exploits vulnerabilities in a household's smart meters. As already mentioned, this SG device gathers information about energy consumption and monitors user behavior, which is displayed at the IHD interface. In turn, this private information can be linked to real persons on the internet. In our case, we chose a setup where the SM is connected via the HAN to the gateway. In our scenario, the attacker gains via the gateway access to the SM. Afterwards, she is able to interfere with the smart homes' energy consumption in different ways. The use case for this attack is shown in Figure 5.

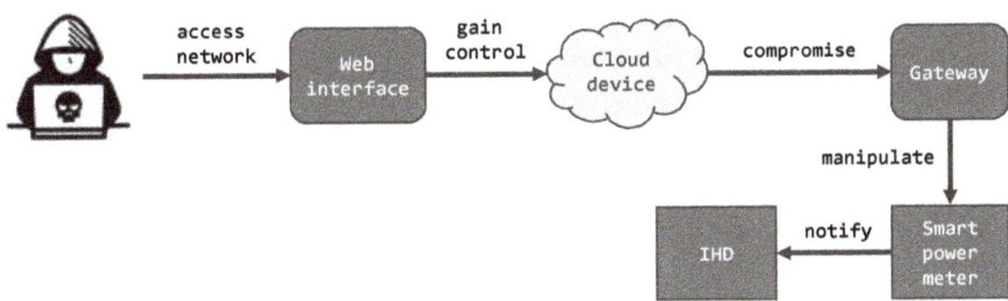

Figure 5. Attack scenario for hijacking of smart meters in smart homes.

6.2.1. Attack Scenario 2.1—Fraud

In this case, the attacker alters the SM of another smart home in the local network. Subsequently, she increases a neighbor's power consumption and simultaneously decreases her own. By balancing the energy expenditure in the local network, the attacker conceals the committed fraud at the cost of the neighbor.

6.2.2. Attack Scenario 2.2—Decrease Bill

If the attacker compromises her own SM or concentrator, she might decrease her recorded energy consumption. In this way, the attacker decreases her bill without changing her consumption habits.

6.2.3. Attack Scenario 2.3—Increase Bill

The attacker compromises the SM that is responsible for collecting information about the energy consumption of another user. Therefore, the expenditure of the unsuspecting victim is increased for the time being.

6.2.4. Attack Scenario 2.4—Increase Bill, no Alarm

Similar to the previous case, the attacker hijacks a victim's SM and alters its consumption for the worst. However, in this scenario, the victim is never alerted about this issue, eventually causing long-lasting consequences.

6.3. Smart Grid and Black-Out Attack Scenarios

The next use case (UC3) differs from the above two scenarios since the attacker targets the main distribution grid. The DMS executes control over a number of distinctive SG devices, which manage and measure energy distribution and exchange information between individual grid components. The goal of this highly sophisticated attack is to disrupt the functionality or to cause a breakdown of the parts designed for islanding and which are supported by storage units. In the most severe case, the attack leads to a complete black-out in an area of the grid network. In order to achieve this ambitious goal, the attacker must get control over the decentralized energy generation and local storage devices. In addition, the attacker must access the switchgear devices to disconnect this section from the grid as well as to disable the monitoring devices to remain unnoticed by the grid control center. In general, Figure 6 depicts an abstract overview of this attack scenario.

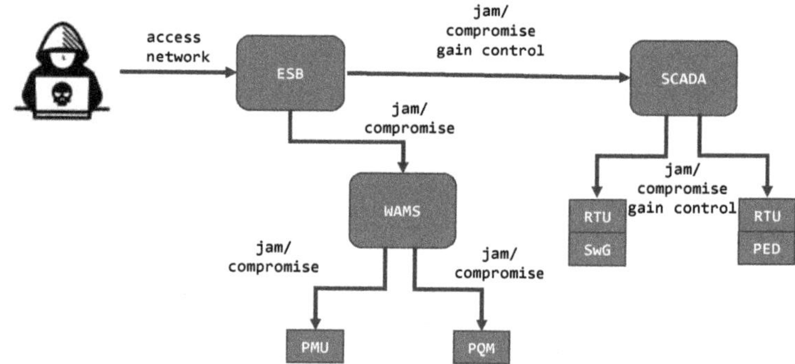

Figure 6. Attack scenario for black-outs in an SG.

6.3.1. Attack Scenario 3.1

This scenario covers the aspect of manipulating various sensor values without attacking the implemented control functionality itself. After the attacker gained access to the ESB, she proceeds further by trying to confiscate the WAMS and, respectively, use it as gateway to perform a lateral movement to the connected PMUs and PQMs. Besides that, the same appears for SCADA and the connected components RTU, SwG and PED. By doing so, the possibility is given that the SCADA initiates the island mode in order to ensure further functionality. From this moment on, the attacker can proceed in one of the following direction.

- Sub scenario 3.1.1
  The attacker manipulates the PED to send a message to the SCADA system that it is fully charged, while in fact, the storage units are empty. The metering devices, PMU and PQM, are also targeted, and they report a power undervoltage or bad quality. This will then lead to the attacker's goal–the SCADA system will initiate islanding in that section by opening the SwG.
- Sub scenario 3.1.2
  The first steps of this scenario is the same as in the previous scenario. Instead of manipulation of the PMUs and PQMs sensor values, this scenario includes the manipulation of the WAMS functionality, where the power undervoltage telegrams are delivered to the control center.
- Sub scenario 3.1.3

In this scenario, the attacker manipulates the RTU of a PED in order to report that the latter is fully charged. The rest of the attack proceeds similarly to scenario 3.1.1, where the PMU and PQM both report undervoltage, and SCADA issues a command to open the SwG.

- Sub scenario 3.1.4
  The first steps of this scenario are the same as in the scenario 3.1.3. Instead of manipulation of the PMUs and PQMs sensor values, the WAMS functionality is manipulated, and the power undervoltage telegrams are delivered to the control center.

6.3.2. Attack Scenario 3.2

The first few steps of this scenario are similar to the previous example. After the attacker gains access to the ESB network, she proceeds further to the SCADA system and to the connected components RTUs and PED.

- Sub scenario 3.2.1
  Here, the attacker gathers information about the status of PED by sniffing the Modbus TCP/IP. Then she interrupts the connection to the PMU2, PQM2 and PED by jamming the communication lines so that no status info is transferred to the control center. In the aftermath, the compromised RTU issues a command to open the SwG when the battery of the PED is empty.
- Sub scenario 3.2.2
  Another option after obtaining information about the status of the PED would be to jam the SCADA system altogether. The RTU will send a SwG open command when the battery is empty.
- Sub scenario 3.2.3
  Instead, to access the Modbus TCP/IP, the attacker can focus on the IEC 60870_5_104 in order to gather information on PED's status. In the aftermath, the attacker is capable of disabling the status transfer from PMU2, PQM2 and PED by jamming them. The RTU connected to the SwG behaves in the same way as in the above examples.
- Sub scenario 3.2.4
  This case is similar to the previous example with the difference that the attacker jams the SCADA system directly instead of the metering and PED control devices.

6.3.3. Attack Scenario 3.3

The first few steps of this scenario are similar to the previous examples. After the attacker gains access to the ESB network, she proceeds further to the WAMs and the SCADA system and the connected components, the PMUs and PQMs, the RTUs and the PED. However, in this case, the attacker is waiting for an undervoltage event by the grid itself. If such an event occurs, the attacker switch of the PV and triggers the battery loading. In this scenario, the attacker gathers information about the status of the grid and the status of the PED by sniffing.

- Sub Scenario 3.3.1 In this scenario, the attacker gathers information about the status of PED by sniffing the Modbus TCP/IP. Additionally, the attacker gathers the information about the status of the grid itself by sniffing the PMU2 and PQM2 measurement values. If the attacker detects an undervoltage event and the possibility to load the batteries, the attacker will trigger the battery load process.
- Sub Scenario 3.3.2 Instead, to access the Modbus TCP/IP, the attacker can focus on the IEC 60870_5_104 in order to gather information on the PED's status, and the stability info about the grid is directly collected via the WAMS.

6.3.4. Attack Scenario 3.4

Instead of targeting the different devices, the attacker launch a direct attack on the SCADA system. By doing so, the attacker can trigger via the SCADA system the event to open the SwG and send the battery load command to the PED.

6.4. Exploitation Probability Assessment

In order to effectively address the different threats, and to model detected vulnerabilities, an assessment of the system components from the exploitation probability point of view is very useful. This is reflected in the fact that different vulnerabilities have different security protection mechanisms in place, and that probability to exploit such vulnerabilities differs. The risk assessment of system components outlined by Marksteiner et al. [42] was obtained by the organization responsible for the specific component. In similar approaches proposed by Mohsin et al. [61] and Krivokuca et al. [56] exploitation probabilities were defined through extensive literature research. Their work proposes a new methodology for quantifying the likelihood of the compromise of a cyber component of SG, and includes risk assessment based on direct or indirect impact of the breach.

In our approach, the CVSS tool is adopted for the calculation of exploitation probabilities, with certain methodology modifications compared to the existing approaches. The first step is a mapping of the components in use and their interrelation at the evaluated environment, based on the threats obtained by the threat modeling process of the specified use cases.

The CVSS tool have several modules available, where the base score module, according to Wadhawan et al., provides a good basis for the calculation of the probability of an attack. Base score module have three sub-modules that can be used for calculating probabilities: *exploitability*, *scope* and *impact* sub-modules. Exploitability sub-module reflects how easily a vulnerability can be exploited, while the scope and impact sub-modules quantify the consequences of a successful exploit.

In contrast to [62], our work takes into account, apart from exploitability of the sub-module, also the core and impact sub-modules, because in penetrating the smart grid, these impact-related parameters are also in the focus of an attacker and, therefore, should be part of the calculation.

The *exploitability* sub-module takes into account:

- The attack vector—reflects the context by which a vulnerability can be exploited;
- The attack complexity—outlines how much effort in the preparation or execution of the attack against a vulnerable component the attacker have to invest;
- Privileges required—denote the level of privileges an attacker must obtain in order to successfully attack a component;
- User interaction—if another human user must be involved for a successful attack.

The *scope* sub-module rates if a vulnerability in one asset affects other assets, which are outside of this security authority.

The *impact* sub-module takes into account the CIA Triade:

- Confidentiality—to what extent confidentiality is affected;
- Integrity—refers to the trustworthiness and correctness;
- Availability—rates the impact of a successful attack to the availability of the affected component.

For less complex attacks, the base metric score is higher because such an attack have a higher likelihood. The score obtained from the CVSS system in range 1–10, and is normalised to the range 0–1 for further calculations. In Tables 3–5, the exploitation probabilities related to the different threats and for the different use cases are listed.

Table 3. UC1 Exploitation probabilities.

Threat Id	Title	Expl. Prob.
1005	An adversary may gain elevated privileges and execute malicious code on HEMS host	0.57
1013	An adversary may execute unknown code on heating/cooling	0.59
1028	Potential excessive resource consumption for IoT gateway or thermometer	0.65
1111	Elevation by changing the execution flow in IoT gateway	0.59

Table 4. UC2 Exploitation probabilities.

Threat Id	Title	Expl. Prob.
857	Elevation by changing the execution flow in DCU	0.57
880	Spoofing the in-house display external entity	0.68
903	An adversary may block access to the application or API hosted on in-house display through a denial of service attack	0.65

Table 5. UC3 Exploitation probabilities.

Threat Id	Title	Expl. Prob.
51	Elevation by changing the execution flow in WAMS	0.59
187	PQM may be subject to elevation of privilege using remote code execution	0.59
230	Data flow MODBUS TCP IP secure is potentially interrupted	0.75
353	Data flow IEEE C37.118 over IEC 61850-8-1 is potentially interrupted	0.75
478	Data flow IEEE C37.118 over IEC 61850-8-1 is potentially interrupted	0.75
593	Data flow MODBUS TCP IP secure is potentially interrupted	0.75
618	Data flow IEC 60870-5-104 is potentially interrupted	0.65
621	Elevation by changing the execution flow in RTU	0.59
640	Data flowing sniffing across MODBUS TCP IP	0.53
647	Elevation by changing the execution flow in PED	0.57
661	Elevation by changing the execution flow in SCADA system	0.64
672	Data flow IEC 60870-5-104 is potentially interrupted	0.65
686	Elevation by changing the execution flow in RTU	0.59
1052	Elevation by changing the execution flow in PMU	0.59
1057	Data flow sniffing	0.53

7. Risk Analysis Using Formal Methods

One of the methods to detect weaknesses and possible vulnerabilities at an early stage is formal verification. Available probabilistic formal verification tools, that found their application in risk analysis, are described at the beginning of this section.

Selected model checker, PRISM, is applied on the three different use cases, in order to perform formal risk analysis and to obtain an indication on how safety and security requirements can be fulfilled within a given environment. Within each use case, several example attack scenarios are modeled. The modeled attacks scenarios are selected from the scenarios outlined in Section 6. This section includes a description of modeled use cases and scenarios, and resulting attack probabilities.

7.1. Formal Methods Overview

Formal verification relies on a diverse set of mathematical and logical methods. These methods can be used to check different parts of a system, including the functional correctness of implementations, programming bugs, hardware Trojans, and security properties, and they can provide both a qualitative and quantitative analysis [63–66].

A survey of probabilistic model checking, including the main probabilistic models, algorithms and abstraction techniques is given in [67], while more details can also be found in [68,69]. In general, there are several different formalisms with difference in the notion of time. Those formalisms can be either non-deterministic or deterministic. When it comes to the notion of time, discrete-time models and continuous-time models have to be distinguished. The basic probabilistic model is the discrete-time Markov chain (DTMC), a deterministic model for discrete time. Its continuous version is called continuous-time Markov chain (CTMC). The Markov decision process (MDP) extends the discrete-time Markov chain with non-determinism. Similarly, its continuous version is called continuous-time Markov decision process (CTMDP). Originally, the MDP was introduced as probabilistic automata (PA). MDP and PA are very similar, with the small difference, that PA allows internal non-determinism, while MDP does not. Probabilistic timed automata (PTA) can be considered as MDP with clock variables. The final probabilistic model, Markov Automata (MA), is a generalization of interactive Markov chains and MDPs, and it is able to express discrete randomness. The selection of probabilistic model depends on the nature of modeled system and processes.

There is a variety of formal verification tools available, covering different aspects. While classic verification gives answers to questions such as "Will a given assertion ever be violated?", and focus on *qualitative properties*, the evaluation of dependability aspects such as, e.g., reliability, availability and performance need a focus on *quantitative* properties and models, including probabilistic behavior and real-time aspects [70]. An extensive overview of quantitative verification, including its different formalisms, modeling languages, properties and verification approaches is given in [70]. The main approaches for quantitative verification are probabilistic model checking and statistical model checking. There is a wide variety of tools with such a focus, including FACT (https://www-users.cs.york.ac.uk/~cap/FACT/, accessed on 31 May 2021), MODEST (http://www.modestchecker.net/, accessed on 31 May 2021), MRMC (http://mrmc-tool.org/, accessed on 31 May 2021), PASS (https://depend.cs.uni-saarland.de/tools/pass/, accessed on 31 May 2021), PARAM (https://depend.cs.uni-saarland.de/tools/param/, accessed on 31 May 2021), PRISM (http://www.prismmodelchecker.org/, accessed on 31 May 2021), UPPAAL (http://www.uppaal.org/, accessed on 31 May 2021), STORM (https://www.stormchecker.org/, accessed on 31 May 2021). The different tools focus on different parts of qualitative model checking. This section describes in more detail the most commonly used tools in the literature–UPPAAL and PRISM, and the most recent tool STORM, and discusses the reasons for the selection of PRISM for the following experiments.

7.1.1. PRISM

The *probabilistic model checker PRISM*, developed at the University of Birmingham, focuses on quantitative verification in a wide area of application domains including wireless communication protocols, quantum cryptography and systems biology [65,71,72]. It supports probabilistic models such as DTMCs, CTMCs, MDPs, PAs, and PTAs, including extensions of these models with costs and rewards. Its state-based input language is based on the Reactive Models formalism of Alur and Henzinger [73].

The tool uses symbolic data structures and algorithms based on Binary Decision Diagrams and Multi-Terminal Binary Decision Diagrams. It has a discrete-event simulation engine providing support for approximation/statistical model checking including different analysis techniques as quantitative abstraction refinement and symmetry reduction.

7.1.2. UPPAAL

The *toolbox UPPAAL* was developed by the Department of Information Technology at Uppsala University in Sweden, in cooperation with the Department of Computer Science in Aalborg University in Denmark for the verification of real-time systems [74,75]. It is intended for systems modeled as a collection of non-deterministic processes with finite control structure and real-valued clocks, communicating through channels or shared variables. The model checker is based on the theory of timed automata especially suited for checking invariant and reachability properties. UPPAAL SMC (http://people.cs.aau.dk/~adavid/smc/, accessed on 31 May 2021) [76,77] (UPPAAL Statistic Model Checking) is an extension of UPPAAL, used for performance property analysis of networks of priced timed automata. Priced timed automata are timed automata whose clocks can be modeled with different rates in different locations.

7.1.3. STORM

The *probabilistic model checker STORM* has been developed at the RWTH Aachen University, Germany for the analysis of systems involving random or probabilistic phenomena [78,79]. It is especially suited for checking quantitative aspects of models, e.g., security in randomized key generation, systems biology or embedded systems. STORM supports several different input languages, and has a modular set-up allowing to exchange solvers and symbolic engines easily. STORM supports several models as DTMCs, CTMCs, MDPs and MAs. It focuses on reachability queries and its supports, including probabilistic computation tree logic, continuous stochastic logic, expected rewards, long-run average rewards, conditional probabilities and multi-objective model checking.

7.1.4. Comparison of Probabilistic Model Checkers

A comparison of the described model checkers' capabilities, and their technical details are given in Table 6. A comparison of features is presented in Table 7.

An extensive comparison between the tools UPPAAL and PRISM is given in [80]. The authors used timed automata and a benchmark study for the modeling and verification in both tools. Another comparison of several tools for probabilistic model checking is given in [70]. This study also includes the tools PRISM and STORM. The authors state that STORM is clearly the most versatile tool. STORM covered all formalism considered in the most of competition. The authors conclude that PRISM and STORM tools support the widest range of properties, comparing to wide variety of available tools. Moreover, they have a wide range of algorithms implemented. On the other hand, the authors state that the PRISM is the most commonly used tool in research, due to its extensive online documentation, the graphical user interface and since it is independent on the platform used.

Table 6. Comparison of probabilistic model checking tools.

	PRISM	**UPPAAL**	**STORM**
Operating System	Windows, Linux, Mac	Windows, Linux, Mac	Linux (min Debian 9, Ubuntu 16.10), Mac (min 10.12)
Last Version	version 4.7 (March 2021)	official release 4.0.15 (Nov. 2019)	version 1.6.3 (Nov. 2020)
Licence	GNU GPL 2	free for non-commercial applications in academia only	GNU GPL 3
Type of input	PRISM language	XTA and XML	PRISM, JANI, GSPNs, DFTs, cpGCL, explicit
Simulator	yes	yes	no
GUI	✓	✓	✗
Case Studies	✓	✓	✗

Table 7. Probabilistic model checkers' feature comparison.

	PRISM	**UPPAAL**	**STORM**
Statistical model checking	✓	✓	✗
Probabilistic model checking	✓	✗	✓
DTMC	✓	✗	✓
CTMC	✓	✗	✓
MDP	✓	✗	✓
MA	✗	✗	✓
PA	✓	✗	✗
PTA	✓	✗	✗
priced TA	✗	✓	✗

Considering all discussed aspects in this section, the PRISM tool is selected for the following experiments.

7.2. UC1: Smart Home and HVAC Hijacking

7.2.1. Probabilistic Model Generation

Figure 7 shows the configuration analyzed in UC1. The system is modeled in the PRISM model checker as MDP (Markov Decision Process) because of the non-deterministic nature of cyber-attacks.

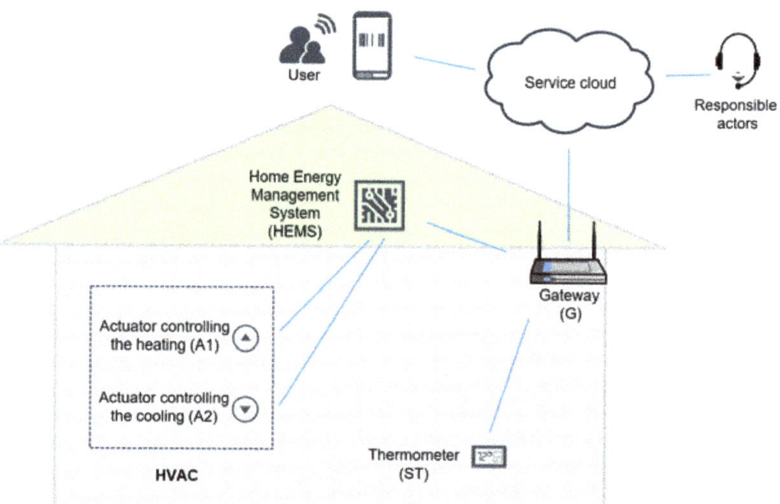

Figure 7. Smart home and HVAC hijacking use-case architecture.

An attack in the modeled system usually leaves traces in a form of unusual values. The response of the system is defined through service policies:

- *A1 policy*: If controller HEMS receives a low temperature reading from sensor ST, it sends a command to actuator A1 to start the heating process;
- *A2 policy*: If controller HEMS receives a high temperature reading from sensor ST, it sends a command to actuator A2 to start the cooling process;
- *ST policy*: ST sends regularly temperature readings to HEMS.

In the modeling process it is assumed that an attacker actively attacks the system, and has skills and means to perform certain attacks by exploiting existing vulnerabilities. An attacker's skills are measured through the maximum number of vulnerabilities that he can try to exploit in one attack scenario—the *cost* value. For example, the attacker's skill level in range one to five ($cost = (1 : 5)$) means that a less skilled attacker is able to exploit only one vulnerability in one attack scenario, while a more skilled attacker can exploit up to five vulnerabilities at the same time in one attack scenario.

The threat modeling and attack modeling processes identified the minimum set of vulnerabilities that can be exploited in order to perform a HVAC hijack attack. Table 8 presents the list of vulnerabilities, which are linked to the threats from Table 2. It is assumed that the attacker is aware of these vulnerabilities, and that the exploitation of at least one vulnerability is necessary in order to successfully carry out an attack. The model also implies that in one iteration the choice of vulnerabilities that are going to be exploited is random. All listed vulnerabilities have defined exploitation probabilities, described in more detail in Section 6. The exploitation probability is the likelihood that an attempt to exploit that particular vulnerability is going to be performed and successful, under previously given assumptions.

The next step, after the system modeling is completed, is the definition of the attack properties. These properties are the formal definition and precondition of the successful attempt of an attack within the modeled system. The properties are defined by using the Probabilistic Computation Tree Logic (PCTL) [81], embedded in the PRISM model checker. The formal verification of the defined attack properties results in the maximum likelihoods of successful attack attempt–risk exposure scores. Table 9 describes the modeled attack properties.

Table 8. Vulnerabilities, exploitation probabilities and associated threats for UC1.

Vulnerability	Associated Threat	Threat Id	Expl. Prob.
ST–G1 link	Link jammed: denied transmission	1028	0.65
HVAC (actuation A1/A2)	Tailored context/incorrect actuation	1013	0.59
G (gain control)	Tailored context/incorrect reading	1111	0.59
HEMS (actuation A1/A2)	Tailored context/incorrect actuation	1005	0.57

Table 9. Modeled attacks, their impact and properties for UC1.

Attack	Impact	Attack Properties
Attack scenario 1.1: Hijack heating	High	Attacker takes control over heating, high or optimal temperature is detected, heating is switched on, resulting in damage
Attack scenario 1.2: Hijack HVAC	High	Attacker takes control over HVAC, optimal temperature is detected, both cooling and heating are switched on, resulting in damage and high power consumption

7.2.2. Results

The results–risk exposure scores are presented in Figure 8.

Table 10 presents the obtained values of risk exposure scores, under the previously described assumptions.

Table 10. Smart home and HVAC hijack use case risk exposure scores.

Attack	$cost = 1$	$cost = 2$	$cost = 3$	$cost = 4$	$cost = 5$
1.1 Hijack heating	0.0000	0.3481	0.4908	0.5473	0.5473
1.2 Hijack HVAC	0.0000	0.0000	0.2054	0.3709	0.4365

The presented results show the maximum likelihood of a successful attempt of the attack, considering different *cost* values. Considering that the goal of an attacker is to carry out an attack by exploiting the smallest amount of vulnerabilities, the presented analysis considers cases where the number of exploited vulnerabilities is in range one to five ($cost = (1:5)$).

The results show that the attack 1.2 (hijack HVAC) is less likely to be successful comparing to the attack 1.1 (hijack heating). This is due to the fact that in order to successfully conduct attack 1.2 more vulnerabilities need to be exploited. The attack 1.1 requires exploitation of two vulnerabilities, while the attack 1.2 requires exploitation of at least three vulnerabilities to be successful.

7.3. UC2: Smart Home and Smart Meter Hijacking

7.3.1. Probabilistic Model Generation

The UC2 configuration is presented in Figure 9. Similarly to the previous use case, the system is modeled in the PRISM model checker as an MDP process.

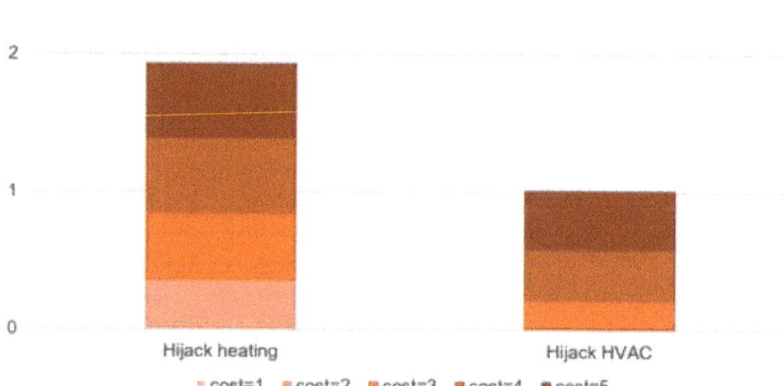

Figure 8. Smart home and HVAC hijack use case results.

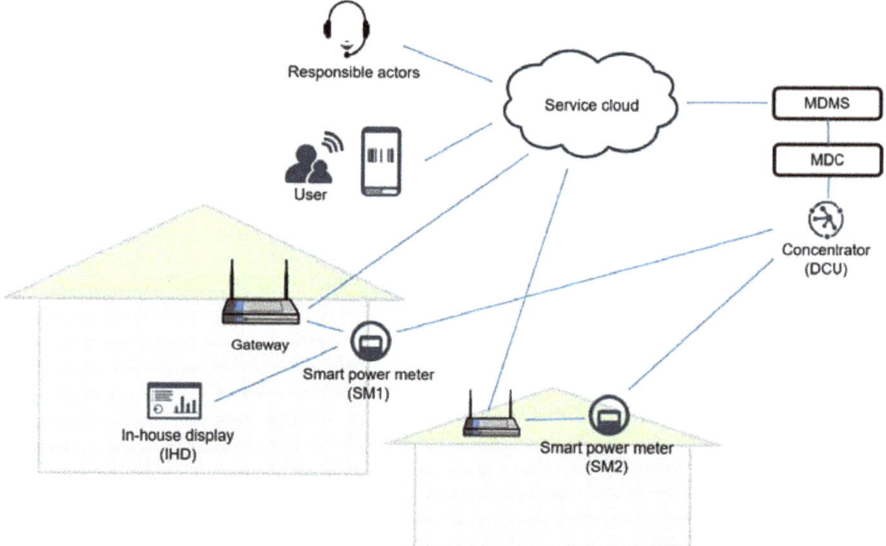

Figure 9. Smart home and smart meter hijacking use-case architecture.

The following service policies, reflecting the system response, permeate through the system model:

- *SM1/2 policy*: Smart meters SM1 and SM2 measure the power consumption and notify IHD in near real time, and MDMS periodically (every 15 min);
- *MDMS policy*: The MDMS is periodically updated, and checks the consistency of the consumption/billing in the neighborhood;
- *Service Cloud policy*: The MDMS is periodically polled with the credentials provided by customer;
- *User policy*: The service cloud periodically notifies the user about the power consumption;
- *Maintenance policy*: The service cloud notifies responsible actors in case of a consumption inconsistency.

Similarly to the first use case, it is assumed that an attacker actively attacks the system, and that he has skills and means to perform a certain attack by exploiting existing vulnerabilities.

The minimum set of vulnerabilities that can be exploited in order to perform a described attack, identified by the threat modeling and attack modeling processes, is presented in Table 11. The system is modeled with the assumptions similar to UC1 – the attacker identified vulnerabilities, and the exploitation of one or more vulnerabilities is necessary in order to successfully carry out an attack. Exploitation probabilities are defined for all listed vulnerabilities, and presented in Table 11.

Table 11. Vulnerabilities, exploitation probabilities and associated threats for UC2.

Vulnerability	Associated Threat	Threat Id	Expl. Prob.
IHD–SM link	Link jammed: denied transmission	903	0.65
SM1, SM2 (reading)	Tailored context/incorrect reading	880	0.68
C (reading SM1/SM2)	Tailored context/incorrect reading	857	0.57
G (gain control)	Tailored context/incorrect reading	1111	0.59

The attack properties, similarly to UC1, are defined using the PCTL. Table 12 describes modeled attack properties, with different impacts (medium and high).

Table 12. Modeled attacks, their impact and properties for UC2.

Attack	Impact	Attack Properties
Attack scenario 2.1: Fraud	High	The user takes control over their and neighbor's smart meter or concentrator, decreases their own power consumption, increases their neighbor's power consumption, responsible actors are not alerted
Attack scenario 2.2: Decrease bill	Medium	The attacker takes control over their own smart meter or concentrator, and decreases the power consumption
Attack scenario 2.3: Increase bill	Medium	The attacker takes control over the user's smart meter or concentrator, and increases the power consumption
Attack scenario 2.4: Increase bill, no alarm	Medium	The attacker takes control over the user's smart meter or concentrator, increases the power consumption, the user is not alerted

7.3.2. Results

Figure 10 presents the results–risk exposure scores for the modeled system and attack properties.

Table 13 presents the resulting values of the risk exposure score.

Table 13. Smart home and smart meter hijacking use case risk exposure scores.

Attack	$cost = 1$	$cost = 2$	$cost = 3$	$cost = 4$	$cost = 5$
2.1 Fraud	0.0000	0.0000	0.2287	0.4098	0.4851
2.2 Decrease bill	0.0000	0.4012	0.5390	0.6467	0.6467
2.3 Increase bill	0.0000	0.4012	0.6467	0.6467	0.6467
2.4 Increase bill, no alarm	0.0000	0.3363	0.5063	0.5063	0.5063

Similarly to UC1, the presented analysis considers cases where the number of exploited vulnerabilities is in the range one to five ($cost = (1 : 5)$).

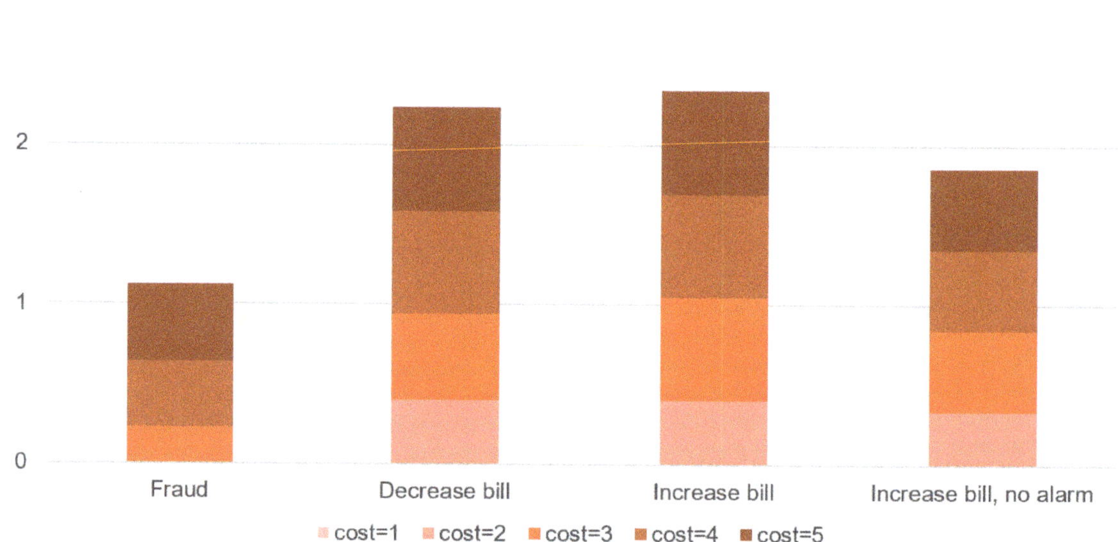

Figure 10. Smart home and smart meter hijacking use case results.

The results show that the attack 2.1 (fraud) has the least likelihood to be successful, under given assumptions. Attack 2.1 requires the exploitation of at least three vulnerabilities, while the other three scenarios considered in this use case can be conducted by exploiting only two vulnerabilities. Attack 2.1 requires the exploitation of at least five vulnerabilities for maximum likelihood, attack 2.2 requires exploitation of three, while attacks 2.3 and 2.4 only require the exploitation of two vulnerabilities for the maximum likelihood. It has to be emphasized that although both attacks 2.3 (or 2.2) and 2.4 require the exploitation of a minimum of two vulnerabilities to be successful, the risk exposure scores of 2.4 are lower because this attack requires the exploitation of parts of the system that have a higher security level (reflected in the lower exploitation probability).

7.4. UC3: Smart Grid and Black-Out Attack

7.4.1. Probabilistic Model Generation

Configuration analyzed in UC3 is presented on Figure 11.
The system response is defined through the following service policies:

- *PQM/PMU policy*: PQM/PMU send regularly Grid Stability (GS) status to WAMS;
- *WAMS policy*: Based on PQM/PMU readings WAMS sends GS value to SCADA; GS value is considered available when at least one PQM and at least on PMU readings are available;
- *PED policy*: PED sends its status to SCADA via RTU, upon request;
- *SCADA policy*: SCADA initiates island mode for the network segment based on GS and PED readings;
- *SwG actuator policy*: SwG opens when island mode is initiated by SCADA.

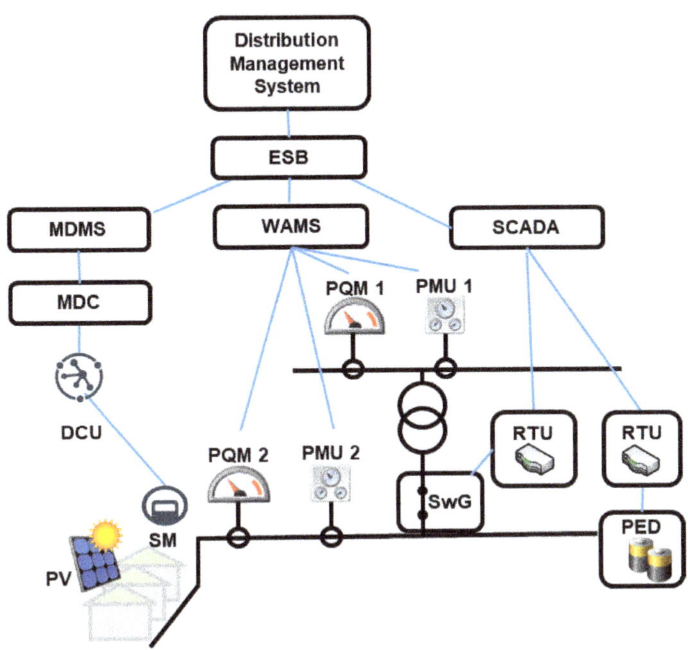

Figure 11. Smart grid and black-out attack use case architecture.

Table 14 presents the minimum set of vulnerabilities that can be exploited in order to perform described attacks, identified by the threat modeling and attack modeling processes.

Table 14. Vulnerabilities, exploitation probabilities and associated threats for UC3.

Vulnerability	Associated Threat	Threat Id	Expl. Prob.
PQM1/2–SCADA link	Link jammed: denied transmission	230/353	0.75
PMU1/2–SCADA link	Link jammed: denied transmission	478/593	0.75
WAMS	Tailored context/incorrect reading	51	0.59
SCADA–RTU(SwG) link	Link jammed: denied transmission	618	0.65
RTU(PED)–SCADA link	Link jammed: denied transmission	672	0.65
RTU(SwG)	Tailored context/incorrect actuation	621	0.59
RTU(PED)	Tailored context/incorrect reading	686	0.59
PED	Tailored context/incorrect reading	647	0.57
PED sniffing	Sniffing	1057	0.53
PQM/PMU sniffing	Sniffing	640	0.53
PMU	Tailored context/incorrect reading	1052	0.59
PQM	Tailored context/incorrect reading	187	0.59
SCADA	Highjacking	672	0.64

The modeled attack properties are described in Table 15. Modeled attacks include four attack scenarios selected from the attack scenarios described in Section 6 as the most representative ones. One additional attack scenario, *Grid stability unavailable/jammed*, is modeled because it is a part of an other described scenarios as middle step (e.g., 3.2.1, 3.2.3), but it has final outcome and impact, completely masking GS value for the rest of the system.

Table 15. Modeled attacks, their impact and properties for UC3.

Attack	Impact	Attack Properties
Attack scenario 3.1.1: Black-out	High	The attacker causes network segment black-out by manipulating status reading on PED, and GS reading on WAMS;
Attack scenario 3.1.2: Black-out	High	The attacker causes network segment black-out by manipulating PED status reading on RTU, and GS reading on WAMS;
Attack scenario 3.2.1: Black-out	High	The attacker causes network segment black-out while PED is discharged by sniffing PED status, and jamming PMU2, PQM2 and PED communication lines;
Attack scenario 3.2.2: Black-out	High	The attacker causes network segment black-out while PED is discharged by sniffing PED status, and jamming SCADA communication lines;
Attack scenario: Grid stability jammed	Medium	Grid stability (GS) value is jammed, and there is no reliable sensor reading for that network segment.

7.4.2. Results

The resulting risk exposure scores are graphically presented on Figure 12, and given in Table 16.

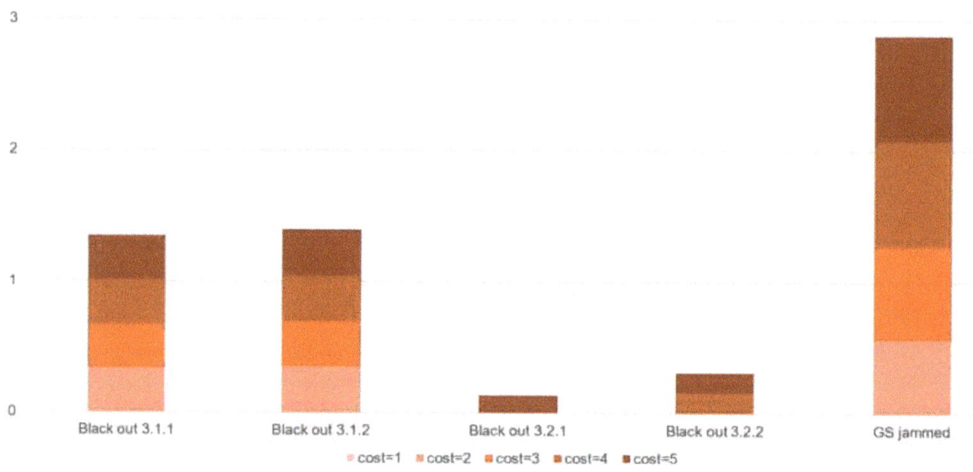

Figure 12. Smart grid and black-out attack use case results.

Table 16. Smart grid and black-out attack use case risk exposure scores.

Attack	*cost* = 1	*cost* = 2	*cost* = 3	*cost* = 4	*cost* = 5
3.1.1 Black out	0.0000	0.3363	0.3363	0.3363	0.3363
3.1.2 Black out	0.0000	0.3481	0.3481	0.3481	0.3481
3.2.1 Black out	0.0000	0.0000	0.0000	0.0000	0.1337
3.2.2 Black out	0.0000	0.0000	0.0000	0.1545	0.1545
GS jammed	0.0000	0.5625	0.7031	0.8086	0.8086

The results show that none of the considered attacks can be successfully conducted by exploiting only one vulnerability. Attacks 3.1.1, 3.1.2 and 3.3 need at least two vulnerabilities exploitations, while attack 3.2.2 requires four and attack 3.2.1 requires five vulnerabilities exploitations. The attack *GS jammed* has the highest likelihood, due to the fact that it is more simple from the attacker point of view. Additionally, the modeled attack sub scenarios from scenario 3.2 have lower risk exposure scores comparing 3.1 sub scenarios because these attacks require the exploitation of parts of the system that have a higher security level, such as SCADA, reflected in the lower exploitation probability.

8. Conclusions and Discussion

In this paper, we addressed persistent security issues in modern power energy systems. For this sake, we presented a methodology for conducting a risk assessment for cyber attacks in SGs. The assessment takes into consideration the technological aspects of the SG architecture. According to that, we developed a threat model from our research templates from the energy and IoT domains, respectively. Based on the identified threats, we introduced several attack scenarios, which exploit the inferred vulnerabilities. Therefore, we estimated the probability of such exploitations based on CVSS. Afterwards, a formal model of the system was implemented in the form of a Markov Decision Process. This model integrates the attacker behavior and vulnerabilities with the exploitation probabilities. In turn, we applied a probabilistic model checker to verify our model. The obtained results yield to the risk exposure scores for different attack scenarios.

With our methodology, we were able to identify 1137 possible threats that fit our scenarios. It is important to note that our calculations are drawn from just 2% of that number. Furthermore, our results provide information on the exposure of individual devices in the SG system. In this way, prevention mechanisms can be applied to prevent such scenarios. This is especially important in dynamic systems that are under limited control of their owners. With this paper, we tackle an important problem in the field of SG security, namely, the lack of general and automatized approaches. For this sake, we provide a reusable methodology for the calculation of exploitation probabilities in complex systems, as one of the novelties of our approach. An additional contribution of our paper is an extensive analysis of the relevant literature in the smart grid domain, including stochastic modeling of cyber attacks, cyber security and formal verification. We also include a detailed comparison between existing formal verification tools.

Our results show that in a smart home environment, an attacker needs to exploit at least two vulnerabilities to successfully perform an attack, and that more complex attack scenarios requires successful exploitation of at least three vulnerabilities. The results of the risk analysis of the smart grid use case also show that none of the considered attacks can be successfully conducted by exploiting only one vulnerability. Some attack scenarios need to exploit at least two vulnerabilities, while some requires at least five vulnerabilities to be exploited. Additionally, the modeled attack scenarios in a smart grid system have lower risk exposure scores comparing to the smart home scenarios because these attacks require the exploitation of parts of the system that have a higher security level, like SCADA, reflected in the lower exploitation probability. These results clearly indicate that all use cases could benefit from a layered security approach, that includes several protection mechanisms in place.

In the future, we want to introduce automation to the individual layers of our methodology. In practice, processes at each level in the hierarchy of the SG system should be automatized. This can be established, for example, by coupling with other applications, like security scanners or testing tools. Furthermore, the conducted assessment should be extended to comprehend other identified threats. The coupling of our approach with common vulnerability exploitation approach represents another challenge . Since our scenarios do not consider implemented countermeasures, applying defense strategies would lead to other assessment results. In addition to that, the methodology provides guidance to address bigger and more complex scenarios. For example, it can be applied in order to address security in smart cities. Since the formal risk analysis differs with respect to applied model checkers, gaining results from other checkers is of interest as well.

Besides technological challenges for secure SG systems, regulatory implications must be considered as well. Initially, the centralized structure of the electric system left the grid exposed to damage in terms of security and safety [23]. The progressing use of IT results in greater diversity and a decentralized system structure [27]. In this way, security issues are diminished by distributing the danger to multiple devices. However, different devices are owned by non-utility stakeholders with different technology choices. This means that security of the overall grid is not imposed by a single organization with a common policy [3]. In fact, the existence of different regulative standards causes confusion with regard to best practices in handling security issues [82]. For this sake, a common framework for critical infrastructure is needed, which provides common guidelines for individual organizational profiles [83].

Author Contributions: Conceptualization, H.V., B.S., J.B. and K.H.-S.; methodology, B.S., H.V. and J.B.; software, B.S.; validation, H.V., B.S. and K.H.-S.; formal analysis, B.S.; investigation, J.B., K.H.-S., B.S. and H.V.; data curation, H.V. and B.S.; writing—original draft preparation, H.V., B.S., J.B. and K.H.-S.; visualization, H.V., B.S. and J.B.; project administration, H.V., B.S., J.B. and K.H.-S.; funding acquisition, H.V. and B.S. All authors have read and agreed to the published version of the manuscript.

Funding: This research has received funding from the European Union's Horizon 2020 research and innovation programme under grant agreement No. 773715, Project RESOLVD H2020-LCE-2017-SGS, and under grant agreement No. 830892, Project SPARTA–H2020-SU-ICT-2018-2.

Institutional Review Board Statement: Not applicable.

Informed Consent Statement: Not applicable.

Conflicts of Interest: The authors declare no conflict of interest. The funders had no role in the design of the study; in the collection, analysis, or interpretation of data; in the writing of the manuscript, or in the decision to publish the results.

References

1. Dharmesh, F.; Nestoras, C.; Vlachou, S.; Kalopoulou, O.; Leandros, M. Cybersecurity in smart grids, challenges and solutions. *AIMS Electron. Electr. Eng.* **2021**, *5*, 24–37.
2. NIST Framework and Roadmap for Smart Grid Interoperability Standards, Release 1.0. *NIST Spec. Publ.* **2010**, *1108*, 1–145.
3. Gopstein, A.; Nguyen, C.; O'Fallon, C.; Wollman, D.; Hasting, N. NIST Framework and Roadmap for Smart Grid Interoperability Standards, Release 4.0. *NIST Spec. Publ.* **2021**, *1108r4*, 8.
4. Global Smart Grid Projections for 2020-IEEE Innovation at Work. Available online: https://innovationatwork.ieee.org/global-smart-grid-projections-for-2020/ (accessed on 9 March 2021).
5. Butt, O.M.; Zulqarnain, M.; Butt, T.M. Recent advancement in smart grid technology: Future prospects in the electrical power network. *Ain Shams Eng. J.* **2021**, *12*, 687–695 [CrossRef]
6. Metke, A.R.; Ekl, R.L. Smart Grid security technology. In Proceedings of the Innovative Smart Grid Technologies (ISGT), Gaithersburg, Maryland, USA, 19–21 January 2010.
7. El Mrabet, Z.; Kaabouch, N.; El Ghazi, H.; El Ghazi, H. Cyber-security in smart grid: Survey and challenges. *Comput. Electr. Eng.* **2018**, *67*, 469–482. [CrossRef]
8. Lee, R.M.; Assante, M.J.; Conway, T. *Analysis of the Cyber Attack on the Ukrainian Power Grid*; SANS Industrial Control Systems: Washington, DC, USA, 2016.
9. Goel, S.; Hong, Y. Security Challenges in Smart Grid Implementation. In *Smart Grid Security*; SpringerBriefs in Cybersecurity; Springer: London, UK, 2015.
10. Paté-Cornell, M.E.; Kuypers, M.; Smith, M.; Keller, P. Cyber Risk Management for Critical Infrastructure: A Risk Analysis Model and Three Case Studies. *Risk Anal.* **2018**, *38*, 226–241 [CrossRef] [PubMed]
11. Di Pinto, A.; Dragoni, Y.; Carcano, A. TRITON: The First ICS Cyber Attack on Safety Instrument Systems. *Black Hat USA* **2018**, *2018*, 1–26.
12. Khan, R.; Maynard, P.; McLaughlin, K.; Laverty, D.; Sezer, S. Threat Analysis of BlackEnergy Malware for Synchrophasor based Real-time Control and Monitoring in Smart Grid. In Proceedings of the 4th International Symposium for ICS & SCADA Cyber Security Research 2016 (ICS-CSR), Belfast, UK, 23–25 August 2016.
13. Kao, D.Y.; Hsiao, S.C. The Dynamic Analysis of WannaCry Ransomware. In Proceedings of the International Conference on Advanced Communications Technology (ICACT), Chuncheon, Korea, 11–14 February 2018; pp. 159–166
14. Wannacry Ransomware. Available online: https://www.europol.europa.eu/wannacry-ransomware (accessed on 9 March 2021).
15. Han, Q.; Molinaro, C.; Picariello, A.; Sperli, G.; Subrahmanian, V.S.; Xiong, Y. Generating Fake Documents using Probabilistic Logic Graphs. *IEEE Trans. Dependable Secur. Comput.* **2021**. [CrossRef]
16. Esposito, C.; Moscato, V.; Sperlí, G. Trustworthiness Assessment of Users in Social Reviewing Systems. *IEEE Trans. Syst. Man Cybern. Syst.* **2021**. [CrossRef]
17. Common Vulnerability Scoring System SIG. Available online: https://www.first.org/cvss/ (accessed on 17 May 2021).

18. Langer, L.; Smith, P.; Hutle, M. Smart grid cybersecurity risk assessment. In Proceedings of the International Symposium on Smart Electric Distribution Systems and Technologies (EDST), Vienna, Austria, 8–11 September 2015; pp. 475–482

19. Jauhar, S.; Chen, B.; Temple, W.G.; Dong, X.; Kalbarczyk, Z.; Sanders, W.H.; Nicol, D.M. Model-Based Cybersecurity Assessment with NESCOR Smart Grid Failure Scenarios. In Proceedings of the 21st Pacific Rim International Symposium on Dependable Computing (PRDC), Zhangjiajie, China, 18–20 November 2015; pp. 319–324. [CrossRef]

20. Lee, A. *Electric Sector Failure Scenarios and Impact Analyses-Version 3.0*; National Electric Sector Cybersecurity Organization Resource (NESCOR) Technical Working Group: Palo Alto, CA, USA, 2015.

21. Pillitteri, V.Y.; Brewer, T.L. Guidelines for Smart Grid Cybersecurity. *NIST Interagency/Internal Report (NISTIR)-7628 Rev 1*; National Institute of Standards and Technology: Gaithersburg, MD, USA, 2014.

22. Salehi Dobakhshari, A.; Ranjbar, A.M. A Novel Method for Fault Location of Transmission Lines by Wide-Area Voltage Measurements Considering Measurement Errors. *IEEE Trans. Smart Grid* **2015**, *6*, 874–884. [CrossRef]

23. Rao, N.S.V.; Poole, S.W.; Ma, C.Y.T.; He, F.; Zhuang, J.; Yau, D.K.Y. Defense of Cyber Infrastructures Against Cyber-Physical Attacks Using Game-Theoretic Models. *Risk Anal.* **2016**, *36*, 694–710. [CrossRef] [PubMed]

24. Gao, J.; Bai, H.; Wang, D.; Wang, L.; Huo, C.; Hou, Y. Rapid Security Situation Prediction of Smart Grid Based on Markov Chain. In Proceedings of the 3rd Information Technology Networking, Electronic and Automation Control Conference (ITNEC), Chengdu, China, 15–17 March 2019; pp. 2386–2389. [CrossRef]

25. Hao, J.; Kang, E.; Sun, J.; Wang, Z.; Meng, Z.; Li, X.; Ming, Z. An Adaptive Markov Strategy for Defending Smart Grid False Data Injection From Malicious Attackers. *IEEE Trans. Smart Grid* **2018**, *9*. [CrossRef]

26. Leszczyna, R. Standards on cyber security assessment of smart grid. *Int. J. Crit. Infrastruct. Prot.* **2018**, *22*, 70–89. [CrossRef]

27. Sun, C.C.; Hahn, A.; Liu, C.C. Cyber security of a power grid: State-of-the-art. *Int. J. Electr. Power Energy Syst.* **2018**, *99*, 45–56. [CrossRef]

28. Soltan, S.; Mittal, P.; Poor, H.V. BlackIoT: IoT Botnet of High Wattage Devices Can Disrupt the Power Grid. In Proceedings of the 27th USENIX Security Symposium, Baltimore, MD, USA, 15–17 August 2018; pp. 15–32.

29. Pliatsios, D.; Sarigiannidis, P.; Lagkas, T.; Sarigiannidis, A.G. A Survey on SCADA Systems: Secure Protocols, Incidents, Threats and Tactics. *IEEE Commun. Surv. Tutor.* **2020**, *22*, 1942–1976 [CrossRef]

30. Nazir, S.; Patel, S.; Patel, D. Assessing and augmenting SCADA cyber security: A survey of techniques. *Comput. Secur.* **2017**, *70*, 436–454. [CrossRef]

31. Irmak, E.; Erkek, I. An overview of cyber-attack vectors on SCADA systems. In Proceedings of the International Symposium on Digital Forensic and Security (ISDFS), Antalya, Turkey, 22–25 March 2018; pp. 1–5. [CrossRef]

32. Ghosh, S.; Sampalli, S. A Survey of Security in SCADA Networks: Current Issues and Future Challenges. *IEEE Access* **2019**, *7*, 135812–135831. [CrossRef]

33. Antón, S.D.; Fraunholz, D.; Lipps, C.; Pohl, F.; Zimmermann, M.; Schotten, H.D. Two Decades of SCADA Exploitation: A Brief History. In Proceedings of the 2017 IEEE Conference on Application, Information and Network Security (AINS), Miri, Malaysia, 13–14 November 2017; pp. 98–104. [CrossRef]

34. McLaughlin, S.; Konstantinou, C.; Wang, X.; Davi, L.; Sadeghi, A.R.; Maniatakos, M.; Karri, R. The Cybersecurity Landscape in Industrial Control Systems. *Proc. IEEE* **2016**, *104*, 1039–1057 [CrossRef]

35. Garcia, L.; Brasser, F.; Cintuglu, M.H.; Sadeghi, A.R.; Mohammed, O.; Zonouz, S.A. Hey, My Malware Knows Physics! Attacking PLCs with Physical Model Aware Rootkit. In *Network and Distributed System Security Symposium (NDSS)*; Internet Society: San Diego, CA, USA, 2017.

36. Spenneberg, R.; Brüggemann, M.; Schwartke H., PLC-Blaster: A Worm Living Solely in the PLC. Available online: https://www.blackhat.com/docs/asia-16/materials/asia-16-Spenneberg-PLC-Blaster-A-Worm-Living-Solely-In-The-PLC-wp.pdf (accessed on 31 May 2021).

37. Klick, J.; Lau, S.; Marzin, D.; Malchow, J.O.; Roth, V. Internet-facing PLCs-A New Back Orifice. *Black Hat USA* **2015**, *2015*, 22–26.

38. Amini, S.; Mohsenian-Rad, H.; Pasqualetti, F. Dynamic Load Altering Attacks in Smart Grid. In Proceedings of the IEEE Power & Energy Society Innovative Smart Grid Technologies Conference (ISGT), Washington, DC, USA, 18–20 February 2015; pp. 1–5. [CrossRef]

39. Dvorkin, Y.; Garg, S. IoT-enabled Distributed Cyber-attacks on Transmission and Distribution Grids. In Proceedings of the 2017 Annual North-American Power Symposium (NAPS), Morgantown, WV, USA, 17–19 September 2017; pp. 1–6. [CrossRef]

40. Li, S.; Yılmaz, Y.; Wang, X. Quickest Detection of False Data Injection Attack in Wide-Area Smart Grids. *IEEE Trans. Smart Grid* **2015**, *6*, 2725–2735. [CrossRef]

41. Cui, S.; Han, Z.; Kar, S.; Kim, T.T.; Poor, H.V.; Tajer, A. Coordinated Data-Injection Attack and Detection in the smart grid: A detailed look at enriching detection solutions. *IEEE Signal Process. Mag.* **2012**, *29*, 106–115.

42. Marksteiner, S.; Vallant, H.; Nahrgang, K. Cyber security requirements engineering for low-voltage distribution smart grid architectures using threat modeling. *J. Inf. Secur. Appl.* **2019**, *49*, 102389. [CrossRef]

43. Li, X.; Liang, X.; Lu, R.; Shen, X.; Lin, X.; Zhu, H. Securing Smart Grid: Cyber Attacks, Countermeasures, and Challenges. *IEEE Commun. Mag.* **2012**, *50*, 38–45. [CrossRef]

44. Rawat, D.B.; Chandra, B. Cyber Security for Smart Grid Systems: Status, Challenges and Perspectives. In Proceedings of the IEEE SoutheastCon, Fort Lauderdale, FL, USA, 9–12 April 2015; pp. 1–6. [CrossRef]

45. Shapsough, S.; Qatan, F.; Aburukba, R.; Aloul, F.; Al Ali, A.R. Smart Grid Cyber Security: Challenges and Solutions. In Proceedings of the International Conference on Smart Grid and Clean Energy Technologies, Offenburg, Germany, 20–23 October 2015; pp. 170–175. [CrossRef]

46. Gunduz, M.Z.; Das, R. Analysis of cyber-attacks on smart grid applications. In Proceedings of the International Artificial Intelligence and Data Processing Symposium (IDAP), Malatya, Turkey, 28–30 September 2018; pp. 1–5. [CrossRef]

47. Yan, Y.; Qian, Y.; Sharif, H.; Tipper, D. A Survey on Cyber Security for Smart Grid Communications. *IEEE Commun. Surv. Tutor.* **2012**, *14*, 998–1010. [CrossRef]

48. Rashid, A.; Hasan, O.; Saghar, K. Formal analysis of a ZigBee-based routing protocol for smart grids using UPPAAL. In Proceedings of the 2015 12th International Conference on High-capacity Optical Networks and Enabling/Emerging Technologies (HONET), Islamabad, Pakistan, 21–23 December 2015; pp. 1–5. [CrossRef]

49. Odelu, V.; Das, A.K.; Wazid, M.; Conti, M. Provably Secure Authenticated Key Agreement Scheme for Smart Grid. *IEEE Trans. Smart Grid* **2018**, *9*, 1900–1910. [CrossRef]

50. Naseem, S.A.; Eslampanah, R.; Uddin, R. Probability estimation for the fault detection and isolation of pmu-based transmission line system of smart grid. In Proceedings of the 2018 5th International Conference on Electrical and Electronic Engineering (ICEEE), Istanbul, Turkey, 3–5 May 2018; pp. 284–288. [CrossRef]

51. Uddin, R.; Naseem, S.A.; Iqbal, Z. Formal reliability analyses of power line communication network-based control in smart grid. *Int. J. Control. Autom. Syst.* **2019**, *17*, 3047–3057. [CrossRef]

52. Hamman, S.T.; Hopkinson, K.M.; Fadul, J.E. A Model Checking Approach to Testing the Reliability of Smart Grid Protection Systems. *IEEE Trans. Power Deliv.* **2017**, *32*, 2408–2415. [CrossRef]

53. Garlapati, S.K.R. Enabling Communication and Networking Technologies for Smart Grid. Ph.D. Thesis, Virginia Tech, Blacksburg, VA, USA, 2014.

54. Bashar, A.; Muhammad, S.; Mohammad, N.; Khan, M. Modeling and Analysis of MDP-based Security Risk Assessment System for Smart Grids. In Proceedings of the 2020 Fourth International Conference on Inventive Systems and Control (ICISC), Coimbatore, India, 8–10 January 2020; pp. 25–30. [CrossRef]

55. Diovu, R.C.; Agee, J.T. Quantitative analysis of firewall security under DDoS attacks in smart grid AMI networks. In Proceedings of the 2017 IEEE 3rd International Conference on Electro-Technology for National Development (NIGERCON), Owerri, Nigeria, 7–10 November 2017; pp. 696–701. [CrossRef]

56. Krivokuća, S.; Stojanović, B.; Hofer-Schmitz, K.; Nešković, N.; Nešković, A. Smart Water Distribution System Communication Architecture Risk Analysis Using Formal Methods. In Proceedings of the 2020 28th Telecommunications Forum (TELFOR), Belgrade, Serbia, 24–25 November 2020; pp. 1–4. [CrossRef]

57. Shostack, A. *Threat Modeling: Designing for Security*; John Wiley & Sons: Indianapolis, IN, USA, 2014.

58. Microsoft Threat Modeling Tool. Available online: https://docs.microsoft.com/en-us/azure/security/develop/threat-modeling-tool (accessed on 26 April 2021).

59. LeBlanc, D.; Howard, M. *Writing Secure Code*; Microsoft Press: Redmond, DC, USA, 2014.

60. Mirsky, Y.; Guri, M.; Elovici, Y. HVACKer: Bridging the Air-Gap by Attacking the Air Conditioning System. *arXiv* **2017**, arXiv:1703.10454.

61. Mohsin, M.; Sardar, M.U.; Hasan, O.; Anwar, Z. IoTRiskAnalyzer: A Probabilistic Model Checking Based Framework for Formal Risk Analytics of the Internet of Things. *IEEE Access* **2017**, *5*, 5494–5505. [CrossRef]

62. Wadhawan, Y.; AlMajali, A.; Neuman, C. A Comprehensive Analysis of Smart Grid Systems against Cyber-Physical Attacks. *Electronics* **2018**, *7*, 249. [CrossRef]

63. Keerthi, K.; Roy, I.; Hazra, A.; Rebeiro, C. Formal Verification for Security in IoT Devices. In *Security and Fault Tolerance in Internet of Things*; Springer: Cham, Switzerland, 2019; pp. 179–200.

64. Basin, D.; Cremers, C.; Meadows, C. Model Checking Security Protocols. In *Handbook of Model Checking*; Springer: Cham, Switzerland, 2018; pp. 727–762.

65. Kwiatkowska, M.; Norman, G.; Parker, D. PRISM: Probabilistic Symbolic Model Checker. In *International Conference on Modelling Techniques and Tools for Computer Performance Evaluation*; Springer: Berlin/Heidelberg, Germany, 2002; pp. 200–204.

66. Hofer-Schmitz, K.; Stojanović, B. Towards formal verification of IoT protocols: A Review. *Comput. Netw.* **2020**, *174*, 107233. [CrossRef]

67. Katoen, J.P. The Probabilistic Model Checking Landscape. In Proceedings of the 31st Annual ACM/IEEE Symposium on Logic in Computer Science LICS'16, New York, NY, USA, 5–8 July 2016; Association for Computing Machinery: New York, NY, USA, 2016; pp. 31–45. [CrossRef]

68. Bartels, F.; Sokolova, A.; de Vink, E. A hierarchy of probabilistic system types. *Theor. Comput. Sci.* **2004**, *327*, 3–22. [CrossRef]

69. Hartmanns, A.; Hermanns, H. In the quantitative automata zoo. Fundamentals of Software Engineering (selected papers of FSEN 2013). *Sci. Comput. Program.* **2015**, *112*, 3–23. [CrossRef]

70. Hahn, E.M.; Hartmanns, A.; Hensel, C.; Klauck, M.; Klein, J.; Křetínský, J.; Parker, D.; Quatmann, T.; Ruijters, E.; Steinmetz, M. The 2019 comparison of tools for the analysis of quantitative formal models. In *International Conference on Tools and Algorithms for the Construction and Analysis of Systems*; Springer: Cham, Switzerland, 2019; pp. 69–92.

71. Hinton, A.; Kwiatkowska, M.; Norman, G.; Parker, D. PRISM: A Tool for Automatic Verification of Probabilistic Systems. In *Tools and Algorithms for the Construction and Analysis of Systems*; Hermanns, H., Palsberg, J., Eds.; Springer: Berlin/Heidelberg, Germany, 2006; pp. 441–444.

72. Kwiatkowska, M.; Norman, G.; Parker, D. PRISM 4.0: Verification of Probabilistic Real-Time Systems. In *International Conference on Computer Aided Verification*; Springer: Berlin/Heidelberg, Germany, 2011; pp. 585–591.

73. Alur, R.; Henzinger, T.A. Reactive Modules. *Form. Methods Syst. Des.* **1999**, *15*, 7–48. [CrossRef]

74. Bengtsson, J.; Larsen, K.; Larsson, F.; Pettersson, P.; Yi, W. *UPPAAL—A Tool Suite for Automatic Verification of Real-Time Systems*; Hybrid, S., III, Alur, R., Henzinger, T.A., Sontag, E.D., Eds.; Springer: Berlin/Heidelberg, Germany, 1996; pp. 232–243.

75. Behrmann, G.; David, A.; Larsen, K.G. *A Tutorial on UPPAAL 4.0*; Department of Computer Science, Aalborg University: Aalborg, Denmark, 2006.

76. David, A.; Larsen, K.G.; Legay, A.; Mikučionis, M.; Poulsen, D.B.; Van Vliet, J.; Wang, Z. Statistical model checking for networks of priced timed automata. In *International Conference on Formal Modeling and Analysis of Timed Systems*; Springer: Berlin/Heidelberg, Germany, 2011; pp. 80–96.

77. Ma, X.; Rinast, J.; Schupp, S.; Gollmann, D. Evaluating Online Model Checking in UPPAAL-SMC using a laser tracheotomy case study. In *5th Workshop on Medical Cyber-Physical Systems*; Schloss Dagstuhl–Leibniz-Zentrum für Informatik: Wadern, Germany, 2014; pp. 100–112.

78. Dehnert, C.; Junges, S.; Katoen, J.P.; Volk, M. A storm is coming: A modern probabilistic model checker. In *International Conference on Computer Aided Verification*; Springer: Cham, Switzerland, 2017; pp. 592–600.

79. Hensel, C.; Junges, S.; Katoen, J.P.; Quatmann, T.; Volk, M. The Probabilistic Model Checker Storm. *arXiv* **2020**, arXiv:2002.07080.

80. Naeem, A.; Azam, F.; Amjad, A.; Anwar, M.W. Comparison of Model Checking Tools Using Timed Automata-PRISM and UPPAAL. In Proceedings of the 2018 IEEE International Conference on Computer and Communication Engineering Technology (CCET), Beijing, China, 18–20 August 2018; pp. 248–253. [CrossRef]

81. Hansson, H.; Jonsson, B. A logic for reasoning about time and reliability. *Form. Asp. Comput.* **1994**, *6*, 512–535. [CrossRef]

82. Glenn, C.; Sterbentz, D.; Wright, A. Cyber Threat and Vulnerability Analysis of the U.S. Electric Sector. 2021. Available online: https://www.osti.gov/servlets/purl/1337873/ (accessed on 31 May 2021). [CrossRef]

83. *Framework for Improving Critical Infrastructure Cybersecurity: Version 1.1*; National Institute of Standards and Technology: Gaithersburg, MD, USA, 2018.

Article

Detecting Vulnerabilities in Critical Infrastructures by Classifying Exposed Industrial Control Systems Using Deep Learning

Pablo Blanco-Medina [1,2,*], Eduardo Fidalgo [1,2], Enrique Alegre [1,2], Roberto A. Vasco-Carofilis [1,2], Francisco Jañez-Martino [1,2] and Victor Fidalgo Villar [2]

1 Department of Electrical, Systems and Automation, Universidad de León, 24071 León, Spain; eduardo.fidalgo@unileon.es (E.F.); enrique.alegre@unileon.es (E.A.); rvasc@unileon.es (R.A.V.-C.); fjanm@unileon.es (F.J.-M.)

2 INCIBE (Spanish National Cybersecurity Institute), 24005 León, Spain; victor.fidalgo@incibe.es

* Correspondence: pblanm@unileon.es

Featured Application: We present a deep-learning-based pipeline to solve a novel problem in Cybersecurity and Industry 4.0. Our proposal, which automatically classifies screenshots of industrial control systems, might support the task of an industrial monitoring tool for detecting vulnerable or exposed industrial control systems on the internet, which might be related to critical infrastructures.

Abstract: Industrial control systems depend heavily on security and monitoring protocols. Several tools are available for this purpose, which scout vulnerabilities and take screenshots of various control panels for later analysis. However, they do not adequately classify images into specific control groups, which is crucial for security-based tasks performed by manual operators. To solve this problem, we propose a pipeline based on deep learning to classify snapshots of industrial control panels into three categories: internet technologies, operation technologies, and others. More specifically, we compare the use of transfer learning and fine-tuning in convolutional neural networks (CNNs) pre-trained on ImageNet to select the best CNN architecture for classifying the screenshots of industrial control systems. We propose the critical infrastructure dataset (CRINF-300), which is the first publicly available information technology (IT)/operational technology (OT) snapshot dataset, with 337 manually labeled images. We used the CRINF-300 to train and evaluate eighteen different pipelines, registering their performance under CPU and GPU environments. We found out that the Inception-ResNet-V2 and VGG16 architectures obtained the best results on transfer learning and fine-tuning, with F1-scores of 0.9832 and 0.9373, respectively. In systems where time is critical and the GPU is available, we recommend using the MobileNet-V1 architecture, with an average time of 0.03 s to process an image and with an F1-score of 0.9758.

Keywords: deep learning; image classification; transfer learning; industrial control system; fine-tuning

Citation: Blanco-Medina, P.; Fidalgo, E.; Alegre, E.; Vasco-Carofilis, R.A.; Jañez-Martino, F.; Villar, V.F. Detecting Vulnerabilities in Critical Infrastructures by Classifying Exposed Industrial Control Systems Using Deep Learning. *Appl. Sci.* **2021**, *11*, 367. https://doi.org/10.3390/app11010367

Received: 29 November 2020
Accepted: 28 December 2020
Published: 1 January 2021

Publisher's Note: MDPI stays neutral with regard to jurisdictional claims in published maps and institutional affiliations.

1. Introduction

Interconnection between electronic devices connected to the internet has become necessary to ensure the control, communication, and monitoring of multiple systems. Those systems, which are exposed online, should be deployed under various security measures to avoid potential attacks [1].

In critical infrastructures, such as healthcare, transportation, or manufacturing, a system shutdown or restart would lead to severe economic and social consequences, as well as significant time costs [2]. Furthermore, the threat of a potential security breach can range from an information leak to system overtaking, which entails high risks in environments such as industrial control systems (ICSs) [1]. Due to this, these systems must rely on constant surveillance [1] to guarantee robustness and stability.

Appl. Sci. **2021**, *11*, 367. https://doi.org/10.3390/app11010367   https://www.mdpi.com/journal/applsci

Supervisory control and data acquisition (SCADA) systems are used to control physical equipment and ICS infrastructures. SCADA systems are commonly referred to as operational technology (OT) systems, which directly control and monitor specific devices. Other industrial systems used to control software, including management, storage, and delivery of data, are known as information technology (IT) systems [3].

To monitor these exposed assets, law enforcement agencies (LEAs) use open-source intelligence (OSINT) tools [4]. In particular, specialized tools, such as Shodan [5], known as metasearchers, monitor the open ports of a network, and the services of the devices that are exposed to the internet. For services that include a graphical user interface (GUI), specific metasearchers usually take screenshots to log relevant information graphically.

The classification of these assets is useful for determining the types of compromised devices. Therefore, the screenshots taken help to discover vulnerabilities, classify the devices based on the images taken, and analyze the obtained information afterwards. However, these metasearchers may not correctly classify images as belonging to IT or OT, thus requiring a manual classification. Due to the large number of devices connected to the internet and the multiple monitoring options existing in the metasearchers, this manual process can be an arduous task for a human operator. Moreover, the dynamic environment and continual updates of these systems may increase the difficulty of classifying these images.

In our work, we propose a pipeline based on existing deep learning models to solve a novel problem related to cybersecurity. More specifically, we use convolutional neural networks (CNNs) pre-trained on an ImageNet dataset to automatically classify screenshots of ICSs that could be linked to critical infrastructures. The ICS snapshots are taken during the monitoring of open ports and devices exposed to the internet through OSINT sources. We designed and evaluated several pipelines based on transfer learning and fine-tuning, which first take and resize a screenshot of an ICS as input. On the one hand, in the transfer learning approach, we use a pre-trained CNN architecture for extracting features. Then, they are fed to a classifier trained with images labeled in three categories, i.e., OT, IT, and others. On the other hand, we fine-tuned the same set of CNN architectures to automatically classify the three categories mentioned above. We consider architectures used in similar studies [6], but we also included others available in the field of computer vision.

We trained and evaluated our pipelines in the critical infrastructure classification dataset (CRINF-300), a dataset of 337 samples crawled from Shodan and manually labeled with IT and OT categories. Even if it is a small dataset for image classification, to the best of our knowledge, it is the first to contain snapshots of real and exposed IT and OT systems. The requirement for a large amount of training data is alleviated due to transfer learning and data augmentation techniques capable of constructing a robust and powerful classifier [7]. Thanks to our collaboration with the Spanish National Cybersecurity Institute (INCIBE) (https://www.incibe.es/en), we applied the pipeline that we present in this paper to recognize industrial control systems based on screenshots in order to support the task of detecting vulnerable systems exposed on the internet in real-time.

The rest of the paper is organized as follows. Section 2 presents a summary of the state-of-the-art image classification approaches and architectures. In Section 3, we introduce the methodology followed. Section 4 discusses our experimental settings and the obtained results. Lastly, in Section 5, we present our conclusions and future lines of work.

2. State of the Art

Image classification is the task of assigning a label to an image. Traditionally, hand-crafted features [8] were extracted from the images and used for training classifiers. CNNs have been established among the best learning algorithm for image-based tasks [9] by achieving the best results on the ILSVRC (ImageNet Large-Scale Visual Recognition Challenge) [10].

Despite this, there are cases where the number of images for training a model is scarce, or the classification tasks are challenging. In these cases, manually crafted feature extraction can outperform the results obtained by CNNs [11,12].

Their parameter optimization and their capability of changing their structures to fit different problems [9,13] have allowed the improvement of CNNs' performance over the years. Technological advances like the use of graphics processing units (GPUs) have also made their progress possible [14].

As seen in [15], CNNs can be divided into seven different categories according to their architecture: spatial exploitation, feature-map exploitation, depth, width, multi-path, channel boosting, and attention. Multiple networks can appear in various categories.

However, these networks need to be trained on a large amount of data, and data gathering and annotation can be a complex, tedious, and time-consuming process. Furthermore, these datasets may soon become outdated, needing the addition of new data [16].

Transfer learning is a technique that allows one to take a model trained for a specific application and apply it to a similar or related task [17]. This approach is used to retain the features obtained from bigger datasets and use them for training a new model on a smaller, similar dataset. Several works have studied the use of transfer learning applied to CNNs for the task of image classification in different fields, but often omit the most recent architectures [6,18], such as NasNetLarge [19]. A brief summary of the most notable image classification CNNs can be seen in Table 1.

Table 1. Image classification architectures' accuracy scores.

Architecture	Top-5 Accuracy (%)	Top-1 Accuracy (%)	Dataset
LeNet [20]	99.80	-	MNIST [20]
DenseNet [21]	93.88	77.85	CIFAR-10 [22]
AlexNet [23]	84.60	63.30	ImageNet
ZFNet [24]	84.00	62.50	ImageNet
GoogleNet [25]	89.90	69.80	ImageNet
VGG16 [26]	91.90	74.40	ImageNet
ResNet [27]	94.29	78.57	ImageNet
ResNeXt-101 [28]	95.60	80.90	ImageNet
Inception-V3 [29]	94.40	78.80	ImageNet
SENet [30]	96.20	82.70	ImageNet
MobileNet-V1 [31]	90.92	71.56	ImageNet
MobileNet-V2 [32]	-	74.70	ImageNet
MobileNet-V3 [33]	-	75.20	ImageNet
EfficientNet [34]	97.00	84.30	ImageNet
Xception [35]	94.50	79.00	ImageNet
Inception-ResNet-V2 [36]	95.10	80.10	ImageNet
NasNetLarge [19]	96.20	82.70	ImageNet

Hussain et al. [17] studied the application of transfer learning on the InceptionV3 [29] architecture pre-trained on the ImageNet dataset [37] and re-trained it on the CIFAR-10 [22] dataset, obtaining 70.1% accuracy and surpassing CNNs trained from scratch on this dataset.

Sharma et al. [38] applied transfer learning on AlexNet [23], GoogLeNet [25], and ResNet50 [27]. They replaced the last three layers with a fully connected layer, a softmax layer, and a classification output layer for each network. Afterwards, they trained the networks on the CIFAR-10 dataset, obtaining classification accuracy per image category. The resulting average performance for each network was 71.67% for GoogLeNet, 78.10% for ResNet50, and 36.12% for AlexNet.

Extensive architecture research and engineering are required to improve neural network classification [39]. To fix this problem, Bello et al. [40] proposed an approach called neural architecture search that helps in optimizing the architecture configuration and improving the classification performance and training time on the CIFAR-10 dataset.

However, training architectures on large datasets, such as ImageNet, causes techniques like this to have a high computational cost. Therefore, Zoph et al. [19] proposed the use of a smaller dataset, CIFAR-10, as a proxy, and then transferred the learned architecture to the ImageNet dataset. The resulting architecture, called NASNet, was compared to multiple CNNs, such as MobileNet-224 [31], Inception-ResnetV2 [36], and Xception [35], on image classification on the ImageNet dataset, comparing both the number of parameters and the accuracy. The proposed solution achieved state-of-the-art results on the ImageNet dataset.

3. Methodology

3.1. Critical Infrastructure Dataset

For our proposal, we used ICS images provided by the Spanish National Cybersecurity Institute (INCIBE), which were retrieved using multiple metasearchers, such as Shodan (https://www.shodan.io). We manually labeled the 337 snapshots in two categories: 74 IT and 263 OT images, and we named the resulting dataset the Critical Infrastructure (CRINF-300). Although our small dataset may be insufficient for training a classifier from scratch, we used data augmentation techniques to increase the number of training images by five times when we performed fine-tuning and transfer learning, which allowed us to properly train a robust classifier. Figure 1 presents four screenshots of IT and OT systems to appreciate their differences.

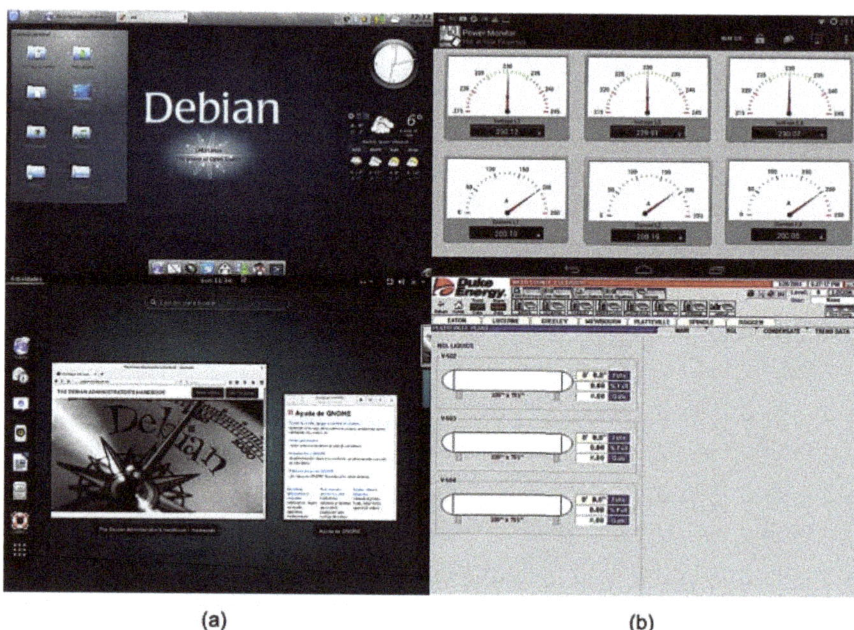

(a)   (b)

Figure 1. Information technology (IT) (column **a**) and operational technology (OT) (column **b**) sample images. IT systems focus on software management and data control, while OT systems help directly monitor device values.

3.2. Proposed Pipeline

We modified and evaluated several pipelines based on transfer learning and fine-tuning, which use resized screenshots of ICSs as input. On the one hand, in transfer learning, we used a pre-trained CNN architecture for extracting features, which were then fed to a classifier trained in three image categories: OT, IT, and others. On the other hand, we performed fine-tuning on a newly defined head of the model of the CNN architecture to automatically classify an image into the three previous categories. Using nine different CNNs, we trained three classifiers applying both transfer learning and fine-tuning techniques. After obtaining the classifier, we used it to label images in three categories: IT, OT, and others, according to the classifier's confidence score. If it was below a certain threshold, 0.9 in our experiments, the images were classified as Others. This third category helps identify additional clusters retrieved from the metasearchers, such as Internet of Things (IoT) images, which could be useful for adding future labels. For our experiments with the selected architectures, we only classified images as IT or OT, omitting the third category for performance purposes. Figure 2 presents an overview of the proposed systems for classifying screenshots.

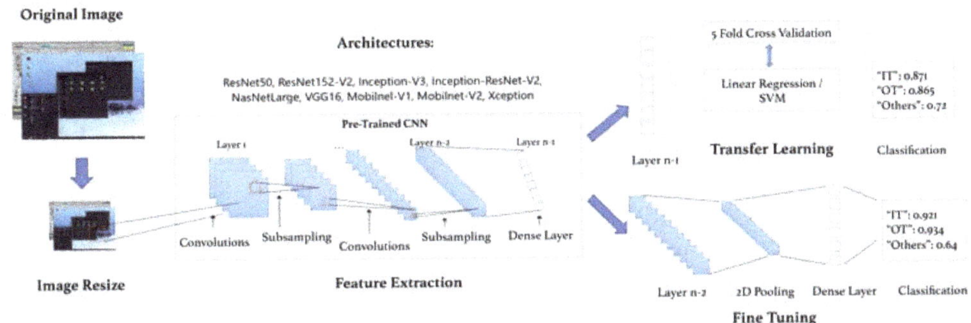

Figure 2. Transfer learning and fine-tuning pipeline. Transfer learning and fine-tuning were used to train classifiers and output the final labels.

3.2.1. Transfer Learning and Fine-Tuning

Due to the limited number of images available, our initial approach was focused on the implementation of transfer learning, instead of re-training the architectures from scratch. For the experiments, we selected a total of nine architectures to apply to our image classification problem: Inception-V3 [29], MobileNet-V1 [31], MobileNet-V2 [32], ResNet50, ResNet152v2 [27], VGG16 [26], NasNetLarge [19], Inception-ResNet-V2 [36], and Xception [35].

We froze the final layer for each network using the pre-trained ImageNet weights and obtained the features before classification. Then, we used those features to train two separate classifiers based on a logistic regression model and a support vector machine (SVM) with a linear kernel, which would output the final labels.

In addition to transfer learning, we also propose fine-tuning in the previously selected architectures for our classification problem based on their F1-score.

3.2.2. Architectures

We selected the given architectures due their common use in similar problems that use transfer learning, and computer vision techniques [18,39]. Additionally, specific architectures, such as MobilenetV1, were chosen due to the real-time-based nature of the given task, which focuses on mobile, lightweight deployment.

VGG [26] is an architecture focused on spatial exploitation, which attempts to consider various filter sizes to analyze both low- and high-level details [15]. It is known for its vast number of parameters, which complicates deployment and increases computational costs, but achieves good results in image classification and object localization tasks.

ResNet [27] introduced the concept of residual learning in CNNs by creating connections between layers that speed up network convergence. It comprises 152 layers, eight times more than the original VGG, while being less complicated. Due to these connection types and its focus on depth, it is classified as both depth- and multi-path-based. The network variations using 50 and 101 layers, respectively, have shown remarkable results in tasks of image classification [15]. ResNet152v2's [27] main difference when compared to the original is that it uses batch normalization before each of the weight layers.

Inception-V3 [29] is a depth- and width-based CNN focused on reducing the computational cost of these types of networks. However, it has a sophisticated design and lacks homogeneity. Inception-ResNet-V2 [36] was introduced as a higher-cost variant of the Inception network, with a higher object recognition performance.

Xception [35] attempts to regulate computational complexity, making learning more efficient and improving the architecture's performance, but has the disadvantage of a high computational cost compared to other architectures.

NasNetLarge [19] designs a search space to reduce architecture complexity and make it independent of image size or network depth, finding the best possible network on a smaller dataset before scaling it to bigger datasets. The resulting architecture can be further modified to fit a larger scale, achieving state-of-the-art results.

Lastly, MobileNet-V1 [31] is an architecture oriented towards mobile device applications, and is designed to have a light architecture for ease of deployment. MobileNet-V2 [32] improves the original proposal by introducing linear bottlenecks between the layers and shortcut connections between them, reducing operation complexity and the parameters needed.

4. Experimental Results and Discussion

4.1. Experimental Settings

We evaluated our proposal on an Intel Xeon E5 v3 computer with 128 GB of RAM using an NVIDIA Titan Xp GPU for both training and testing. All of the CNN architectures were implemented using Python3 under the Keras library [41] and using Tensorflow as the backend.

To measure the performance of both our approaches, we used the F1-score and the accuracy metrics, as can be seen in Equations (3) and (4). We chose these metrics to represent the robustness of our classifiers. The F1-score is the harmonic mean of the precision and recall measures, which are detailed in Equations (1) and (2), respectively. In these equations, TP and FP represent true and false positives, respectively, while TN and FN indicate true and false negatives.

$$P = \frac{TP}{TP + FP} \tag{1}$$

$$R = \frac{TP}{TP + FN} \tag{2}$$

$$F1 = 2 \cdot \frac{P \cdot R}{P + R} \tag{3}$$

$$Accuracy = \frac{TP + TN}{TP + TN + FP + FN} \tag{4}$$

In image classification problems, a true positive is when an image is assigned its correct label. For a particular class, a false positive is considered for each image that has been labeled as belonging to a given class, but belongs to a different one. At the same time, false negatives are considered as the images from the class that have been assigned other labels.

We selected the 0.9 value for the confidence threshold of the pipeline empirically. Following the requirements of the monitoring service where the proposal would be integrated, we initially evaluated high threshold values of −0.8 and 0.9− over a small set of images. We visually inspected the results of the classification under both thresholds and concluded that 0.9 obtains a higher generalization capability with a lower number of failures, meeting the requirements of the monitoring tool.

We also measured the time needed for feature extraction and classification in both the CPU and GPU, retrieving both the mean and standard deviation in each task.

4.1.1. Transfer Learning Settings

To fit the architecture's input size, each image was resized to the required value. For VGG16, ResNet50, and both MobileNet architectures, images were fixed to a size of 224×224. On the Xception, ResNet152v2, and Inception-based architectures, they were scaled to 299×299, while NasNetLarge used a size of 331×331. We fed the resized images to these pre-trained networks, extracted the features, and trained our classifiers using them. Both the logistic regression classifier and the SVM classifier were implemented using the Scikit-learn Python library [42].

Since there is a difference between the available images per category, we implemented five-fold cross-validation, generating five models per architecture. This technique helps reduce model bias when compared to other approaches, such as the training–testing split. The dataset was divided into five folds, four of which were used to fit the model, whereas the last one is used for validation. This process was repeated until every fold was used to test the proposed model.

4.1.2. Fine-Tuning Settings

Using the previously selected architectures, we performed fine-tuning by dropping the last layer of each model, adding a new set of layers that consisted of an average 2D pooling of 7×7, a Flatten layer, a Dense layer with ReLu activation and 256 outputs, a 0.5 dropout layer, and a final Dense layer with softmax activation and two outputs.

For fine-tuning, we used 70% of our images for training, 20% for testing, and 10% for validation. For a fair comparison, we added the same layers with the same parameters on all the architectures. We trained these added layers with a batch size of 26 images and a learning rate of 1×10^{-4} over 20 epochs, we evaluated them with the same set of images. To increase the number of images for training our architectures, we used data augmentation in order to learn more general and robust features, improving the performance of our models.

We used the ImageDataGenerator function from Keras to create these new images. We set the rotation range parameter to 25, zoom range to 0.1, width shift range to 0.1, height and shift range to 0.1, shear range to 0.2, horizontal flip to True, and fill mode to "nearest". These new slightly altered images were used for training and consisted of five times the number of images of the original training batch.

4.2. Discussion of Results

The results of our transfer learning experiments can be seen in Tables 2 and 3. In both of the proposed approaches, Inception-ResNet-V2 obtained the highest F1-score and accuracy. Using the logistic regression classifier, this architecture obtained a 98.32 F1-score and 97.33 accuracy, with a variance of 0.70 and 1.12, respectively. On the SVM classifier, Inception-ResNet-V2 obtained slightly lower results, with a 98.13 F1-score and 97.03 accuracy. VGG16 and ResNet50 obtained the lowest F1-score, with 87.67 in both approaches. ResNet152v2 obtained a 10% F1-score over ResNet50, but with a computational time three times higher, highlighting the enhancements of the batch normalization.

NasNetLarge was the slowest-performing architecture in all our given approaches in both the CPU and GPU, which can be attributed to the large size of the network. When combining transfer learning with an SVM classifier, NasNetLarge took an average of 0.54 s in the CPU and 0.27 s in the GPU to process a single image. In the CPU, it was followed by ResNet152v2 with 0.33 s and 0.17 s in the GPU. NasNetLarge took an average of 0.49 s in the CPU and 0.26 s in GPU to process a single image, the fastest time recorded for this architecture in all of our approaches. In the CPU, it was followed by ResNet152v2 with 0.49 s and by Inception-ResNet-V2 in the GPU with 0.12 s.

Table 2. F1-score and accuracy results of our transfer learning strategy using a logistic regression classifier on critical infrastructure classification dataset (CRINF-300) with five-fold cross-validation. The time indicates the average computational cost per individual image. Bold highlights the best results.

Architecture	F1-Score (%)	Accuracy (%)	CPU (s)	GPU (s)
ResNet50	87.67 (+/−0.35)	78.04 (+/−0.55)	0.10 (+/−0.10)	0.05 (+/−0.18)
VGG16	87.67 (+/−0.35)	78.04 (+/−0.55)	0.16 (+/−0.03)	**0.03 (+/−0.07)**
Xception	89.46 (+/−0.67)	81.60 (+/−1.24)	0.14 (+/−0.07)	0.05 (+/−0.15)
Inception-V3	97.02 (+/−1.50)	95.25 (+/−2.40)	0.10 (+/−0.16)	0.07 (+/−0.26)
Mobilenet-V1	97.58 (+/−0.70)	96.15 (+/−1.17)	**0.06 (+/−0.05)**	0.04 (+/−0.12)
Mobilenet-V2	97.55 (+/−0.49)	96.14 (+/−0.73)	0.08 (+/−0.09)	0.05 (+/−0.17)
NasNetLarge	96.44 (+/−1.29)	94.36 (+/−1.99)	0.49 (+/−0.65)	0.16 (+/−0.96)
Inception-ResNet-V2	**98.32 (+/−0.70)**	**97.33 (+/−1.12)**	0.21 (+/−0.41)	0.12 (+/−0.63)
ResNet152v2	97.41 (+/−0.90)	95.85 (+/−1.46)	0.26 (+/−0.32)	0.10 (+/−0.46)

Table 3. F1-score and accuracy results of our transfer learning strategy using a support vector machine (SVM) classifier on CRINF-300 with five-fold cross-validation. The time indicates the average computational cost per individual image. Bold highlights the best results.

Architecture	F1-Score (%)	Accuracy (%)	CPU (s)	GPU (s)
ResNet50	87.67 (+/−0.35)	78.04 (+/−0.55)	0.10 (+/−0.11)	0.06 (+/−0.18)
VGG16	87.67 (+/−0.35)	78.04 (+/−0.55)	0.14 (+/−0.03)	**0.03 (+/−0.07)**
Xception	90.08 (+/−1.51)	82.77 (+/−2.83)	0.14 (+/−0.07)	0.05 (+/−0.14)
Inception-V3	97.20 (+/−1.31)	95.54 (+/−2.12)	0.14 (+/−0.16)	0.12 (+/−0.26)
Mobilenet-V1	97.56 (+/−0.75)	96.14 (+/−1.19)	**0.08 (+/−0.05)**	0.06 (+/−0.11)
Mobilenet-V2	97.92 (+/−0.74)	96.73 (+/−1.12)	0.10 (+/−0.09)	0.07 (+/−0.16)
NasNetLarge	96.78 (+/−1.04)	94.96 (+/−1.52)	0.54 (+/−0.66)	0.27 (+/−0.81)
Inception-ResNet-V2	**98.13 (+/−0.84)**	**97.03 (+/−1.33)**	0.27 (+/−0.40)	0.15 (+/−0.55)
ResNet152v2	97.59 (+/−0.94)	96.14 (+/−1.52)	0.33 (+/−0.32)	0.17 (+/−0.42)

The MobileNet architecture obtained the second-best performance. On the logistic approach, V1 obtained scores of 97.58 and 96.15 against the 97.55 and 96.14 obtained by V2. When using the SVM classifier, MobileNet-V2 yielded slightly better results, with a 97.92 F1-score and 96.73 accuracy compared to the 97.56 and 96.14 results reported by V1. Despite these slight differences, MobileNet-V1 always achieved faster CPU and GPU times than V2, which included less variance in the time required to process the images.

When comparing the logistic regression approach against the SVM classifier, we saw small improvements in both the F1-score and accuracy in Xception, InceptionV3, NasNetLarge, ResNet152V2, and MobileNet-V2. However, both MobileNet-V1 and Inception-ResNet-V2 obtained slightly lower F1 results, decreasing to 97.56 and 98.13, respectively. Lastly, ResNet50 and VGG16 remained with the same scores performance-wise.

As for performance variance, ResNet50 and VGG16 obtained the least variance in both approaches for F1-score and accuracy, with 0.35 and 0.55, respectively. In our logistic classifier, InceptionV3 obtained the largest variance of F1-score and accuracy with 1.50 and 2.40, respectively, followed by NasNetLarge with 1.29 and 1.99. On our SVM approach, the Xception architecture surpassed the variance of these two models, with 1.51 F1-score and 2.83 accuracy variances.

Regarding the computational cost, VGG16 and MobileNet-V1 were the best-performing architectures in both types of transfer learning. MobileNet-V1 was faster on the CPU, with 0.06 s against the 0.16 obtained by VGG16, but was surpassed on the GPU by 0.03 s.

On average, the SVM classifier took slightly more time than the logistic regression approach, with the highest average increase seen in the NasNetLarge architecture from 0.49 to 0.54 on the CPU and 0.16 to 0.27 on the GPU. While the differences in most architectures between the two proposed transfer learning approaches are minimal, they can be critical in real-time systems.

Our fine-tuned results are presented in Table 4. VGG16 obtained the best results with an F1-score of 93.73 and an accuracy of 95.59, which improved significantly from the transfer learning results of 87.67 F1-score and 78.04 accuracy. ResNet50 was the second-best-performing architecture, with a 92.16 F1-score and 94.12 accuracy.

Table 4. Results using the fine-tuning strategy over the nine selected architectures, using the F1-score and accuracy. The time indicates the average computational cost per individual image. Bold highlights the best results.

Architecture	F1-Score (%)	Accuracy (%)	CPU (s)	GPU (s)
ResNet50	92.16	94.12	0.27 (+/−0.25)	0.20 (+/−0.25)
VGG16	**93.73**	**95.59**	**0.21 (+/−0.07)**	**0.14 (+/−0.10)**
Xception	80.98	88.24	0.23 (+/−0.08)	0.15 (+/−0.12)
Inception-V3	82.01	88.24	0.28 (+/−0.18)	0.18 (+/−0.17)
MobileNet-V1	86.51	91.18	**0.21 (+/−0.14)**	0.18 (+/−0.19)
MobileNet-V2	48.78	76.47	0.25 (+/−0.19)	0.18 (+/−0.15)
NasNetLarge	81.20	83.82	0.68 (+/−0.44)	0.32 (+/−0.43)
Inception-ResNet-V2	76.22	85.29	0.36 (+/−0.17)	0.22 (+/−0.27)
ResNet152v2	71.22	83.82	0.34 (+/−0.15)	0.19 (+/−0.22)

While ResNet50 and VGG16 obtained significant improvements in both F1-score and accuracy, they did not surpass the best results yielded by the transfer learning approach. Furthermore, they increased the computational time required by up to three times when compared to the transfer-based models. However, VGG16 remained as the architecture with the least time variance and fastest processing time, followed by MobileNet-V1 on the CPU and Xception on the GPU. These enhanced scores suggest that fine-tuning of the layers and parameters is more beneficial to these architectures than simply freezing the final layers, despite our low number of images.

Although our fine-tuning additions perform well on top of ResNet50 and VGG16, they do not achieve good performance on the rest of the architectures, obtaining lower results than those of the transfer-learning-based models. In particular, MobileNet-V2 obtained an F1-score of 48.78 from the original 97.55. This drop in performance can also be attributed to our new head of the model and the low number of training images.

Given the real-time processing requirement of monitoring ICSs and the particular focus on light deployment, we recommend the pipeline with the Mobilenet-V1 architecture in order to have the best speed–performance trade-off. Although VGG16 is the fastest architecture on the GPU for our fine-tuning approach, and with the least variance across all proposals, its performance is much lower than that of Mobilenet-V1 in transfer learning. Figures 3 and 4 graphically illustrate our results, highlighting the best-performing architectures across all three approaches.

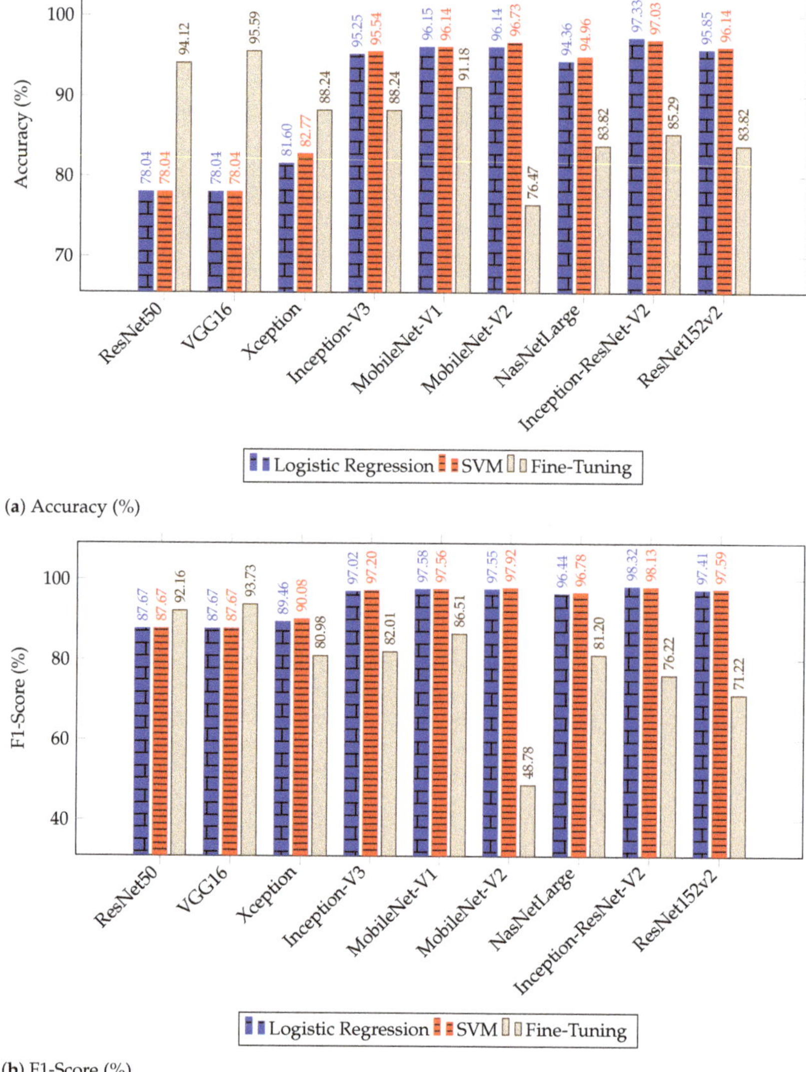

(**a**) Accuracy (%)

(**b**) F1-Score (%)

Figure 3. Accuracy (**a**) and F1-score (**b**) comparison of the results obtained in our proposed pipeline using transfer learning and fine-tuning on CRINF-300.

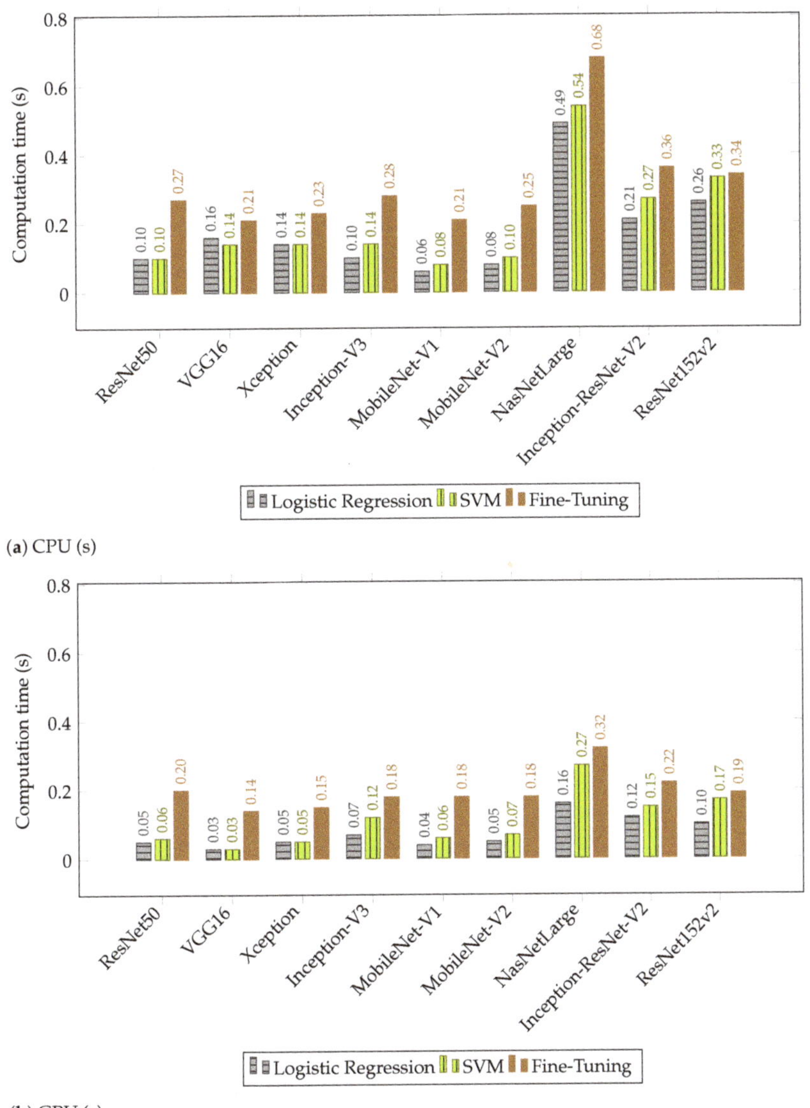

(**a**) CPU (s)

(**b**) GPU (s)

Figure 4. Time comparison of the results obtained in our proposed pipeline using the CPU (**a**) and the GPU (**b**).

5. Conclusions

This paper has presented a pipeline for classifying SCADA images as belonging to IT or OT systems using transfer-learning- and fine-tuning-based approaches on a custom SCADA dataset named CRINF-300. Moreover, we have applied it to a real-case scenario where we classify images to enhance surveillance tasks on critical infrastructure systems to detect and prevent potential security breaches.

We have analyzed nine different CNN architectures using their pre-trained weights on the ImageNet dataset to train three different approaches for image classifiers: transfer learning with a logistic regression classifier or with an SVM classifier and fine-tuning. We also register the average time in the classification of each image on both the CPU and GPU.

For transfer learning, we validated our approach using the F1-score, and accuracy, and five-fold cross-validation during training. We trained and tested the architectures on a 337 image dataset provided by INCIBE, containing 74 IT images and 263 OT images.

We proposed a new model head after freezing the last layers of the selected architectures and adding a custom set of layers for fine-tuning. We enhanced our images with data augmentation and trained our model using training–testing validation.

Our results show that the best CNN architectures for our problem are Inception-ResNet-V2, as it is the best architecture performance-wise with a 98.32 and 98.13 F1-score on our transfer-learning-based approach, and MobileNet-V1, as it has the best performance-speed trade-off, with an F1-score of 97.58 and a speed of 0.06 s on the CPU. Out of the two classifiers proposed for transfer learning, although close in performance, the logistic regression model seems to be better suited for our problem, with higher average scores, reduced performance variance, and decreased computational time.

We conclude that transfer learning is the better approach for our problem based on the performance of the Inception-ResNet-V2 and MobileNet-V1 networks. While VGG16 obtained a noticeable improvement when fine-tuning the architecture, it did not surpass MobileNet's original score, even when increasing our total number of images by up to five times using data augmentation.

Our future work will be focused on further improving the proposed solution, increasing our training images, and adding new layers on top of the given architectures based on our VGG16 results. Another possibility is to extend the current study further to include different architectures, such as VGG19 [26]. The inclusion of more detailed classes is also a future line of investigation for this image classification task.

Finally, after classification is performed, images can be further analyzed by retrieving information found within them [8], such as brand or company names, using text spotting [43,44].

Author Contributions: Conceptualization, P.B.-M., E.F., and E.A.; methodology, P.B.-M., E.F., and E.A; software, P.B.-M.; validation, P.B.-M., E.F., and E.A.; formal analysis, P.B.-M., E.F., E.A., R.A.V.-C., F.J.-M., and V.F.V.; investigation, P.B.-M.; resources, E.F. and E.A.; data curation, P.B.-M.; writing—original draft preparation, P.B.-M.; writing—review and editing, P.B.-M., E.F., E.A., R.A.V.-C., F.J.-M., and V.F.V.; visualization, P.B.-M.; supervision, E.F. and E.A.; project administration, E.F. and E.A.; funding acquisition, E.A. All authors have read and agreed to the published version of the manuscript.

Funding: This research was supported by the grant "Ayudas para la realización de estudios de doctorado en el marco del programa propio de investigación de la Universidad de León Convocatoria 2018" and by the framework agreement between Universidad de León and INCIBE (Spanish National Cybersecurity Institute) under Addendum 01.

Institutional Review Board Statement: Not applicable.

Informed Consent Statement: Not applicable.

Data Availability Statement: Data sharing is not applicable in this article.

Acknowledgments: This work was supported by the framework agreement between the Universidad de León and INCIBE (Spanish National Cybersecurity Institute) under Addendum 01. We acknowledge the NVIDIA Corporation for the donation of the TITAN Xp and Tesla K40 GPUs used for this research.

Conflicts of Interest: The authors declare no conflict of interest. The funders had no role in the design of the study.

References

1. Wolf, M.; Serpanos, D. Safety and security in cyber-physical systems and internet-of-things systems. *Proc. IEEE* **2017**, *106*, 9–20. [CrossRef]
2. Cherdantseva, Y.; Burnap, P.; Blyth, A.; Eden, P.; Jones, K.; Soulsby, H.; Stoddart, K. A review of cyber security risk assessment methods for SCADA systems. *Comput. Secur.* **2016**, *56*, 1–27. [CrossRef]
3. Conklin, W.A. IT vs. OT security: A time to consider a change in CIA to include resilienc. In Proceedings of the 2016 49th Hawaii International Conference on System Sciences (HICSS), Koloa, HI, USA, 5–8 January 2016; IEEE: Piscataway, NJ, USA, 2016; pp. 2642–2647.
4. Lee, S.; Shon, T. Open source intelligence base cyber threat inspection framework for critical infrastructures. In Proceedings of the 2016 Future Technologies Conference (FTC), San Francisco, CA, USA, 6–7 December 2016; IEEE: Piscataway, NJ, USA, 2016; pp. 1030–1033.
5. Genge, B.; Enăchescu, C. ShoVAT: Shodan-based vulnerability assessment tool for Internet-facing services. *Secur. Commun. Networks* **2016**, *9*, 2696–2714. [CrossRef]
6. Liu, Q.; Feng, C.; Song, Z.; Louis, J.; Zhou, J. Deep Learning Model Comparison for Vision-Based Classification of Full/Empty-Load Trucks in Earthmoving Operations. *Appl. Sci.* **2019**, *9*, 4871. [CrossRef]
7. Han, D.; Liu, Q.; Fan, W. A new image classification method using CNN transfer learning and web data augmentation. *Expert Syst. Appl.* **2018**, *95*, 43–56. [CrossRef]
8. Fidalgo, E.; Alegre, E.; Fernández-Robles, L.; González-Castro, V. Fusión temprana de descriptores extraídos de mapas de prominencia multi-nivel para clasificar imágenes. *Rev. Iberoam. Automática E Informática* **2019**, *16*, 358–368. [CrossRef]
9. Rawat, W.; Wang, Z. Deep convolutional neural networks for image classification: A comprehensive review. *Neural Comput.* **2017**, *29*, 2352–2449. [CrossRef]
10. Russakovsky, O.; Deng, J.; Su, H.; Krause, J.; Satheesh, S.; Ma, S.; Huang, Z.; Karpathy, A.; Khosla, A.; Bernstein, M.; et al. Imagenet large scale visual recognition challenge. *Int. J. Comput. Vis.* **2015**, *115*, 211–252. [CrossRef]
11. Fidalgo, E.; Alegre, E.; Fernández-Robles, L.; González-Castro, V. Classifying suspicious content in tor darknet through Semantic Attention Keypoint Filtering. *Digit. Investig.* **2019**, *30*, 12–22. [CrossRef]
12. Fidalgo, E.; Alegre, E.; Gonzalez-Castro, V.; Fernández-Robles, L. Boosting image classification through semantic attention filtering strategies. *Pattern Recognit. Lett.* **2018**, *112*, 176–183. [CrossRef]
13. Sun, Y.; Xue, B.; Zhang, M.; Yen, G.G.; Lv, J. Automatically Designing CNN Architectures Using the Genetic Algorithm for Image Classification. *IEEE Trans. Cybern.* **2020**, *50*, 3840–3854. [CrossRef] [PubMed]
14. Ma, B.; Li, X.; Xia, Y.; Zhang, Y. Autonomous deep learning: A genetic DCNN designer for image classification. *Neurocomputing* **2020**, *379*, 152–161. [CrossRef]
15. Khan, A.; Sohail, A.; Zahoora, U.; Qureshi, A.S. A survey of the recent architectures of deep convolutional neural networks. *arXiv* **2019**, arXiv:1901.06032.
16. Tan, C.; Sun, F.; Kong, T.; Zhang, W.; Yang, C.; Liu, C. A survey on deep transfer learning. In Proceedings of the International Conference on Artificial Neural Networks, Rhodes, Greece, 4–7 October 2018; Springer: Berlin/Heidelberg, Germany, 2018; pp. 270–279.
17. Hussain, M.; Bird, J.J.; Faria, D.R. A study on cnn transfer learning for image classification. In *UK Workshop on Computational Intelligence*; Springer: Cham, Switzerland, 2018; pp. 191–202.
18. Xiao, Z.; Tan, Y.; Liu, X.; Yang, S. Classification Method of Plug Seedlings Based on Transfer Learning. *Appl. Sci.* **2019**, *9*, 2725. [CrossRef]
19. Zoph, B.; Vasudevan, V.; Shlens, J.; Le, Q.V. Learning transferable architectures for scalable image recognition. In Proceedings of the IEEE Conference on Computer Vision and Pattern Recognition, Salt Lake City, UT, USA, 18–22 June 2018; pp. 8697–8710.
20. LeCun, Y.; Bottou, L.; Bengio, Y.; Haffner, P. Gradient-based learning applied to document recognition. *Proc. IEEE* **1998**, *86*, 2278–2324. [CrossRef]
21. Huang, G.; Liu, Z.; Van Der Maaten, L.; Weinberger, K.Q. Densely connected convolutional networks. In Proceedings of the IEEE Conference on Computer Vision and Pattern Recognition, Honolulu, HI, USA, 21–26 July 2017; pp. 4700–4708.
22. Krizhevsky, A.; Hinton, G. Learning multiple layers of features from tiny images. *Citeseer* **2009**, *7*, 1–58.
23. Krizhevsky, A.; Sutskever, I.; Hinton, G.E. Imagenet classification with deep convolutional neural networks. *Adv. Neural Inf. Process. Syst.* **2012**, 1097–1105. [CrossRef]
24. Zeiler, M.D.; Fergus, R. Visualizing and understanding convolutional networks. In Proceedings of the European Conference on Computer Vision, Zurich, Switzerland, 6–12 September 2014; Springer: Cham, Switzerland, 2014; pp. 818–833.
25. Szegedy, C.; Liu, W.; Jia, Y.; Sermanet, P.; Reed, S.; Anguelov, D.; Erhan, D.; Vanhoucke, V.; Rabinovich, A. Going deeper with convolutions. In Proceedings of the IEEE Conference on Computer Vision and Pattern Recognition, Boston, MA, USA, 7–12 June 2015; pp. 1–9.
26. Simonyan, K.; Zisserman, A. Very Deep Convolutional Networks for Large-Scale Image Recognition. *arXiv* **2014**, arXiv:1409.1556.
27. He, K.; Zhang, X.; Ren, S.; Sun, J. Deep residual learning for image recognition. In Proceedings of the IEEE Conference on Computer Vision and Pattern Recognition, Las Vegas, NV, USA, 26 June–1 July 2016; pp. 770–778.
28. Xie, S.; Girshick, R.; Dollár, P.; Tu, Z.; He, K. Aggregated residual transformations for deep neural networks. In Proceedings of the IEEE Conference on Computer Vision and Pattern Recognition, Honolulu, HI, USA, 21–26 July 2017; pp. 1492–1500.

29. Szegedy, C.; Vanhoucke, V.; Ioffe, S.; Shlens, J.; Wojna, Z. Rethinking the inception architecture for computer vision. In Proceedings of the IEEE Conference on Computer Vision and Pattern Recognition, Las Vegas, NV, USA, 26 June–1 July 2016; pp. 2818–2826.
30. Hu, J.; Shen, L.; Sun, G. Squeeze-and-excitation networks. In Proceedings of the IEEE Conference on Computer Vision and Pattern Recognition, Salt Lake City, UT, USA, 18–22 June 2018; pp. 7132–7141.
31. Howard, A.G.; Zhu, M.; Chen, B.; Kalenichenko, D.; Wang, W.; Weyand, T.; Andreetto, M.; Adam, H. Mobilenets: Efficient convolutional neural networks for mobile vision applications. *arXiv* **2017**, arXiv:1704.04861.
32. Sandler, M.; Howard, A.; Zhu, M.; Zhmoginov, A.; Chen, L.C. Mobilenetv2: Inverted residuals and linear bottlenecks. In Proceedings of the IEEE Conference on Computer Vision and Pattern Recognition, Salt Lake City, UT, USA, 18–22 June 2018; pp. 4510–4520.
33. Howard, A.; Sandler, M.; Chu, G.; Chen, L.C.; Chen, B.; Tan, M.; Wang, W.; Zhu, Y.; Pang, R.; Vasudevan, V.; et al. Searching for mobilenetv3. In Proceedings of the IEEE International Conference on Computer Vision, Seoul, Korea, 27 October–2 November 2019; pp. 1314–1324.
34. Tan, M.; Le, Q.V. Efficientnet: Rethinking model scaling for convolutional neural networks. *arXiv* **2019**, arXiv:1905.11946.
35. Chollet, F. Xception: Deep learning with depthwise separable convolutions. In Proceedings of the IEEE Conference on Computer Vision and Pattern Recognition, Honolulu, HI, USA, 21–26 July 2017; pp. 1251–1258.
36. Szegedy, C.; Ioffe, S.; Vanhoucke, V.; Alemi, A.A. Inception-v4, inception-resnet and the impact of residual connections on learning. In Proceedings of the Thirty-First AAAI Conference on Artificial Intelligence, San Francisco, CA, USA, 4–9 February 2017.
37. Deng, J.; Dong, W.; Socher, R.; Li, L.J.; Li, K.; Fei-Fei, L. Imagenet: A large-scale hierarchical image database. In Proceedings of the 2009 IEEE Conference on Computer Vision and Pattern Recognition, Miami, FL, USA, 20–25 June 2009; IEEE: Piscataway, NJ, USA, 2009; pp. 248–255.
38. Sharma, N.; Jain, V.; Mishra, A. An analysis of convolutional neural networks for image classification. *Procedia Comput. Sci.* **2018**, *132*, 377–384. [CrossRef]
39. Taormina, V.; Cascio, D.; Abbene, L.; Raso, G. Performance of Fine-Tuning Convolutional Neural Networks for HEp-2 Image Classification. *Appl. Sci.* **2020**, *10*, 6940. [CrossRef]
40. Bello, I.; Zoph, B.; Vasudevan, V.; Le, Q.V. Neural optimizer search with reinforcement learning. In Proceedings of the 34th International Conference on Machine Learning, Sydney, Australia, 6–11 August 2017; Volume 70, pp. 459–468.
41. Chollet, F. Keras. 2015. Available online: https://keras.io (accessed on 29 November 2020)
42. Pedregosa, F.; Varoquaux, G.; Gramfort, A.; Michel, V.; Thirion, B.; Grisel, O.; Blondel, M.; Prettenhofer, P.; Weiss, R.; Dubourg, V.; et al. Scikit-learn: Machine Learning in Python. *J. Mach. Learn. Res.* **2011**, *12*, 2825–2830.
43. Blanco-Medina, P.; Alegre, E.; Fidalgo, E.; Al-Nabki, M.; Chaves, D. Enhancing text recognition on Tor Darknet images. *XL Jornadas Autom.* **2019**, 828–835. [CrossRef]
44. Blanco-Medina, P.; Fidalgo, E.; Alegre, E.; Jáñez Martino, F. Improving Text Recognition in Tor darknet with Rectification and Super-Resolution techniques. In Proceedings of the 9th International Conference on Imaging for Crime Detection and Prevention (ICDP-2019), London, UK, 16–18 December 2019; pp. 32–37.

Article

Cybersecurity against the Loopholes in Industrial Control Systems Using Interval-Valued Complex Intuitionistic Fuzzy Relations

Abdul Nasir [1], Naeem Jan [1], Abdu Gumaei [2,*], Sami Ullah Khan [1] and Fahad R. Albogamy [3]

1   Department of Mathematics, Institute of Numerical Sciences, Gomal University, Dera Ismail Khan 29050, Pakistan; theabdulnasir@gmail.com (A.N.); naeem.phdma73@iiu.edu.pk (N.J.); gomal85@gmail.com (S.U.K.)
2   STC's Artificial Intelligence Chair, Department of Information Systems, College of Computer and Information Sciences, King Saud University, Riyadh 11543, Saudi Arabia
3   Computer Sciences Program, Turabah University College, Taif University, P.O. Box 11099, Taif 21944, Saudi Arabia; f.alhammdani@tu.edu.sa
*   Correspondence: agumaei.c@ksu.edu.sa

Abstract: Technology is rapidly advancing and every aspect of life is being digitalized. Since technology has made life much better and easier, so organizations, such as businesses, industries, companies and educational institutes, etc., are using it. Despite the many benefits of technology, several risks and serious threats, called cyberattacks, are associated with it. The method of neutralizing these cyberattacks is known as cybersecurity. Sometimes, there are uncertainties in recognizing a cyberattack and nullifying its effects using righteous cybersecurity. For that reason, this article introduces interval-valued complex intuitionistic fuzzy relations (IVCIFRs). For the first time in the theory of fuzzy sets, we investigated the relationships among different types of cybersecurity and the sources of cyberattacks. Moreover, the Hasse diagram for the interval-valued complex intuitionistic partial order set and relation is defined. The concepts of the Hasse diagram are used to inspect different cybersecurity techniques and practices. Then, using the properties of Hasse diagrams, the most beneficial technique is identified. Furthermore, the best possible selection of types of cybersecurity is made after putting some restrictions on the selection. Lastly, the advantages of the proposed methods are illuminated through comparison tests.

Keywords: cybercrime; cybersecurity; Hasse diagram; interval-valued complex intuitionistic fuzzy relations; interval-valued complex intuitionistic fuzzy sets

Citation: Nasir, A.; Jan, N.; Gumaei, A.; Khan, S.U.; Albogamy, F.R. Cybersecurity against the Loopholes in Industrial Control Systems Using Interval-Valued Complex Intuitionistic Fuzzy Relations. *Appl. Sci.* **2021**, *11*, 7668. https://doi.org/10.3390/app11167668

Received: 10 June 2021
Accepted: 18 August 2021
Published: 20 August 2021

1. Introduction

Mathematical modeling is a process of transforming practical problems and real life events into a mathematical form. Human opinions and views are generally imprecise and unclear. Similarly, experimental errors, inaccuracies and miscalculations lead to inexact results. Hence, uncertainty is an inevitable part of our lives. Information is often complex, unknowable and puzzling. Modeling uncertainty was almost impossible before 1965 when Zadeh [1] introduced the world to a life changing theory of fuzzy sets (FS) and fuzzy logic. This theory opened up new gates for mathematicians because it could model uncertainty such as complexity in human opinions and errors in scientific experimentations. An FS is a set that associates a mapping $\acute{m}$ to each of its elements, where $\acute{m}$ is called the degree of membership that attains values from the [0, 1] interval.

The concept of relations between crisp sets was defined by Klir [2]. Crisp sets are limited to only two possibilities; the elements are either a member of a set or not a member of a set. In other words, crisp sets and their relations only work with yes-and-no-type problems. This theory of sets deals with exact information and cannot be used to model uncertainty. Mendel [3] invented the concepts of relations for FSs, which are called fuzzy relations (FRs). FRs are not just limited to respond in yes-or-no, but they can also specify the strength, grade and level of good relations between any pair of FSs. This depends on its degree of membership: if the value of the degree of membership is near one then the relation is said to be a strong relation, while values near zero are the sign of weak relationship. FRs generalize crisp relations because, if the values of the degrees of membership of each element of an FR are restricted to zero (0) and one (1), then the FR becomes a crisp relation. In 1975, Zadeh [4] generalized FSs and introduced the idea of interval-valued fuzzy sets (IVFSs). An IVFS replaces the single value of the degree of membership in an FS by an interval, whose extremes belong to a $[0, 1]$ interval, i.e., the degree of membership is a subinterval of a $[0, 1]$ interval. The concept of relation was introduced to IVFSs by Bustince and Burillo [5], called the interval-valued fuzzy relation (IVFR). Goguen [6] gave a system of axioms for a relatively simple form of FS theory; Zywica [7] modeled the medical uncertainties using FSs; Roman-Flores et al. [8] proposed a note on Zadeh's extensions; Dubious and Prade [9] discussed uncertainty, bipolarity and gradualness through FSs; Gehrke et al. [10] commented on IVFSs; Bustince [11] applied IVFSs to approximate reasoning; Turksen [12] entitled his work *IVFSs and compensatory AND*.

In 2002, Ramot et al. [13] came up with an innovative idea of switching to complex valued mappings as the degrees of membership of a set. They introduced the complex fuzzy set (CFS) in which the degree of membership takes on values from a unit disk of a complex plane. Since the degrees of membership in a CFS are complex numbers, thus, they involve two parts: the real and imaginary parts. Each of these parts corresponds to a different entity. The real part is known as the amplitude term of the degree of membership, while the imaginary part is said to be the phase term of the degree of membership. Most often, CFSs are used to model phase altering problems. When the phase term or the imaginary part of the degree of membership is set to zero in the CFS, it becomes an FS. Further, Ramot et al. [14] defined complex fuzzy relations (CFRs), which are used to find the relationships between CFSs. Greenfield et al. [15] changed the degree of membership of a CFS from a single number to an interval, thus bringing up the notion of interval-valued complex fuzzy set (IVCFS). Recently, Nasir et al. [16] defined interval-valued complex fuzzy relations as being used for studying the relationships between two or more IVCFSs. Chen et al. [17] studied a neuro-fuzzy architecture employing CFSs; Yazdanbakhsh and Dick [18] reviewed the CFSs; Tamir et al. [19] proposed some applications of CFSs; Dai et al. [20] formulated distance measures between IVCFSs; Greenfield et al. [21] defined the join and meet operations for IVCFSs.

According to the definition of an FS, it only discusses the degree of membership $\acute{m}$, but it also covers the degree of non-membership $\underline{n}$, which is the complement of the degree of membership with respect to unit interval, i.e., $\underline{n} = 1 - \acute{m}$. This hidden feature of FSs led to the definition of the intuitionistic fuzzy set (IFS), proposed by Atanassov [22]. An IFS associates two mappings to each of its elements, which are called the degree of membership and the degree of non-membership. Both of these mappings take on values from $[0, 1]$ interval given that their sum also lies within the range of this interval. The idea of the intuitionistic fuzzy relation (IFR) was concocted by Burillo et al. [23]. In 1999, Atanassov [24] expressed the degree of membership and the degree of non-membership of an IFS in the form of intervals and, hence, developed a new concept called the interval-valued intuitionistic fuzzy set (IVIFS). Complex valued mappings were brought in to IFS theory by Alkouri et al. [25] in 2012, who introduced the notion of the complex intuitionistic fuzzy set (CIFS). Garg and Rani [26] put forward the interval-valued complex intuitionistic fuzzy relation (IVCIFS). Li [27] used IFSs for multiattribute decision-making (MD) models and methods; De et al. [28] applied IFSs in medical diagnosis; Vlachos and Sergiadis [29] applied IFSs to pattern recognition; Lee et al. [30] compared IVFSs, IFSs and bipolar-valued FSs; Grzegorzewski [31] used Hausdorff metric for finding the distances between IVFSs and IFSs; Nasir et al. [32–35] applied complex relations to analysis of economic relationships. Ali et al. [36] studied complex intuitionistic fuzzy classes; Liu and Jiang [37] proposed a new distance measure of IVIFSs and applied it in decision making; Bustince and Burillo [38] devised the correlation of IVIFSs; Nayagam and Sivaraman [39] proposed a method of ranking IVIFSs. Otero et al. [40] proposed the application of fuzzy logic in assessing the security control for organizations; Tariq et al. [41,42] also used fuzzy theory to rank and prioritize the security controls for cloud wireless sensor networks and computing networks; Mokhtari et al. [43] used fuzzy multiple-attribute decision-making techniques for selecting the best control system.

In this paper, the Cartesian product (CP) of two IVCIFSs is introduced. Furthermore, the innovative conception of an interval-valued complex intuitionistic fuzzy relation (IV-CIFR) is defined by using the concept of the CP of IVCIFSs. Moreover, the types of IVCIFRs have been defined, including the interval-valued complex intuitionistic converse fuzzy relation, interval-valued complex intuitionistic equivalence fuzzy relation, interval-valued complex intuitionistic partial order fuzzy relation, interval-valued complex intuitionistic total order fuzzy relation, interval-valued complex intuitionistic composite fuzzy relation and many more. Every definition is held by the examples. In addition, some results have been proved for the type of IVCIFRs. Besides these, the Hasse diagrams for interval-valued complex intuitionistic partial order fuzzy sets and relations have also been presented. Furthermore, the ideas related to a Hasse diagram: maximum element, minimum element, maximal element, minimal element, supremum, infimum, upper and lower bounds are discussed. The novel concepts introduced in this study, i.e., IVCIFSs and IVCIFRs are superior to the pre-existing frameworks of FSs, IFSs, IVFSs, IVIFSs, CFSs, CIFSs and IVCFSs. Since IVCIFRs analyze the relationships among the IVCIFSs, so they are composed of the degrees of membership and non-membership, which are in the form of intervals with complex numbers. These notions can handle uncertainty much better than other mentioned concepts. The advantage of the interval-values is that they cover the little mistakes, ambiguities and errors made by decision makers and experts. In addition, their intuitionistic-type structure talks about the degree of membership and the degree of non-membership. Due to the complex valued mappings, they are also capable of handling information with multiple variables.

Every now and then, enterprises, corporations, industries and business companies shift their structures and organizations to a digital system. These computerized systems are extremely beneficial, but they are also vulnerable to many risky attacks and threats, which are known as cybercrimes. Different techniques, practices, tools and methods called cybersecurities, are used to tackle cybercrimes. At times, it can be challenging to detect the type of cybercrime, thus leaving many uncertainties. Similarly, there are also hesitations in employing the right cybersecurity techniques to rescue the business or companies from attacks as there are so many options out there to overcome these threats and risks. The selection of the most appropriate security technique may be difficult due to uncertainties and an inability to make the right decision. Therefore, in order to overcome all these uncertainties, we used the fuzzy theory. This article mathematically analyzes the relationships among cybersecurities and the sources of cyberattacks in an industrial control system (ICS), such as the effectiveness and ineffectiveness of cybersecurity against a certain source. Moreover, the current article also proposes a method to inspect different types of cybersecurity and choose the best one for an organization or network. This innovative method is based on the concepts of Hasse diagrams and interval-valued complex intuitionistic partial order fuzzy relations. Furthermore, the proposed methods are compared with other similar methods that pre-exist in the literature. The dominance and the reliability of the introduced methods are verified using numerical problems as the complex relations have not yet been discovered in fuzzy set theory. Henceforth, there is a great opportunity for potential research work to be carried out to explore these structures.

The arrangement of the remaining the paper is as follows:

Section 2 reviews some predefined concepts of fuzzy set theory that are used as the basis for the paper. Section 3 introduces the innovative conceptions of IVCIFRs, CPs between two IVCIFSs, the types of IVCIFRs and some theorems are also proved. In Section 4, we deliberate on the Hasse diagram for interval-valued complex intuitionistic partial order fuzzy sets and relations, including some useful properties and definitions related to Hasse diagrams. Section 5 offers two applications of IVCIFSs and IVCIFRs. The first application investigates cybercrimes, cyber-securities and the sources of penetration in an industrial control system (ICS). The second application uses Hasse diagrams and interval-valued complex intuitionistic partial order fuzzy relations to find the best cybersecurity technique. The existing structures in the field of fuzzy set theory are compared with the proposed structures in Section 6. Finally, the paper ends with the conclusion.

2. Preliminaries

In this section, some basic definitions and their examples are presented such as fuzzy set (FS), interval-valued fuzzy set (IVFS), complex fuzzy set (CFS), interval-valued complex fuzzy set (IVCFS), Cartesian product (CP) of two IVCFSs, interval-valued complex fuzzy relation (IVCFR), intuitionistic fuzzy set (IFS), complex intuitionistic fuzzy set (CIFS), and interval-valued complex intuitionistic fuzzy set (IVCIFS).

Definition 1. *Ref.* [1] *A fuzzy set (FS) $\check{E}$ in a universe T is of the following form*

$$\check{E} = \{(x, \acute{m}(x)) : x \in T\}$$

where $\acute{m} : \check{E} \to [0,\ 1]$ is a mapping called degree of membership.

Example 1. $\check{E} = \{(s, 0), (t, 0.63), (u, 0.97), (v, 0.26), (w, 0.18), (x, 1),\ (y, 0.85), (z, 0.35)\}$ *is an FS.*

Definition 2. *Ref.* [4] *An interval-valued fuzzy set (IVFS) $\check{E}$ in a universe T is of the following form*

$$\check{E} = \{(x, [\acute{m}^-(x), \acute{m}^+(x)]) : x \in T\}$$

where $\acute{m}^- : \check{E} \to [0,\ 1]$ and $\acute{m}^+ : \check{E} \to [0,\ 1]$ are the mappings called lower and upper degrees of membership, respectively.

Example 2. $\check{E} = \left\{ \begin{array}{l} (u, [0.32, 0.68]), (v, [0.16, 0.26]), (w, [0.58, 0.63]), \\ (x, [0.55, 0.66]), (y, [0.23, 0.49]), (z, [0.87, 1]) \end{array} \right\}$ *is an IVFS.*

Definition 3. *Ref.* [13] *A complex fuzzy set (CFS) $\check{E}$ in a universe T is of the following form*

$$\check{E} = \{(x, \acute{m}_{\mathbb{C}}(x)) : x \in T\}$$

where $\acute{m}_{\mathbb{C}} : \check{E} \to \mathcal{Z} \ni 0 \leq |\mathcal{Z}| \leq 1$ and $\mathcal{Z}$ is a complex number. The mapping $\acute{m}_{\mathbb{C}}$ is called degree of membership. Equivalently, the CFS can also be represented in the following form

$$\check{E} = \left\{ \left(x, \alpha(x)e^{i2\rho(x)\pi}\right) : x \in T \right\}$$

where $\alpha : \check{E} \to [0,\ 1]$ and $\rho : \check{E} \to [0,\ 1]$ are mappings called amplitude term and phase term of degree of membership and $i = \sqrt{-1}$.

Example 3. $\check{E} = \left\{ \begin{array}{l} \left(u, 0.15e^{i2(0.58)\pi}\right), \left(v, 0.75e^{i2(0.23)\pi}\right), \left(w, 0.65e^{i2(0.42)\pi}\right), \\ \left(x, 0e^{i2(1)\pi}\right), \left(y, 1e^{i2(0)\pi}\right), \left(z, 0.93e^{i2(0.37)\pi}\right) \end{array} \right\}$ *is a CFS.*

Definition 4. *Ref.* [15] *An interval-valued complex fuzzy set (IVCFS) $\check{E}$ in a universe T is of the following form*

$$\check{E} = \{(x, [\acute{m}_{\mathbb{C}}^-(x), \acute{m}_{\mathbb{C}}^+(x)]) : x \in T\}$$

where $\acute{m}_{\mathbb{C}}^- : \breve{E} \to \mathcal{Z}$ and $\acute{m}_{\mathbb{C}}^+ : \breve{E} \to \mathcal{Z} \ni 0 \leq |\mathcal{Z}| \leq 1$ and $\mathcal{Z}$ is a complex number. The mappings $\acute{m}_{\mathbb{C}}^-$ and $\acute{m}_{\mathbb{C}}^+$ are called the lower and upper degrees of membership, respectively. Equivalently, the IVCFS can also be represented in the following form

$$\breve{E} = \left\{ \left(x, \left[\alpha^-(x), \alpha^+(x) \right] e^{i2[\rho^-(x), \rho^+(x)]\pi} \right) : x \in T \right\}$$

where $\alpha^- : \breve{E} \to [0, 1]$, $\alpha^+ : \breve{E} \to [0, 1]$, $\rho^- : \breve{E} \to [0, 1]$ and $\rho^+ : \breve{E} \to [0, 1]$ are mappings called the lower amplitude term, upper amplitude term, lower phase term and upper phase term of degree of membership and $i = \sqrt{-1}$.

Example 4. $\breve{E} = \left\{ \begin{array}{l} \left(w, [0.28, 0.36] e^{i2[0.43, 0.61]\pi} \right), \left(x, [0.17, 0.23] e^{i2[0.67, 0.81]\pi} \right), \\ \left(y, [0.45, 0.50] e^{i2[0.04, 0.21]\pi} \right), \left(z, [0.89, 0.98] e^{i2[0, 0.16]\pi} \right) \end{array} \right\}$ is an IVCFS.

Definition 5. *Ref.* [16] *Let* $\breve{E} = \left\{ \left(x, \left[\alpha_{\breve{E}}^-(x), \alpha_{\breve{E}}^+(x) \right] e^{i2[\rho_{\breve{E}}^-(x), \rho_{\breve{E}}^+(x)]\pi} \right) : x \in T \right\}$ *and* $\dot{F} = \left\{ \left(y, \left[\alpha_{\dot{F}}^-(y), \alpha_{\dot{F}}^+(y) \right] e^{i2[\rho_{\dot{F}}^-(y), \rho_{\dot{F}}^+(y)]\pi} \right) : y \in T \right\}$ *be two IVCFSs in a universe T, then their Cartesian product is given as*

$$\breve{E} \times \dot{F} = \left\{ \left((x, y), \left[\alpha_{\breve{E} \times \dot{F}}^-(x, y), \alpha_{\breve{E} \times \dot{F}}^+(x, y) \right] e^{i2[\rho_{\breve{E} \times \dot{F}}^-(x, y), \rho_{\breve{E} \times \dot{F}}^+(x, y)]\pi} \right) : x \in \breve{E}, y \in \dot{F} \right\}$$

where $\alpha_{\breve{E} \times \dot{F}}^-(x, y) = \min\left\{ \alpha_{\breve{E}}^-(x), \alpha_{\dot{F}}^-(y) \right\}$, $\alpha_{\breve{E} \times \dot{F}}^+(x, y) = \min\left\{ \alpha_{\breve{E}}^+(x), \alpha_{\dot{F}}^+(y) \right\}$ $\rho_{\breve{E} \times \dot{F}}^-(x, y) = \min\left\{ \rho_{\breve{E}}^-(x), \rho_{\dot{F}}^-(y) \right\}$ and $\rho_{\breve{E} \times \dot{F}}^+(x, y) = \min\left\{ \rho_{\breve{E}}^+(x), \rho_{\dot{F}}^+(y) \right\}$.

Definition 6. *Ref.* [16] *The interval-valued complex fuzzy relation (IVCFR) is a subset of the Cartesian product of any two IVCFSs, i.e.,* $\bar{R} \subseteq \breve{E} \times \dot{F}$, *where* $\breve{E}$ *and* $\dot{F}$ *are IVCFSs and* $\bar{R}$ *denotes the IVCFR.*

Example 5. *The Cartesian product of two IVCFSs*

$$\breve{E} = \left\{ \begin{array}{l} \left(s, [0.28, 0.36] e^{i2[0.43, 0.61]\pi} \right), \left(t, [0.17, 0.23] e^{i2[0.67, 0.81]\pi} \right), \\ \left(u, [0.45, 0.50] e^{i2[0.04, 0.21]\pi} \right), \left(v, [0.89, 0.98] e^{i2[0, 0.16]\pi} \right) \end{array} \right\} \text{ and}$$

$$\dot{F} = \left\{ \begin{array}{l} \left(w, [0.53, 0.64] e^{i2[0.29, 0.37]\pi} \right), \left(x, [0.73, 0.79] e^{i2[0.43, 0.52]\pi} \right), \\ \left(y, [0.19, 0.35] e^{i2[0.24, 0.35]\pi} \right), \left(z, [0.39, 0.55] e^{i2[0.45, 0.55]\pi} \right) \end{array} \right\} \text{ is}$$

$$\breve{E} \times \dot{F} = \left\{ \begin{array}{l} \left((s, w), [0.28, 0.36] e^{i2[0.29, 0.37]\pi} \right), \left((s, x), [0.28, 0.36] e^{i2[0.43, 0.52]\pi} \right), \\ \left((s, y), [0.19, 0.35] e^{i2[0.24, 0.36]\pi} \right), \left((s, z), [0.28, 0.36] e^{i2[0, 0.16]\pi} \right), \\ \left((t, w), [0.17, 0.23] e^{i2[0.29, 0.37]\pi} \right), \left((t, x), [0.17, 0.23] e^{i2[0.43, 0.52]\pi} \right), \\ \left((t, y), [0.17, 0.23] e^{i2[0.24, 0.35]\pi} \right), \left((t, z), [0.17, 0.23] e^{i2[0.45, 0.55]\pi} \right), \\ \left((u, w), [0.45, 0.50] e^{i2[0.04, 0.21]\pi} \right), \left((u, x), [0.45, 0.50] e^{i2[0.04, 0.21]\pi} \right), \\ \left((u, y), [0.19, 0.35] e^{i2[0.04, 0.21]\pi} \right), \left((u, z), [0.39, 0.50] e^{i2[0.04, 0.21]\pi} \right), \\ \left((v, w), [0.53, 0.64] e^{i2[0.29, 0.37]\pi} \right), \left((v, x), [0.73, 0.79] e^{i2[0, 0.16]\pi} \right), \\ \left((v, y), [0.19, 0.35] e^{i2[0, 0.16]\pi} \right), \left((v, z), [0.39, 0.55] e^{i2[0, 0.16]\pi} \right) \end{array} \right\}$$

The IVCFR $\bar{\dot{R}}$. between the IVCFSs $\check{E}$ and $\dot{F}$ is given

$$\bar{\dot{R}} = \left\{ \begin{array}{l} \left((s,w),[0.28,0.36]e^{i2[0.29,0.37]\pi}\right), \left((s,x),[0.28,0.36]e^{i2[0.43,0.52]\pi}\right), \\ \left((t,x),[0.17,0.23]e^{i2[0.43,0.52]\pi}\right), \left((u,w),[0.45,0.50]e^{i2[0.04,0.21]\pi}\right), \\ \left((v,y),[0.19,0.35]e^{i2[0,0.16]\pi}\right), \left((v,z),[0.39,0.55]e^{i2[0,0.16]\pi}\right) \end{array} \right\}.$$

Definition 7. *Ref.* [22] *An intuitionistic fuzzy set (IFS)* $\check{E}$ *in a universe T is of the following form*

$$\check{E} = \{(x,\acute{m}(x),\underline{n}(x)) : x \in T\}$$

where $\acute{m} : \check{E} \to [0,\ 1]$ *and* $\underline{n} : \check{E} \to [0,\ 1]$ *are mappings called the degree of membership and degree of non-membership, respectively, given that* $0 \leq \acute{m}(x) + \underline{n}(x) \leq 1$.

Example 6. $\check{E} = \{(u,0.86,012),(w,0.47,0.36),(x,0.34,0.59),(y,0.61,0.28),(z,0.16,0.73)\}$ *is an IFS.*

Definition 8. *Ref.* [25] *A complex intuitionistic fuzzy set (CIFS)* $\check{E}$ *in a universe T is of the following form*

$$\check{E} = \{(x,\acute{m}_{\mathbb{C}}(x),\underline{n}_{\mathbb{C}}(x)) : x \in T\}$$

where $\acute{m}_{\mathbb{C}} : \check{E} \to \mathcal{Z}$ *and* $\underline{n}_{\mathbb{C}} : \check{E} \to \mathcal{Z} \ni 0 \leq |\mathcal{Z}| \leq 1$ *and* $\mathcal{Z}$ *is a complex number. The mappings* $\acute{m}_{\mathbb{C}}$ *and* $\underline{n}_{\mathbb{C}}$ *are called degree of membership and degree of non-membership, respectively, given that* $0 \leq |\acute{m}_{\mathbb{C}}(x)| + |\underline{n}_{\mathbb{C}}(x)| \leq 1$. *Equivalently, the CIFS can also be represented in the following form*

$$\check{E} = \left\{ \left(x,\alpha_{\acute{m}}(x)e^{i2\rho_{\acute{m}}(x)\pi},\alpha_{\underline{n}}(x)e^{i2\rho_{\underline{n}}(x)\pi}\right) : x \in T \right\}$$

where $\alpha_{\acute{m}} : \check{E} \to [0,\ 1]$, $\alpha_{\underline{n}} : \check{E} \to [0,\ 1]$, $\rho_{\acute{m}} : \check{E} \to [0,\ 1]$ *and* $\rho_{\underline{n}} : \check{E} \to [0,\ 1]$ *are mappings called amplitude term of degree of membership, amplitude term of degree of non-membership, phase term of degree of membership and phase term of degree of non-membership and* $i = \sqrt{-1}$, *given that* $0 \leq \alpha_{\acute{m}}(x) + \alpha_{\underline{n}}(x) \leq 1$ *and* $0 \leq \rho_{\acute{m}}(x) + \rho_{\underline{n}}(x) \leq 1$.

Example 7. $\check{E} = \left\{ \begin{array}{l} \left(w,0.35e^{i2(0.42)\pi},0.26e^{i2(0.31)\pi}\right), \left(x,0.51e^{i2(0.16)\pi},0.38e^{i2(0.84)\pi}\right), \\ \left(y,0.71e^{i2(0.22)\pi},0.19e^{i2(0.49)\pi}\right), \left(z,0.09e^{i2(0.17)\pi},0.03e^{i2(0.75)\pi}\right) \end{array} \right\}$ *is a CIFS.*

Definition 9. *Ref.* [26] *An interval-valued complex intuitionistic fuzzy set (IVCIFS)* $\check{E}$ *in a universe T is of the following form*

$$\check{E} = \left\{ \left(x,[\acute{m}_{\mathbb{C}}^{-}(x),\acute{m}_{\mathbb{C}}^{+}(x)],[\underline{n}_{\mathbb{C}}^{-}(x),\underline{n}_{\mathbb{C}}^{+}(x)]\right) : x \in T \right\}$$

where $\acute{m}_{\mathbb{C}}^{-} : \check{E} \to \mathcal{Z}$, $\acute{m}_{\mathbb{C}}^{+} : \check{E} \to \mathcal{Z}$, $\underline{n}_{\mathbb{C}}^{-} : \check{E} \to \mathcal{Z}$ *and* $\underline{n}_{\mathbb{C}}^{+} : \check{E} \to \mathcal{Z} \ni 0 \leq |\mathcal{Z}| \leq 1$ *and* $\mathcal{Z}$ *is a complex number. The mappings* $\acute{m}_{\mathbb{C}}^{-}$ *and* $\acute{m}_{\mathbb{C}}^{+}$ *are called the lower and upper degrees of membership, respectively, while the mappings* $\underline{n}_{\mathbb{C}}^{-}$ *and are called the lower and upper degrees of non-membership, respectively, given that* $0 \leq |\acute{m}_{\mathbb{C}}^{+}(x)| + |\underline{n}_{\mathbb{C}}^{+}(x)| \leq 1$. *Equivalently, the IVCIFS can also be represented in the following form*

$$\check{E} = \left\{ \left(x,[\alpha_{\acute{m}}^{-}(x),\alpha_{\acute{m}}^{+}(x)]e^{i2[\rho_{\acute{m}}^{-}(x),\rho_{\acute{m}}^{+}(x)]\pi},[\alpha_{\underline{n}}^{-}(x),\alpha_{\underline{n}}^{+}(x)]e^{i2[\rho_{\underline{n}}^{-}(x),\rho_{\underline{n}}^{+}(x)]\pi}\right) : x \in T \right\}$$

where $\alpha_{\acute{m}}^{-} : \check{E} \to [0,\ 1]$, $\alpha_{\acute{m}}^{+} : \check{E} \to [0,\ 1]$, $\rho_{\acute{m}}^{-} : \check{E} \to [0,\ 1]$, $\rho_{\acute{m}}^{\downarrow} : \check{E} \to [0,\ 1]$ *and* $\alpha_{\underline{n}}^{-} : \check{E} \to [0,\ 1]$, $\alpha_{\underline{n}}^{+} : \check{E} \to [0,\ 1]$, $\rho_{\underline{n}}^{-} : \check{E} \to [0,\ 1]$, $\rho_{\underline{n}}^{+} : \check{E} \to [0,\ 1]$ *are mappings called the lower amplitude term, upper amplitude term, lower phase term, upper phase term of degree of membership and lower ampli-*

tude term, upper amplitude term, lower phase term, upper phase term of degree of non-membership, respectively, $i = \sqrt{-1}$. Given that $0 \leq \alpha_{\acute{m}}^{+} + \alpha_{\underline{n}}^{+} \leq 1$ and $0 \leq \rho_{\acute{m}}^{+} + \rho_{\underline{n}}^{+} \leq 1$.

Example 8. $\breve{E} = \left\{ \begin{array}{l} \left(w, [0.28, 0.36]e^{i2[0.43, 0.53]\pi}, [0.48, 0.61]e^{i2[0.32, 0.42]\pi}\right), \\ \left(x, [0.07, 0.12]e^{i2[0.24, 0.29]\pi}, [0.56, 0.66]e^{i2[0.57, 0.71]\pi}\right), \\ \left(y, [0.63, 0.71]e^{i2[0.03, 0.15]\pi}, [0.11, 0.23]e^{i2[0.73, 0.82]\pi}\right) \\ \left(z, [0.37, 0.46]e^{i2[0.15, 0.25]\pi}, [0.32, 0.43]e^{i2[0.53, 0.68]\pi}\right) \end{array} \right\}$ *is an IVCIFS.*

3. Interval-Valued Complex Fuzzy Relations

This section introduces the novel concepts of the Cartesian product of two interval-valued complex intuitionistic fuzzy sets (IVCIFSs), an interval-valued complex intuitionistic fuzzy relation (CIVIFR) and its types. For each definition a suitable example is given. Moreover, some interesting results for IVCIFRs have also been proved.

Definition 10. *Let* $\breve{E} = \left\{ \left(x, \left[\alpha_{(\breve{E})\acute{m}}^{-}(x), \alpha_{(\breve{E})\acute{m}}^{+}(x)\right]e^{i2[\rho_{(\breve{E})\acute{m}}^{-}(x), \rho_{(\breve{E})\acute{m}}^{+}(x)]\pi}, \left[\alpha_{(\breve{E})\underline{n}}^{-}(x), \alpha_{(\breve{E})\underline{n}}^{+}(x)\right]e^{i2[\rho_{(\breve{E})\underline{n}}^{-}(x), \rho_{(\breve{E})\underline{n}}^{+}(x)]\pi} \right) : x \in T \right\}$ and*

$\dot{F} = \left\{ \left(y, \left[\alpha_{(\dot{F})\acute{m}}^{-}(y), \alpha_{(\dot{F})\acute{m}}^{+}(y)\right]e^{i2[\rho_{(\dot{F})\acute{m}}^{-}(y), \rho_{(\dot{F})\acute{m}}^{+}(y)]\pi}, \left[\alpha_{(\dot{F})\underline{n}}^{-}(y), \alpha_{(\dot{F})\underline{n}}^{+}(y)\right]e^{i2[\rho_{(\dot{F})\underline{n}}^{-}(y), \rho_{(\dot{F})\underline{n}}^{+}(y)]\pi} \right) : y \in T \right\}$ *be two IVCFSs in a universe T, then their Cartesian product is given as*

$\breve{E} \times \dot{F} = \left\{ \left((x, y), \left[\alpha_{(\breve{E}\times\dot{F})\acute{m}}^{-}(x, y), \alpha_{(\breve{E}\times\dot{F})\acute{m}}^{+}(x, y)\right]e^{i2[\rho_{(\breve{E}\times\dot{F})\acute{m}}^{-}(x, y), \rho_{(\breve{E}\times\dot{F})\acute{m}}^{+}(x, y)]\pi}, \left[\alpha_{(\breve{E}\times\dot{F})\underline{n}}^{-}(x, y), \alpha_{(\breve{E}\times\dot{F})\underline{n}}^{+}(x, y)\right]e^{i2[\rho_{(\breve{E}\times\dot{F})\underline{n}}^{-}(x, y), \rho_{(\breve{E}\times\dot{F})\underline{n}}^{+}(x, y)]\pi} \right) : x \in \breve{E}, y \in \dot{F} \right\}$

where $\alpha_{(\breve{E}\times\dot{F})\acute{m}}^{-}(x, y) = \min\left\{ \alpha_{(\breve{E})\acute{m}}^{-}(x), \alpha_{(\dot{F})\acute{m}}^{-}(y) \right\}, \alpha_{(\breve{E}\times\dot{F})\acute{m}}^{+}(x, y) = \min\left\{ \alpha_{(\breve{E})\acute{m}}^{+}(x), \alpha_{(\dot{F})\acute{m}}^{+}(y) \right\}$

$\rho_{(\breve{E}\times\dot{F})\acute{m}}^{-}(x, y) = \min\left\{ \rho_{(\breve{E})\acute{m}}^{-}(x), \rho_{(\dot{F})\acute{m}}^{-}(y) \right\}$ *and* $\rho_{(\breve{E}\times\dot{F})\acute{m}}^{+}(x, y) = \min\left\{ \rho_{(\breve{E})\acute{m}}^{+}(x), \rho_{(\dot{F})\acute{m}}^{+}(y) \right\}$

and $\alpha_{(\breve{E}\times\dot{F})\underline{n}}^{-}(x, y) = \max\left\{ \alpha_{(\breve{E})\underline{n}}^{-}(x), \alpha_{(\dot{F})\underline{n}}^{-}(y) \right\}, \alpha_{(\breve{E}\times\dot{F})\underline{n}}^{+}(x, y) = \max\left\{ \alpha_{(\breve{E})\underline{n}}^{+}(x), \alpha_{(\dot{F})\underline{n}}^{+}(y) \right\}$

$\rho_{(\breve{E}\times\dot{F})\underline{n}}^{-}(x, y) = \max\left\{ \rho_{(\breve{E})\underline{n}}^{-}(x), \rho_{(\dot{F})\underline{n}}^{-}(y) \right\}$ *and* $\rho_{(\breve{E}\times\dot{F})\underline{n}}^{+}(x, y) = \max\left\{ \rho_{(\breve{E})\underline{n}}^{+}(x), \rho_{(\dot{F})\underline{n}}^{+}(y) \right\}$.

Definition 11. *The interval-valued complex intuitionistic fuzzy relation (IVCIFR) is a subset of the Cartesian product of any two IVCIFSs, i.e., $\bar{R} \subseteq \breve{E} \times \dot{F}$, where $\breve{E}$ and $\dot{F}$ are IVCIFSs and $\bar{R}$ denotes the IVCIFR.*

Example 9. *The Cartesian product of two IVCFSs*

$\breve{E} = \left\{ \begin{array}{l} \left(u, [0.07, 0.12]e^{i2[0.24, 0.29]\pi}, [0.56, 0.66]e^{i2[0.57, 0.71]\pi}\right), \\ \left(v, [0.63, 0.71]e^{i2[0.03, 0.15]\pi}, [0.11, 0.23]e^{i2[0.73, 0.82]\pi}\right), \\ \left(w, [0.37, 0.46]e^{i2[0.15, 0.25]\pi}, [0.32, 0.43]e^{i2[0.53, 0.68]\pi}\right) \end{array} \right\}$ *and*

$\dot{F} = \left\{ \begin{array}{l} \left(x, [0.26, 0.34]e^{i2[0.15, 0.25]\pi}, [0.38, 0.49]e^{i2[0.63, 0.74]\pi}\right), \\ \left(y, [0.52, 0.58]e^{i2[0.30, 0.42]\pi}, [0.18, 0.37]e^{i2[0.06, 0.13]\pi}\right), \\ \left(z, [0.28, 0.36]e^{i2[0.43, 0.53]\pi}, [0.48, 0.61]e^{i2[0.32, 0.42]\pi}\right), \end{array} \right\}$ *is*

$$\breve{E} \times \dot{F} = \begin{Bmatrix} \left((u,x),\,[0.07,0.12]e^{i2[0.15,0.25]\pi},\,[0.56,0.66]e^{i2[0.63,0.74]\pi}\right), \\ \left((u,y),\,[0.07,0.12]e^{i2[0.24,0.29]\pi},\,[0.56,0.66]e^{i2[0.57,0.71]\pi}\right), \\ \left((u,z),\,[0.07,0.12]e^{i2[0.24,0.29]\pi},\,[0.56,0.66]e^{i2[0.57,0.71]\pi}\right), \\ \left((v,x),\,[0.26,0.34]e^{i2[0.03,0.15]\pi},\,[0.38,0.49]e^{i2[0.73,0.82]\pi}\right), \\ \left((v,y),\,[0.52,0.58]e^{i2[0.03,0.15]\pi},\,[0.18,0.37]e^{i2[0.73,0.82]\pi}\right), \\ \left((v,z),\,[0.28,0.36]e^{i2[0.03,0.15]\pi},\,[0.48,0.61]e^{i2[0.73,0.82]\pi}\right), \\ \left((w,x),\,[0.26,0.34]e^{i2[0.15,0.25]\pi},\,[0.38,0.49]e^{i2[0.63,0.74]\pi}\right), \\ \left((w,y),\,[0.37,0.46]e^{i2[0.15,0.25]\pi},\,[0.32,0.43]e^{i2[0.53,0.68]\pi}\right), \\ \left((w,z),\,[0.28,0.36]e^{i2[0.15,0.25]\pi},\,[0.48,0.61]e^{i2[0.53,0.68]\pi}\right) \end{Bmatrix}$$

The IVCIFR $\bar{R}$ between the IVCFSs $\breve{E}$ and $\dot{F}$ is

$$\bar{R} = \begin{Bmatrix} \left((u,y),\,[0.07,0.12]e^{i2[0.24,0.29]\pi},\,[0.56,0.66]e^{i2[0.57,0.71]\pi}\right), \\ \left((v,x),\,[0.26,0.34]e^{i2[0.03,0.15]\pi},\,[0.38,0.49]e^{i2[0.73,0.82]\pi}\right), \\ \left((v,z),\,[0.28,0.36]e^{i2[0.03,0.15]\pi},\,[0.48,0.61]e^{i2[0.73,0.82]\pi}\right), \\ \left((w,y),\,[0.37,0.46]e^{i2[0.15,0.25]\pi},\,[0.32,0.43]e^{i2[0.53,0.68]\pi}\right), \\ \left((w,z),\,[0.28,0.36]e^{i2[0.15,0.25]\pi},\,[0.48,0.61]e^{i2[0.53,0.68]\pi}\right) \end{Bmatrix}$$

The IVCIFR $\bar{R}$ between the IVCFSs $\breve{E}$ and $\dot{F}$ is given as below (Figure 1),

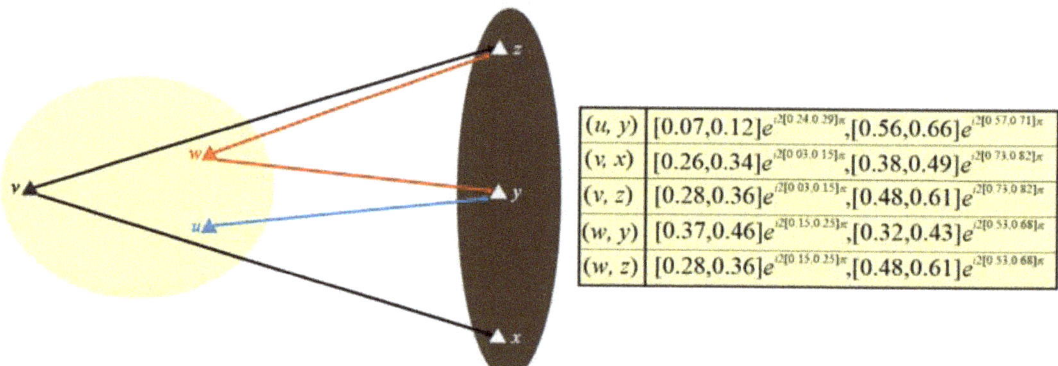

Figure 1. Interval-valued complex intuitionistic fuzzy relation.

NOTE: For convenience, throughout this article, x and (x,y) will be used to denote

$$\left(x,\,\left[\alpha^{-}_{(\breve{E})\dot{m}}(x),\,\alpha^{+}_{(\breve{E})\dot{m}}(x)\right]e^{i2[\rho^{-}_{(\breve{E})\dot{m}}(x),\,\rho^{+}_{(\breve{E})\dot{m}}(x)]\pi},\,\left[\alpha^{-}_{(\breve{E})\underline{n}}(x),\,\alpha^{+}_{(\breve{E})\underline{n}}(x)\right]e^{i2[\rho^{-}_{(\breve{E})\underline{n}}(x),\,\rho^{+}_{(\breve{E})\underline{n}}(x)]\pi}\right)$$

and

$$\left(\begin{array}{l}(x,y),\,\left[\alpha^{-}_{(\breve{E}\times\breve{E})\dot{m}}(x,y),\,\alpha^{+}_{(\breve{E}\times\breve{E})\dot{m}}(x,y)\right]e^{i2[\rho^{-}_{(\breve{E}\times\breve{E})\dot{m}}(x,y),\,\rho^{+}_{(\breve{E}\times\breve{E})\dot{m}}(x,y)]\pi}, \\ \left[\alpha^{-}_{(\breve{E}\times\breve{E})\underline{n}}(x,y),\,\alpha^{+}_{(\breve{E}\times\breve{E})\underline{n}}(x,y)\right]e^{i2[\rho^{-}_{(\breve{E}\times\breve{E})\underline{n}}(x,y),\,\rho^{+}_{(\breve{E}\times\breve{E})\underline{n}}(x,y)]\pi}\end{array}\right),\text{ respectively,}$$

otherwise it will be mentioned.

Definition 12. *Let $Ĕ$ be an IVCIFS in a universe T and $\bar{R}$ be an IVCIFR on $Ĕ$. Then,*

1. *$\bar{R}$ is known as an IVCI reflexive fuzzy relation (IVCI-reflexive-FR) on $Ĕ$ if $(x,x) \in \bar{R}$, $\forall x \in Ĕ$.*
2. *$\bar{R}$ is known as an IVCI irreflexive fuzzy relation (IVCI-irreflexive-FR) on $Ĕ$ if $(x,x) \notin \bar{R}$, $\forall x \in Ĕ$.*
3. *$\bar{R}$ is known as an IVCI symmetric fuzzy relation (IVCI-symmetric-FR) on $Ĕ$ if $\forall x,y \in Ĕ$, $(x,y) \in \bar{R} \Rightarrow (y,x) \in \bar{R}$.*
4. *$\bar{R}$ is known as an IVCI antisymmetric fuzzy relation (IVCI-antisymmetric-FR) on $Ĕ$ if $\forall x,y \in Ĕ$, $(x,y) \in \bar{R}$ and $(y,x) \in \bar{R} \Rightarrow x = y$.*
5. *$\bar{R}$ is known as an IVCI asymmetric fuzzy relation (IVCI-asymmetric-FR) on $Ĕ$ if $\forall x,y \in Ĕ$, $(y,x) \in \bar{R} \Rightarrow (x = y \notin)\bar{R}$.*
6. *$\bar{R}$ is known as an IVCI complete fuzzy relation (IVCI-complete-FR) on $Ĕ$ if $\forall x,y \in \bar{R}$, $(x,y) \in \bar{R}$ or $(y,x) \in \bar{R}$.*
7. *$\bar{R}$ is known as an IVCI transitive fuzzy relation on (IVCI-transitive-FR) $Ĕ$ if $\forall x,y,z \in Ĕ$, $(x,y) \in \bar{R}$ and $(y,z) \in \bar{R} \Rightarrow (x,z) \in \bar{R}$.*
8. *$\bar{R}$ is known as an IVCI equivalence fuzzy relation (IVCI-equivalence-FR) on $Ĕ$ if $\bar{R}$ is IVCI-reflexive-FR, IVCI-symmetric-FR and IVCI-transitive-FR on $Ĕ$.*
9. *$\bar{R}$ is known as an IVCI preorder fuzzy relation (IVCI-preorder-FR) on $Ĕ$ if $\bar{R}$ is IVCI-reflexive-FR and IVCI-transitive-FR on $Ĕ$.*
10. *$\bar{R}$ is known as an IVCI strict order fuzzy relation (IVCI-strict order-FR) on $Ĕ$ if $\bar{R}$ is IVCI-irreflexive-FR and IVCI-transitive-FR on $Ĕ$.*
11. *$\bar{R}$ is known as an IVCI partial order fuzzy relation (IVCI-partial order-FR) on $Ĕ$ if $\bar{R}$ is IVCI-preorder-FR and IVCI-antisymmetric-FR on $Ĕ$.*
12. *$\bar{R}$ is known as an IVCI linear order fuzzy relation (IVCI-linear order-FR) on $Ĕ$ if $\bar{R}$ is IVCI-partial order-FR and IVCI-complete-FR on $Ĕ$.*

Example 10. *For an IVCIFS*
$$
Ĕ = \left\{
\begin{array}{l}
\left(u, [0.07,0.12]e^{i2[0.24,0.29]\pi}, [0.56,0.66]e^{i2[0.57,0.71]\pi}\right), \\
\left(v, [0.63,0.71]e^{i2[0.03,0.15]\pi}, [0.11,0.23]e^{i2[0.73,0.82]\pi}\right), \\
\left(w, [0.37,0.46]e^{i2[0.15,0.25]\pi}, [0.32,0.43]e^{i2[0.53,0.68]\pi}\right)
\end{array}
\right\}
$$
the Cartesian product $Ĕ \times Ĕ$ is

$$
Ĕ \times Ĕ = \left\{
\begin{array}{l}
\left((u,u), [0.07,0.12]e^{i2[0.15,0.25]\pi}, [0.56,0.66]e^{i2[0.63,0.74]\pi}\right), \\
\left((u,v), [0.07,0.12]e^{i2[0.03,0.15]\pi}, [0.56,0.66]e^{i2[0.73,0.82]\pi}\right), \\
\left((u,w), [0.07,0.12]e^{i2[0.15,0.25]\pi}, [0.56,0.66]e^{i2[0.57,0.71]\pi}\right), \\
\left((v,u), [0.07,0.12]e^{i2[0.03,0.15]\pi}, [0.56,0.66]e^{i2[0.73,0.82]\pi}\right), \\
\left((v,v), [0.63,0.71]e^{i2[0.03,0.15]\pi}, [0.11,0.23]e^{i2[0.73,0.82]\pi}\right), \\
\left((v,w), [0.37,0.46]e^{i2[0.03,0.15]\pi}, [0.32,0.43]e^{i2[0.73,0.82]\pi}\right), \\
\left((w,u), [0.07,0.12]e^{i2[0.15,0.25]\pi}, [0.56,0.66]e^{i2[0.57,0.71]\pi}\right), \\
\left((w,v), [0.37,0.46]e^{i2[0.03,0.15]\pi}, [0.32,0.43]e^{i2[0.73,0.82]\pi}\right), \\
\left((w,w), [0.37,0.46]e^{i2[0.15,0.25]\pi}, [0.32,0.43]e^{i2[0.53,0.68]\pi}\right)
\end{array}
\right\}
$$

Then,

1. The IVCI-equivalence-FR $\breve{R}_1$ on $\breve{E}$ is given as

$$\breve{R}_1 = \left\{ \begin{array}{l} \left((u,u), [0.07,0.12]e^{i2[0.15,0.25]\pi}, [0.56,0.66]e^{i2[0.63,0.74]\pi} \right), \\ \left((u,w), [0.07,0.12]e^{i2[0.15,0.25]\pi}, [0.56,0.66]e^{i2[0.57,0.71]\pi} \right), \\ \left((v,v), [0.63,0.71]e^{i2[0.03,0.15]\pi}, [0.11,0.23]e^{i2[0.73,0.82]\pi} \right), \\ \left((w,u), [0.07,0.12]e^{i2[0.15,0.25]\pi}, [0.56,0.66]e^{i2[0.57,0.71]\pi} \right), \\ \left((w,w), [0.37,0.46]e^{i2[0.15,0.25]\pi}, [0.32,0.43]e^{i2[0.53,0.68]\pi} \right) \end{array} \right\}$$

2. The IVCI-preorder-FR $\breve{R}_2$ on $\breve{E}$ is given as

$$\breve{R}_2 = \left\{ \begin{array}{l} \left((u,u), [0.07,0.12]e^{i2[0.15,0.25]\pi}, [0.56,0.66]e^{i2[0.63,0.74]\pi} \right), \\ \left((v,u), [0.07,0.12]e^{i2[0.03,0.15]\pi}, [0.56,0.66]e^{i2[0.73,0.82]\pi} \right), \\ \left((v,v), [0.63,0.71]e^{i2[0.03,0.15]\pi}, [0.11,0.23]e^{i2[0.73,0.82]\pi} \right), \\ \left((w,w), [0.37,0.46]e^{i2[0.15,0.25]\pi}, [0.32,0.43]e^{i2[0.53,0.68]\pi} \right) \end{array} \right\}$$

3. The IVCI-strict order-FR $\breve{R}_3$ on $\breve{E}$ is given as

$$\breve{R}_3 = \left\{ \begin{array}{l} \left((v,u), [0.07,0.12]e^{i2[0.03,0.15]\pi}, [0.56,0.66]e^{i2[0.73,0.82]\pi} \right), \\ \left((w,u), [0.07,0.12]e^{i2[0.15,0.25]\pi}, [0.56,0.66]e^{i2[0.57,0.71]\pi} \right), \\ \left((w,v), [0.37,0.46]e^{i2[0.03,0.15]\pi}, [0.32,0.43]e^{i2[0.73,0.82]\pi} \right) \end{array} \right\}$$

4. The IVCI-partial order-FR $\breve{R}_4$ on $\breve{E}$ is given as

$$\breve{R}_4 = \left\{ \begin{array}{l} \left((u,u), [0.07,0.12]e^{i2[0.15,0.25]\pi}, [0.56,0.66]e^{i2[0.63,0.74]\pi} \right), \\ \left((v,u), [0.07,0.12]e^{i2[0.03,0.15]\pi}, [0.56,0.66]e^{i2[0.73,0.82]\pi} \right), \\ \left((v,v), [0.63,0.71]e^{i2[0.03,0.15]\pi}, [0.11,0.23]e^{i2[0.73,0.82]\pi} \right), \\ \left((w,u), [0.07,0.12]e^{i2[0.15,0.25]\pi}, [0.56,0.66]e^{i2[0.57,0.71]\pi} \right), \\ \left((w,w), [0.37,0.46]e^{i2[0.15,0.25]\pi}, [0.32,0.43]e^{i2[0.53,0.68]\pi} \right) \end{array} \right\}$$

5. The IVCI-linear order-FR $\breve{R}_5$ on $\breve{E}$ is given as

$$\breve{R}_5 = \left\{ \begin{array}{l} \left((u,u), [0.07,0.12]e^{i2[0.15,0.25]\pi}, [0.56,0.66]e^{i2[0.63,0.74]\pi} \right), \\ \left((v,u), [0.07,0.12]e^{i2[0.03,0.15]\pi}, [0.56,0.66]e^{i2[0.73,0.82]\pi} \right), \\ \left((v,v), [0.63,0.71]e^{i2[0.03,0.15]\pi}, [0.11,0.23]e^{i2[0.73,0.82]\pi} \right), \\ \left((w,u), [0.07,0.12]e^{i2[0.15,0.25]\pi}, [0.56,0.66]e^{i2[0.57,0.71]\pi} \right), \\ \left((w,v), [0.37,0.46]e^{i2[0.03,0.15]\pi}, [0.32,0.43]e^{i2[0.73,0.82]\pi} \right), \\ \left((w,w), [0.37,0.46]e^{i2[0.15,0.25]\pi}, [0.32,0.43]e^{i2[0.53,0.68]\pi} \right) \end{array} \right\}$$

Definition 13. *For an IVCIFR $\breve{R}$, the converse relation $\breve{R}^c$ of $\breve{R}$ is defined as,* $\breve{R}^c = \{ (y,x) : (x,y) \in \breve{R} \}$.

Example 11. *If* $\breve{R} =$

$$
\left\{
\begin{array}{l}
\Big((u,y), [0.07,0.12]e^{i2[0.24,0.29]\pi}, [0.56,0.66]e^{i2[0.57,0.71]\pi}\Big), \\
\Big((v,x), [0.26,0.34]e^{i2[0.03,0.15]\pi}, [0.38,0.49]e^{i2[0.73,0.82]\pi}\Big), \\
\Big((v,z), [0.28,0.36]e^{i2[0.03,0.15]\pi}, [0.48,0.61]e^{i2[0.73,0.82]\pi}\Big), \\
\Big((w,y), [0.37,0.46]e^{i2[0.15,0.25]\pi}, [0.32,0.43]e^{i2[0.53,0.68]\pi}\Big), \\
\Big((w,z), [0.28,0.36]e^{i2[0.15,0.25]\pi}, [0.48,0.61]e^{i2[0.53,0.68]\pi}\Big)
\end{array}
\right\}
$$ *is an*

IVCIFR between IVCIFSs $\breve{E} =$
$$
\left\{
\begin{array}{l}
\Big(u, [0.07,0.12]e^{i2[0.24,0.29]\pi}, [0.56,0.66]e^{i2[0.57,0.71]\pi}\Big), \\
\Big(v, [0.63,0.71]e^{i2[0.03,0.15]\pi}, [0.11,0.23]e^{i2[0.73,0.82]\pi}\Big), \\
\Big(w, [0.37,0.46]e^{i2[0.15,0.25]\pi}, [0.32,0.43]e^{i2[0.53,0.68]\pi}\Big)
\end{array}
\right\}
$$ *and*

$$
\breve{F} =
\left\{
\begin{array}{l}
\Big(x, [0.26,0.34]e^{i2[0.15,0.25]\pi}, [0.38,0.49]e^{i2[0.63,0.74]\pi}\Big), \\
\Big(y, [0.52,0.58]e^{i2[0.30,0.42]\pi}, [0.18,0.37]e^{i2[0.06,0.13]\pi}\Big), \\
\Big(z, [0.28,0.36]e^{i2[0.43,0.53]\pi}, [0.48,0.61]e^{i2[0.32,0.42]\pi}\Big),
\end{array}
\right\}
$$, *then the converse relation* $\breve{R}^c$

of $\breve{R}$ *is*

$$
\breve{R}^c =
\left\{
\begin{array}{l}
\Big((x,v), [0.26,0.34]e^{i2[0.03,0.15]\pi}, [0.38,0.49]e^{i2[0.73,0.82]\pi}\Big), \\
\Big((y,u), [0.07,0.12]e^{i2[0.24,0.29]\pi}, [0.56,0.66]e^{i2[0.57,0.71]\pi}\Big), \\
\Big((y,w), [0.37,0.46]e^{i2[0.15,0.25]\pi}, [0.32,0.43]e^{i2[0.53,0.68]\pi}\Big), \\
\Big((z,v), [0.28,0.36]e^{i2[0.03,0.15]\pi}, [0.48,0.61]e^{i2[0.73,0.82]\pi}\Big), \\
\Big((z,w), [0.28,0.36]e^{i2[0.15,0.25]\pi}, [0.48,0.61]e^{i2[0.53,0.68]\pi}\Big)
\end{array}
\right\}
$$

The IVCI-equivalence-FRs sets off the concept of IVCIF-equivalences classes, which are defined as follows.

Definition 14. *For an IVCI-equivalence-FR* $\breve{R}$*, the IVCIF equivalence class of* x *modulo* $\breve{R}$ *is defined as,* $\breve{R}[x] = \{y \,|\, (y,x) \in \breve{R}\}$.

Example 12. *If* $\breve{R} =$

$$
\left\{
\begin{array}{l}
\Big((u,u), [0.07,0.12]e^{i2[0.15,0.25]\pi}, [0.56,0.66]e^{i2[0.63,0.74]\pi}\Big), \\
\Big((u,w), [0.07,0.12]e^{i2[0.15,0.25]\pi}, [0.56,0.66]e^{i2[0.57,0.71]\pi}\Big), \\
\Big((v,v), [0.63,0.71]e^{i2[0.03,0.15]\pi}, [0.11,0.23]e^{i2[0.73,0.82]\pi}\Big), \\
\Big((w,u), [0.07,0.12]e^{i2[0.15,0.25]\pi}, [0.56,0.66]e^{i2[0.57,0.71]\pi}\Big), \\
\Big((w,w), [0.37,0.46]e^{i2[0.15,0.25]\pi}, [0.32,0.43]e^{i2[0.53,0.68]\pi}\Big),
\end{array}
\right\}
$$ *is an*

IVCIFR on an IVCIFS $\breve{E} =$
$$
\left\{
\begin{array}{l}
\Big(u, [0.07,0.12]e^{i2[0.24,0.29]\pi}, [0.56,0.66]e^{i2[0.57,0.71]\pi}\Big), \\
\Big(v, [0.63,0.71]e^{i2[0.03,0.15]\pi}, [0.11,0.23]e^{i2[0.73,0.82]\pi}\Big), \\
\Big(w, [0.37,0.46]e^{i2[0.15,0.25]\pi}, [0.32,0.43]e^{i2[0.53,0.68]\pi}\Big)
\end{array}
\right\}
$$, *then*

the IVCIF-equivalences class of

1. x *modulo* $\breve{R}$ *is given as*

$$
\breve{R}[u] =
\left\{
\begin{array}{l}
\Big(u, [0.07,0.12]e^{i2[0.24,0.29]\pi}, [0.56,0.66]e^{i2[0.57,0.71]\pi}\Big), \\
\Big(w, [0.37,0.46]e^{i2[0.15,0.25]\pi}, [0.32,0.43]e^{i2[0.53,0.68]\pi}\Big)
\end{array}
\right\}
$$

2. y *modulo* $\breve{R}$ *is given as*

$$
\breve{R}[v] = \left\{ \Big(v, [0.63,0.71]e^{i2[0.03,0.15]\pi}, [0.11,0.23]e^{i2[0.73,0.82]\pi}\Big) \right\}
$$

3.   *z modulo* $\dot{\bar{R}}$ is given as

$$\dot{\bar{R}}[w] = \left\{ \begin{array}{l} \left(u, [0.07, 0.12]e^{i2[0.24, 0.29]\pi}, [0.56, 0.66]e^{i2[0.57, 0.71]\pi}\right), \\ \left(w, [0.37, 0.46]e^{i2[0.15, 0.25]\pi}, [0.32, 0.43]e^{i2[0.53, 0.68]\pi}\right) \end{array} \right\}$$

Definition 15. *For an IVCIFR $\dot{\bar{R}}$ on an IVCIFS $\breve{E}$, then the IVCI-composite-FR $\bar{R} \circ \dot{\bar{R}}$ is defined as: for each $(x, y) \in \dot{\bar{R}}$ and $(y, z) \in \dot{\bar{R}} \Rightarrow (x, z) \in \dot{\bar{R}} \circ \dot{\bar{R}}, \forall x, y, z \in T.$*

Example 13. *For some IVCIFRs*

$$\dot{\bar{R}}_1 = \left\{ \begin{array}{l} \left((u, v), [0.07, 0.12]e^{i2[0.03, 0.15]\pi}, [0.56, 0.66]e^{i2[0.73, 0.82]\pi}\right), \\ \left((v, u), [0.07, 0.12]e^{i2[0.03, 0.15]\pi}, [0.56, 0.66]e^{i2[0.73, 0.82]\pi}\right), \\ \left((w, v), [0.37, 0.46]e^{i2[0.03, 0.15]\pi}, [0.32, 0.43]e^{i2[0.73, 0.82]\pi}\right) \end{array} \right\} \textit{and}$$

$$\dot{\bar{R}}_2 = \left\{ \begin{array}{l} \left((u, w), [0.07, 0.12]e^{i2[0.15, 0.25]\pi}, [0.56, 0.66]e^{i2[0.57, 0.71]\pi}\right), \\ \left((v, u), [0.07, 0.12]e^{i2[0.03, 0.15]\pi}, [0.56, 0.66]e^{i2[0.73, 0.82]\pi}\right), \\ \left((w, v), [0.37, 0.46]e^{i2[0.03, 0.15]\pi}, [0.32, 0.43]e^{i2[0.73, 0.82]\pi}\right) \\ \left((w, w), [0.37, 0.46]e^{i2[0.15, 0.25]\pi}, [0.32, 0.43]e^{i2[0.53, 0.68]\pi}\right) \end{array} \right\} \textit{on an IVCIFS.}$$

$$\breve{E} = \left\{ \begin{array}{l} \left(u, [0.07, 0.12]e^{i2[0.24, 0.29]\pi}, [0.56, 0.66]e^{i2[0.57, 0.71]\pi}\right), \\ \left(v, [0.63, 0.71]e^{i2[0.03, 0.15]\pi}, [0.11, 0.23]e^{i2[0.73, 0.82]\pi}\right), \\ \left(w, [0.37, 0.46]e^{i2[0.15, 0.25]\pi}, [0.32, 0.43]e^{i2[0.53, 0.68]\pi}\right) \end{array} \right\}, \textit{then the IVCI-composite-}$$

FR $\dot{\bar{R}}_1 \circ \dot{\bar{R}}_2$ is given as,

$$\dot{\bar{R}}_1 \circ \dot{\bar{R}}_2 = \left\{ \begin{array}{l} \left((u, u), [0.07, 0.12]e^{i2[0.15, 0.25]\pi}, [0.56, 0.66]e^{i2[0.63, 0.74]\pi}\right), \\ \left((v, w), [0.37, 0.46]e^{i2[0.03, 0.15]\pi}, [0.32, 0.43]e^{i2[0.73, 0.82]\pi}\right), \\ \left((w, u), [0.07, 0.12]e^{i2[0.15, 0.25]\pi}, [0.56, 0.66]e^{i2[0.57, 0.71]\pi}\right), \end{array} \right\}$$

Theorem 1. *An IVCIFR $\dot{\bar{R}}$ is an IVCI-symmetric-FR on an IVCIFS $\breve{E}$ iff $\dot{\bar{R}} = \dot{\bar{R}}^c$.*

Proof. Assume that $\dot{\bar{R}} = \dot{\bar{R}}^c$, then
$(u, v) \in \dot{\bar{R}} \Rightarrow (v, u) \in \dot{\bar{R}}^c \Rightarrow (v, u) \in \dot{\bar{R}}.$
Thus, $\dot{\bar{R}}$ is an IVCI-symmetric-FR on an IVCIFS $\breve{E}$.
Conversely, suppose that $\dot{\bar{R}}$ is an IVCI-symmetric-FR on an IVCIFS $\breve{E}$, then
$(u, v) \in \dot{\bar{R}} \Rightarrow (v, u) \in \dot{\bar{R}}.$
However, $(v, u) \in \dot{\bar{R}}^c \Rightarrow \dot{\bar{R}} = \dot{\bar{R}}^c. \square$

Theorem 2. *An IVCIFR $\dot{\bar{R}}$ is an IVCI-transitive-FR on an IVCIFS $\breve{E}$ if f $\dot{\bar{R}} \circ \dot{\bar{R}} \subseteq \dot{\bar{R}}$.*

Proof. Assume that $\dot{\bar{R}}$ is an IVCI-transitive-FR on an IVCIFS $\breve{E}$.
Let $(u, w) \in \dot{\bar{R}} \circ \dot{\bar{R}}$,
Then, by the definition of IVCI-transitivity-FR,
$(u, v) \in \dot{\bar{R}}$ and $(v, w) \in \dot{\bar{R}} \Rightarrow (u, w) \in \dot{\bar{R}} \Rightarrow \dot{\bar{R}} \circ \dot{\bar{R}} \subseteq \dot{\bar{R}}.$
Conversely assume that $\dot{\bar{R}} \circ \dot{\bar{R}} \subseteq \dot{\bar{R}}$, then
for $(u, v) \subset \dot{\bar{R}}$ and $(v, w) \in \dot{\bar{R}} \Rightarrow (u, w) \in \dot{\bar{R}} \circ \dot{\bar{R}} \subseteq \dot{\bar{R}} \Rightarrow (u, w) \in \dot{\bar{R}}.$
Thus, $\dot{\bar{R}}$ is an IVCI-transitive-FR on $\breve{E}$. $\square$

Theorem 3. *Suppose $\dot{\bar{R}}$ is an IVCI-equivalence-FR on an IVCIFS $\breve{E}$, then $\dot{\bar{R}} \circ \dot{\bar{R}} = \dot{\bar{R}}$.*

Proof. Assume that $(u, v) \in \dot{\bar{R}}$,

then by the definition of IVCI-symmetric-FR,

$(v, u) \in \dot{\bar{R}}$.

Now, by using the definition of IVCI-transitive-FR,

$(u, u) \in \dot{\bar{R}}$.

However, by the definition of IVCI-composite-FR,

$(u, u) \in \dot{\bar{R}} \circ \dot{\bar{R}}$.

Thus,

$$\dot{\bar{R}} \subseteq \dot{\bar{R}} \circ \dot{\bar{R}} \tag{1}$$

Conversely, assume that $(u, v) \in \dot{\bar{R}} \circ \dot{\bar{R}}$, then $\exists\, w \in S \ni (u, w) \in \dot{\bar{R}}$ and $(w, v) \in \dot{\bar{R}}$.

However, it is given that $\dot{\bar{R}}$ is an IVCI-equivalence-FR on $\breve{E}$, so $\dot{\bar{R}}$ is also an IVCI-transitive-FR. Therefore,

$$(u, v) \in \dot{\bar{R}} \Rightarrow \dot{\bar{R}} \circ \dot{\bar{R}} \subseteq \dot{\bar{R}} \tag{2}$$

Thus, by (1) and (2),

$\dot{\bar{R}} \circ \dot{\bar{R}} = \dot{\bar{R}}$. $\square$

Theorem 4. *Suppose $\dot{\bar{R}}$ is an IVCI-partial order-FR on an IVCIFS $\breve{E}$, then the converse relation $\dot{\bar{R}}^c$ of $\dot{\bar{R}}$ is also an IVCI-partial order-FR on $\breve{E}$.*

Proof. In order to prove the assertion, it is sufficient to show that the converse of a complex intuitionistic partial order fuzzy relation R^c satisfies the three properties of a complex intuitionistic partial order fuzzy relation.

By using the properties of IVCI-partial order-FR $\dot{\bar{R}}$, we prove the statement.

i.   It is given that $\dot{\bar{R}}$ is an IVCI-reflexive-FR. Therefore, for any $u \in S$, $(u, u) \in \dot{\bar{R}} \Rightarrow (u, u) \in \dot{\bar{R}}^c$. Thus, $\dot{\bar{R}}^c$ is an IVCI-reflexive-FR.

ii.   Assume that $(u, u) \in \dot{\bar{R}}^c$ and $(v, u) \in \dot{\bar{R}}^c$, then, $(u, v) \in \dot{\bar{R}}$ and $(v, u) \in \dot{\bar{R}}$. However, $\dot{\bar{R}}$ is an IVCI-antisymmetric-FR. Therefore, $(u, v) = (v, u)$. Thus, $\dot{\bar{R}}$ is also an IVCI-antisymmetric-FR.

iii.   Suppose that $(u, v) \in \dot{\bar{R}}^c$ and $(v, w) \in \dot{\bar{R}}^c$, then, $(w, v) \in \dot{\bar{R}}$ and $(v, u) \in \dot{\bar{R}}$. However, it is given that $\dot{\bar{R}}$ is an IVCI-transitive-FR. Therefore, $(w, u) \in \dot{\bar{R}} \Rightarrow (u, w) \in \dot{\bar{R}}^c$. Thus, $\dot{\bar{R}}^c$ is also an IVCI-transitive-FR.

From i, ii and iii, the converse relation $\dot{\bar{R}}^c$ of an IVCI-partial order-FR $\dot{\bar{R}}$ is proved to be an IVCI-partial order-FR too. $\square$

Theorem 5. *Suppose $\dot{\bar{R}}$ is an IVCI-equivalence-FR on an IVCIFS $\breve{E}$, then $(x, y) \in \dot{\bar{R}}$, if $\dot{\bar{R}}[u] = \dot{\bar{R}}[v]$.*

Proof. Assume that $(u, v) \in \dot{\bar{R}}$ and $w \in \dot{\bar{R}}[u]\, \dot{\bar{R}}(w, u) \in \dot{\bar{R}}$.

Now, by using the fact that an IVCI-equivalence-FR is also an IVCI-transitive-FR, so

$(w, v) \in \dot{\bar{R}} \Rightarrow w \in \dot{\bar{R}}[v]$.

Thus,

$$\dot{\bar{R}}[u] \subseteq R[v] \tag{3}$$

As $(u, v) \in \dot{\bar{R}}$, by using the fact that an IVCI-equivalence-FR is also an IVCI-symmetric-FR, so

$(v, u) \in \dot{\bar{R}}$.

Additionally, assume that $w \in \dot{\bar{R}}[v] \Rightarrow (w, v) \in \dot{\bar{R}}$.

Now, again by using the fact that an IVCI-equivalence-FR is also an IVCI-transitive-FR, so

$(w, u) \in \dot{\bar{R}} \Rightarrow w \in \dot{\bar{R}}[u]$.

Thus,

$$\dot{\bar{R}}[v] \subseteq \dot{\bar{R}}[u] \tag{4}$$

Therefore, (3) and (4) infer that $\dot{\bar{R}}[v] = \dot{\bar{R}}[u]$.

Conversely, assume that $\ddot{R}[v] = \ddot{R}[u]$, $w \in \ddot{R}[u]$ and $w \in \ddot{R}[v] \Rightarrow (w, v) \in \ddot{R}$ and $(w, u) \in \ddot{R}$.

Again, by using the fact that an IVCI-equivalence-FR is also an IVCI-symmetric-FR, so $(w, u) \in \ddot{R} \Rightarrow (u, w) \in \ddot{R}$.

Now, by the definition of IVCI-transitive-FR,

$(u, w) \in \ddot{R}$ and $(w, v) \in \ddot{R} \Rightarrow (u, v) \in \ddot{R}$,

which completes the proof. $\square$

4. Hasse Diagram for IVCI-Partial Order-FRs

In this section, the Hasse diagram for IVCI-partial order-FR is defined. Moreover, some important ideas related to Hasse diagrams, such as maximum element, minimum element, maximal element, minimal element, supremum, infimum, upper and lower bounds, are defined.

Definition 16. *The graphical representation of an IVCI-partial order-FS is called the Hasse diagram. The diagram is made up of dots and lines, which are called the vertices and edges, respectively. Each vertex represents an element of an IVCI-partial order-FS, while each edge represents a relationship between some pair of these elements. There are certain rules in the construction of a Hasse diagram, that are discussed below:*

i. *The elements are arranged in an ascending order to distinguish the lower ranks and higher ranks. In an ordered pair of an IVCI-partial order-FR, the preceding element is considered to be smaller than the element appearing second in the pair. For example, in the ordered pair (t, x), the element t is smaller than the element x. t will appear higher than x in the diagram.*

ii. *There are no self-loops for IVCI-reflexive-FRs. In a Hasse diagram, the self-relation is not represented by any edge, it is just assumed to be there.*

iii. *There are no directional edges. The directional edges indicate the order of the element in the ordered pair. For example, in the ordered pair (t, x), the relation in normal diagrams, would be represented by a directional edge with an arrow head pointing towards element x. Thus, ranking x higher than t. But in a Hasse diagram, the stepwise ascending arrangement of the elements automatically distinguishes the higher ranked and lower ranked elements.*

iv. There are no redundant edges, for instance the edge for an IVCI-transitive-FRs and IVCI-reflexive-FRs. Consider (t, x) and then IVCI-transitive-FR $\ddot{R}(t, k)$. In normal diagrams, there would be three edges representing the above three relations. However, in a Hasse diagram, only two edges are constructed, i.e., from *t to x and another from to k. The indirect link between t and k via x is represented by the two edges. The IVCI-transitive-FR is intuitively understood.*

Example 14. *Consider an IVCIFS*

$$\breve{E} = \left\{ \begin{array}{l} \left(u, [0.07, 0.12]e^{i2[0.24, 0.29]\pi}, [0.56, 0.66]e^{i2[0.57, 0.71]\pi}\right), \\ \left(v, [0.63, 0.71]e^{i2[0.03, 0.15]\pi}, [0.11, 0.23]e^{i2[0.73, 0.82]\pi}\right), \\ \left(w, [0.37, 0.46]e^{i2[0.15, 0.25]\pi}, [0.32, 0.43]e^{i2[0.53, 0.68]\pi}\right), \\ \left(x, [0.26, 0.34]e^{i2[0.15, 0.25]\pi}, [0.38, 0.49]e^{i2[0.63, 0.74]\pi}\right), \\ \left(y, [0.52, 0.58]e^{i2[0.30, 0.42]\pi}, [0.18, 0.37]e^{i2[0.06, 0.13]\pi}\right), \\ \left(z, [0.28, 0.36]e^{i2[0.43, 0.53]\pi}, [0.48, 0.61]e^{i2[0.32, 0.42]\pi}\right), \end{array} \right\} \quad in \ a \ universe \ S.$$

The Cartesian product $\breve{E} \times \breve{E}$ is

$$\breve{E}_s \times \breve{E}_s = \begin{bmatrix}
\left((u,u), [0.07,0.12]e^{i2[0.15,0.25]\pi}, [0.56,0.66]e^{i2[0.63,0.74]\pi}\right), \\
\left((u,v), [0.07,0.12]e^{i2[0.03,0.15]\pi}, [0.56,0.66]e^{i2[0.73,0.82]\pi}\right), \\
\left((u,w), [0.07,0.12]e^{i2[0.15,0.25]\pi}, [0.56,0.66]e^{i2[0.57,0.71]\pi}\right) \\
\left((u,x), [0.07,0.12]e^{i2[0.15,0.25]\pi}, [0.56,0.66]e^{i2[0.63,0.74]\pi}\right), \\
\left((u,y), [0.07,0.12]e^{i2[0.24,0.29]\pi}, [0.56,0.66]e^{i2[0.57,0.71]\pi}\right), \\
\left((u,z), [0.07,0.12]e^{i2[0.24,0.29]\pi}, [0.56,0.66]e^{i2[0.57,0.71]\pi}\right), \\
\left((v,u), [0.07,0.12]e^{i2[0.03,0.15]\pi}, [0.56,0.66]e^{i2[0.73,0.82]\pi}\right), \\
\left((v,v), [0.63,0.71]e^{i2[0.03,0.15]\pi}, [0.11,0.23]e^{i2[0.73,0.82]\pi}\right), \\
\left((v,w), [0.37,0.46]e^{i2[0.03,0.15]\pi}, [0.32,0.43]e^{i2[0.73,0.82]\pi}\right), \\
\left((v,x), [0.26,0.34]e^{i2[0.03,0.15]\pi}, [0.38,0.49]e^{i2[0.73,0.82]\pi}\right), \\
\left((v,y), [0.52,0.58]e^{i2[0.03,0.15]\pi}, [0.18,0.37]e^{i2[0.73,0.82]\pi}\right), \\
\left((v,z), [0.28,0.36]e^{i2[0.03,0.15]\pi}, [0.48,0.61]e^{i2[0.73,0.82]\pi}\right), \\
\left((w,u), [0.07,0.12]e^{i2[0.15,0.25]\pi}, [0.56,0.66]e^{i2[0.57,0.71]\pi}\right), \\
\left((w,v), [0.37,0.46]e^{i2[0.03,0.15]\pi}, [0.32,0.43]e^{i2[0.73,0.82]\pi}\right), \\
\left((w,w), [0.37,0.46]e^{i2[0.15,0.25]\pi}, [0.32,0.43]e^{i2[0.53,0.68]\pi}\right), \\
\left((w,x), [0.26,0.34]e^{i2[0.15,0.25]\pi}, [0.38,0.49]e^{i2[0.63,0.74]\pi}\right), \\
\left((w,y), [0.37,0.46]e^{i2[0.15,0.25]\pi}, [0.32,0.43]e^{i2[0.53,0.68]\pi}\right), \\
\left((w,z), [0.28,0.36]e^{i2[0.15,0.25]\pi}, [0.48,0.61]e^{i2[0.53,0.68]\pi}\right), \\
\left((x,u), [0.07,0.12]e^{i2[0.15,0.25]\pi}, [0.56,0.66]e^{i2[0.63,0.74]\pi}\right), \\
\left((x,v), [0.26,0.34]e^{i2[0.03,0.15]\pi}, [0.38,0.49]e^{i2[0.73,0.82]\pi}\right), \\
\left((x,w), [0.26,0.34]e^{i2[0.15,0.25]\pi}, [0.38,0.49]e^{i2[0.63,0.74]\pi}\right), \\
\left((x,x), [0.26,0.34]e^{i2[0.15,0.25]\pi}, [0.38,0.49]e^{i2[0.63,0.74]\pi}\right), \\
\left((x,y), [0.26,0.34]e^{i2[0.15,0.25]\pi}, [0.38,0.49]e^{i2[0.63,0.74]\pi}\right), \\
\left((x,z), [0.26,0.34]e^{i2[0.15,0.25]\pi}, [0.38,0.49]e^{i2[0.63,0.74]\pi}\right), \\
\left((y,u), [0.07,0.12]e^{i2[0.24,0.29]\pi}, [0.56,0.66]e^{i2[0.57,0.71]\pi}\right), \\
\left((y,v), [0.52,0.58]e^{i2[0.03,0.15]\pi}, [0.18,0.37]e^{i2[0.73,0.82]\pi}\right), \\
\left((y,w), [0.37,0.46]e^{i2[0.15,0.25]\pi}, [0.32,0.43]e^{i2[0.53,0.68]\pi}\right), \\
\left((y,x), [0.26,0.34]e^{i2[0.15,0.25]\pi}, [0.38,0.49]e^{i2[0.63,0.74]\pi}\right), \\
\left((y,y), [0.52,0.58]e^{i2[0.30,0.42]\pi}, [0.18,0.37]e^{i2[0.06,0.13]\pi}\right), \\
\left((y,z), [0.26,0.34]e^{i2[0.43,0.53]\pi}, [0.48,0.61]e^{i2[0.63,0.74]\pi}\right), \\
\left((z,u), [0.07,0.12]e^{i2[0.24,0.29]\pi}, [0.56,0.66]e^{i2[0.57,0.71]\pi}\right), \\
\left((z,v), [0.28,0.36]e^{i2[0.03,0.15]\pi}, [0.48,0.61]e^{i2[0.73,0.82]\pi}\right), \\
\left((z,w), [0.28,0.36]e^{i2[0.15,0.25]\pi}, [0.48,0.61]e^{i2[0.53,0.68]\pi}\right), \\
\left((z,x), [0.26,0.34]e^{i2[0.15,0.25]\pi}, [0.38,0.49]e^{i2[0.63,0.74]\pi}\right), \\
\left((z,y), [0.37,0.46]e^{i2[0.15,0.25]\pi}, [0.32,0.43]e^{i2[0.53,0.68]\pi}\right), \\
\left((z,z), [0.28,0.36]e^{i2[0.43,0.53]\pi}, [0.48,0.61]e^{i2[0.32,0.42]\pi}\right)
\end{bmatrix}$$

An IVCI-partial order-FR $\breve{R}$ is

$$\breve{R} = \left\{ \begin{array}{l}
\left((u,v), [0.07,0.12]e^{i2[0.15,0.25]\pi}, [0.56,0.66]e^{i2[0.63,0.74]\pi}\right), \\
\left((u,y), [0.07,0.12]e^{i2[0.24,0.29]\pi}, [0.56,0.66]e^{i2[0.57,0.71]\pi}\right), \\
\left((u,z), [0.07,0.12]e^{i2[0.24,0.29]\pi}, [0.56,0.66]e^{i2[0.57,0.71]\pi}\right), \\
\left((v,v), [0.63,0.71]e^{i2[0.03,0.15]\pi}, [0.11,0.23]e^{i2[0.73,0.82]\pi}\right), \\
\left((v,z), [0.28,0.36]e^{i2[0.03,0.15]\pi}, [0.48,0.61]e^{i2[0.73,0.82]\pi}\right), \\
\left((w,w), [0.37,0.46]e^{i2[0.15,0.25]\pi}, [0.32,0.43]e^{i2[0.53,0.68]\pi}\right), \\
\left((w,y), [0.37,0.46]e^{i2[0.15,0.25]\pi}, [0.32,0.43]e^{i2[0.53,0.68]\pi}\right), \\
\left((w,z), [0.28,0.36]e^{i2[0.15,0.25]\pi}, [0.48,0.61]e^{i2[0.53,0.68]\pi}\right), \\
\left((x,u), [0.07,0.12]e^{i2[0.15,0.25]\pi}, [0.56,0.66]e^{i2[0.63,0.74]\pi}\right), \\
\left((x,v), [0.26,0.34]e^{i2[0.03,0.15]\pi}, [0.38,0.49]e^{i2[0.73,0.82]\pi}\right), \\
\left((x,w), [0.26,0.34]e^{i2[0.15,0.25]\pi}, [0.38,0.49]e^{i2[0.63,0.74]\pi}\right), \\
\left((x,x), [0.26,0.34]e^{i2[0.15,0.25]\pi}, [0.38,0.49]e^{i2[0.63,0.74]\pi}\right), \\
\left((x,y), [0.26,0.34]e^{i2[0.15,0.25]\pi}, [0.38,0.49]e^{i2[0.63,0.74]\pi}\right), \\
\left((x,z), [0.26,0.34]e^{i2[0.15,0.25]\pi}, [0.38,0.49]e^{i2[0.63,0.74]\pi}\right), \\
\left((y,y), [0.52,0.58]e^{i2[0.30,0.42]\pi}, [0.18,0.37]e^{i2[0.06,0.13]\pi}\right), \\
\left((y,z), [0.26,0.34]e^{i2[0.43,0.53]\pi}, [0.48,0.61]e^{i2[0.63,0.74]\pi}\right), \\
\left((z,z), [0.28,0.36]e^{i2[0.43,0.53]\pi}, [0.48,0.61]e^{i2[0.32,0.42]\pi}\right)
\end{array} \right\}$$

Figure 2 depicts the Hasse diagram of the IVCI-partial order-FR $\breve{R}$.

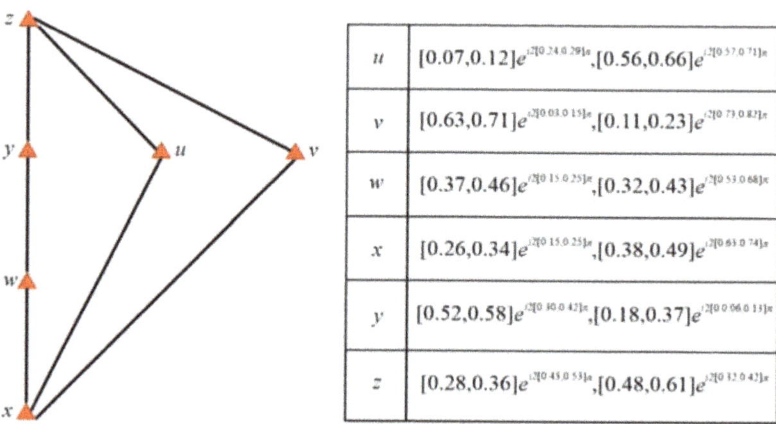

Figure 2. Hasse diagram for an IVCI-partial order-FR $\breve{R}$

Definition 17. *Suppose that an IVCI-partial order-FR is represented by a Hasse diagram, then an element is known as:*

1. *The maximal element if it appears at the top of the diagram.*
2. *The minimal if it appears at the bottom of the diagram.*
3. *The maximum or greatest element if every element related to it is smaller than it.*
4. *The minimum or least element if every element related to it is greater than it.*

Definition 18. *Suppose $\dot{F}$ be the subset of an IVCI-partial order-FS $\check{E}$, then an element $x \in \dot{R} \subseteq \check{E} \times \check{E}$ is known as the:*

1. *Upper bound of $\dot{F}$ if $(v, u) \in \bar{R}, \forall v \in \dot{F}$.*
2. *Lower bound of $\dot{F}$ if $(u, v) \in \bar{R}, \forall v \in \dot{F}$.*
3. *Supremum of $\dot{F}$ if it is the least upper bound of $\dot{F}$.*
4. *Infimum of $\dot{F}$ if it is the greatest upper bound of $\dot{F}$.*

Example 15. *Assume that $\check{E} = \{f, g, h, k, l, o, p, q, r, s, t, u\}$ is a complex intuitionistic partial order fuzzy set $\check{E}$. For ease, we close your eyes to the degree of membership and degree of non-membership. Figure 3 is the illustration of the Hasse diagram of set $\check{E}$.*

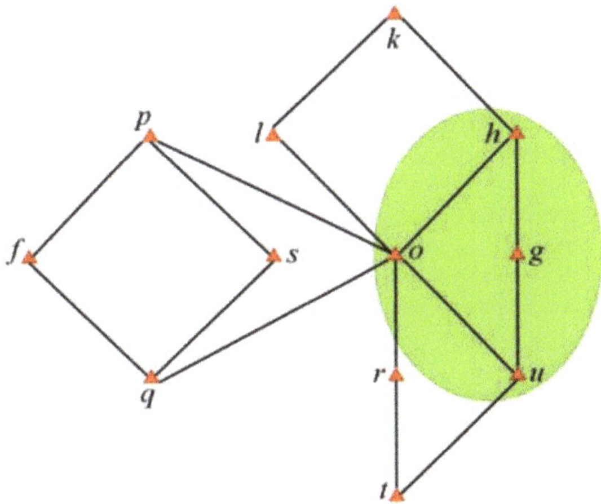

Figure 3. Hasse Diagram for set $\check{E}$.

In the above diagram,

a. Maximal element is k.
b. Minimal element is t.
c. Maximum elements are k and p.
d. Minimum elements are q and t.

Now considering a subset $\dot{F} = \{g, h, o, u\}$ of set $\check{E}$ whose elements are enclosed in the green oval.

e. Upper bounds of $\dot{F}$ are h and k.
f. Lower bounds of $\dot{F}$ are u and t.
g. Supremum of $\dot{F}$ is h.
h. Infimum of $\dot{F}$ is u.

5. Application

As the name suggests, in this section the applications of the proposed concepts are presented. We applied the introduced relations and their different types in the fields of computer technology, more precisely: cybersecurity and cybercrime in the industrial control system (ICS).

5.1. Security Measures

Over the past few decades, the automation of industrial systems has been steadily gaining momentum. Business demands continuous improvements in the efficiency of the production process, thus, the depth of IT penetration and system connectivity grows continuously; industrial facilities are connected to corporate networks and are frequently managed remotely over the network. However, along with their benefits, these new technologies have brought new threats into the world of industrial automation, and these new threats came as a surprise. The Industrial Control Systems (ICS) in place today were designed to operate for decades, and many of them were developed without any serious regard to IT security.

The stable operation of today's industrial networks can be disrupted, not only by a failure at a production unit or an operator's error but also by a software error, an accidental infection of workstations with malware or a deliberate cybercriminal attack.

Some security measures are given below. Figure 4 portrays the flowchart for the process being followed in the application.

I. **Default-deny as a standard policy:** In default-deny mode, the ICS works in a protected environment that only allows programs to run that are required for the technological process to function. All unknown and unwanted applications, including malicious programs, are blocked. Thus, a secure running environment is created with minimum load on system resources.

II. **Proactive protection** against unknown malicious programs and automatic protection against exploits. The technology scans executable programs, assessing the security of each application by monitoring its activities when in operation.

III. **Device Control technology** helps to manage removable devices (USB storages, GPRS modems, smartphones, USB network cards) and creates limited lists of permitted devices and the users who can access them.

IV. **All-in-one IT security console** helps to monitor and control all solutions to ensure IT security. With the single management console, admins can install, configure and manage security, and access reports.

V. **Integration with SIEM** (using special connectors) allows admins to export information about security incidents at protected nodes of the technological network into the corporate SIEM system.

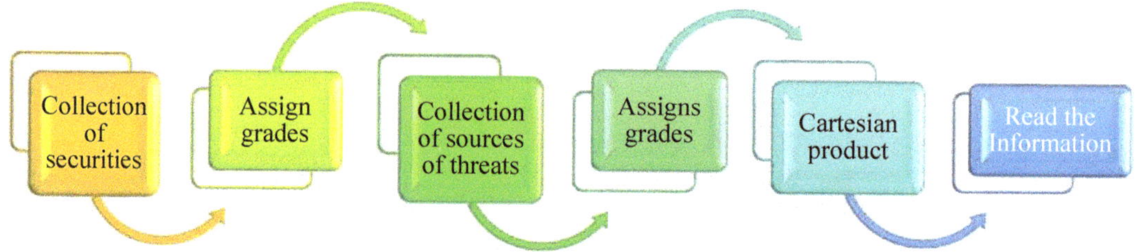

Figure 4. Flowchart for the process being followed.

The above security measures are abbreviated and assigned degrees of membership and degrees of non-membership values. Table 1 contains the details of their abbreviations and values of degrees.

Table 1. Details of security measures.

Security Measures	Abbreviations	Membership	Non-Membership
Default-deny as a standard policy	DD	$[0.550, 0.600]e^{i16[0.500,0.750]\pi}$	$[0.300, 0.400]e^{i16[0.125,0.313]\pi}$
Proactive protection	PP	$[0.475, 0.525]e^{i16[0.375,0.438]\pi}$	$[0.250, 0.375]e^{i16[0.125,0.250]\pi}$
Device Control technology	DCT	$[0.375, 0.525]e^{i16[0.438,0.563]\pi}$	$[0.200, 0.350]e^{i16[0.188,0.250]\pi}$
All-in-one IT Security Console	$ITSC$	$[0.400, 0.500]e^{i16[0.313,0.438]\pi}$	$[0.250, 0.350]e^{i16[0.250,0.313]\pi}$
Integration with SIEM	$SIEM$	$[0.300, 0.375]e^{i16[0.563,0.613]\pi}$	$[0.275, 0.325]e^{i16[0.188,0.313]\pi}$

5.2. Sources of Code Penetration

In a computer system there are certain weaknesses. Cybercriminals are always looking for opportunities. They inject their codes using media that are sources of code penetration. Following are the eight sources of code penetration:

I. Mobile Devices
II. Via USB Ports
III. Via Remote Access
IV. Wi-Fi
V. HMI Interface
VI. Internet Connections
VII. Outside Contractors
VIII. Via Corporate Networks

Table 2 contains the details of the above sources such as the abbreviations used in calculations and the degree of membership and degree of non-membership.

Table 2. Details of code penetration sources.

Source	Abbreviation	Membership	Non-Membership
Mobile Devices	MD	$[0.275, 0.375]e^{i16[0.375,0.500]\pi}$	$[0.350, 0.400]e^{i16[0.250,0.313]\pi}$
USB Ports	USB	$[0.375, 0.450]e^{i16[0.250,0.375]\pi}$	$[0.275, 0.325]e^{i16[0.125,0.250]\pi}$
Remote Access	RA	$[0.350, 0.400]e^{i16[0.375,0.438]\pi}$	$[0.325, 0.350]e^{i16[0.188,0.313]\pi}$
Wi-Fi	$WiFi$	$[0.450, 0.550]e^{i16[0.313,0.500]\pi}$	$[0.325, 0.400]e^{i16[0.188,0.250]\pi}$
HMI Interface	HMI	$[0.300, 0.375]e^{i16[0.438,0.563]\pi}$	$[0.200, 0.300]e^{i16[0.063,0.188]\pi}$
Internet Connections	IC	$[0.475, 0.575]e^{i16[0.500,0.613]\pi}$	$[0.250, 0.350]e^{i16[0.125,0.313]\pi}$
Outside Contractors	OC	$[0.500, 0.550]e^{i16[0.375,0.500]\pi}$	$[0.225, 0.300]e^{i16[0.063,0.250]\pi}$
Corporate Networks	CN	$[0.200, 0.475]e^{i16[0.313,0.613]\pi}$	$[0.250, 0.375]e^{i16[0.125,0.188]\pi}$

5.3. Calculations

Next, the relationships are analyzed, i.e., the effectiveness and ineffectiveness of each cybersecurity method against every cybercrime. We carry out the following mathematics.

We present the following two IVCIFSs $\breve{E}$ and $\dot{F}$ representing the set of securities and the set of sources of threats, respectively.

$$\breve{E} = \left\{ \begin{array}{l} \left(DD, [0.550, 0.600]e^{i16[0.500,0.750]\pi}, [0.300, 0.400]e^{i16[0.125,0.313]\pi} \right), \\ \left(PP, [0.475, 0.525]e^{i16[0.375,0.438]\pi}, [0.250, 0.375]e^{i16[0.125,0.250]\pi} \right), \\ \left(DCT, [0.375, 0.525]e^{i16[0.438,0.563]\pi}, [0.200, 0.350]e^{i16[0.188,0.250]\pi} \right), \\ \left(ITSC, [0.400, 0.500]e^{i16[0.313,0.438]\pi}, [0.250, 0.350]e^{i16[0.250,0.313]\pi} \right), \\ \left(SIEM, [0.300, 0.375]e^{i16[0.563,0.613]\pi}, [0.275, 0.325]e^{i16[0.188,0.313]\pi} \right) \end{array} \right\}$$

$$\dot{F} = \left\{ \begin{array}{l} \left(MD, [0.275, 0.375]e^{i16[0.375,0.500]\pi}, [0.350, 0.400]e^{i16[0.250,0.313]\pi} \right), \\ \left(USB, [0.375, 0.450]e^{i16[0.250,0.375]\pi}, [0.275, 0.325]e^{i16[0.125,0.250]\pi} \right), \\ \left(RA, [0.350, 0.400]e^{i16[0.375,0.438]\pi}, [0.325, 0.350]e^{i16[0.188,0.313]\pi} \right), \\ \left(WiFi, [0.450, 0.550]e^{i16[0.313,0.500]\pi}, [0.325, 0.400]e^{i16[0.188,0.250]\pi} \right), \\ \left(HMI, [0.300, 0.375]e^{i16[0.438,0.563]\pi}, [0.200, 0.300]e^{i16[0.063,0.188]\pi} \right), \\ \left(IC, [0.475, 0.575]e^{i16[0.500,0.613]\pi}, [0.250, 0.350]e^{i16[0.125,0.313]\pi} \right), \\ \left(OC, [0.500, 0.550]e^{i16[0.375,0.500]\pi}, [0.225, 0.300]e^{i16[0.063,0.250]\pi} \right), \\ \left(CN, [0.200, 0.475]e^{i16[0.313,0.613]\pi}, [0.250, 0.375]e^{i16[0.125,0.188]\pi} \right) \end{array} \right\}$$

To find out the efficacy of certain security measures against particular sources of code penetration, we use the Cartesian product. Thus, finding the Cartesian product between the IVCIFSs $\breve{E}$ and $\dot{F}$

$$\breve{R} = \breve{E} \times \dot{F}$$

$$
= \begin{bmatrix}
\left((DD,MD), [0.275,0.375]e^{i16[0.375,0.500]\pi}, [0.350,0.400]e^{i16[0.250,0313]\pi} \right), \\
\left((DD,USB), [0.375,0.450]e^{i16[0.2500.375]\pi}, [0.300,0.400]e^{i16[0.125,031:3]\pi} \right), \\
\left((DD,RA), [0.350,0.400]e^{i16[0.375,0.438]\pi}, [0.325,0.400]e^{i16[0.188,0.313]\pi} \right), \\
\left((DD,WiFi), [0.450,0.550]e^{i16[0.313,0.500]\pi}, [0.325,0.400]e^{i16[0.188,0.313]\pi} \right), \\
\left((DD,HMI), [0.300,0.375]e^{i16[0.438,0.563]\pi}, [0.300,0.400]e^{i16[0.125,0.313]\pi} \right), \\
\left((DD,IC), [0.475,0.575]e^{i16[0.500,0.613]\pi}, [0.350,0.400]e^{i16[0.125,0.313]\pi} \right), \\
\left((DD,OC), [0.500,0.550]e^{i16[0.375,0.500]\pi}, [0.300,0.400]e^{i16[0.125,0.313]\pi} \right), \\
\left((DD,CN), [0.200,0.475]e^{i16[0.313,0.613]\pi}, [0.300,0.400]e^{i16[0.125,0.313]\pi} \right), \\
\left((PP,MD), [0.275,0.375]e^{i16[0.375,0.438]\pi}, [0.350,0.400]e^{i16[0.250,0.313]\pi} \right), \\
\left((PP,USB), [0.375,0.450]e^{i16[0.250,0.375]\pi}, [0.275,0.375]e^{i16[0.125,0.250]\pi} \right), \\
\left((PP,RA), [0.350,0.400]e^{i16[0.375,0.438]\pi}, [0.325,0.375]e^{i16[0.188,0.313]\pi} \right), \\
\left((PP,WiFi), [0.450,0.525]e^{i16[0.313,0.438]\pi}, [0.325,0.400]e^{i16[0.188,0.250]\pi} \right), \\
\left((PP,HMI), [0.300,0.375]e^{i16[0.375,0.438]\pi}, [0.250,0.375]e^{i16[0.125,0.250]\pi} \right), \\
\left((PP,IC), [0.475,0.525]e^{i16[0.375,0.438]\pi}, [0.250,0.375]e^{i16[0.125,0.313]\pi} \right), \\
\left((PP,OC), [0.475,0.525]e^{i16[0.375,0.438]\pi}, [0.250,0.375]e^{i16[0.125,0250]\pi} \right), \\
\left((PP,CN), [0.200,0.475]e^{i16[0.313,0.438]\pi}, [0.250,0.375]e^{i16[0.125,0.188]\pi} \right), \\
\left((DCT,MD), [0.275,0.375]e^{i16[0.375,0.500]\pi}, [0.350,0.400]e^{i16[0.250,0.313]\pi} \right), \\
\left((DCT,USB), [0.375,0.450]e^{i16[0.250,0.375]\pi}, [0.275,0.350]e^{i16[0.188,0.250]\pi} \right), \\
\left((DCT,RA), [0.350,0.400]e^{i16[0.375,0.438]\pi}, [0.325,0.350]e^{i16[0.188,0.313]\pi} \right), \\
\left((DCT,WiFi), [0.375,0.525]e^{i16[0.313,0.500]\pi}, [0.325,0.400]e^{i16[0.188,0.250]\pi} \right), \\
\left((DCT,HMI), [0.300,0.375]e^{i16[0.438,0.563]\pi}, [0.200,0.350]e^{i16[0.188,0.250]\pi} \right), \\
\left((DCT,IC), [0.375,0.525]e^{i16[0.438,0.563]\pi}, [0.250,0.350]e^{i16[0.125,0.313]\pi} \right), \\
\left((DCT,OC), [0.375,0.525]e^{i16[0.375,0.500]\pi}, [0.225,0.350]e^{i16[0.188,0.250]\pi} \right), \\
\left((DCT,CN), [0.200,0.475]e^{i16[0.313,0.563]\pi}, [0.250,0.375]e^{i16[0.188,0.250]\pi} \right), \\
\left((ITSC,MD), [0.275,0.375]e^{i16[0.313,0.438]\pi}, [0.350,0.400]e^{i16[0.250,0.313]\pi} \right), \\
\left((ITSC,USB), [0.375,0.450]e^{i16[0.250,0.375]\pi}, [0.275,0.350]e^{i16[0.250,0.313]\pi} \right), \\
\left((ITSC,RA), [0.350,0.400]e^{i16[0.313,0.438]\pi}, [0.325,0.350]e^{i16[0.250,0.313]\pi} \right), \\
\left((ITSC,WiFi), [0.400,0.500]e^{i16[0.313,0.438]\pi}, [0.325,0.400]e^{i16[0.250,0.313]\pi} \right), \\
\left((ITSC,HMI), [0.300,0.375]e^{i16[0.313,0.438]\pi}, [0.250,0.350]e^{i16[0.250,0.313]\pi} \right), \\
\left((ITSC,IC), [0.400,0.500]e^{i16[0.313,0.438]\pi}, [0.250,0.350]e^{i16[0.250,0.313]\pi} \right), \\
\left((ITSC,OC), [0.400,0.500]e^{i16[0.313,0.438]\pi}, [0.250,0.350]e^{i16[0.250,0.313]\pi} \right), \\
\left((ITSC,CN), [0.200,0.475]e^{i16[0.313,0.438]\pi}, [0.250,0.375]e^{i16[0.250,0.313]\pi} \right), \\
\left((SIEM,MD), [0.275,0.375]e^{i16[0.375,0.500]\pi}, [0.350,0.400]e^{i16[0.250,0.313]\pi} \right), \\
\left((SIEM,USB), [0.300,0.375]e^{i16[0.250,0.375]\pi}, [0.275,0.325]e^{i16[0.188,0.313]\pi} \right), \\
\left((SIEM,RA), [0.300,0.375]e^{i16[0.375,0.438]\pi}, [0.325,0.350]e^{i16[0.188,0.313]\pi} \right), \\
\left((SIEM,WiFi), [0.300,0.375]e^{i16[0.313,0.500]\pi}, [0.325,0.400]e^{i16[0.188,0.313]\pi} \right), \\
\left((SIEM,HMI), [0.300,0.375]e^{i16[0.438,0.563]\pi}, [0.275,0.3250]e^{i16[0.188,0.313]\pi} \right), \\
\left((SIEM,IC), [0.300,0.375]e^{i16[0.500,0.613]\pi}, [0.275,0.350]e^{i16[0.188,0.313]\pi} \right), \\
\left((SIEM,OC), [0.300,0.375]e^{i16[0.375,0.500]\pi}, [0.275,0.325]e^{i16[0.188,0.313]\pi} \right), \\
\left((SIEM,CN), [0.200,0.375]e^{i16[0.313,0.613]\pi}, [0.275,0.375]e^{i16[0.188,0.313]\pi} \right)
\end{bmatrix}
$$

Every member of $\breve{E} \times \dot{F}$ is an order pair, which characterizes the connection between that pair, i.e., the influences and impacts of the first parameter on the second, in an ordered pair. The degrees of membership indicate the efficacy of a security measure to overcome a particular source of vulnerability with respect to time. In contrast, the degrees of non-membership give the inefficiency or ineptness of a certain security measure against a specific source of threat injection. For instance, the ordered pair $\left((DD, IC), [0.475, 0.575]e^{i16[0.500,0.613]\pi}, [0.350, 0.400]e^{i16[0.125,0.313]\pi}\right)$ expresses that the default-deny mode can successfully tackle the risks incoming through internet connections. Further, the numbers explain that the level of inefficiency is low. More specifically, the degree values are translated as: the grade of security that a default-deny mode provides against the vulnerabilities injected through internet connections is 47.5% to 57.5%, with respect to 8 to 10 time units and the chances of a cyberattack via internet connections bypassing the default-deny mode is 35% to 40%, with respect to 2 to 5 time units. As far as the security is concerned, the longer duration of time in the degree of membership is considered better, while the smaller time frame in the degree of non-membership is better.

5.4. Cyber-Security Techniques and Practices

An industry's digital system faces threats of all shapes and sizes. Thus, they should be ready to defend, identify and respond to a full range of attacks. The security strategy needs to be able to address the various methods these cybercriminals might employ. The process of the method is depicted in Figure 5.

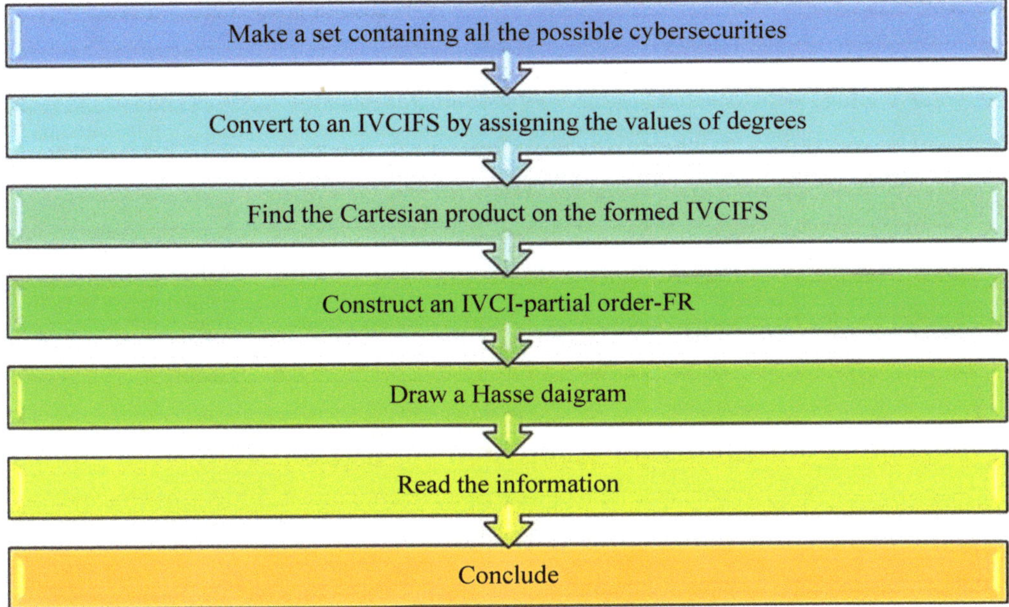

Figure 5. Flowchart of the process for selecting the best cyber-security.

Following are fourteen different security practices and techniques:

I. **Access control (*AC*)** If the cyberattacker is unable to access your industrial network, then they will do very limited harm. *AC* limits user access according to their responsibilities, which increases the security, and especially, internal breaches are restricted.

$$\left(AC, [0.455, 0.575]e^{i[0.305,0.540]\pi}, [0.100, 0.180]e^{i[0.009,0.270]\pi}\right)$$

II. **Anti-malware software (*AMS*)** Viruses, Trojans, worms, key-loggers and spyware are all malwares, which are used to infect digital systems. Anti-malware software identifies risky programs and files, then prevents them from spreading.

$$\left(AMS, [0.560, 0.610]e^{i[0.350, 0.460]\pi}, [0.350, 0.390]e^{i[0.300, 0.450]\pi} \right)$$

III. **Anomaly detection (*AD*)** Identifying anomalies is a difficult task. Henceforth, anomaly detection engines (ADE) are designed that allow the analysis of an industrial network. This alerts the authorities when breaches occur so that they can respond at the right times.

$$\left(AD, [0.295, 0.385]e^{i[0.390, 0.480]\pi}, [0.270, 0.350]e^{i[0.250, 0.330]\pi} \right)$$

IV. **Application security (*AS*)** An *AS* establishes security parameters for any applications that are relevant to industrial security.

$$\left(AS, [0.250, 0.340]e^{i[0.300, 0.375]\pi}, [0.280, 0.400]e^{i[0.220, 0.300]\pi} \right)$$

V. **Data loss prevention (DLP)** *DLP* equipment and strategies protect employees and users from ill-use, such as giving away sensitive data.

$$\left(DLP, [0.360, 0.425]e^{i[0.290, 0.380]\pi}, [0.310, 0.385]e^{i[0.200, 0.300]\pi} \right)$$

VI. **Email security (*ES*)** An *ES* system basically identifies risky emails. These phishing emails are usually very convincing because their target is to trick people. Further, this system also stops cyberattacks and prevents the sharing of important data.

$$\left(ES, [0.405, 0.505]e^{i[0.650, 0.780]\pi}, [0.050, 0.100]e^{i[0.060, 0.100]\pi} \right)$$

VII. **Endpoint security (*EPS*)** These days, the difference between personal and business devices is nearly non-existent. Unfortunately, personal devices are targeted to attack businesses. Endpoint security is a defensive layer between business networks and such remote devices.

$$\left(EPS, [0.375, 0.440]e^{i[0.350, 0.450]\pi}, [0.450, 0.560]e^{i[0.350, 0.450]\pi} \right)$$

VIII. **Firewalls (*FW*)** *FW* act as gateways in a network, used to secure the borders between the internet and local networks. They are used to manage network traffic by allowing approved traffic and blocking non-authorized traffic.

$$\left(FW, [0.450, 0.530]e^{i[0.500, 0.580]\pi}, [0.220, 0.290]e^{i[0.130, 0.210]\pi} \right)$$

IX. **Intrusion prevention systems (*IPS*)** Different types of cyberattacks are identified and then rapidly responded to by *IPS* after scanning and analyzing the traffic. These systems use databases of well-known cyberattack approaches; therefore, they immediately recognize threats.

$$\left(IPS, [0.275, 0.375]e^{i[0.375, 0.500]\pi}, [0.350, 0.400]e^{i[0.250, 0.350]\pi} \right)$$

X. **Network segmentation (*NS*)** *NS* restricts the traffic from suspicious sources that carries risky threats, and allows the authorized and right traffic.

$$\left(NS, [0.500, 0.580]e^{i[0.400, 0.500]\pi}, [0.200, 0.280]e^{i[0.200, 0.300]\pi} \right)$$

XI. **Security information and event management (SIEM)** $SIEM$ is a field of cyber-security that provides instantaneous analysis of security warnings spawned by applications and network hardware.

$$\left(SIEM, [0.500, 0.600]e^{i[0.650,0.700]\pi}, [0.040, 0.080]e^{i[0.050,0.120]\pi}\right)$$

XII. **Virtual private network (VPN)** The communication between secure networks and an endpoint device is authenticated by using VPN tools. They block other parties from spying by creating an encrypted line.

$$\left(VPN, [0.320, 0.450]e^{i[0.200,0.300]\pi}, [0.350, 0.410]e^{i[0.375,0.500]\pi}\right)$$

XIII. **Web security (WS)** WS is an extensive word that describes the security measures taken by businesses to ensure a harmless web experience when connected to an internal network. This prevents web-based cyberthreats from using browsers as access points to get into the network.

$$\left(WS, [0.325, 0.375]e^{i[0.365,0.490]\pi}, [0.335, 0.400]e^{i[0.275,0.395]\pi}\right)$$

XIV. **Wireless security (WiS)** Traditional networks are generally more secure than wireless networks. Thus, severe types of WiS measures are essential to certify that cybercriminals are not gaining access.

$$\left(WiS, [0.450, 0.560]e^{i[0.330,0.460]\pi}, [0.285, 0.390]e^{i[0.230,0.315]\pi}\right)$$

XV. **Encryption (E)** The encryption of data is an exceptionally effective security technique. The term "encryption" means the transformation of data to a code language or cipher text that cannot be read by a human. Special keys are used to decrypt these codes and cipher text back to a readable format. The complex encryption algorithms keep the information safe. Some of these algorithms are; Twofish algorithm, Rivest–Shamir–Adleman (RSA) algorithm and triple data encryption algorithm. Figure 6 illustrates the process of data encryption.

$$\left(E, [0.650, 0.810]e^{i[0.660,0.790]\pi}, [0.085, 0.190]e^{i[0.040,0.130]\pi}\right)$$

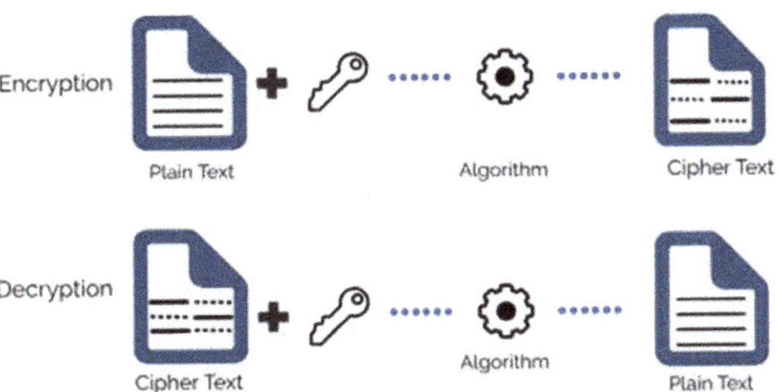

Figure 6. Data Encryption.

Let us assign the degree of membership and the degree of non-membership to each of the security measures and construct an IVCIFS $\breve{E}$

$$
\breve{E} = \left\{
\begin{aligned}
&\left(AC, [0.455, 0.575]e^{i[0.305,0.540]\pi}, [0.100, 0.180]e^{i[0.009,0.270]\pi} \right), \\
&\left(AMS, [0.560, 0.610]e^{i[0.350,0.460]\pi}, [0.350, 0.390]e^{i[0.300,0.450]\pi} \right), \\
&\left(AD, [0.295, 0.385]e^{i[0.390,0.480]\pi}, [0.270, 0.350]e^{i[0.250,0.330]\pi} \right), \\
&\left(AS, [0.250, 0.340]e^{i[0.300,0.375]\pi}, [0.280, 0.400]e^{i[0.220,0.300]\pi} \right), \\
&\left(DLP, [0.360, 0.425]e^{i[0.290,0.380]\pi}, [0.310, 0.385]e^{i[0.200,0.300]\pi} \right), \\
&\left(E, [0.650, 0.810]e^{i[0.660,0.790]\pi}, [0.085, 0.190]e^{i[0.040,0.130]\pi} \right), \\
&\left(ES, [0.405, 0.505]e^{i[0.650,0.780]\pi}, [0.050, 0.100]e^{i[0.060,0.100]\pi} \right), \\
&\left(EPS, [0.375, 0.440]e^{i[0.350,0.450]\pi}, [0.450, 0.560]e^{i[0.350,0.450]\pi} \right), \\
&\left(FW, [0.450, 0.530]e^{i[0.500,0.580]\pi}, [0.220, 0.290]e^{i[0.130,0.210]\pi} \right), \\
&\left(IPS, [0.275, 0.375]e^{i[0.375,0.500]\pi}, [0.350, 0.400]e^{i[0.250,0.350]\pi} \right), \\
&\left(NS, [0.500, 0.580]e^{i[0.400,0.500]\pi}, [0.200, 0.280]e^{i[0.200,0.300]\pi} \right), \\
&\left(SIEM, [0.500, 0.600]e^{i[0.650,0.700]\pi}, [0.040, 0.080]e^{i[0.050,0.120]\pi} \right), \\
&\left(VPN, [0.320, 0.450]e^{i[0.200,0.300]\pi}, [0.350, 0.410]e^{i[0.375,0.500]\pi} \right), \\
&\left(WS, [0.325, 0.375]e^{i[0.365,0.490]\pi}, [0.335, 0.400]e^{i[0.275,0.395]\pi} \right), \\
&\left(WiS, [0.450, 0.560]e^{i[0.330,0.460]\pi}, [0.285, 0.390]e^{i[0.230,0.315]\pi} \right),
\end{aligned}
\right\}
$$

Table 3 contains the descriptions of an interval-valued complex intuitionistic partial order-FR $\breve{R}$ that is obtained from the CP $\breve{E} \times \breve{E}$.

Table 3. IVCI-partial order-FR $\breve{R} \subseteq \breve{E} \times \breve{E}$.

Ordered Pairs ($\breve{R}=\breve{E}\times\breve{E}$)	Degrees of Membership ($\acute{m}_C$)	DegreesofNon-Membership($\underline{n}_C$)
(AC, AC)	$[0.455, 0.575]e^{i[0.305,0.540]\pi}$	$[0.100, 0.180]e^{i[0.009,0.270]\pi}$
$(AC, SIEM)$	$[0.455, 0.575]e^{i[0.305,0.540]\pi}$	$[0.100, 0.180]e^{i[0.050,0.270]\pi}$
(AMS, AMS)	$[0.560, 0.610]e^{i[0.350,0.460]\pi}$	$[0.350, 0.390]e^{i[0.300,0.450]\pi}$
(AMS, E)	$[0.560, 0.610]e^{i[0.350,0.460]\pi}$	$[0.350, 0.390]e^{i[0.300,0.450]\pi}$
$(AMS, SIEM)$	$[0.500, 0.600]e^{i[0.350,0.460]\pi}$	$[0.350, 0.390]e^{i[0.300,0.450]\pi}$
(AD, AD)	$[0.295, 0.385]e^{i[0.390,0.480]\pi}$	$[0.270, 0.350]e^{i[0.250,0.330]\pi}$
(AD, AMS)	$[0.295, 0.385]e^{i[0.350,0.460]\pi}$	$[0.350, 0.390]e^{i[0.300,0.450]\pi}$
(AD, E)	$[0.295, 0.385]e^{i[0.390,0.480]\pi}$	$[0.270, 0.350]e^{i[0.250,0.330]\pi}$
$(AD, SIEM)$	$[0.295, 0.385]e^{i[0.390,0.480]\pi}$	$[0.270, 0.350]e^{i[0.250,0.330]\pi}$
(AS, AMS)	$[0.250, 0.340]e^{i[0.300,0.375]\pi}$	$[0.350, 0.400]e^{i[0.300,0.450]\pi}$
(AS, AS)	$[0.250, 0.340]e^{i[0.300,0.375]\pi}$	$[0.280, 0.400]e^{i[0.220,0.300]\pi}$
(AS, E)	$[0.650, 0.810]e^{i[0.660,0.790]\pi}$	$[0.280, 0.400]e^{i[0.220,0.300]\pi}$
$(AS, SIEM)$	$[0.250, 0.340]e^{i[0.300,0.375]\pi}$	$[0.280, 0.400]e^{i[0.220,0.300]\pi}$
(DLP, AC)	$[0.360, 0.425]e^{i[0.290,0.380]\pi}$	$[0.310, 0.385]e^{i[0.200,0.300]\pi}$
(DLP, DLP)	$[0.360, 0.425]e^{i[0.290,0.380]\pi}$	$[0.310, 0.385]e^{i[0.200,0.300]\pi}$
(DLP, EPS)	$[0.360, 0.425]e^{i[0.290,0.380]\pi}$	$[0.450, 0.560]e^{i[0.350,0.450]\pi}$
$(DLP.SIEM)$	$[0.360, 0.425]e^{i[0.290,0.380]\pi}$	$[0.310, 0.385]e^{i[0.200,0.300]\pi}$
(E, E)	$[0.360, 0.425]e^{i[0.290,0.380]\pi}$	$[0.085, 0.190]e^{i[0.040,0.130]\pi}$
$(F,, SIEM)$	$[0.500, 0.600]e^{i[0.650,0.700]\pi}$	$[0.085, 0.190]e^{i[0.050,0.130]\pi}$
(ES, AC)	$[0.405, 0.505]e^{i[0.305,0.540]\pi}$	$[0.100, 0.180]e^{i[0.060,0.270]\pi}$
(ES, DLP)	$[0.360, 0.425]e^{i[0.290,0.380]\pi}$	$[0.310, 0.385]e^{i[0.200,0.300]\pi}$
(ES, ES)	$[0.405, 0.505]e^{i[0.650,0.780]\pi}$	$[0.050, 0.100]e^{i[0.060,0.100]\pi}$
(ES, EPS)	$[0.375, 0.440]e^{i[0.350,0.450]\pi}$	$[0.450, 0.560]e^{i[0.350,0.450]\pi}$
$(ES, SIEM)$	$[0.405, 0.505]e^{i[0.650,0.700]\pi}$	$[0.050, 0.100]e^{i[0.060,0.120]\pi}$

Table 3. *Cont.*

Ordered Pairs ($\breve{R}=\breve{E}\times\breve{E}$)	Degrees of Membership ($\acute{m}_{\mathbb{C}}$)	Degrees of Non-Membership ($\underline{n}_{\mathbb{C}}$)
(EPS, EPS)	$[0.375, 0.440]e^{i[0.350,0.450]\pi}$	$[0.450, 0.560]e^{i[0.350,0.450]\pi}$
$(EPS, SIEM)$	$[0.375, 0.440]e^{i[0.350,0.450]\pi}$	$[0.450, 0.560]e^{i[0.350,0.450]\pi}$
(FW, AMS)	$[0.450, 0.530]e^{i[0.350,0.460]\pi}$	$[0.350, 0.390]e^{i[0.300,0.450]\pi}$
(FW, AD)	$[0.295, 0.385]e^{i[0.390,0.480]\pi}$	$[0.270, 0.350]e^{i[0.250,0.330]\pi}$
(FW, E)	$[0.450, 0.530]e^{i[0.500,0.580]\pi}$	$[0.220, 0.290]e^{i[0.130,0.210]\pi}$
(FW, FW)	$[0.450, 0.530]e^{i[0.500,0.580]\pi}$	$[0.220, 0.290]e^{i[0.130,0.210]\pi}$
$(FW, SIEM)$	$[0.450, 0.530]e^{i[0.500,0.580]\pi}$	$[0.220, 0.290]e^{i[0.130,0.210]\pi}$
(IPS, AMS)	$[0.275, 0.375]e^{i[0.375,0.500]\pi}$	$[0.350, 0.400]e^{i[0.300,0.450]\pi}$
(IPS, AD)	$[0.275, 0.375]e^{i[0.375,0.480]\pi}$	$[0.350, 0.400]e^{i[0.250,0.350]\pi}$
(IPS, AS)	$[0.250, 0.340]e^{i[0.300,0.375]\pi}$	$[0.350, 0.400]e^{i[0.250,0.350]\pi}$
(IPS, E)	$[0.275, 0.375]e^{i[0.375,0.500]\pi}$	$[0.350, 0.400]e^{i[0.250,0.350]\pi}$
(IPS, FW)	$[0.275, 0.375]e^{i[0.375,0.500]\pi}$	$[0.350, 0.400]e^{i[0.250,0.350]\pi}$
(IPS, IPS)	$[0.275, 0.375]e^{i[0.375,0.500]\pi}$	$[0.350, 0.400]e^{i[0.250,0.350]\pi}$
$(IPS, SIEM)$	$[0.275, 0.375]e^{i[0.375,0.500]\pi}$	$[0.350, 0.400]e^{i[0.250,0.350]\pi}$
(NS, AC)	$[0.455, 0.575]e^{i[0.305,0.500]\pi}$	$[0.200, 0.280]e^{i[0.200,0.300]\pi}$
(NS, AMS)	$[0.500, 0.580]e^{i[0.350,0.460]\pi}$	$[0.350, 0.390]e^{i[0.300,0.450]\pi}$
(NS, AD)	$[0.295, 0.385]e^{i[0.390,0.480]\pi}$	$[0.270, 0.350]e^{i[0.250,0.330]\pi}$
(NS, AS)	$[0.250, 0.340]e^{i[0.300,0.375]\pi}$	$[0.280, 0.400]e^{i[0.220,0.300]\pi}$
(NS, DLP)	$[0.360, 0.425]e^{i[0.290,0.380]\pi}$	$[0.310, 0.385]e^{i[0.200,0.300]\pi}$
(NS, E)	$[0.500, 0.580]e^{i[0.400,0.500]\pi}$	$[0.200, 0.280]e^{i[0.200,0.300]\pi}$
(NS, ES)	$[0.405, 0.505]e^{i[0.400,0.500]\pi}$	$[0.200, 0.280]e^{i[0.200,0.300]\pi}$
(NS, EPS)	$[0.375, 0.440]e^{i[0.350,0.450]\pi}$	$[0.450, 0.560]e^{i[0.350,0.450]\pi}$
(NS, FW)	$[0.450, 0.530]e^{i[0.400,0.500]\pi}$	$[0.220, 0.290]e^{i[0.200,0.300]\pi}$
(NS, IPS)	$[0.275, 0.375]e^{i[0.375,0.500]\pi}$	$[0.350, 0.400]e^{i[0.250,0.350]\pi}$
(NS, NS)	$[0.500, 0.580]e^{i[0.400,0.500]\pi}$	$[0.200, 0.280]e^{i[0.200,0.300]\pi}$
$(NS, SIEM)$	$[0.500, 0.580]e^{i[0.400,0.500]\pi}$	$[0.200, 0.280]e^{i[0.200,0.300]\pi}$
$(SIEM, SIEM)$	$[0.500, 0.600]e^{i[0.650,0.700]\pi}$	$[0.040, 0.080]e^{i[0.050,0.120]\pi}$
(VPN, AC)	$[0.320, 0.450]e^{i[0.200,0.300]\pi}$	$[0.350, 0.410]e^{i[0.375,0.500]\pi}$
(VPN, AMS)	$[0.320, 0.450]e^{i[0.200,0.300]\pi}$	$[0.350, 0.410]e^{i[0.375,0.500]\pi}$
(VPN, AD)	$[0.295, 0.385]e^{i[0.200,0.300]\pi}$	$[0.350, 0.410]e^{i[0.375,0.500]\pi}$
(VPN, AS)	$[0.250, 0.340]e^{i[0.200,0.300]\pi}$	$[0.350, 0.410]e^{i[0.375,0.500]\pi}$
(VPN, DLP)	$[0.320, 0.425]e^{i[0.200,0.300]\pi}$	$[0.350, 0.410]e^{i[0.375,0.500]\pi}$
(VPN, E)	$[0.320, 0.450]e^{i[0.200,0.300]\pi}$	$[0.350, 0.410]e^{i[0.375,0.500]\pi}$
(VPN, ES)	$[0.320, 0.450]e^{i[0.200,0.300]\pi}$	$[0.350, 0.410]e^{i[0.375,0.500]\pi}$
(VPN, EPS)	$[0.320, 0.450]e^{i[0.200,0.300]\pi}$	$[0.450, 0.560]e^{i[0.375,0.500]\pi}$
(VPN, FW)	$[0.320, 0.450]e^{i[0.200,0.300]\pi}$	$[0.350, 0.410]e^{i[0.375,0.500]\pi}$
(VPN, IPS)	$[0.275, 0.375]e^{i[0.200,0.300]\pi}$	$[0.350, 0.410]e^{i[0.375,0.500]\pi}$
(VPN, NS)	$[0.320, 0.450]e^{i[0.200,0.300]\pi}$	$[0.350, 0.410]e^{i[0.375,0.500]\pi}$
$(VPN, SIEM)$	$[0.320, 0.450]e^{i[0.200,0.300]\pi}$	$[0.350, 0.410]e^{i[0.375,0.500]\pi}$
(VPN, VPN)	$[0.320, 0.450]e^{i[0.200,0.300]\pi}$	$[0.350, 0.410]e^{i[0.375,0.500]\pi}$
(VPN, WiS)	$[0.320, 0.450]e^{i[0.200,0.300]\pi}$	$[0.350, 0.410]e^{i[0.375,0.500]\pi}$
(WS, AC)	$[0.325, 0.375]e^{i[0.305,0.490]\pi}$	$[0.335, 0.400]e^{i[0.275,0.395]\pi}$
(WS, AMS)	$[0.325, 0.375]e^{i[0.350,0.460]\pi}$	$[0.350, 0.400]e^{i[0.300,0.450]\pi}$
(WS, AD)	$[0.295, 0.385]e^{i[0.390,0.480]\pi}$	$[0.335, 0.400]e^{i[0.275,0.395]\pi}$
(WS, AS)	$[0.250, 0.340]e^{i[0.300,0.375]\pi}$	$[0.335, 0.400]e^{i[0.275,0.395]\pi}$
(WS, DLP)	$[0.325, 0.375]e^{i[0.290,0.380]\pi}$	$[0.335, 0.400]e^{i[0.275,0.395]\pi}$
(WS, E)	$[0.325, 0.375]e^{i[0.365,0.490]\pi}$	$[0.335, 0.400]e^{i[0.275,0.395]\pi}$
(WS, ES)	$[0.325, 0.375]e^{i[0.365,0.490]\pi}$	$[0.335, 0.400]e^{i[0.275,0.395]\pi}$
(WS, EPS)	$[0.325, 0.375]e^{i[0.350,0.450]\pi}$	$[0.450, 0.560]e^{i[0.350,0.450]\pi}$
(WS, FW)	$[0.325, 0.375]e^{i[0.365,0.490]\pi}$	$[0.335, 0.400]e^{i[0.275,0.395]\pi}$
(WS, IPS)	$[0.275, 0.375]e^{i[0.365,0.490]\pi}$	$[0.350, 0.400]e^{i[0.275,0.395]\pi}$
(WS, NS)	$[0.325, 0.375]e^{i[0.365,0.490]\pi}$	$[0.335, 0.400]e^{i[0.275,0.395]\pi}$
$(WS, SIEM)$	$[0.325, 0.375]e^{i[0.365,0.490]\pi}$	$[0.335, 0.400]e^{i[0.275,0.395]\pi}$

Table 3. *Cont.*

Ordered Pairs ($\bar{R}=\breve{E}\times\breve{E}$)	Degrees of Membership ($\acute{m}_C$)	DegreesofNon-Membership($\underline{n}_C$)
(WS, WS)	$[0.325, 0.375]e^{i[0.365,0.490]\pi}$	$[0.335, 0.400]e^{i[0.275,0.395]\pi}$
(WS, WiS)	$[0.325, 0.375]e^{i[0.330,0.460]\pi}$	$[0.335, 0.400]e^{i[0.275,0.395]\pi}$
(WiS, AC)	$[0.450, 0.560]e^{i[0.305,0.460]\pi}$	$[0.285, 0.390]e^{i[0.230,0.315]\pi}$
(WiS, AMS)	$[0.450, 0.560]e^{i[0.330,0.460]\pi}$	$[0.350, 0.390]e^{i[0.200,0.450]\pi}$
(WiS, AD)	$[0.295, 0.385]e^{i[0.330,0.460]\pi}$	$[0.285, 0.390]e^{i[0.250,0.330]\pi}$
(WiS, AS)	$[0.250, 0.340]e^{i[0.300,0.375]\pi}$	$[0.280, 0.400]e^{i[0.220,0.300]\pi}$
(WiS, DLP)	$[0.360, 0.425]e^{i[0.290,0.380]\pi}$	$[0.310, 0.385]e^{i[0.230,0.315]\pi}$
(WiS, WiS)	$[0.450, 0.560]e^{i[0.330,0.460]\pi}$	$[0.285, 0.390]e^{i[0.230,0.315]\pi}$
(WiS, ES)	$[0.405, 0.505]e^{i[0.330,0.460]\pi}$	$[0.285, 0.390]e^{i[0.230,0.315]\pi}$
(WiS, EPS)	$[0.375, 0.440]e^{i[0.330,0.460]\pi}$	$[0.450, 0.560]e^{i[0.350,0.450]\pi}$
(WiS, FW)	$[0.450, 0.530]e^{i[0.330,0.460]\pi}$	$[0.285, 0.390]e^{i[0.230,0.315]\pi}$
(WiS, IPS)	$[0.275, 0.375]e^{i[0.330,0.460]\pi}$	$[0.350, 0.400]e^{i[0.250,0.350]\pi}$
(WiS, NS)	$[0.450, 0.560]e^{i[0.330,0.460]\pi}$	$[0.285, 0.390]e^{i[0.230,0.315]\pi}$
$(WiS, SIEM)$	$[0.450, 0.560]e^{i[0.330,0.460]\pi}$	$[0.285, 0.390]e^{i[0.230,0.315]\pi}$
(WiS, WiS)	$[0.450, 0.560]e^{i[0.330,0.460]\pi}$	$[0.285, 0.390]e^{i[0.230,0.315]\pi}$

Figure 7 portrays the Hasse diagram for the above interval-valued complex intuitionistic partial order fuzzy relation. For ease, the degrees of membership and non-membership are kept hidden in the following diagram of $\bar{R}$. In Figure 7, one can clearly tell that the best security technique of these fourteen tools is the $SIEM$ because it is the maximum as well as the maximal element. On the contrary side, the VPN and WS are considered to provide the least protection as they appear at the bottom of the diagram.

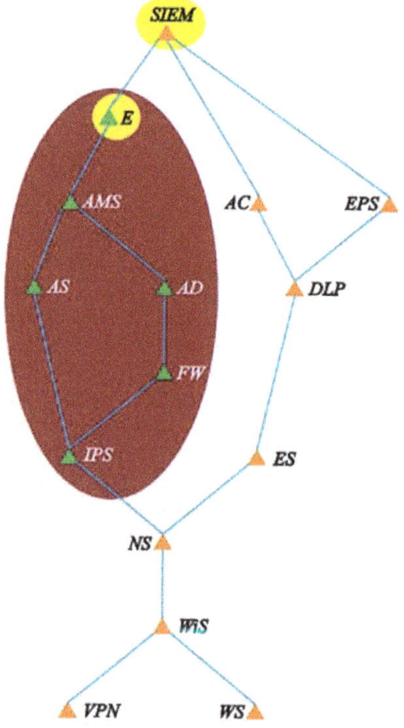

Figure 7. Hasse diagram for IVC-partial order-FR $\bar{R}$

If we consider that an industry has to choose from among the following shortlisted security techniques because of some restrictions, these nominated security techniques and practices are listed in the following subset $\dot{F}$.

$$\dot{F} = \left\{ \begin{array}{l} \left(AMS, [0.560, 0.610]e^{i[0.350,0.460]\pi}, [0.350, 0.390]e^{i[0.300,0.450]\pi} \right), \\ \left(AD, [0.295, 0.385]e^{i[0.390,0.480]\pi}, [0.270, 0.350]e^{i[0.250,0.330]\pi} \right), \\ \left(AS, [0.250, 0.340]e^{i[0.300,0.375]\pi}, [0.280, 0.400]e^{i[0.220,0.300]\pi} \right), \\ \left(E, [0.650, 0.810]e^{i[0.660,0.790]\pi}, [0.085, 0.190]e^{i[0.040,0.130]\pi} \right), \\ \left(FW, [0.450, 0.530]e^{i[0.500,0.580]\pi}, [0.220, 0.290]e^{i[0.130,0.210]\pi} \right), \\ \left(IPS, [0.275, 0.375]e^{i[0.375,0.500]\pi}, [0.350, 0.400]e^{i[0.250,0.350]\pi} \right) \end{array} \right\}$$

The business seeks to select the best security technique from among the shortlisted techniques in set $\dot{F}$. In order to do so, they must look out for the upper bounds and supremum. In the following diagram, the elements of $\dot{F}$ are enclosed in the dark red circle. Here, the upper bounds are $\{E, SIEM\}$. The supremum is the least upper bound, therefore, encryption is the supremum in this case. Amongst the members of the subset $\dot{F}$, encryption (E) is the best choice for coping with the cybersecurity reasons.

6. Comparative Analysis

In this section, the reliability of the proposed framework of IVCIFRs is verified through comparing it with the pre-existing structures such as CFRs or IFRs and IVIFRs.

6.1. Comparison with FRs, CFRs, IVFRs and IVCFRs

The leading limitation of FR, CFR, IVFRs and IVCFR as compared to IVCIFRs is that these notions discuss the degrees of membership only and do not talk about the degrees of non-membership, while an IVCIFR argues about both of the degrees. Thus, an IVCIFR has the power to analyze both the strength and the weakness of any relationship.

Moreover, the complex structure of IVCIFRs can model multivariable problems, whereas the frameworks of FRs and IFRs are unable to model such problems. As an IVCFR is the greatest structure among the aforementioned contestants, so a detailed comparison between IVCIFRs and IVCFRs is given below.

We investigate the problem discussed in Section 5.3 by using IVCFRs and thinking of the following two IVCFSs $\breve{E}$ and $\dot{F}$ representing the set of securities and the set of sources of threats, respectively. In order to minimize the amount of calculation and to conclude the comparative analysis, some of the securities and the sources of threats are omitted.

$$\breve{E} = \left\{ \begin{array}{c} \left(DD, [0.550, 0.600]e^{i16[0.500,0.750]\pi} \right), \left(PP, [0.475, 0.525]e^{i16[0.375,0.438]\pi} \right) \\ \left(SIEM, [0.300, 0.375]e^{i16[0.563,0.613]\pi} \right) \end{array} \right\}$$

$$\dot{F} = \left\{ \begin{array}{l} \left(USB, [0.375, 0.450]e^{i16[0.250,0.375]\pi} \right), \left(RA, [0.350, 0.400]e^{i16[0.375,0.438]\pi} \right), \\ \left(IC, [0.475, 0.575]e^{i16[0.500,0.613]\pi} \right), \left(OC, [0.500, 0.550]e^{i16[0.375,0.500]\pi} \right) \end{array} \right\}$$

Table 4 contains the details of abbreviations used in the above sets.

Table 4. Abbreviations.

Abbreviations	Full Names
DD	Default-Deny as a standard policy
PP	Proactive Protection
SIEM	Integration with SIEM
USB	USB Ports
RA	Remote Access
IC	Internet Connections
OC	Outside Contractors

The IVCFR $\check{\underline{R}}$ between $\underline{\check{E}}$ and $\dot{F}$ is

$$
\check{\underline{R}} = \left\{
\begin{array}{l}
\left((DD, USB), [0.375, 0.450]e^{i16[0.250,0.375]\pi} \right), \\
\left((DD, RA), [0.350, 0.400]e^{i16[0.375,0.438]\pi} \right), \\
\left((DD, IC), [0.475, 0.575]e^{i16[0.500,0.613]\pi} \right), \\
\left((DD, OC), [0.500, 0.550]e^{i16[0.375,0.500]\pi} \right), \\
\left((PP, USB), [0.375, 0.450]e^{i16[0.250,0.375]\pi} \right), \\
\left((PP, RA), [0.350, 0.400]e^{i16[0.375,0.438]\pi} \right), \\
\left((PP, IC), [0.475, 0.525]e^{i16[0.375,0.438]\pi} \right), \\
\left((PP, OC), [0.475, 0.525]e^{i16[0.375,0.438]\pi} \right), \\
\left((SIEM, USB), [0.300, 0.375]e^{i16[0.250,0.375]\pi} \right), \\
\left((SIEM, RA), [0.300, 0.375]e^{i16[0.375,0.438]\pi} \right), \\
\left((SIEM, IC), [0.300, 0.375]e^{i16[0.500,0.613]\pi} \right), \\
\left((SIEM, OC), [0.300, 0.375]e^{i16[0.375,0.500]\pi} \right)
\end{array}
\right\}
$$

It can be seen that the above IVCFR $\check{\underline{R}}$ only gives the information for the degree of membership. That is, it only gives the effectiveness of a cybersecurity technique against a source of penetration and fails to reveal the ineffectiveness of these securities versus the source of penetration's relations because the structure of an IVCFR does not have any degree of non-membership. Hence, these structures have certain limitations and, thus, they give limited information.

6.2. Comparison with IFRs, CIFRs and IVIFRs

An IVCIFR argues about both the degree of membership and non-membership, as do the structures of IFRs, CIFRs and IVIFRs. However, IFRs and IVIFRs involve only real numbers, which limits them to single variable problems. They cannot express the problems involving time (periodic) or with phase changes.

The structure of CIFRs is based on complex numbers and, thus, consists of amplitude and phase terms. However, the advantage of an interval-valued structure of IVCIFRs over the crisp valued structure of CIFRs is that the interval represents a set, thus covering the uncertainties, including mistakes made by the person, experimental errors or computer approximations that lead to fuzziness. A detailed comparison between IVCIFRs and IVIFRs is given below.

We look to solve the problem discussed in Section 5.3 by using IVIFRs. We present the following two IVIFSs $\underline{\check{E}}$ and $\dot{F}$ representing the set of securities and the set of sources of threats, respectively. We omit some of the securities and the sources in order to minimize the amount of calculation. The details of abbreviations used in the following sets are in Table 4.

$$
\underline{\check{E}} = \left\{
\begin{array}{l}
(DD, [0.550, 0.600], [0.300, 0.400]), (PP, [0.475, 0.525], [0.250, 0.375]), \\
(SIEM, [0.300, 0.375], [0.275, 0.325])
\end{array}
\right\}
$$

$$
\dot{F} = \left\{
\begin{array}{l}
(USB, [0.375, 0.450], [0.275, 0.325]), (RA, [0.350, 0.400], [0.325, 0.350]), \\
(IC, [0.475, 0.575], [0.250, 0.350]), (OC, [0.500, 0.550], [0.225, 0.300])
\end{array}
\right\}
$$

The IVIFR $\bar{\dot{R}}$ between $\breve{\mathrm{E}}$ and $\dot{F}$ is

$$
\bar{\dot{R}} = \left\{
\begin{array}{l}
((DD, USB), [0.375, 0.450], [0.300, 0.400]), \\
((DD, RA), [0.350, 0.400], [0.325, 0.400]), \\
((DD, IC), [0.475, 0.575], [0.350, 0.400]), \\
((DD, OC), [0.500, 0.575], [0.300, 0.400]), \\
((PP, USB), [0.375, 0.450], [0.275, 0.375]), \\
((PP, RA), [0.350, 0.400], [0.325, 0.375]), \\
((PP, IC), [0.475, 0.525], [0.250, 0.375]), \\
((PP, OC), [0.475, 0.525], [0.250, 0.375]), \\
((SIEM, USB), [0.300, 0.375], [0.275, 0.325]), \\
((SIEM, RA), [0.300, 0.375], [0.325, 0.350]), \\
((SIEM, IC), [0.300, 0.375], [0.275, 0.350]), \\
((SIEM, OC), [0.300, 0.375], [0.275, 0.325])
\end{array}
\right\}
$$

From the above IVCFR $\bar{\dot{R}}$, it is observed that it explains the effectiveness and ineffectiveness of securities against the sources of penetration. However, unlike IVCIFR, it does not involve the time frame. We are also interested in the time durations for which certain security can successfully handle the vulnerability. Therefore, the complex valued degrees must be involved to achieve the needed information.

6.3. Cons of Alternative Methods

I.   The structure of FRs, IVFRs, CFRs and IVCFRs lack the degree of non-membership.

II.   The IFR, CIFR and IVIFR methods discuss the degree of membership as well as the degree of no-membership, but they have certain limitations.

- IFR, with its single valued degrees, does not cope with uncertainty as efficiently as interval-valued structures. Moreover, it cannot model multivariable problems.
- Though CIFR is capable of modeling multivariable problems, it lags behind in handling uncertainty due to its single valued degrees.
- An IVIFR can grip the uncertainty quite well with its interval-valued structure, but it is only limited to one-dimensional problems.

6.4. Pros of IVCIFR

I.   The structure is composed of the degrees of membership and non-membership.

II.   Interval values cover the mistakes and errors made by the expert or that occur during the survey or experiments.

III.   Complex valued memberships and non-memberships can be used to cope with multidimensional variables.

Table 5 provides a summary of the characteristics of eight different structures in fuzzy set and logic theory. From Table 5, the grand structure of IVCIFRs is verified as it ticks all four characteristics, while the rest of the competitors have limitations in their structures.

Table 5. Comparison on the basis of structural properties.

Structure	Membership	Non-Membership	Multidimensional Variables	Interval-Values
FR	✓	✗	✗	✗
CFR	✓	✗	✓	✗
IVFR	✓	✗	✗	✓
IVCFR	✓	✗	✓	✓
IFR	✓	✓	✗	✗
CIFR	✓	✓	✓	✗
IVIFR	✓	✓	✗	✓
IVCIFR	✓	✓	✓	✓

7. Conclusions

This article introduced the innovative concepts of interval-valued complex intuitionistic fuzzy relation (IVCIFR) and the Cartesian product (CP) between two interval-valued complex intuitionistic fuzzy sets (IVCIFSs). Further, various types of IVCIFRs are also defined, including the interval-valued complex intuitionistic equivalence fuzzy relation (IVCI-equivalence-FR), IVCI-partial order-FR, IVCI-total order-FR, IVCI-composite-FR and many more. Moreover, the Hasse diagram has been introduced for the IVCI-partial order-FR and IVCI-partial order-FS. The concepts and ideas related to the Hasse diagram have also been defined. Suitable examples are given for each of the definitions and some results are proved for the different types of IVCIFRs. Furthermore, the proposed ideas are utilized to investigate the relationships between different types of cybersecurity and cybercrimes and their sources. The section titled *comparative analysis* verifies the omnipotence of IVCIFRs by its head-to-head comparison with other alternative mathematical techniques. In addition, it also sums up the vast structure of IVCIFRs and the limitations of pre-existing frameworks. The flaws of IVCIFRs include the absence of a degree of abstinence as well as the restrictions and constraints on the sum of the degrees of membership and non-membership. In the future, these concepts can be extended to the generalizations of fuzzy sets, which will give rise to many interesting structures with a vast range of applications.

Author Contributions: Conceptualization, A.N., N.J., A.G. and F.R.A.; Data curation, A.N., N.J. and S.U.K.; Formal analysis, A.N., N.J., A.G., S.U.K. and F.R.A.; Funding acquisition, A.N., A.G. and F.R.A.; Investigation, A.N., N.J., A.G., S.U.K. and F.R.A.; Methodology, A.N., N.J., A.G., S.U.K. and F.R.A.; Project administration, N.J. and A.G.; Resources, A.N., N.J., A.G. and S.U.K.; Software, A.N., N.J., A.G., S.U.K. and F.R.A.; Supervision, A.N., N.J., A.G., S.U.K. and F.R.A.; Validation, A.N., N.J., S.U.K. and F.R.A.; Visualization, A.N. and N.J.; Writing—original draft, A.N. and N.J.; Writing—review & editing, A.N., N.J., A.G., S.U.K. and F.R.A. All authors have read and agreed to the published version of the manuscript.

Funding: This research received no external funding.

Institutional Review Board Statement: Not applicable.

Informed Consent Statement: Not applicable.

Data Availability Statement: There is no data supported this study.

Acknowledgments: The authors are grateful to the Deanship of Scientific Research, King Saud University for funding through Vice Deanship of Scientific Research Chairs.

Conflicts of Interest: All the authors declare that they do not have conflicts in the publication of this article.

References

1. Zadeh, L.A. Fuzzy sets. *Inf. Control.* **1965**, *8*, 338–353. [CrossRef]
2. Klir, G.J.; Folger, T.A. *Fuzzy Sets, Uncertainty, and Information*; Prentice Hall: Englewood Cliffs, NJ, USA, 1988.
3. Mendel, J.M. Fuzzy logic systems for engineering: A tutorial. *Proc. IEEE* **1995**, *83*, 345–377. [CrossRef]
4. Zadeh, L.A. The concept of a linguistic variable and its application to approximate reasoning—I. *Inf. Sci.* **1975**, *8*, 199–249. [CrossRef]
5. Bustince, H.; Burillo, P. Mathematical analysis of interval-valued fuzzy relations: Application to approximate reasoning. *Fuzzy Sets Syst.* **2000**, *113*, 205–219. [CrossRef]
6. Goguen, J.A., Jr. Concept representation in natural and artificial languages: Axioms, extensions and applications for fuzzy sets. *Int. J. Man-Mach. Stud.* **1974**, *6*, 513–561. [CrossRef]
7. Żywica, P. Modelling medical uncertainties with use of fuzzy sets and their extensions. In Proceedings of the International Conference on Information Processing and Management of Uncertainty in Knowledge-Based Systems, Cádiz, Spain, 18 May 2018; Springer: Cham, Germany, 2018; pp. 369–380. [CrossRef]
8. Román-Flores, H.; Barros, L.C.; Bassanezi, R.C. A note on Zadeh's extensions. *Fuzzy Sets Syst.* **2001**, *117*, 327–331. [CrossRef]
9. Dubois, D.; Prade, H. Gradualness, uncertainty and bipolarity: Making sense of fuzzy sets. *Fuzzy Sets Syst.* **2012**, *192*, 3–24. [CrossRef]
10. Gehrke, M.; Walker, C.; Walker, E. Some comments on interval valued fuzzy sets! *Structure* **1996**, *1*, 2. [CrossRef]
11. Bustince, H. Indicator of inclusion grade for interval-valued fuzzy sets. Application to approximate reasoning based on interval-valued fuzzy sets. *Int. J. Approx. Reason.* **2000**, *23*, 137–209. [CrossRef]

12. Turksen, I.B. Interval-valued fuzzy sets and 'compensatory AND'. *Fuzzy Sets Syst.* **1992**, *51*, 295–307. [CrossRef]
13. Ramot, D.; Milo, R.; Friedman, M.; Kandel, A. Complex fuzzy sets. *IEEE Trans. Fuzzy Syst.* **2002**, *10*, 171–186. [CrossRef]
14. Ramot, D.; Friedman, M.; Langholz, G.; Kandel, A. Complex fuzzy logic. *IEEE Trans. Fuzzy Syst.* **2003**, *11*, 450–461. [CrossRef]
15. Greenfield, S.; Chiclana, F.; Dick, S. Interval-valued complex fuzzy logic. In Proceedings of the 2016 IEEE International Conference on Fuzzy Systems (FUZZ-IEEE), Vancouver, BC, Canada, 24–29 July 2016; pp. 2014–2019.
16. Nasir, A.; Jan, N.; Gumaei, A.; Khan, S.U. Medical diagnosis and life span of sufferer using interval valued complex fuzzy relations. *IEEE Access* **2021**, *9*, 93764–93780. [CrossRef]
17. Chen, Z.; Aghakhani, S.; Man, J.; Dick, S. ANCFIS: A neurofuzzy architecture employing complex fuzzy sets. *IEEE Trans. Fuzzy Syst.* **2010**, *19*, 305–322. [CrossRef]
18. Yazdanbakhsh, O.; Dick, S. A systematic review of complex fuzzy sets and logic. *Fuzzy Sets Syst.* **2018**, *338*, 1–22. [CrossRef]
19. Tamir, D.E.; Rishe, N.D.; Kandel, A. Complex fuzzy sets and complex fuzzy logic an overview of theory and applications. *Fifty Years Fuzzy Log. Its Appl.* **2015**, *326*, 661–681.
20. Dai, S.; Bi, L.; Hu, B. Distance measures between the interval-valued complex fuzzy sets. *Mathematics* **2019**, *7*, 549. [CrossRef]
21. Greenfield, S.; Chiclana, F.; Dick, S. Join and meet operations for interval-valued complex fuzzy logic. In Proceedings of the 2016 Annual Conference of the North American Fuzzy Information Processing Society (NAFIPS), El Paso, TX, USA, 31 October–4 November 2016; pp. 1–5.
22. Atanassov, K.T. Intuitionistic fuzzy sets. *Fuzzy Sets Syst.* **1986**, *20*, 87–96. [CrossRef]
23. Burillo, P.; Bustince, H. Intuitionistic fuzzy relations (Part I). *Mathw. Soft Comput.* **2016**, *2*, 5–38.
24. Atanassov, K.T. Interval valued intuitionistic fuzzy sets. In *Intuitionistic Fuzzy Sets*; Physica: Heidelberg, Germany, 1999; pp. 139–177.
25. Alkouri, A.S.; Salleh, A.R. Complex intuitionistic fuzzy sets. *AIP Conf. Proc.* **2012**, *1482*, 464.
26. Garg, H.; Rani, D. Complex interval-valued intuitionistic fuzzy sets and their aggregation operators. *Fundam. Inform.* **2019**, *164*, 61–101. [CrossRef]
27. Li, D.F. Multiattribute decision making models and methods using intuitionistic fuzzy sets. *J. Comput. Syst. Sci.* **2005**, *70*, 73–85. [CrossRef]
28. De, S.K.; Biswas, R.; Roy, A.R. An application of intuitionistic fuzzy sets in medical diagnosis. *Fuzzy Sets Syst.* **2001**, *117*, 209–213. [CrossRef]
29. Vlachos, I.K.; Sergiadis, G.D. Intuitionistic fuzzy information–applications to pattern recognition. *Pattern Recognit. Lett.* **2007**, *28*, 197–206. [CrossRef]
30. Lee, K.M.; LEE, K.M.; CIOS, K.J. Comparison of interval-valued fuzzy sets, intuitionistic fuzzy sets, and bipolar-valued fuzzy sets. In *Computing and Information Technologies: Exploring Emerging Technologies*; World Scientific: Hackensack, NJ, USA, 2001; pp. 433–439. [CrossRef]
31. Grzegorzewski, P. Distances between intuitionistic fuzzy sets and/or interval-valued fuzzy sets based on the Hausdorff metric. *Fuzzy Sets Syst.* **2004**, *148*, 319–328. [CrossRef]
32. Nasir, A.; Jan, N.; Yang, M.-S.; Khan, S.U. Complex T-spherical fuzzy relations with their applications in economic relationships and international trades. *IEEE Access* **2021**, *9*, 66115–66131. [CrossRef]
33. Khan, S.U.; Nasir, A.; Jan, N.; Ma, Z.H. Graphical Analysis of Covering and Paired Domination in the Environment of Neutrosophic Information. *Math. Probl. Eng.* **2021**, *2021*. [CrossRef]
34. Nasir, A.; Jan, N.; Gumaei, A.; Khan, S.U.; Al-Rakhami, M. Evaluation of the Economic Relationships on the Basis of Statistical Decision-Making in Complex Neutrosophic Environment. *Complexity* **2021**, *2021*. [CrossRef]
35. Jan, N.; Rehman, S.U.; Nasir, A.; Aydi, H.; Khan, S.U. Analysis of Economic Relationship Using the Concept of Complex Pythagorean Fuzzy Information. *Secu. Comm. Nets.* **2021**, *2021*. [CrossRef]
36. Ali, M.; Tamir, D.E.; Rishe, N.D.; Kandel, A. Complex intuitionistic fuzzy classes. In Proceedings of the 2016 IEEE International Conference on Fuzzy Systems (FUZZ-IEEE), Vancouver, BC, Canada, 24–29 July 2016; pp. 2027–2034.
37. Liu, Y.; Jiang, W. A new distance measure of interval-valued intuitionistic fuzzy sets and its application in decision making. *Soft Comput.* **2019**, *24*, 1–17. [CrossRef]
38. Bustince, H.; Burillo, P. Correlation of interval-valued intuitionistic fuzzy sets. *Fuzzy Sets Syst.* **1995**, *74*, 237–244. [CrossRef]
39. Nayagam, V.L.G.; Sivaraman, G. Ranking of interval-valued intuitionistic fuzzy sets. *Appl. Soft Comput.* **2011**, *11*, 3368–3372. [CrossRef]
40. Otero, A.R.; Tejay, G.; Otero, L.D.; Ruiz-Torres, A.J. A fuzzy logic-based information security control assessment for organizations. In Proceedings of the 2012 IEEE Conference on Open Systems, Kuala Lumpur, Malaysia, 21–24 October 2012; pp. 1–6.
41. Tariq, M.I.; Ahmed, S.; Memon, N.A.; Tayyaba, S.; Ashraf, M.W.; Nazir, M.; Hussain, A.; Balas, V.E.; Balas, M.M. Prioritization of information security controls through fuzzy AHP for cloud computing networks and wireless sensor networks. *Sensors* **2020**, *20*, 1310. [CrossRef] [PubMed]
42. Tariq, M.I.; Tayyaba, S.; Ali Mian, N.; Sarfraz, M.S.; De-la-Hoz-Franco, E.; Butt, S.A.; Santarcangelo, V.; Rad, D.V. Combination of AHP and TOPSIS methods for the ranking of information security controls to overcome its obstructions under fuzzy environment. *J. Intell. Fuzzy Sys.* **2020**, *38*, 6075–6088. [CrossRef]
43. Mokhtari, S.M.; Alinejad-Rokny, H.; Jalalifar, H. Selection of the best well control system by using fuzzy multiple-attribute decision-making methods. *J. Appl. Stat.* **2014**, *41*, 1105–1121. [CrossRef]

Article

Privacy Preserving Face Recognition in Cloud Robotics: A Comparative Study

Chiranjeevi Karri [1], Omar Cheikhrouhou [2,3,*], Ahmed Harbaoui [4], Atef Zaguia [5] and Habib Hamam [6,7]

1 C4-Cloud Computing Competence Centre, University of Beira Interior, 6200-506 Covilhã, Portugal; karri.chiranjeevi@ubi.pt
2 CES Laboratory, National School of Engineers of Sfax, University of Sfax, Sfax 3038, Tunisia
3 Higher Institute of Computer Science of Mahdia, University of Monastir, Monastir 5019, Tunisia
4 Faculty of Computing and Information Technology, King AbdulAziz University, Jeddah 21589, Saudi Arabia; aharbaoui@kau.edu.sa
5 Department of Computer Science, College of Computers and Information Technology, Taif University, P.O. BOX 11099, Taif 21944, Saudi Arabia; zaguia.atef@tu.edu.sa
6 Faculty of Engineering, Université de Moncton, Moncton, NB E1A3E9, Canada; habib.hamam@umoncton.ca
7 Department of Electrical and Electronic Engineering Science, University of Johannesburg, Johannesburg 2006, South Africa
* Correspondence: omar.cheikhrouhou@isetsf.rnu.tn

Abstract: Real-time robotic applications encounter the robot on board resources' limitations. The speed of robot face recognition can be improved by incorporating cloud technology. However, the transmission of data to the cloud servers exposes the data to security and privacy attacks. Therefore, encryption algorithms need to be set up. This paper aims to study the security and performance of potential encryption algorithms and their impact on the deep-learning-based face recognition task's accuracy. To this end, experiments are conducted for robot face recognition through various deep learning algorithms after encrypting the images of the ORL database using cryptography and image-processing based algorithms.

Keywords: cloud robotics; image face recognition; deep learning algorithms; security; encryption algorithms

Citation: Karri, C.; Cheikhrouhou, O.; Harbaoui, A.; Zaguia, A.; Hamam, H. Privacy Preserving Face Recognition in Cloud Robotics: A Comparative Study. *Appl. Sci.* **2021**, *11*, 6522. https://doi.org/10.3390/app11146522

Academic Editors: Leandros Maglaras and Ioanna Kantzavelou

Received: 2 April 2021
Accepted: 9 July 2021
Published: 15 July 2021

Publisher's Note: MDPI stays neutral with regard to jurisdictional claims in published maps and institutional affiliations.

1. Introduction

Advancements in the robotics field have led to the emergence of a diversity of robot-based applications and favored the integration of robots in the automation of several applications of our daily life. Multi-robot systems, where several robots collaborate in the achievement of a task [1], are now used in several applications including smart transportation [2], smart healthcare [3], traffic management [4], disaster management [5], and face recognition [6,7]. Although robots' resources have been improving in terms of energy, computation power, and storage, they still cannot satisfy the need of emerging applications [8]. As a solution, researchers focused on solutions that leverage the use of cloud computing [9]. A new paradigm has emerged, namely cloud robotics [8]. Cloud robotics resulted from the integration of advancement in the robotics field with the progress made in the cloud computing field.

Cloud robotics have several advantages compared to traditional robotics systems, including large storage, remote data availability, and more computing power.

The need for computing power is also motivated by the emergence of a new generation of applications using artificial intelligence and learning algorithms to analyze and interpret data. Thanks to these resource powered robotics systems, complex problems that have long been considered very difficult, such as speech and face recognition, can now be executed on robots and have also achieved very promising results. More precisely, it is possible nowadays to design a robot with limited resources and to execute facial recognition tasks

279

using convolutional neural network (CNN) algorithms by simply connecting to a cloud service [7].

However, this solution faces security problems. Indeed, it is essential to ensure the security of the facial images to be sent through the network. An interesting solution for this problem is the use of a cryptographic system allowing for avoiding network attacks able to recover these data.

In the present work, we focus on the two environments: robot and cloud. Certainly, the confidentiality of robots is essential. This is because private information is shared in public clouds. As a result, there is a clear risk of abuse or at least misuse of private data. The robot contains and uses private information. They should be safeguarded and treated with respect for confidentiality and privacy.

The contribution of this paper is threefold:

- We provide a security analysis of the potential encryption algorithms that can be used to encrypt images stored on the cloud.
- We present a comparative and experimental study of several CNN based secure robotic facial recognition solutions.
- We study the impact of encryption algorithms on the performance of the CNN based robot face recognition models.
  The experimental done in this paper includes several combinations of various encryption algorithms and deep learning algorithms that have been tested and have shown an improvement in recognition speed and accuracy without impacting privacy issues when executed on cloud compared to their execution in robot environment.

The remainder of this paper is structured as follows: Section 2 presents an overview of the main CNN based robot face recognition models. Then, Section 3 highlights the different encryption techniques that can be used for images encryption. Section 4, provides a security analysis of the encryption algorithms studied. The performance of the CNN based robot face recognition and the impact of the encryption algorithms on their performance were presented and discussed in Section 5. Later, The benefit of outsourcing computation to the cloud for face recognition algorithms is shown before concluding the paper.

2. CNN Models for Robot Face Recognition

The evaluation of convolution neural network (CNN) tremendously changed the researchers thought process towards the applications of computer vision like object recognition, semantic segmentation, image fusion and so on. It play key role in machine learning or deep learning algorithms. The major difference between these two is in their structure. In machine learning architecture, features are extracted with various CNN layers and classification is done with other classification algorithm whereas in deep-learning both feature extraction and classification are available in same architecture [10]. The artificial neural network (ANN) are feedback networks and CNN are feed-forward networks which are inspired by the process of neurons in human brain. It (ANN's) has mostly one input, output and one hidden layers depends on the problem one can increase the hidden layers. In general, a CNN has a convolution layer, an activation layer, a pooling layer, and a fully connected layer. Convolution layer has a number of filters of different sizes (such as 3×3, 5×5, 7×7) to perform convolution operation on input image aiming to extract the image features. To detect features, these filters are sliding over the image and perform a dot product, and these features are given to an activation layer. In the activation layer, activation function decides the outcome. The main activation functions are: binary step, linear activation, Sigmoid, and Rectified linear unit (ReLu). Especially in our article, we preferred ReLu activation function and its respective neuron outcome becomes one if summed multiplied inputs and weights exceeds certain threshold value or else it becomes zero. In certain region it obeys the linearity rule between input and output of respective neuron. Once features are extracted with various kernels, for dimensionality reduction, outcome of CNN are passed through pooling layer. In present research, there are various ways pool the layer. Average pooling is one simple approach in which average of feature map is consider, in max pooling, consider maximum among the feature map. Finally, a fully connected layer resulting from the pooled feature map is converted to a single long continuous linear vector as shown in Figure 1. In what follows, we discuss the main CNN models including LeNet, Alexnet, VGG16Net, GoogLeNet, ResNet, DenseNet, MobileFaceNet, EffNet, and ShuffleNet.

Convolutional neural networks (CNN or ConvNet) present a category of deep neural networks, which are most often used in visual image analysis [10]. The model of connectivity between the CNN neurons is inspired from the organization of the animal visual cortex. A CNN is generally composed of a convolution layer, an activation layer, a pooling layer, and a fully connected layer. The convolution layer includes a set of filters of different sizes (e.g., $3 \times 3, 5 \times 5, 7 \times 7$). These filters are applied in a convolution operation on the input image in order to extract the image features. To detect these features, the input image is scanned by these filters and a scalar product is performed. The obtained features presents the input of the activation layer which decides on the outcome.

The main activation functions are: binary step, linear activation, Sigmoid and ReLu. We opted, in this work, for a rectified linear unit transformation function (ReLu). The ReLu transformation function activates a node only if the input is greater than a certain threshold. If the input is less than zero, the output is zero. On the other hand, when the input exceeds a certain threshold, the activation function becomes a linear relationship with the independent variable. Then, the rectified features go through a pooling layer. Pooling is a downsampling operation reducing the dimensionality of the feature map. Pooling can be average pooling, which calculates the average value for each patch on the feature map, or max pooling, which calculates the maximum value for each patch on the feature map. In the last step, a fully connected layer incorporating all features is converted into one single vector, as shown in Figure 1.

Let us now discuss some commonly used CNN models including LeNet, Alexnet, VGG16Net, GoogLeNet, ResNet, DenseNet, MobileFaceNet, EffNet, and ShuffleNet.

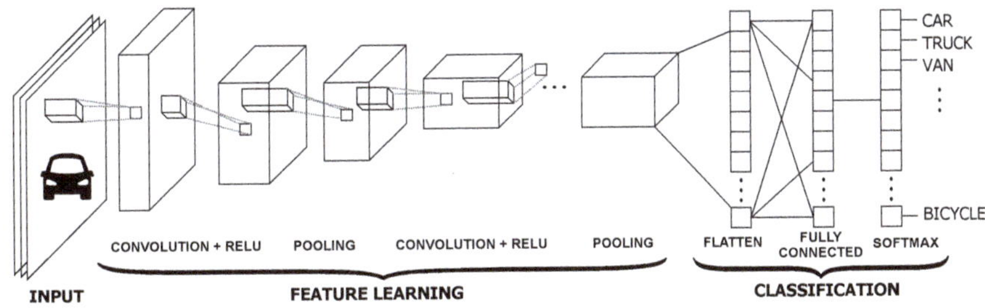

Figure 1. CNN architecture.

2.1. LeNet

In 1989, the LeNet was introduced by the researcher named Y. LeCun and was used for object recognition especially in images with low resolution. Initially, the input image is pre-processed to size of $3 \times 32 \times 32$ and with six kernels or filter, image features are extracted of size $6 \times 28 \times 28$. These features are passed through average pooling and its outcome is reduced version of original of size $6 \times 14 \times 14$. In next stage of filtering (feature extraction), the same cascaded operations are performed with 16 kernels which leads to size of $6 \times 10 \times 10$. In last stage of feature extraction, two cascaded polling layers applied which leads to size of $16 \times 5 \times 5$ and $120 \times 1 \times 1$ respectively. In classification stage, two fully connected layers (FCN) are used which outcome is of size 4096 [11]. The Figure 2 replicates the hole architecture of LeNet.

2.2. AlexNet

AlexNet is a cnn that has had a significant impact on deep learning, particularly in the training and testing process to machine vision. It successfully winning the 2012 ImageNet LSVRC-2012 contest by a significant margin (15.3 percent mistake rates versus 26.2 percent error rates in second position). The network's design was quite similar to that of LeNet, although it was richer, with much more filters per layer and cascaded convolution layers.The AlexNet has eight weighted layers, the first five of which are convolutional and the last three of which are completely linked. The last fully-connected layer's output is sent into a 1000-way softmax, which generates a distribution across the 1000 class labels. The network aims to maximise the multi-variable logistic regression goal, which is really the mean of the log-probability of the right label underneath the forecast distributions throughout all training cases as in Figure 2. Only those kernel mappings in the preceding layer that are on the same GPU are interconnected to the filter or kernels of the 2nd, 4th, and 5th convolutional layers. All kernel/filters mappings in the 2nd layer are connected to the filters of 3rd layer of convolutional. Every neurons throughout the preceding or previous layer are connected to FCN layer of neurons. It supports parallel training on two GPU's because of group convolution compatibility [12]. The architecture of AlexNet is replicated in Figure 2.

2.3. Visual Geometry Group (VGG16Net)

This convolution net was invited in 2004 by simonyan and it is available in two forms, one named VGG16 has 16 layers and VGG19 has 19 layers with filter size of 3×3. It also won the first prize with 93 percent accuracy in training and testing. It takes input image of size (224, 224, 3). The first two layers share the same padding and have 64 channels with 3*3 filter sizes. Following a stride (2, 2) max pool layer, two layers with 256 filter size and filter size convolution layers are added (3, 3). Following that is a stride (2, 2) max pooling layer, which is identical to the preceding layer. There are then two convolution layers with filter sizes of 3 and 3 and a 256 filter. Depends on the requirement of the user one can increase the number of layers for deeper features for better classification Figure 2. It has 138 million hyper-parameters for tuning. It offers parallel processing for reduction in computational time and uses max-pooling which obeys sometime non-linearity [13].

2.4. GoogleNet

It is invited and won the prize in ILSVRC in 2014. Its architecture are as similar to other nets In contrast, dropout regularisation is used in the FCN layer, and ReLU activation function is used in all convolution operation. This network, however, is significantly bigger and longer than AlexNet, with 22 overall layers and a far fewer number of hyper-parameters. For computational expanses reduction, in GoogleNet, in-front of 3×3 and 5×5 they used 1×1. This layer is called bottleneck layers as in Figure 3. It optimizes the feature maps based on back propagation algorithm with bearing the increment in computational cost and kernel size. This computational cost is reduced by bottleneck layer which transform the feature maps to matrix of smaller size. This reduction is along the volume direction, so feature map depth becomes less. It has 5 million hyper-parameters and obtained accuracy is near around 93 percent [14].

2.5. ResNet

It also own the prize in ILSVRC—2015 and was developed by He et al. In this architecture, performance metric (accuracy) becomes saturated as the features are deeper (number of convolution layers) and after certain extent of deep features its performance is drastically degrading. This architecture offers skip connections for improvement of performance and to get rapid convergence of network. In training phase of the network, these skip connections allows the data to move in another flexible path for better gradient. The ResNet architecture is somewhat complex than VGGNet. It comprises 65 million hyper-parameters and accuracy is near arround 96 to 96.4 percent, this value is better than human accuracy (94.9 percent). A deeper network can be made from a shallow network by copying weights in a shallow network and setting other layers in the deeper network to be identity mapping. This formulation indicates that the deeper model should not produce higher training errors than the shallow counterpart [15]. From Figure 4, ResNet, layers learn a residual mapping with reference to the layer inputs $F(x) := H(x) - x$ rather than directly learning a desired underlying mapping $H(x)$ to ease the training of very deep networks (up to 152 layers). The original mapping is recast into $F(x) + x$ and can be realized by shortcut connections.

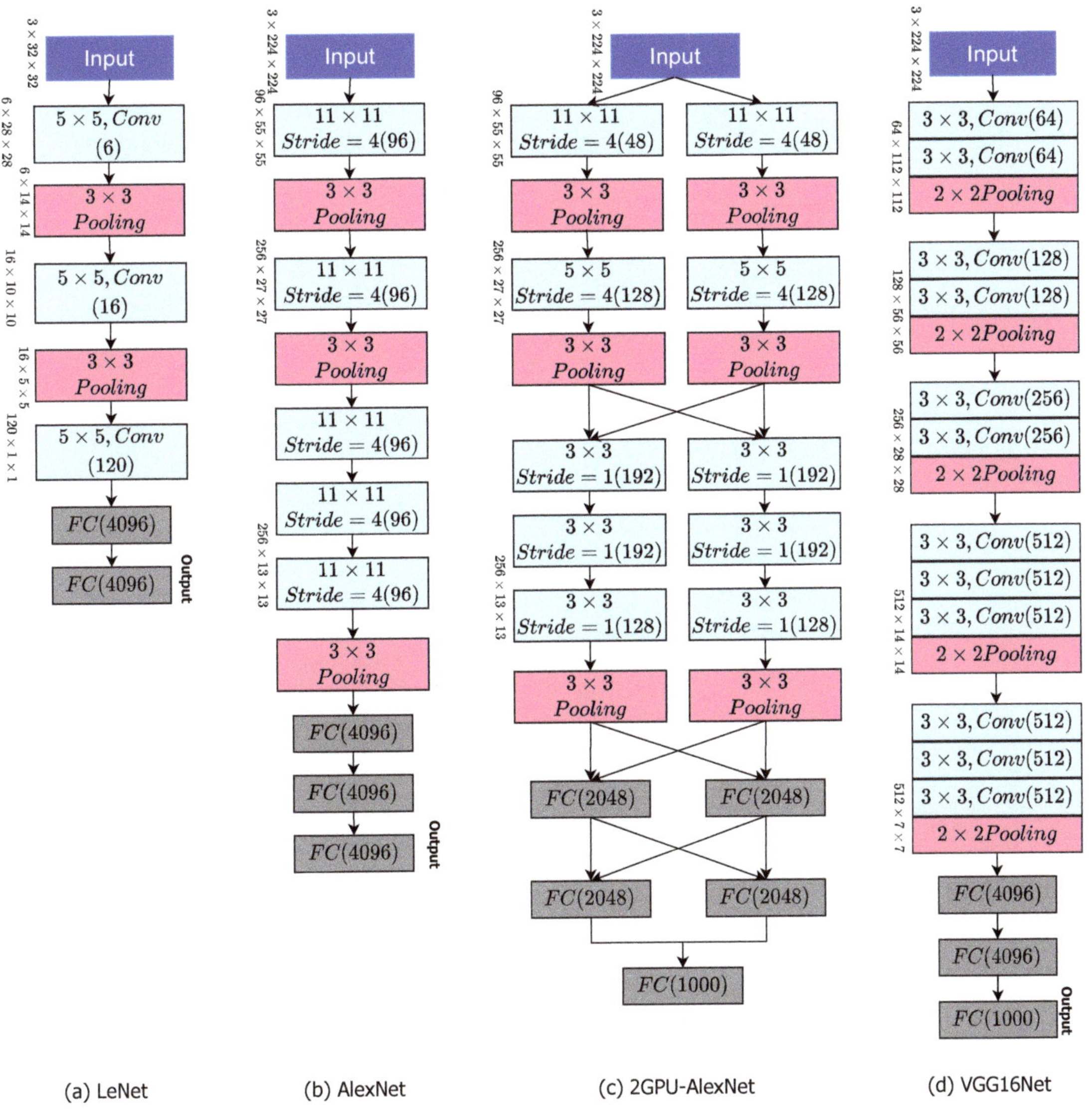

Figure 2. CNN Models: Architecture for LeNet, AlexNet, 2GPU-AlexNet, and VGG16Net.

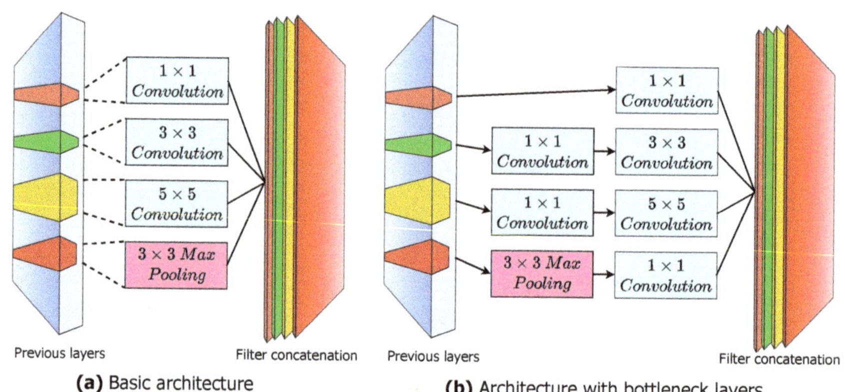

Figure 3. GoogleNet architecture.

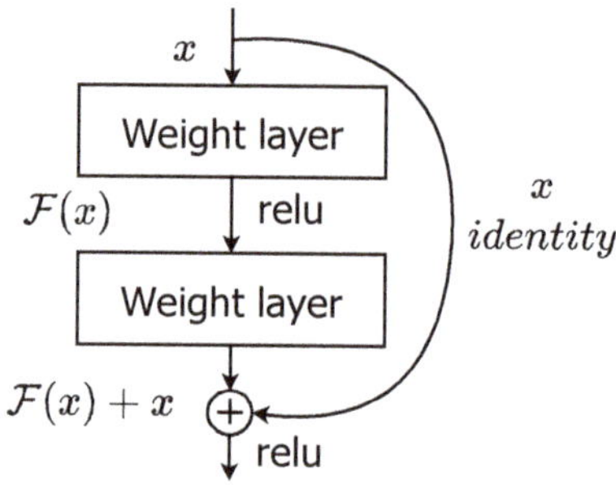

Figure 4. ResNet architecture.

2.6. DenseNet

DenseNet is a contemporary CNN architecture for visual object recognition that has achieved state-of-the-art performance while requiring fewer parameters. DenseNet is fairly similar to ResNet but for a few key differences. DenseNet uses its concatenates (.) attributes to mix the previous layer output with a future layer, whereas ResNet blends the previous layer with the future layers using an additive attribute (+). The training process of ResNet CNN becomes harder because of lack of gradient when deeper into the deeper features. This problem is resolved in DensNet by establishing a shortest path between one layer and its successor layer. This architecture establish shortest path between all layers, a architecture with L layers establish Z connections which is equal to $\frac{L(L+1)}{2}$. In this architecture, every layer carries a limited tuning parameters and for each layers are convoluted with 12 filters. Implicit deep supervision character improves the flow of the gradient through the network. The outcome features of all layers can directly pass though loss function and its gradients are available for all the layers as shown in Figure 5. In the architecture of DenseNet, each layer outputs k feature maps, where k is the growth factor. The bottleneck layer of 1×1 followed by 3×3 convolutions and 1×1 convolutions, output $4k$ feature maps, for ImageNet, the initial convolution layer, outputs $2k$ feature maps [16].

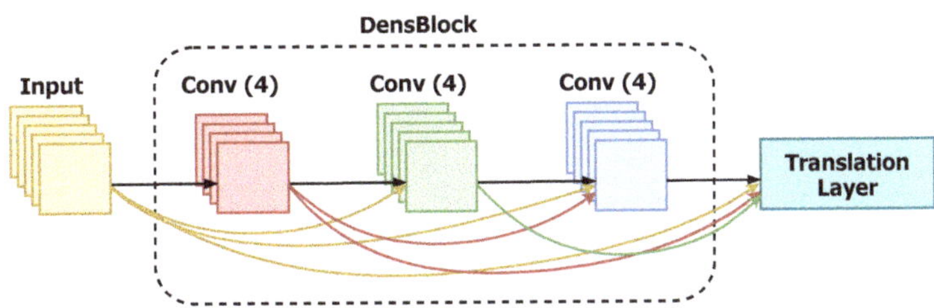

Figure 5. DenseNet architecture.

2.7. MobileFaceNet

This network architecture uses newly introduced convolution called depth-wise separable convolution which allows the network tune with less hyper-parameters as in Figure 6. It also provides flexibility in the selection of a right sized model dependence on the application of designers or users by introducing two simple global hyperparameters i.e., Width Multiplier (Thinner models) and Resolution Multiplier (Reduced representation). In standard convolution, the application of filters across all input channels and the combination of these values is done in a single step—whereas, in depthwise convolution, convolution is performed in two stages: depthwise convolution—filtering stage and pointwise convolution—combination stage. Efficiency has a trade-off with accuracy [17]. Depthwise convolution reduces the number of computations because of the fact that multiplication is an expansive operation compared to addition.

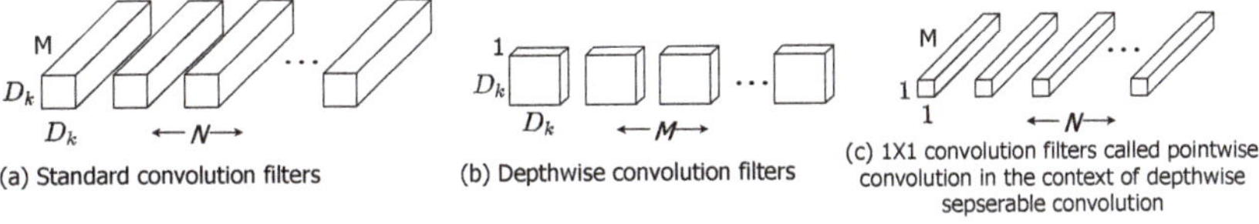

Figure 6. Depthwise separable convolution.

2.8. ShuffleNet

The ShuffleNet has advantages over MobileNet, it has higher accuracy with less computational time and less hyper-parameters. It is used for many applications like cloud robotics, drones and mobile phones. It overcome the drawback of expansive point-wise convolution by introducing group convolution point-wise and to override side effects by introducing channel shuffler. Practical experiment shows that ShuffleNet has an accuracy of 7.8 percent higher than MobileNet and its computation time is much faster (13 times) than MobileNet. The group convolution already explained in ResNet or AlexNet [18]. Figure 7a is a bottleneck unit with depth-wise seperable convolution (3 × 3 DWConv). Figure 7b is a ShuffleNet architecture with point-wise group convolution with channel shuffler. In figure, second level of group convolution in point-wise is to provide the shortest path for reduction of channel dimension. Figure 7c gives a ShuffleNet architecture including stride of size 2. The shufflent performance is better because of the group convolution and shuffling process in channel.

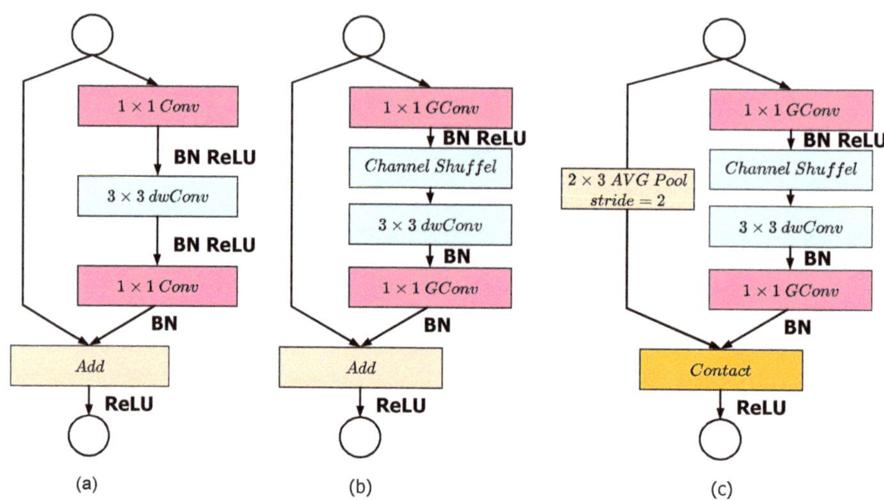

Figure 7. ShuffleNet architecture: (**a**) bottleneck unit with depthwise convolution (DWConv); (**b**) ShuffleNet unit with pointwise group convolution (GConv) and channel shuffle; (**c**) ShuffleNet unit with stride = 2.

2.9. EffNet

Unlike other previously discussed networks, spatial separable convolution was introduced in EffNet and is almost similar to depth-wise separable convolution which is used in MobileNet. In separable convolution, filter or kernel matrix *k* partitioned into two vector of size *k*1 and *k*2 (indirectly multiplication of matrix *k*1 and *k*2). This partition allows convolution of two one dimensional is equivalent to convolution of one two dimensional. For the sake of better understanding lets take s simple example,

$$\begin{pmatrix} 3 & 6 & 9 \\ 4 & 8 & 12 \\ 5 & 10 & 15 \end{pmatrix} = \begin{pmatrix} 3 \\ 4 \\ 5 \end{pmatrix} \begin{pmatrix} 1 & 2 & 3 \end{pmatrix}$$

$$k = k1 * k2$$

Now, perform convolution with *k*1 followed by convolution with *k*2 which leads reduction in number of multiplication. With this simple concept, the of multiplication required to perform convolution comes down to 6 (each one has 3 multiplication) in place of 9 multiplications. This reduction in multiplication leads to improvement in training speed and reduce the computational complexity of the network as in Figure 8. The major drawback with this process is that, possibility of portioning is not possible for all the kernels. This drawback is the bottleneck of Effnet particularly while in training process [19].

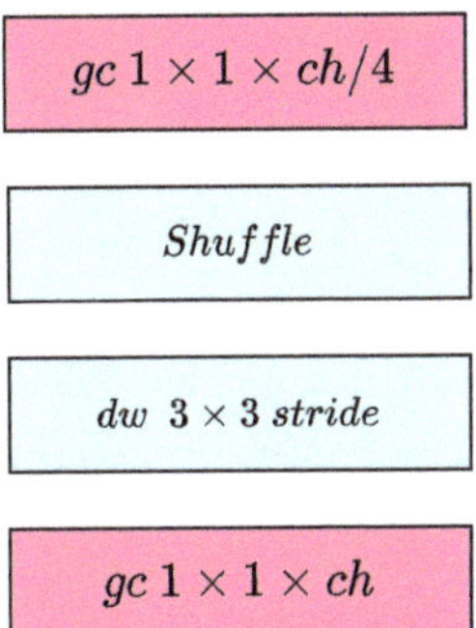

Figure 8. EffNet architecture.

3. Overview of the Encryption Algorithms under Study

In this section, we give an overview of the encryption algorithms that we will compare between them.

3.1. DNA Algorithm

The confusion matrix generated from the chaotic crypto system is applied on input image, and these encrypted image pixels are further diffused in accordance with the transformation of nucleotides into respective base pairs. Images of any size to be encrypted are reshaped to specific sizes and rearranged as an array which follows a strategy generated by a chaotic logistic map. In the second stage, according to a DNA inbuilt process, the confused pixels are shuffled for further encryption. In the third stage, every DNA nucleotide is transformed to a respective base pair by means of repetitive calculations that follow Chebyshev's chaotic map [20].

3.2. AES Algorithm

Here, images are transformed from gray-levels to respective strings; upon the converted strings, the AES algorithm performs image encryption. It is a symmetric key block cipher to safeguard the confidential information. AES is used to encrypt the images and was implemented on board a circuit or in a programmatic form in a computer. AES comes with three forms with key sizes of 128, 192, and 256, respectively. In the case of an AES-128 bit key, for both encryption and decryption of strings, it uses a key length of 128 bits. It is symmetric cryptography, also known as private key cryptography, and encryption and decryption uses the same key; therefore, the transmitter and recipient both should know and utilize a certain secret key. The sensitive and security levels can be protected with any key length. Key lengths of 192 or 256 bits are required for a high level of security. In this article, we used an AES-512 scheme for encryption of images; in addition, a bit wise shuffling was applied [21]. AES performance depends on key size and number of iterations chosen for shuffling and is proportional to size of the key and iterations. In the encryption stage, in general, 10, 12, or 14 iterations are used. Except for the last round, each round comprises four steps, shown in Algorithm 1:

Algorithm 1: AES encryption flow.

---

1. KeyExpansion: Using the AES key schedule, round keys are produced from cipher keys. Each round of AES requires a distinct 128-bit round key block, plus one extra.
2. SubBytes: By means of an "s-box" nonlinear lookup table, one byte is substituted with another byte [21].
3. ShiftRows: Byte transposition occurs, one, two, or three bytes are used to cyclically change the state matrix's of the second, third, and fourth rows to the left.
4. The final step is to find final state matrix calculation by multiplying fixed polynomial and current state matrixes.

3.3. Genetic Algorithm (GA)

The genetic algorithm encryption follows natural reproduction of genetics of humans or animals. In this, images are encrypted in three levels: first stage—reproduction, second stage—crossover, and third stage—mutation.

Reproduction: The production of new solution/encrypted image is obtained by fusion of two images (one original and another reference image); sometimes, it is called offspring and the algorithm security depends on the best selection of offspring/reference image. [22].

Crossover: The images to be encrypted are represented with eight bits per pixel; these eight bits of each pixel are equally portioned. For crossover operation, select any two pixels and its equivalent binary bits; now, swap the first four bits of one pixel with the last four bits of another pixel. The swapped pixel and its equivalent intensity generate a new value. This process is repeated for all the pixels in an image to get an encrypted image.

Mutation: This stage is not mandatory in general, but, for some reason, we perform this stage. It is simply a complement/invert (making 1 to 0 or 0 to 1) of any bit of any pixel, and its key depends on the position of the selected bit for complement.

3.4. Bit Slicing Algorithm

This image encryption scheme is simple, flexible, and faster, the security of encryption and randomness in genetic algorithm are improved with image bit slicing and rotating slices at any preferred angles 90, 180, and 270 degrees. The gray scale image is sliced into eight slices because it takes eight bits to represent any intensity [23]. The algorithm is given in Algorithm 2:

Algorithm 2: BSR image encryption flow.

---

1. In an image, all pixels are represented with their respective 8-bit binary equivalents.
2. Segregate the image into eight parts based on bits starting from MSB to LSB.
3. For each slice, now apply rotation operation with predefined angles.
4. Perform the three above for specified iterations and, after that, convert encrypted bits to gray scale intensity to get encrypted images.

3.5. Chaos Algorithm

The authors were at first enthusiastic in employing simple chaotic maps including the tent map and the logistic map since the quickness of the crypto algorithm is always a key aspect in evaluating the efficiency of a cryptography algorithm. However, new image encryption algorithms based on more sophisticated chaotic maps shown in 2006 and 2007 that using a higher-dimensional chaotic map could raise the effectiveness and security of crypto algorithms. To effectively apply chaos theory in encryption, chaotic maps must be built in such a way that randomness produced by the map may induce the requisite confusion matrix and diffusion matrix. In obtaining the confusion matrix, the pixel positions are changed in a specific fashion and ensure no change in pixel intensity levels.

Similarly, a diffusion matrix is obtained by modifying the pixel values sequentially in accordance with the order of sequence which are generated by chaotic crypto systems [24].

3.6. RSA Algorithm

Ron Rivest, Adi Shamir, and Leonard Adleman (RSA) created this algorithm back in 1977. The RSA algorithm is used to encrypt and decrypt the information. The public-key cryptosystems include the RSA cryptosystem. RSA is a cryptographic algorithm that is commonly used to transmit safe information. The RSA technique is used to encrypt the images, with two huge prime integers and a supplementary value chosen as the public key. The prime numbers are kept hidden from the public. The public key is used for image encryption, while the private key is used to decrypt them. It is not only used for image but also for text encryption. Consider two large prime numbers r and s as public and private keys respectively for encryption of images. Take two integers f and e in such a way that $f \times e \bmod \phi(m) = 1$. With these four selected integers, images are encrypted with the simple formula $D = q^f \bmod \phi(m)$, where q is an original input image, f is a publicly available key, and $\phi(m)$ is a product of $(r\text{-}1)$ and $(s\text{-}1)$ ensures that $gcd(f, \phi(m)) = 1$ i.e., f and $\phi(m)$ are co-primes. D is the encrypted image after encryption through RSA. To get back the original image, use $R = D^e \bmod \phi(m)$, where e is a key available only for private people [25] (Show in Algorithm 3).

Algorithm 3: Image encryption using RSA.

---

1. Initially access the original gray-scale image of any size (R)
2. Now, select two different large prime numbers r and s.
3. Measure the m value which is equal to $m = rs$.
4. Now, calculate $\Phi(m) = \Phi(r)\Phi(s) = (r - 1)(s - 1) = m - (r + s - 1)$, where function Φ is Euler's totient function.
5. Select another integer f (public key) in such a way that $1 < f < \Phi(m)$;
and ;
$gcd(f, \Phi(m)) = 1$, in which f and $\Phi(m)$ are co-primes.
6. Calculate e as $e = (f - 1) \bmod \Phi(m)$; i.e., e is the multiplicative modular inverse of $f\ (modulo\ \Phi(m))$.
7. Get the encrypted image, $D = Sf \bmod \Phi(m)$.
8. For gray-scale images, perform $D = D \bmod 256$, since image contains the pixel intensity levels between 0 to 255.
9. To decrypt the encrypted image, perform, $S = De \bmod 256$.
10. Then, the original input image $R = S \bmod \Phi(n)$.

3.7. Rubik's Cube Principle (RCP)

This encryption scheme follows the strategy of Rubik's Cube Principle. In this, the gray-scale image pixels are encrypted by changing the position of the individual pixels, and this change follows the principle of Rubik's Cube. Initially, two keys are generated randomly; with these keys, image pixels are gone through the XOR operation bit by bit between odd row-wise and columns-wise. In the same way, image pixels are gone through the XOR operation bit by bit between even row-wise and columns-wise with flipped versions of two secret keys. This process is repeated until the maximum or termination criteria reached [26]. Algorithm 4 shows the detailed description of the RCP algorithm for image encryption.

Algorithm 4: Rubik's Cube flow.

---

1. Let us assume an image I_o of size $M \times N$, and assume it is represented with α-bits. Now, two randomly generated vectors L_S and L_D of size M and N, respectively. Each elements in $L_S(i)$ and $L_D(j)$ can take a random number between a set A of range between 0 to $2\alpha - 1$.
2. Pre-define the maximum iterations count (itrmax), and set itr to zero.
3. For every iterations, itr is incremented by one: itr = itr + 1.
4. For every row of image I_o,
(a) calculate addition of all pixels in ith row, and is calculated by

$$\alpha(i) = \sum_{j=1}^{N} I_0(i,j), i = 1, 2, 3, 4, ..., M, \tag{1}$$

(b) Now, calculate $M\alpha(i)$ by doing modulo 2 of α (i),
(c) The ith row is shifted right or left or circular by $L_S(i)$ positions (pixels of images are moved to right or left direction by $K_R(i)$; after this operation, the first pixel becomes the last, and the last pixel becomes first), as per the following equation:

$$if \begin{cases} M\alpha(i) & = 0 \quad right \quad circular \quad shift \\ M\alpha(i) & \neq 0 \quad left \quad circular \quad shift \end{cases} \tag{2}$$

5. Similarly, for every column of image I_o,
(a) calculate addition of all pixels $\beta(i)$ in jth column, and is calculated by

$$\beta(i) = \sum_{i=1}^{N} I_0(i,j), i = 1, 2, 3, 4, 5....., M, \tag{3}$$

(b) now, calculate $M\beta(j)$ by doing modulo 2 of $\beta(j)$.
(c) the jth column of image is shifted up or circular or down by $L_D(i)$ positions, by following equations:

$$if \begin{cases} M\beta(i) & = 0 \quad up \quad circular \quad shift \\ M\beta(i) & \neq 0 \quad down \quad circular \quad shift \end{cases} \tag{4}$$

These steps (4 & 5) create a scrambled image $I_{SCR}(i)$.
6. With the help of vector L_D, a bit-wise XOR operation is performed on each row of scrambled image $I_{SCR}(i)$ by means of following equation:

$$I_1(2i-1,j) = I_{SCR}(2i-1,j) \oplus L_D(j), \tag{5}$$

$$I_1(2i,j) = I_{SCR}(2i,j) \oplus rot180(L_D(j)). \tag{6}$$

In the above equation, $\oplus$ shows XOR operation (bit-wise) and $rot180(L_D)$ shows a flip operation on vector K_C from right to left.
7. With the help of vector L_S, the XOR operation (bit-wise) is performed on the column of scrambled image by means of following equation:

$$I_{ENC}(i,2j-1) = I_1(i,2j-1) \oplus (L_S(j)), \tag{7}$$

$$I_{ENC}(i,2j) = I_1(i,2j)rot180 \oplus (L_S(j)), \tag{8}$$

where $rot180(L_S(j))$ shows the flip operation from left to right with respect to vector K_R.
8. Repeat step 1 to step 7 until itr = itrmax. 9. Finally, encrypted image I_{ENC} is generated and process is terminated; otherwise, the algorithm moves to step 3.

3.8. Hill Cipher Algorithm

It also comes, under symmetric key encryption algorithm, in which one can retrieve the decryption key by means of very simple transformation or repetitive computations. Encryption of images with this is very simple, easy, and a limited time-consuming process. It just replaced the original image pixels of size m with the Hill cipher image pixels of the same size. Let us take a simple example, let us assume three pixels with names as Q_1, Q_2, Q_3, and assume the Hill cipher algorithm and replace them with C_1, C_2, C_3, as per the following procedure:

$$C_1 = (L_{11}Q_1 + L_{12}Q_2 + L_{13}Q_3)mod26 \tag{9}$$

$$C_2 = (L_{21}Q_1 + L_{22}Q_2 + L_{23}Q_3)mod26 \tag{10}$$

$$C_3 = (L_{31}Q_1 + L_{32}Q_2 + L_{33}Q_3)mod26 \tag{11}$$

This can be represented by matrices as follows:

$$\begin{bmatrix} C_1 \\ C_2 \\ C_3 \end{bmatrix} = \begin{bmatrix} L_{11} & L_{12} & L_{13} \\ L_{21} & L_{22} & L_{23} \\ L_{31} & L_{32} & L_{33} \end{bmatrix} \begin{bmatrix} Q_1 \\ Q_2 \\ Q_3 \end{bmatrix} \tag{12}$$

The above can also be represented as: $C = LQ$, where Q is plain pixels and C is Hill cipher pixels. L is a key matrix that performs the transformation from original to hill cipher and is reversible. To get back the the original image, simply use $Q = L^{-1}C$. In this algorithm, the generation of reversible key matrix handles complex computations and is very hard task [27].

4. Security Analysis of the Studied Encryption Algorithms

To compare the studied encryption algorithms, we conceived experiments illustrating the effect of the image encryption on robot facial recognition and security issues in cloud environments. These experiments also aim at measuring the time needed to recognize faces in cloud and robotic environments. Our analysis of the results led us to compare the accuracy of the robot's facial recognition on two levels:

- facial recognition of the robot. This includes the cloud and excluding the cloud. It should be noted that the recognition algorithm is executed in the on-board robot;
- the effect of the encryption algorithms on the accuracy of the robot's facial recognition.

We took a dataset composed of 92×112 grayscale pgm images to assess the performance of the encryption algorithms [28]. For each of the 40 different subjects, 10 separate images were provided. To ensure some variability in the image acquisition conditions, images were taken at different times with varying lighting, for some subject. Moreover, the images were taken for different facial expressions (eyes open/closed, smiling/not smiling), different details of the face (glasses/no glasses), and poses of the head with tilts and rotations up to 20 degrees. We took a dark background when capturing all the images. For each of the 40 subjects, four images with different poses are used as the training dataset, while the remaining six images for each are used as the test database. This leads to 160 images for training and 240 images for testing. Likewise, experiments were carried out by varying the number of training images per face. In addition, we have undertaken robot face recognition experiments. To do this, we encrypted training and test images with the algorithms mentioned above.

4.1. Parameters Used for Security Analysis

Attacks such as the Statistical Attack, Plain Text Attack, Cipher Text Attack, Differential Attack, and Brute Force Attack present serious practical difficulties when implementing the algorithms operating them [29]. In this article, we used the following security measure parameters.

4.1.1. Histogram

The histogram is a graph indicating the number of repetitions of each of the gray levels when going through all the pixels of the given image. On the *x*-axis, we find the gray levels from 0 to 255, as shown in Figure 9 for the input image. Certainly, when the histogram of the original image is significantly different from that of the encrypted image, better security is ensured. It is expected that the histogram of the encrypted image is uniformly distributed. In this case, it becomes difficult to divide the content of the original image [30].

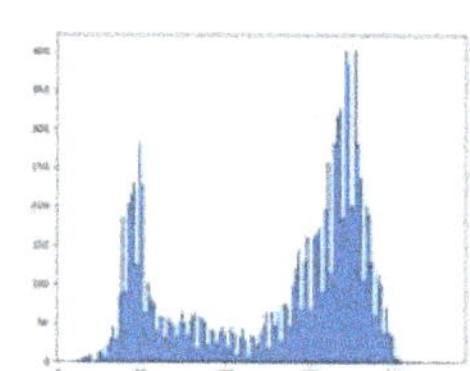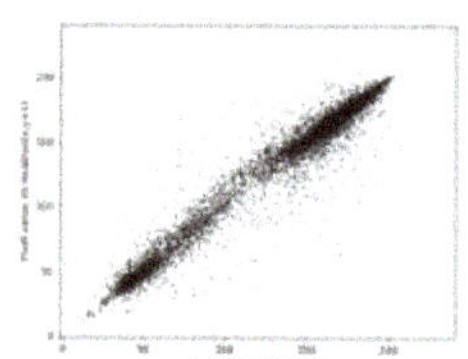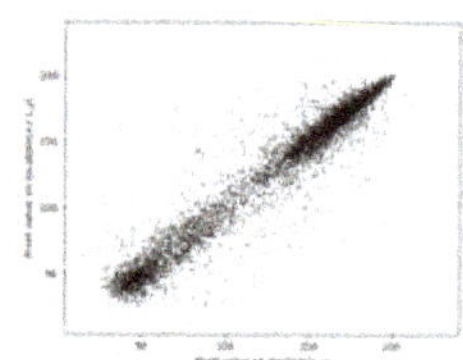

Figure 9. From left to right: Input image, Histogram, Horizontal scatter plot and Vertical scatter plot

4.1.2. Correlation Coefficient (CC)

This coefficient gives us a quantitative indication of the similarity between the original image and the encrypted one [31]. It is calculated as follows:

$$C_{xy} = \frac{cov(x,y)}{\sqrt{D(x)}\sqrt{D(y)}} \tag{13}$$

The variables *x* and *y* respectively denote the gray values of two adjacent horizontal and vertical, or diagonal pixels in the given image. We apply the following formulas:

$$cov(x,y) = \frac{1}{N}\sum_{i=1}^{N}(x_i - E(x))(y_i - E(y)) \tag{14}$$

$$E(x) = \frac{1}{N}\sum_{i=1}^{N}x_i, \quad E(y) = \frac{1}{N}\sum_{i=1}^{N}y_i \tag{15}$$

$$D(x) = \frac{1}{N}\sum_{i=1}^{N}(x_i - E(x))^2 \; D(y) = \frac{1}{N}\sum_{i=1}^{N}(y_i - E(y))^2 \tag{16}$$

where *N* stands for the number of used pairs of pixels.

4.1.3. Scatter Plot

It is a graphical representation of the correlation between two adjacent pixels. While the original image gives us a perfect positive correlation, the encrypted image generally shows no correlation. When the plot is a straight line, this means that the correlation between the two measured pixels is high. If the image contains random or quasi-random values, no correlation appears, as depicted in Figure 9.

4.1.4. Number of Pixels Change Rate (NPCR)

This parameter provides a quantitative indication for the pixel change rates in the encrypted image when a pixel change is introduced at the original image [32]. Let us consider the input image *T1* and its encrypted version *T2*, where *T1* has only one pixel difference. The NPCR metric of two images is defined as follows:

$$NPCR = \frac{\sum_{i,j} s(i,j)}{M * N} * 100 \tag{17}$$

where $s(i,j) = \begin{cases} 0 & if \quad T1(i,j) = T2(i,j) \\ 1 & if \qquad otherwise \end{cases}$

Here, M is the number of rows in the given image, N is the number of columns in this image, and (i,j) represent the position of one pixel.

4.1.5. Unified Averaged Changed Intensity (UACI)

This parameter determines the average intensity difference between original image and its encrypted version [31]:

$$UACI = \frac{1}{M*N}\left[\sum_{i,j} \frac{s(i,j)}{255}\right] * 100 \tag{18}$$

4.1.6. Mean Square Error (MSE)

The mean square error (MSE) is the measurement of difference between the original and cipher images [33]. The high value of MSE is related to a high amount of difference between original image and cipher image. It can be calculated by the following equation:

$$MSE = \frac{1}{M*N} \sum_{i=0}^{M-1} \sum_{j=0}^{N-1} [I(i,j) - K(i,j)]^2 \tag{19}$$

where I and K represent the plain image and the encrypted image.

4.1.7. Peak Signal-to-Noise Ratio (PSNR)

The peak signal-to-noise ratio (PSNR) measures the conformity between the plain and cipher images [33]:

$$PSNR = 10log\frac{255^2}{MSE}(db) \tag{20}$$

where MSE is the Mean Square Error.

4.2. Security Analysis and Discussion

To compare the studied encryption algorithms, we proceed as follows. After choosing a grayscale image from the given database, we apply the encryption algorithms to measure the security parameters and the recognition accuracy in the cloud and robot environments. Figure 9 shows the test image as well as the corresponding histogram and scatter plot. It turned out that the correlation coefficient of the test image is slightly smaller than 1, and more precisely equal to 0.955. Table 1 shows the security parameters of the studied encryption algorithms. Moreover, Figure 10 shows the histogram and the horizontal and vertical scatter point of the encrypted image, for the studied encryption algorithms. It turns out that, if the histogram of the encrypted image is uniform, then it becomes very difficult to break the security key. Indeed, the image with uniform values does not present relevant information. For the range of algorithms used here and the histogram of the image encrypted with Hill Cypher, the Rubik's Cube and DNA methods are found to be more consistent. Therefore, they are more resistant to statistical analysis attacks compared to other methods. It is worth noting that an encryption algorithm is more secure when the correlation between neighboring pixels inside the encrypted image is low. According to Table 1, the correlation coefficient of the encrypted image is almost zero. By applying the DNA method, we obtained a very small correlation coefficient, which is almost zero. In Table 2, we see that the encryption is fast for the chaos algorithm, while the encryption time is bigger for the AES. It should be noted that we carried out our experiments in robot (desktop) and cloud environments. We used the specifications of Table 3. It is clear from the Table 1 that the DNA method produces the highest NPCR value, whereas the GA method gives the highest UACI value. In addition, the GA method turned out to be producing the lowest peak signal-to-noise ratio (PSNR) [34] value. All our observations converge on the fact that, with the exception of bit slicing and GA, the various algorithms generally offer a relatively high level of security for the encrypted image.

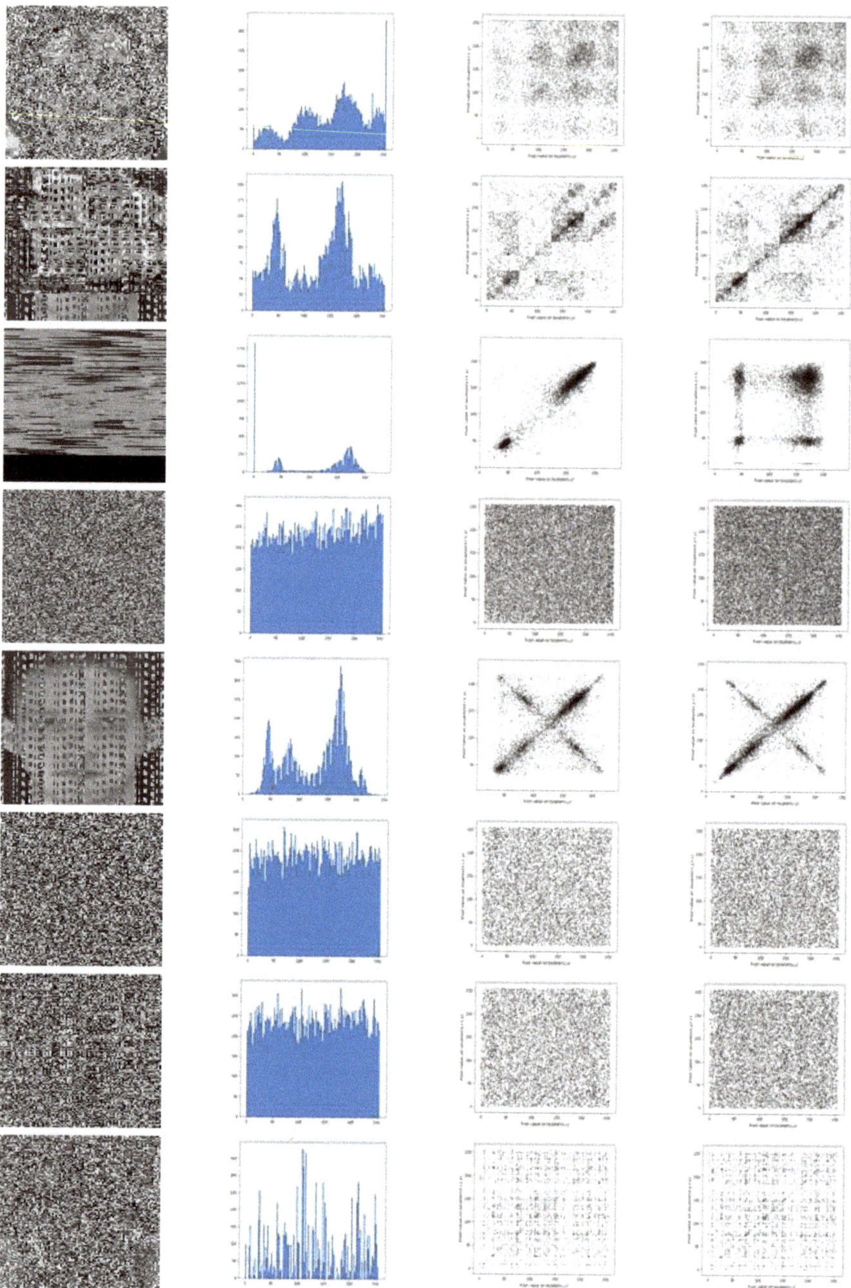

Figure 10. Security measures with proposed encryption algorithms. From left to right: encrypted image histogram, horizontal, vertical scatter plot. From top to bottom: AES, Bit-slice, Chaos, DNA, GA, Hill cypher, RCP, and RSA.

Table 1. Security measures of the studied encryption algorithms.

Algorithm	NPCR	CC	MSE	PSNR	UACI
DNA	99.446	**0.0615**	6372.06	10.088	24.6696
AES	99.572	0.1188	7033.57	9.6590	26.5178
GA	**34.297**	0.3828	**4293.95**	**11.802**	**9.7257**
Bit slicing	86.403	0.4029	6483.72	10.0125	22.4563
Chaos	99.116	0.9529	7262.96	9.5196	26.2900
RSA	99.996	0.0678	7432.85	9.4192	29.8552
RCP	99.621	0.0285	7542.40	9.3557	29.4176
Hill-Cypher	99.631	0.0328	7541.06	9.3564	29.4601

Table 2. Encryption time, average, and standard deviation to encrypt 40 images with various encryption algorithms.

Algorithm	Time	Avg. time	Std. Time
DNA	4.8877	4.9674	0.89898
AES	131.11	66.092	37.7128
GA	1.4454	1.4376	0.00937
Bit slicing	2.3361	2.2671	0.02760
Chaos	**0.0478**	**0.0473**	**0.00097**
RSA	1.4402	1.4357	0.03938
RCP	1.2341	1.3879	0.98765
Hill-Cypher	2.8552	1.4336	0.82310

Table 3. Specifications of cloud and robot.

	Cloud	Robot
Processor	Intel(R) Xeon(R) Silver 4114 CPU @ 2.20GHz	Intel core(R) i7-9700k CPU @ 3.60GHz
RAM	40 GB	32 GB
OS	64-bit Linux	64-bit Windows

5. Performance Analysis of CNN Based Robot Face Recognition Models

In this section, we present the results of the performance of the previously discussed CNN models and the impact of the encryption algorithms on the accuracy of these models. For each CNN model, we measure its accuracy first with clear images and then with encrypted images using one of the studied encryption algorithms. Moreover, we have also evaluated the performance in robot and cloud environment.

To study the impact of the encryption algorithms on the accuracy of the studied CNN models, we proceed as follows. First, we encrypt the images, then the studied CNN model is trained using these encrypted images, and finally, the CNN model is tested using a subset of the encrypted images and consequently, the accuracy is measured.

We studied the performance of the previously nine CNN models namely, Lenet, AlexNet, ResNet, VGG16Net, GoogleNet, DenseNet, MobileFaceNet, ShuffleNet, and EffNEt.

During the pre-processing step, the dataset images are resized to a size suitable for each CNN model and gray scale images are converted to RGB images.

Experimental results are conducted using the ORL database [35]. The Olivetti Research Laboratory (ORL) face dataset contains a set of face images.

5.1. Simulation Settings

In the NVIDIA GEFORCE GTX 1050TI variant, all tests were carried out utilizing the Windows platform with Intel core(R) i7-9700k CPU @ 3.60 GHz with 32 GB RAM in the case of robot environment and Intel(R) Xeon(R) Silver 4114 CPU @ 2.20 GHz with 32 GB RAM in the case of a cloud environment. The Python version 3.7.3 with TENSORFLOW version < 1.15.0 >= 1.14.0 is introduced to test the process and extract the features and for classification of images. As mentioned above, a pre-processing is needed before the formation process is started for the convolution neural network architectures. A re-scale is used for the resized images of data set in order to convert the images to 224 × 224 as AlexNet input and to 224 × 224 as ResNet input. Based on established Quality parameters, the efficiency of the trained neural network is assessed based on accuracy.

5.2. Learning Rate

One of the important hyperparameter of CNN models is the learning rate. The learning rate measures the efficiency of the CNN network weights and more precisely how well are adapted to the loss rate. The lower the rating, the smoother the downhill direction is. Although it could be a smart strategy for us to ensure we do not skip any local minimas (through a low learning rate), it may also imply that we take a long time to converge—especially if we remain on a plateau region. In addition, the learning rate affects how fast our model can converge to a local minimum (the best possible accuracy is achieved). To have it right will mean that we would have less time to learn the algorithm. Given the increasing sophistication of the facial recognition network model, an effective learning rate is especially challenging to achieve, since the scale of various parameters differs considerably and may be changed during the training cycle, which requires a long time.

Thus, experiments are conducted to find the suitable learning rate for each CNN model. Figures 11 and 12 show the learning rate versus accuracy of GoogleNet and Resnet, respectively. For GoogleNet, accuracy is maximum at a value of 0.0005 of learning rate. Regarding ResNet, accuracy is maximum at a value of 0.0007 of learning rate. For other CNN models a learning rate value of 0.0001 was used during experiments.

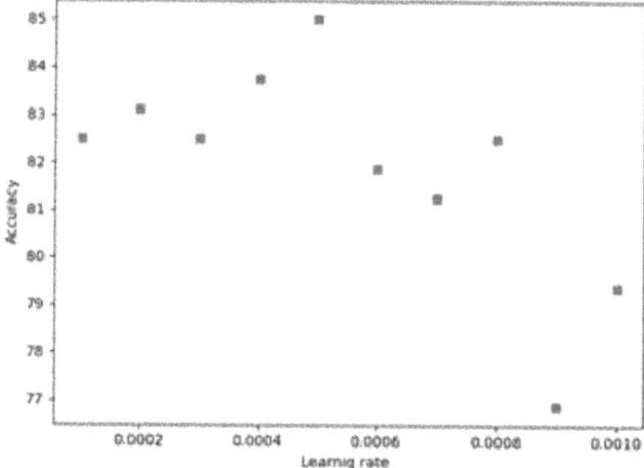

Figure 11. GoogleNet Learning rate.

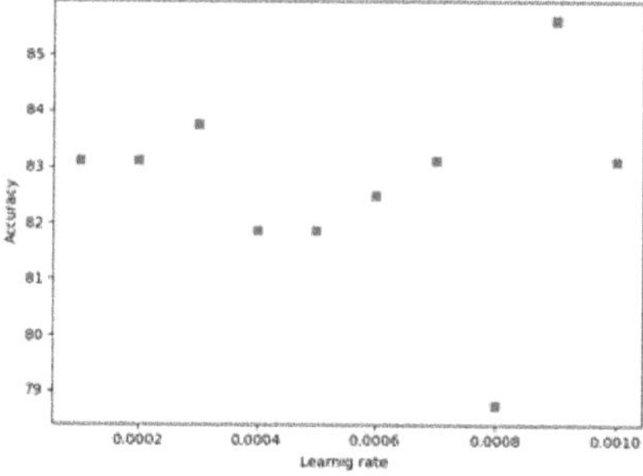

Figure 12. ResNet Learning rate.

5.3. Accuracy

The accuracy of the studied CNN models depends on the efficient feature extraction from encrypted images. Although a CNN model can easily extract features from clear images, it is a challenging task for it to extract features from encrypted images.

The studied CNN models use different filters in convolution layers, pooling layers, and followed different strategies for features extraction, which leads to different accuracy of face recognition. Moreover, images with different poses, illumination effects, and disguise might lead to low face recognition accuracy value.

Regarding encryption algorithms, they introduce some distortion in the images features. Thus, image features are distorted after encryption and consequently CNN models can lead to different results. Moreover, among the studied encryption algorithms some are linear and some are nonlinear. Therefore, they have different effect on the accuracy of the CNN models.

According to the results shown in Table 4 the **Bit Slice, Chacos** and **GA** encryption algorithms preserve the features after encryption, and therefore, give better accuracy compared to other encryption algorithms. More precisely, for the studied CNN based face recognition models, their accuracy obtained using these encrypted algorithms is close to the accuracy obtained using clear images.

Table 4 shows the accuracy obtained with various algorithms including PCA. It It is clearly noted from the table that the face recognition accuracy of CNN models is better than accuracy of PCA for clear images and encrypted images. From the studied CNN models, Effnet gives the best accuracy and outperforms other models with all encrypted algorithms. Compared to PCA, EffNet improves accuracy by 16.2%, 6.16%, 20.4%, 22.1%, 279.88%, 9.56%, 392.1%, 366.32%, and 480 % for plain image, AES, Bit slice, Chacos, DNA, GA, Hill, RSA and RCP, respectively. Additionally, Tables 5 and 6 show the testing and training time of various CNN models. The testing time of all CNN models for one image is less than one second, and training time is 6 to 8 minutes for the used ORL database. Figures 13–17 show the loss and accuracy curves of LeNet, AlexNet, Vgg16Net, ResNet, and DenseNet, respectively. To train the CNN model, Softmax loss function is used in this paper. It is a cascaded combination of softmax function, the last fully connected layer of corresponding CNN model, and cross-entropy loss. In training CNN models, weights wk (where k is number of training class, here 40) of connected layers are normalized to scale parameter s, and, in the same way, outcome features of last convolution layers amplitudes are normalized to s [36]. As a result, softmax loss is calculated from the below equation for a given input vector feature x, and, assuming respective ground truth label y, then

$$L_1 = -log \frac{e^{scos(\theta_{wy},x)}}{e^{scos(\theta_{wy},x)} + \sum_{k\neq y}^{k} e^{scos(\theta_{wy},x)}} \tag{21}$$

where $cos(\theta_{wy}, x) = w_k^T x$ is the cosine similarity and θ_{wy}, x is the angle between w_k and x. The learned features with softmax loss are prone to be separable, rather than to be discriminative for robot face recognition. In addition, the time needed to produce recognition results (with LeNet) for one image for the cloud turned out to be smaller than that required for the robot as depicted in Table 7.

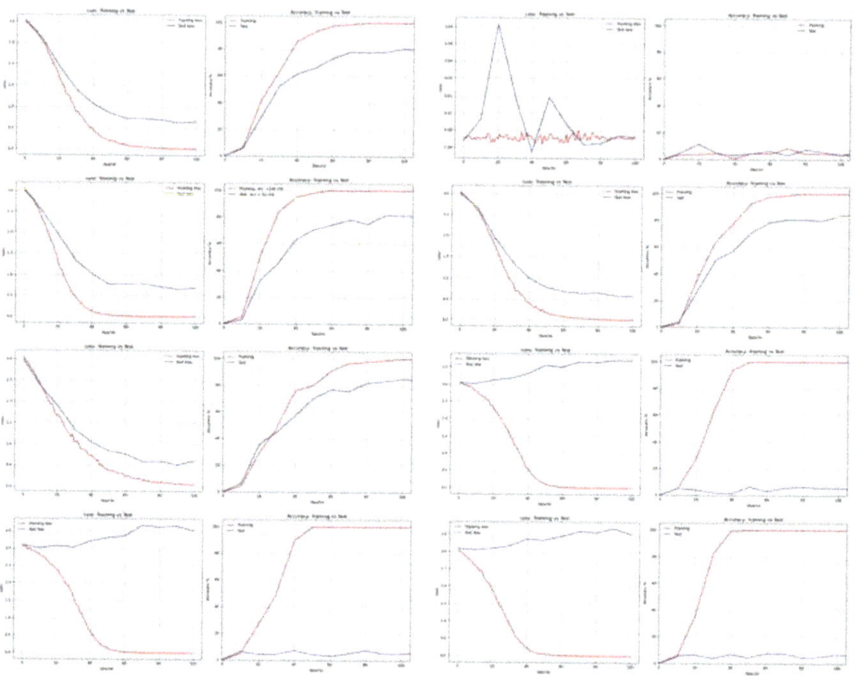

Figure 13. Loss and Accuracy of **AlexNet** from left to right and top to bottom: DNA, AES, Bit Slice, Chaco, GA, Hill, RCP, and RSA.

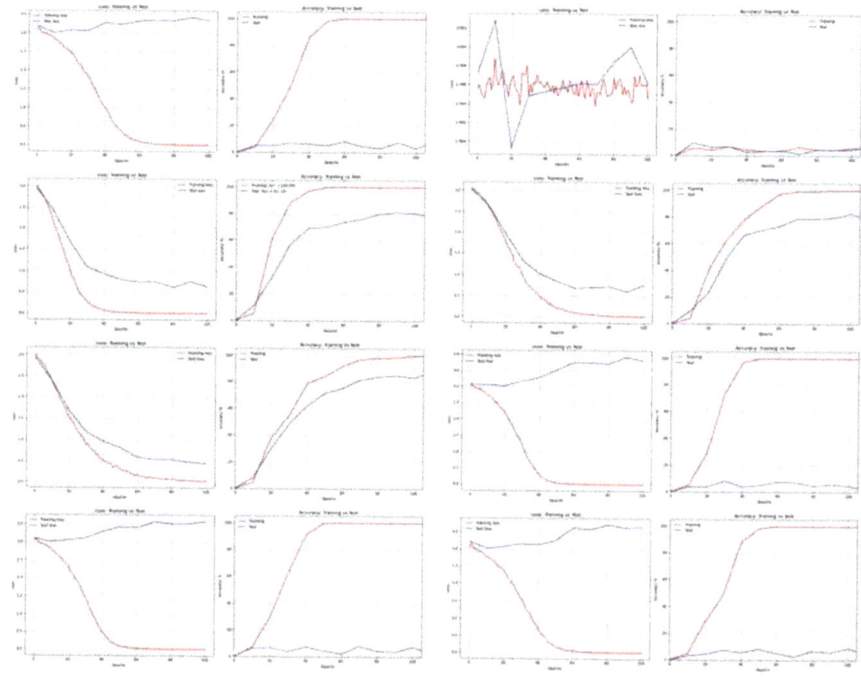

Figure 14. Loss and Accuracy of **Vgg16Net** from left to right and top to bottom: DNA, AES, Bit Slice, Chaco, GA, Hill, RCP, and RSA.

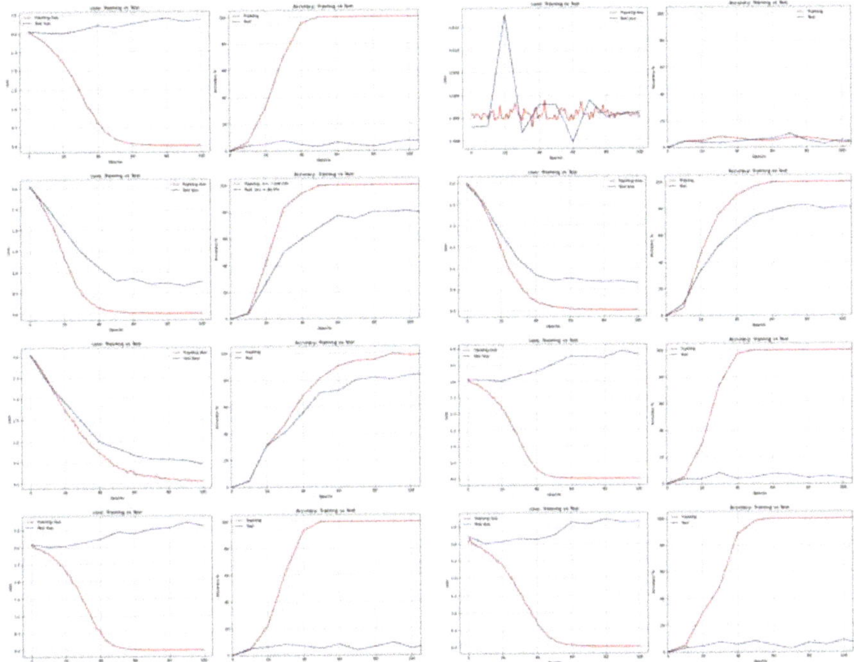

Figure 15. Loss and Accuracy of **GoogleNet** from left to right and top to bottom: DNA, AES, Bit Slice, Chaco, GA, Hill, RCP, and RSA.

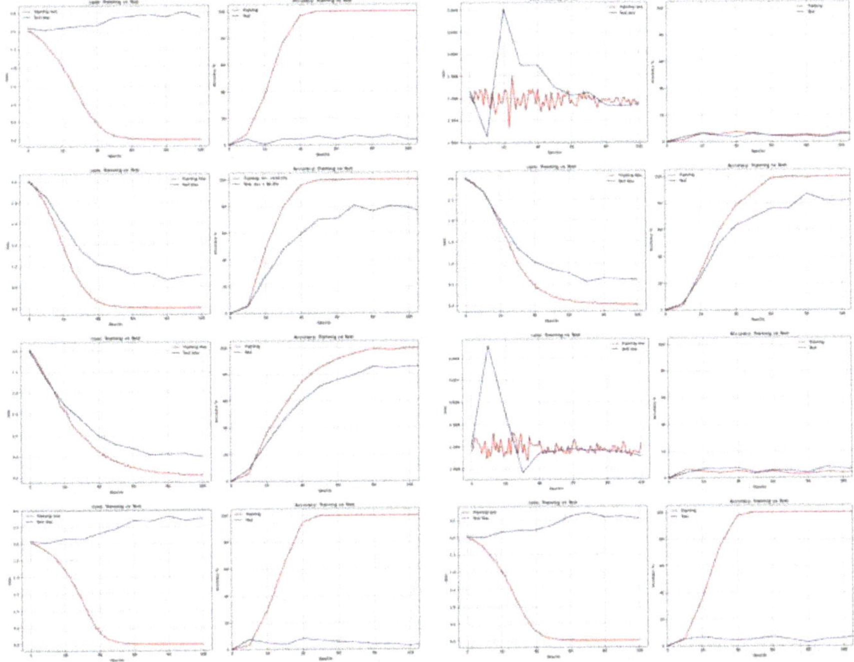

Figure 16. Loss and Accuracy of **ResNet** from left to right and top to bottom: DNA, AES, Bit Slice, Chaco, GA, Hill, RCP, and RSA.

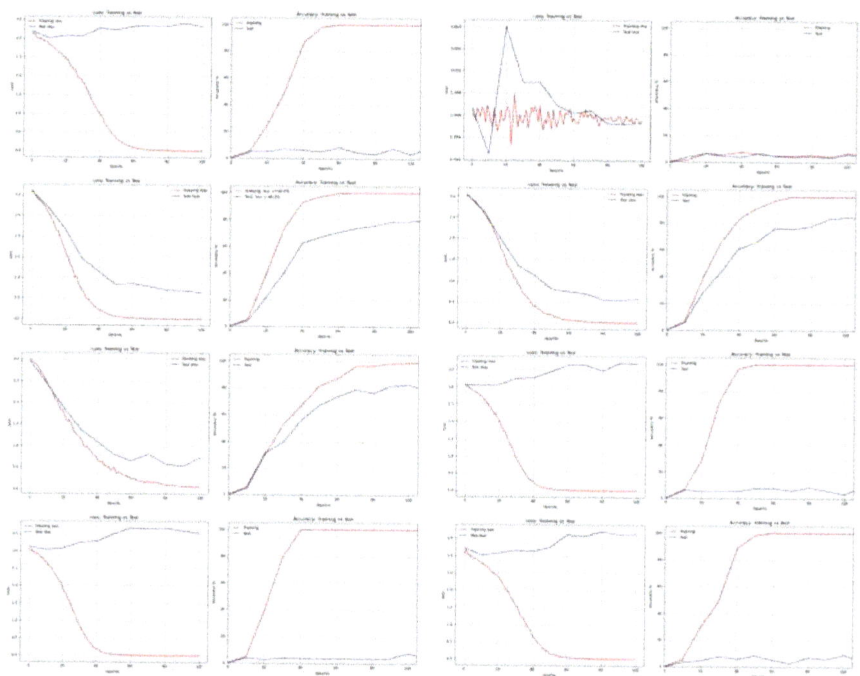

Figure 17. Loss and Accuracy of **DenseNet** from left to right and top to bottom: DNA, AES, Bit Slice, Chaco, GA, Hill, RCP, and RSA.

Table 4. Accuracy of the studied CNN models.

CNN Model	Without Enc.	AES	RCP	Hill	RSA	DNA	Bit Slice	Chacos	GA
PCA	80.4 [37]	2.50	2.51	2.91	2.91	3.33	70.0	72.50	80.4
LeNet	82.5 [38]	6.25	7.50	8.75	9.38	6.25	80.62	84.38	82.50
Alexnet	84.3 [39]	6.88	7.50	6.25	8.75	6.88	82.50	81.88	85.00
VGG16Net	85.6 [13]	7.50	8.12	9.38	8.12	6.88	80.62	86.87	85.62
GoogLeNet	85.7 [14]	8.12	9.38	6.88	8.75	8.24	81.25	82.50	88.75
ResNet	85.3 [39]	9.38	10.00	8.12	8.75	7.50	80.62	82.50	87.50
DenseNet	87.9 [40]	7.50	11.25	8.12	6.88	6.25	80.62	83.13	85.00
MobileFaceNet	88.6 [41]	11.7	12.34	11.34	9.25	10.54	81.35	86.25	87.92
ShuffleNet	90.3 [41]	13.2	15.34	12.31	11.2	11.21	83.34	87.32	87.91
EffNet	93.5 [19]	14.5	17.98	14.32	13.57	12.65	**84.34**	**88.54**	**88.09**

Table 5. Training time (Sec) of CNN models on robots.

CNN Model	Without Enc.	AES	Bit Slice	Chacos	DNA	GA	Hill	RSA	RCP
LeNet	450.84	439.51	458.51	431.033	443.16	441.45	452.97	453.86	435.61
Alexnet	441.77	449.66	435.14	451.181	430.95	438.31	431.39	432.91	454.7
VGG16Net	449.45	436.39	451.82	454.15	457.70	453.40	457.56	455.43	454.68
GoogLeNet	449.17	435.62	457.40	455.15	450.76	448.60	452.73	455.01	457.59
ResNet	444.69	434.37	457.75	451.39	467.23	448.57	448.06	457.42	458.59
DenseNet	443.37	433.25	458.75	459.0	457.64	457.62	454.84	458.95	458.01
MobileFaceNet	449.39	440.73	462.67	455.76	455.12	461.85	450.90	451.40	452.65
ShuffleNet	453.77	458.78	449.67	431.071	455.47	458.02	450.36	452.73	452.29
EffNet	454.44	457.17	433.81	457.401	448.97	432.93	438.35	446.41	458.73

Table 6. Testing time (Sec) of CNN models on robots.

CNN Models	Without Enc.	AES	Bit Slice	Chacos	DNA	GA	Hill	RSA	RCP
LeNet	0.559	0.749	0.98	0.67	0.793	0.612	0.876	0.628	0.753
Alexnet	0.85	0.945	0.98	0.774	0.569	0.575	0.629	0.92	0.627
VGG16Net	0.671	0.671	0.765	0.796	0.656	0.812	0.640	0.796	0.656
GoogLeNet	0.687	0.753	0.781	0.656	0.671	0.562	0.968	0.812	0.578
ResNet	0.538	0.765	0.640	0.968	0.906	0.906	0.671	0.875	0.906
DenseNet	0.527	0.875	0.671	0.781	0.890	1.000	0.875	0.781	0.781
MobileFaceNet	0.765	0.828	0.796	0.828	0.718	0.906	0.781	0.781	0.687
ShuffleNet	0.907	0.622	0.965	0.675	0.598	0.626	0.808	0.737	0.676
EffNet	0.915	0.793	0.775	0.959	0.643	0.879	0.877	0.69	0.784

Table 7. Computational time to recognition one image in cloud and robots.

Source	Time of Execution in Seconds
Robot	0.559
Cloud	0.014

6. Conclusions

We studied several approaches for robot secure face recognition in the cloud environment. We used different algorithms to encrypt a set of images. The encrypted versions were used for training and testing. We trained the robot with various deep learning algorithms. We adopted some tests to measure safety, time complexity, and recognition accuracy. Both cloud and robot environments were considered. The findings showed that some algorithms were well suited for the security criterion. Others were better placed for recognition accuracy. The findings also revealed that, compared to a large range of algorithms, the genetic algorithm is a good candidate for better security and better recognition accuracy. Our study also revealed that the recognition accuracy provided by AES, RSA, and RCP is not reasonably high, although these methods are more secure. The percentages of improvement from PCA to EffNet were 16.2%, 6.16%, 20.4%, 22.1%, 9.56%, 279.88%, 392.1%, 366.32%, and 480% for plain images and encrypted images using AES, Bit slice, Chacos, GA, DNA, Hill, RSA, and RCP encryption algorithms, respectively. By comparing cloud and robot environments, we concluded that the recognition was faster for the cloud. It is advisable to run the same algorithm on the cloud to improve the robot's recognition speed.

Author Contributions: Conceptualization, C.K. and O.C.; methodology, C.K. and O.C.; writing—original draft preparation, C.K.; writing—review and editing, C.K. and O.C. and A.H.; visualization, A.H.; supervision, H.H.; project administration, O.C.; funding acquisition, A.Z. All authors have read and agreed to the published version of the manuscript.

Funding: This research was funded by Taif university Researchers Supporting Project Number (TURSP-2020/114), Taif University, Taif, Saudi Arabia.

Institutional Review Board Statement: Not applicable

Informed Consent Statement: Not applicable.

Data Availability Statement: Publicly available dataset were analyzed in this study. This data can be found here: [https://www.kaggle.com/tavarez/the-orl-database-for-training-and-testing].

Acknowledgments: The authors thank Taif University Research Supporting Project Number (TURSP-2020/114), Taif University, Taif, Saudi Arabia. The authors would like to thank also Asma Cheikhrouhou from University of Carthage, Tunis, Tunisia, for her English editing services.

Conflicts of Interest: The authors declare no conflict of interest.

References

1. Cheikhrouhou, O.; Khoufi, I. A comprehensive survey on the Multiple Traveling Salesman Problem: Applications, approaches and taxonomy. *Comput. Sci. Rev.* **2021**, *40*, 100369. [CrossRef]
2. Jamil, F.; Cheikhrouhou, O.; Jamil, H.; Koubaa, A.; Derhab, A.; Ferrag, M.A. PetroBlock: A blockchain-based payment mechanism for fueling smart vehicles. *Appl. Sci.* **2021**, *11*, 3055. [CrossRef]
3. Ijaz, M.; Li, G.; Lin, L.; Cheikhrouhou, O.; Hamam, H.; Noor, A. Integration and Applications of Fog Computing and Cloud Computing Based on the Internet of Things for Provision of Healthcare Services at Home. *Electronics* **2021**, *10*, 1077. [CrossRef]
4. Allouch, A.; Cheikhrouhou, O.; Koubâa, A.; Toumi, K.; Khalgui, M.; Nguyen Gia, T. UTM-chain: Blockchain-based secure unmanned traffic management for internet of drones. *Sensors* **2021**, *21*, 3049. [CrossRef]
5. Cheikhrouhou, O.; Koubâa, A.; Zarrad, A. A cloud based disaster management system. *J. Sens. Actuator Net.* **2020**, *9*, 6. [CrossRef]
6. Tian, S.; Lee, S.G. An implementation of cloud robotic platform for real time face recognition. In Proceedings of the 2015 IEEE International Conference on Information and Automation, Lijiang, China, 8–10 August 2015; pp. 1509–1514.
7. Masud, M.; Muhammad, G.; Alhumyani, H.; Alshamrani, S.S.; Cheikhrouhou, O.; Ibrahim, S.; Hossain, M.S. Deep learning-based intelligent face recognition in IoT-cloud environment. *Comput. Commun.* **2020**, *152*, 215–222. [CrossRef]
8. Chaari, R.; Cheikhrouhou, O.; Koubâa, A.; Youssef, H.; Hmam, H. Towards a distributed computation offloading architecture for cloud robotics. In Proceedings of the 2019 15th International Wireless Communications & Mobile Computing Conference (IWCMC), Tangier, Morocco, 24–28 June 2019; pp. 434–441.
9. Samriya, J.K.; Chandra Patel, S.; Khurana, M.; Tiwari, P.K.; Cheikhrouhou, O. Intelligent SLA-Aware VM Allocation and Energy Minimization Approach with EPO Algorithm for Cloud Computing Environment. *Math. Probl. Eng.* **2021**, *2021*, 9949995. [CrossRef]
10. Jemal, I.; Haddar, M.A.; Cheikhrouhou, O.; Mahfoudhi, A. Performance evaluation of Convolutional Neural Network for web security. *Comput. Commun.* **2021**, *175*, 58–67. [CrossRef]
11. LeCun, Y.; Jackel, L.; Bottou, L.; Brunot, A.; Cortes, C.; Denker, J.; Drucker, H.; Guyon, I.; Muller, U.; Sackinger, E.; et al. Comparison of learning algorithms for handwritten digit recognition. In Proceedings of the International Conference on Artificial Neural Networks, Perth, Australia, 27 November–1 December 1995; Volume 60, pp. 53–60.
12. Krizhevsky, A.; Sutskever, I.; Hinton, G.E. Imagenet classification with deep convolutional neural networks. *Adv. Neural Inf. Process. Syst.* **2012**, *25*, 1097–1105. [CrossRef]
13. Simonyan, K.; Zisserman, A. Very deep convolutional networks for large-scale image recognition. *arXiv* **2014**, arXiv:1409.1556.
14. Szegedy, C.; Liu, W.; Jia, Y.; Sermanet, P.; Reed, S.; Anguelov, D.; Erhan, D.; Vanhoucke, V.; Rabinovich, A. Going deeper with convolutions. In Proceedings of the IEEE Conference on Computer Vision and Pattern Recognition, Boston, MA, USA, 7–12 June 2015; pp. 1–9.
15. He, K.; Zhang, X.; Ren, S.; Sun, J. Deep residual learning for image recognition. In Proceedings of the IEEE Conference on Computer Vision and Pattern Recognition, Las Vegas, NV, USA, 27–30 June 2016; pp. 770–778.
16. Huang, G.; Liu, Z.; Van Der Maaten, L.; Weinberger, K.Q. Densely connected convolutional networks. In Proceedings of the IEEE Conference on Computer Vision and Pattern Recognition, Honolulu, HI, USA, 21–26 July 2017; pp. 4700–4708.
17. Howard, A.G.; Zhu, M.; Chen, B.; Kalenichenko, D.; Wang, W.; Weyand, T.; Andreetto, M.; Adam, H. Mobilenets: Efficient convolutional neural networks for mobile vision applications. *arXiv* **2017**, arXiv:1704.04861.
18. Zhang, X.; Zhou, X.; Lin, M.; Sun, J. Shufflenet: An extremely efficient convolutional neural network for mobile devices. In Proceedings of the IEEE Conference on Computer Vision and Pattern Recognition, Salt Lake City, UT, USA, 18–23 June 2018; pp. 6848–6856.
19. Freeman, I.; Roese-Koerner, L.; Kummert, A. Effnet: An efficient structure for convolutional neural networks. In Proceedings of the 2018 25th IEEE International Conference on Image Processing (ICIP), Athens, Greece, 7–10 October 2018; pp. 6–10.
20. Liu, H.; Wang, X. Image encryption using DNA complementary rule and chaotic maps. *Appl. Soft Comput.* **2012**, *12*, 1457–1466. [CrossRef]

21. Yahya, A.; Abdalla, A.M.; Arabnia, H.; Daimi, K. An AES-Based Encryption Algorithm with Shuffling. In Proceedings of the 2009 International Conference on Security & Management, SAM 2009, Las Vegas, NV, USA, 13–16 July 2009; pp. 113–116.
22. Zeghid, M.; Machhout, M.; Khriji, L.; Baganne, A.; Tourki, R. A modified AES based algorithm for image encryption. *Int. J. Comput. Sci. Eng.* **2007**, *1*, 70–75.
23. Vijayaraghavan, R.; Sathya, S.; Raajan, N. Security for an image using bit-slice rotation method-image encryption. *Indian J. Sci. Technol.* **2014**, *7*, 1. [CrossRef]
24. Sankpal, P.R.; Vijaya, P. Image encryption using chaotic maps: A survey. In Proceedings of the 2014 Fifth International Conference on Signal and Image Processing, Bangalore, India, 8–10 January 2014; pp. 102–107.
25. Chepuri, S. An RGB image encryption using RSA algorithm. *Int. J. Courrent Trends Eng. Res.* **2017**, *3*, 2455–1392.
26. Loukhaoukha, K.; Chouinard, J.Y.; Berdai, A. A secure image encryption algorithm based on Rubik's cube principle. *J. Electr. Comput. Eng.* **2012**, *2012*. [CrossRef]
27. Acharya, B.; Panigrahy, S.K.; Patra, S.K.; Panda, G. Image encryption using advanced hill cipher algorithm. *Int. J. Recent Trends Eng.* **2009**, *1*, 663–667.
28. Samaria, F.S.; Harter, A.C. Parameterisation of a stochastic model for human face identification. In Proceedings of the 1994 IEEE Workshop on Applications of Computer Vision, Sarasota, FL, USA, 5–7 December 1994; pp. 138–142.
29. Wu, S.; Zhang, Y.; Jing, X. A novel encryption algorithm based on shifting and exchanging rule of bi-column bi-row circular queue. In Proceedings of the 2008 International Conference on Computer Science and Software Engineering, Wuhan, China, 12–14 December 2008; Volume 3, pp. 841–844.
30. Jolfaei, A.; Mirghadri, A. A novel image encryption scheme using pixel shuffler and A5/1. In Proceedings of the 2010 International Conference on Artificial Intelligence and Computational Intelligence, Sanya, China, 23–24 October 2010; Volume 2, pp. 369–373.
31. Ibrahim, S.; Alhumyani, H.; Masud, M.; Alshamrani, S.S.; Cheikhrouhou, O.; Muhammad, G.; Hossain, M.S.; Abbas, A.M. Framework for Efficient Medical Image Encryption Using Dynamic S-Boxes and Chaotic Maps. *IEEE Access* **2020**, *8*, 160433–160449. [CrossRef]
32. Mazloom, S.; Eftekhari-Moghadam, A.M. Color image cryptosystem using chaotic maps. In Proceedings of the 2011 IEEE Symposium On Computational Intelligence For Multimedia, Signal In addition, Vision Processing, Paris, France, 11–15 April 2011; pp. 142–147.
33. Kanwal, S.; Inam, S.; Cheikhrouhou, O.; Mahnoor, K.; Zaguia, A.; Hamam, H. Analytic Study of a Novel Color Image Encryption Method Based on the Chaos System and Color Codes. *Complexity* **2021**, *2021*. [CrossRef]
34. Karri, C.; Jena, U. Fast vector quantization using a Bat algorithm for image compression. *Eng. Sci. Technol. Int. J.* **2016**, *19*, 769–781. [CrossRef]
35. Karri, C. Secure robot face recognition in cloud environments. *Multimed. Tools Appl.* **2021**, *80*, 18611–18626. [CrossRef]
36. Wang, F.; Jiang, M.; Qian, C.; Yang, S.; Li, C.; Zhang, H.; Wang, X.; Tang, X. Residual attention network for image classification. In Proceedings of the IEEE Conference on Computer Vision and Pattern Recognition, Honolulu, HI, USA, 21–26 July 2017; pp. 3156–3164.
37. Fernandes, S.; Bala, J. Performance Analysis of PCA-based and LDA-based Algorithms for Face Recognition. *Int. J. Signal Process. Syst.* **2013**, *1*, 1–6. [CrossRef]
38. Soulie, F.F.; VIENNET, E.; LAMY, B. Multi-modular neural network architectures: Applications in optical character and human face recognition. In *Advances in Pattern Recognition Systems Using Neural Network Technologies*; World Scientific: Singapore 1993; pp. 77–111.
39. Almabdy, S.; Elrefaei, L. Deep convolutional neural network-based approaches for face recognition. *Appl. Sci.* **2019**, *9*, 4397. [CrossRef]
40. Song, J.M.; Kim, W.; Park, K.R. Finger-vein recognition based on deep DenseNet using composite image. *IEEE Access* **2019**, *7*, 66845–66863. [CrossRef]
41. Chen, S.; Liu, Y.; Gao, X.; Han, Z. Mobilefacenets: Efficient cnns for accurate real-time face verification on mobile devices. In *Chinese Conference on Biometric Recognition*; Springer: Berlin/Heidelberg, Germany, 2018; pp. 428–438.

MDPI

St. Alban-Anlage 66

4052 Basel

Switzerland

Tel. +41 61 683 77 34

Fax +41 61 302 89 18

www.mdpi.com

Applied Sciences Editorial Office

E-mail: applsci@mdpi.com

www.mdpi.com/journal/applsci